Renewable Energy Systems

Renewable Energy Systems

A Smart Energy Systems Approach to the Choice and Modeling of Fully Decarbonized Societies

Third edition

Henrik Lund
Department of Sustainability and Planning, Aalborg University, Aalborg, Denmark

ACADEMIC PRESS

An imprint of Elsevier

Academic Press is an imprint of Elsevier
125 London Wall, London EC2Y 5AS, United Kingdom
525 B Street, Suite 1650, San Diego, CA 92101, United States
50 Hampshire Street, 5th Floor, Cambridge, MA 02139, United States

Notices
Knowledge and best practice in this field are constantly changing. As new research and experience broaden
our understanding, changes in research methods, professional practices, or medical treatment may
become necessary.

Practitioners and researchers must always rely on their own experience and knowledge in evaluating
and using any information, methods, compounds, or experiments described herein. In using such
information or methods they should be mindful of their own safety and the safety of others, including
parties for whom they have a professional responsibility.

To the fullest extent of the law, neither the Publisher nor the authors, contributors, or editors, assume
any liability for any injury and/or damage to persons or property as a matter of products liability, negligence
or otherwise, or from any use or operation of any methods, products, instructions, or ideas contained in
the material herein.

ISBN 978-0-443-14137-9

For information on all Academic Press publications visit
our website at https://www.elsevier.com/books-and-journals

Publisher: Megan Ball
Acquisitions Editor: Edward Payne
Editorial Project Manager: Aleksandra Packowska
Production Project Manager: Erragounta Saibabu Rao
Cover Designer: Miles Hitchen

Typeset by STRAIVE, India

Working together
to grow libraries in
developing countries

www.elsevier.com • www.bookaid.org

Contents

About the contributors

Xiliang Zhang is a professor of management science and engineering and director of the Institute of Energy, Environment, and Economy at Tsinghua University. Prof. Zhang is a member of the National Experts Panel on Climate Change and the Chair of the Energy Systems Engineering Committee of the China Energy Research Society. He has been heading the expert group of national carbon market design since 2015. He was the lead author of the fourth and fifth IPCC Climate Change Assessment Report. He was granted the Leading Talent Award by the Ministry of Ecology and Environment and the First Award for Humanity and Social Science Research by the Ministry of Education in 2020. His current research interests include low-carbon energy economy transformation, climate change economics, and climate change policy and mechanism design. Prof. Zhang holds a PhD in Systems Engineering from Tsinghua University.

Willett Kempton is trained as a cognitive anthropologist and electrical engineer. He serves as an associate director of the Center for Research in Wind (CReW) and professor in the College of Earth, Ocean, and Environment and in the Department of Electrical and Computer Engineering at the University of Delaware. With coauthors, Kempton published the first proposal for what is now called Vehicle-to-Grid (V2G) power in 1997 and the fundamental equations for V2G power and V2G markets in 2005. Kempton is cofounder of Nuvve, a company commercializing V2G, with offices in five countries. In addition to his research and scholarship on electric vehicles, Prof. Kempton has peer-reviewed publications on offshore wind power and other energy topics and has advised state governments on procurement of offshore wind power.

Frede Hvelplund is a professor in energy planning at Aalborg University, Denmark. He has a background in economics and social anthropology. Hvelplund has written a comprehensive series of books and articles on the transition to renewable energy systems; among others, *Alternative Energy Plans* was written in interdisciplinary groups together with engineers. Hvelplund is a "concrete institutional economist" and understands the market as a social construction that for decades has been conditioned to support a fossil fuel-based economy. Hence, Hvelplund believes that a transition to a "renewable energy" economy requires fundamental changes of an array of concrete institutional rules, laws, and market conditions. In 2005, Hvelplund obtained the Danish Dr. Techn. Degree, and in December 2008, he received the EUROSOLAR European Solar Prize.

Bernd Möller is a professor in sustainable energy systems management and the director of the MEng program in sustainable energy and development at Flensburg University, Germany. Möller holds a graduate degree in energy systems engineering and a PhD in the use of spatial information and analysis (GIS) in energy planning from Aalborg University, Denmark. Möller's research interest is the application of quantitative geographical analysis in sustainable energy resource economics, technology, and planning, where the distribution, location, and distance influence the feasibility of resources such as wind, biomass, and solar energy, as well as energy efficiency and infrastructures such as district heating and cooling.

Poul Alberg Østergaard is a professor in energy planning at the Department of Sustainability and Planning at Aalborg University, Denmark—listed on the Stanford's list of top 2% scientist in the world. He has worked within the field of energy planning since 1995 with a focus on simulation of energy systems based on high penetrations of renewable energy sources as well as on the design of renewable energy system scenarios. He is the head of the *Study Board of Planning and Land Surveying* at Aalborg University as well as the program director of the MSc program in *Sustainable Energy Planning and Management* at Aalborg University. Østergaard is the editor-in-chief of the *International Journal of Sustainable Energy Planning and Management* as well as coeditor of a series of other energy journals.

Brian Vad Mathiesen is a professor in energy planning at Aalborg University. Mathiesen holds MSc and PhD degrees focusing on fuel cells in future energy systems (2008). His research covers analyses of short-term, well-known transition technologies and 100% renewable energy systems as well as technical energy system analyses and studies of feasibility, public regulation, and technological change. Since 2005, Mathiesen has been involved in research in renewable energy systems as well as technologies for the large-scale integration of wind power. He was responsible for the technical and socioeconomic analyses that formed the basis for a detailed roadmap toward 100% renewable energy in the IDA Climate Plan 2050 (2009) and in the strategic research project CEESA (2011). In 2008 and 2010, he was involved in creating the Heat Plan Denmark, which analyzed future heat options.

David Connolly is the CEO of HeatGrid Ireland, a company which develops, funds, constructs, and operates district heating networks in Ireland. Connolly is also the cofounder and chairperson of the Irish District Energy Association, which represents the district heating and cooling industry across the island of Ireland. Before HeatGrid Ireland, he developed high-temperature heat pumps (up to 200°C) for large industrial users, and he was also the CEO of Wind Energy Ireland, which is Ireland's largest renewable energy association representing onshore and offshore wind across the island of Ireland. Until 2017, Connolly was an associate professor in energy planning at Aalborg University in Copenhagen, Denmark, where his research focused on the design and assessment of 100% renewable energy systems for electricity, heat, and transport (www.dconnolly.net).

Wen Liu is an assistant professor at the Copernicus Institute of Sustainable Development at Utrecht University in the Netherlands. She holds a PhD in energy planning from Aalborg University, Denmark. Liu's research interests are energy system analysis, sustainable heat transition, and seasonal thermal energy storage. Liu has published a series of articles proposing the analysis of renewable energy systems and the evaluation of sustainable heat technologies. Liu was the first researcher to apply the EnergyPLAN computer tool to the Chinese energy system.

Anders N. Andersen is the head of research and development projects and has been responsible for the development of the simulation tool energyPRO, which is used worldwide for simulating and optimizing distributed energy plants equipped with energy stores and participating in wholesale and balancing electricity markets. Andersen holds a PhD in energy system analysis, an MSc in mathematics and physics, and a diploma in business administration and organization.

Iva Ridjan Skov is an associate professor in energy planning and renewable energy systems at Aalborg University, Denmark. She holds a PhD in electrofuels and renewable energy systems. Her research is focused on electrofuel (PtX) pathways for liquid and gaseous fuels for 100% renewable systems and energy system analysis of these pathways from technical, socioeconomic, and policy perspectives. Analyses include demand side, energy system impacts, cross-sectorial integration, and utilization of by-products such as waste heat from electrolysis. She is the program director of the MSc program *Sustainable Cities* and a member of the *Study Board of Planning and Land Surveying* at Aalborg University. She is also an international member of the Croatian Academy of Engineers and a member of International Scientific Council of University of Rijeka.

Peter Sorknæs is an associate professor at the Department of Sustainability and Planning at Aalborg University. He has comprehensive experience with research on energy system analysis both at national and regional levels, with a focus on the operational characteristics of energy systems with 100% renewable energy and on district heating and energy markets. He also has research experience in the operation strategies of energy communities, district heating plants, and technologies in market-based energy systems.

Acknowledgments

First, I would like to thank all my colleagues at the Department of Sustainability and Planning at Aalborg University. You have contributed to the creation of an interdisciplinary environment in which new thoughts are welcome, and useful and fruitful comments are always made on the basis of a variety of different skills and competences. I have used this professional richness many times throughout my career. In particular, I would like to thank Frede Hvelplund for many years of excellent friendship and research partnership, including the joint involvement in most of the cases on which this book is based. Without your help and inspiration, I would never have been able to write this book.

Thanks to Mette Reiche Sørensen from the department for providing excellent and efficient linguistic support, as well as making comments to clarify the discussions in this book.

Thanks to the following former and current members of the research group of Sustainable Energy Planning for important contributions to Chapters 6, 7, and 8: Poul Alberg Østergaard, Bernd Möller, Brian Vad Mathiesen, Peter Sorknæs, Iva Ridjan Skov, Jakob Zinck Thellufsen, Miguel Chang, and David Connolly. Also thanks to other members of the group for many years of good collaboration, as well as helpful comments made on the manuscript: Karl Sperling, Steffen Nielsen, Poul Thøis Madsen, Mikkel Kany, Meng Yuan, and Rasmus Magni Johannsen.

Thanks to Tim Richardson, Aalborg University; Thomas B. Johansson, Lund University; and Olav Hohmeyer, Flensburg University for competent and inspiring discussions during the assessment and defense of the content of the first edition of this book for the senior doctoral degree in April 2009. Thanks to Niels I. Meyer, Andrew Jamison, Bent Flyvbjerg, and Jes Adolphson for contributing with helpful comments on the book. Moreover, thanks to Woody W. Clark II for many years of inspiring collaboration. Thanks to Anders N. Andersen of EMD International for convincing me to convert the EnergyPLAN model into Windows-based Pascal. Also thanks to Ebbe Münster, Henning Mæng, and Leif Tambjerg of PlanEnergi and EMD for helping me design, test, and develop the model during the last many years.

Thanks to Sigurd Lauge Pedersen of the Danish Energy Agency, Jens Pedersen of Energinet.dk, and Hans Henrik Lindboe of EA Consulting for inspiring collaboration on modeling during the work in the expert group on CHP and renewable energy in 2001 and onward. Also thanks to Poul Erik Morthorst and Kenneth Karlsson of

DTU/Risø National Energy Laboratory and Peter Meibom of the Danish Energy Association for providing similar inspiration.

Thanks to the team of the Dubrovnik Conferences, Naim Afgan, Noam Lior, and, in particular, Neven Duić and Goran Krajačić of University of Zagreb, for contributing with inspiring discussions and comparative studies on different energy system analysis models.

Thanks to Willett Kempton of Delaware University for fruitful collaboration on the modeling of V2Gs (vehicle to grid), as well as his contribution to Chapter 5.

Thanks to Brian Elmegaard of the Technical University of Denmark and Axel Hauge Pedersen of DONG Energy as well as Henning Parbo and Kim Behnke of Energinet.dk for contributing with helpful comments on the modeling of CAES.

Thanks to the steering committee of the Danish Society of Engineers' "Energy Year 2006" for inviting me and my colleague Brian Vad Mathiesen to conduct the overall technical and economic analyses of the project: Søren Skibstrup Eriksen, Per Nørgaard, Kurt Emil Eriksen, John Schiøler Andersen, Thomas Sødring, Charles Nielsen, Hans Jørgen Brodersen, Mogens Weel Hansen, and Bjarke Fonnesbech. Also thanks to all participants in "Energy Year 2006" whose inputs and expertise formed the basis for the study.

Thanks to the Danish Society of Engineers' (IDA) ad hoc expert group behind the two reports of the "IDA Climate Response" work, which has formed the basis of Chapter 8 of this edition on energy systems in fully decarbonized societies: Monika Skadborg, Per Homann Jespersen, Niels Brock, Torben Nørgaard, Felicia Fock, Søren Linderoth, Peter Bach, Laura Klitgaard, and Pernille Hagedorn-Rasmussen.

Thanks to the interdisciplinary and inspiring team of the CEESA project including Henrik Wenzel and Lorie Hamelin, University of Southern Denmark; Claus Felby and Niclas Scott Bentsen, Copenhagen University; Peter Karnøe, Per Christensen, Birgitte Bak-Jensen, Mads Pagh Nielsen, Jayakrishnan R. Pillai, and Erik Schaltz, Aalborg University; Thomas Astrup, Davide Tonini, Morten Lind, Kai Heussen, Frits M. Andersen, Marie Münster, and Lise-Lotte P. Hansen, Technical University of Denmark; and Jesper Munksgaard as well as several other colleagues already mentioned above.

Thanks to the team on the Zero Energy Buildings Research Center including Anna Marszal and Per Heiselberg, Aalborg University, and Svend Svendsen, Technical University of Denmark. Also thanks to Urban Persson and Sven Werner, Halmstad University, as well as Robin Wiltshire from the UK Building Research Establishment for a great deal of inspiration regarding the role of future district heating grids.

Thanks to Xiliang Zhang, Tsinghua University in Beijing as well as Wen Liu, Utrecht University, for their contribution to Chapter 7.

Thanks to the team of the Aalborg Energy Office of the early 1980s, especially Poul Bundgaard, for his involvement in the Nordkraft power station case.

Thanks to my fellow students in 1984, Frank Rosager, Henning Mæng, Lars Boye Mortensen, and Sofie Jörby, for designing "Alternative 4" in the Aalborg Heat Planning case. Also thanks to city council member Willy Gregersen for insisting on the involvement of university staff in "real-life" problems and planning procedures both in the Heat Planning case and in the case of Nordjyllandsværket.

Thanks to the Biomass Secretariat of the Danish Energy Agency, in the early 1990s headed by Helge Ørsted Pedersen and Kaare Sandholt, for involving me in analyzing large-scale biogas plants.

Thanks to the many people and organizations that became involved in the case of Nordjyllandsværket. I especially thank Peter Høstgaard Jensen and Flemming Nissen from the power companies for presenting persistent and very competent counterarguments throughout the public debate.

Thanks to all the people and organizations that were involved in the transmission line case in the mid-1990s, including the East Himmerland Energy Office headed by Marianne Bender. Moreover, I would like to thank county council members Thyge Steffensen and Karl Bornhøft for making every possible effort to secure a decent inclusion of relevant alternatives in the decision-making.

Thanks to Ulrich Jochimsen of Netzwerk Dezentrale Energienutzung for initiating our involvement in the Lausitz case in 1992. Also thanks to Niels Winther Knudsen and Annette Grunwald for fruitful collaboration on the design and promotion of an alternative energy strategy.

Thanks to the General Workers' Union, in particular Ole Busck and Sussi Handberg, for inviting me to participate in the making of the Green Energy Plan in the mid-1990s.

Thanks to Ejwin Beuse and Finn Tobiesen from the Danish Organisation for Renewable Energy for involving me and my colleagues in Thai energy planning. Also thanks to the following participants in the workshop in Bangkok in 1999 for initiating and contributing to the Prachuap Khiri Khan power plant case: Decharut Sukkumnoed and S. (Bank) Nunthavorakarn of Kasetsart University and Aroon Lawanprasert and Sumniang Natakuatoong of Thammasat University in Bangkok.

Thanks to Asbjørn Bjerre for initiating my involvement in the design of feasibility studies of Danish wind power, including the case of the Economic Council in 2002. Also thanks to Karl Emil Serup from Aarhus School of Business, Aarhus University, and Carsten Heyn-Johnsen and Erik Christensen from Aalborg University for helpful discussions and comments.

Last, but not least, I would like to express my gratitude to my wife, Søsser Lund, both for bringing me along when she involved herself in kinesiology in the 1980s, which later inspired the term *Choice Awareness*, and for listening and participating in many talks on the subject of this book.

Finally, I would like to thank my two daughters, Olivia and Fanny, for each drawing a picture for this book, one showing a wind turbine illustrating renewable energy and the other illustrating Hobson's Choice: "This horse or none!"

Henrik Lund
July 2023

Abbreviations

Power plant technologies

CAES	compressed air energy storage
CCS	carbon capture and storage
CCU	carbon capture and utilization
CHP	combined heat and power
COP	coefficient of performance (ratio between the output heat and input work/electricity of a heat pump)
HTL	hydrothermal liquefaction
PP	power plant (condensing unit)

Electricity demand and production

CEEP	critical excess electricity production
DSM	demand side management
EEEP	exportable excess electricity production

Renewable energy and fuels

DME	dimethyl ether—the first derivative of methanol
PV	photovoltaic
RES	renewable energy sources

Transportation

BEV	battery electric vehicle
EV	electric vehicle
HFCV	hydrogen fuel cell vehicle
pkm	person kilometer (person transportation)
tkm	ton kilometer (freight transportation)
V2G	vehicle to grid (vehicle supplying power to the public grid)

Buildings and energy infrastructures

4GDC	4th generation district cooling
4GDH	4th generation district heating
DH	district heating
ZEB	zero energy building/zero emissions building

Policy and planning

EIA	Environmental Impact Assessment
GIS	geographical information systems

Economy

DEC	Danish Economic Council
DKK	Danish Kroner
DM	Deutsche Mark
EUR	Euro
GDP	gross domestic product
O&M	operation and maintenance
THB	Thai Baht
USD	U.S. Dollar

Energy and power units

GW	gigawatt (power capacity unit equal to 1 billion watt)
GWh	gigawatt hour (energy unit equal to 1 million kWh)
MW	megawatt (power capacity unit equal to 1 million watt)
MWe	megawatt electric output
MWth	megawatt thermal output
PJ	peta joule (energy unit equal to 1 million billion joule)
TWh	terawatt hour (energy unit equal to 1 billion kWh)

Introduction

How can society convert to 100 percent renewable energy and become fully decarbonized? The answer to that question is the main topic of this book. Two important aspects must be considered. First, from a technical point of view, which technologies can we use to make sure that the resources available meet the demands? To answer this question, this book presents an energy system analysis methodology and a tool for the design of renewable energy systems. This part includes the results of around 20 comprehensive energy system analysis studies with a focus on the implementation of 100 percent renewable energy systems. Moreover, as part of this new edition, a chapter on fully decarbonized societies has been added (Chapter 8).

Second, in terms of politics and social science, how can society implement such a technological change? To answer that question, this book introduces a theoretical framework approach, which aims at understanding how major technological changes, such as the transition to renewable energy, can be implemented at both the national and international levels. This second aspect involves the formulation of the Choice Awareness theory, as well as the analysis of 11 major empirical cases from Denmark and other countries.

Regarding the implementation of the change from fossil fuels to renewable energy, Denmark is an interesting case. Like many other Western countries, Denmark was totally dependent on the import of oil at the time of the first oil crisis in 1973. Almost all transportation and residential heating were based on oil. Furthermore, 85 percent of the electricity supplied in Denmark was produced from oil. Altogether, prior to the oil crisis, more than 90 percent of its primary energy supply was based on oil.

Denmark, like many other countries, was unprepared for the sudden rise in oil prices. Danish energy planning had been based on the principle of supply meeting demand. Power stations were planned and built on a prognosis based on the historical development of needs. Denmark had no minister of energy and no energy department, no action plans in the case of being cut off from oil supplies, and no long-term strategy for the future in case oil resources were depleted.

Nevertheless, now 50 years later, Danish society has proved its ability to implement rather remarkable changes. Fig. 1.1 shows the development of the primary energy supply of Denmark since 1972 and illustrates two important factors: most of the oil consumption has been replaced by other fuels—for example, coal, natural gas, and, to some extent, renewable energy—and Denmark has managed to first stabilize the primary energy supply at the same level as in 1972 and in recent years, even decrease it. This stabilization and decrease are unique compared to other countries, as they have been achieved simultaneously with a "normal western European" economic growth. The primary means have been energy conservation and efficiency improvements in the supply. Buildings have been insulated, and combined heat and power (CHP) production has been expanded. Thus, 50 years later, the primary energy supply for heating has been reduced to two-thirds of the level prior to 1973, even though the

Renewable Energy Systems. https://doi.org/10.1016/B978-0-443-14137-9.00001-2

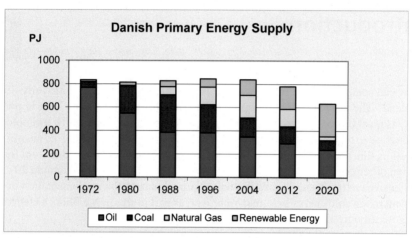

Fig. 1.1 Danish primary energy supply.

heated space area has increased by more than 50 percent in the same period. The renewable energy share of the primary energy supply has increased from around zero in 1972 to more than 40 percent in 2020, and wind power production has grown to approx. a 50 percent share of the electricity demand. Moreover, the share of green gas based on biogas in the natural gas supply has reached 30 percent in 2022 and is increasing rapidly.

Since 2006, the target of the Danish Government has been that Denmark will become completely fossil fuel free. Since the publication of previous editions of this book, Denmark has first defined a target of using 100 percent renewable energy in its energy and transportation sectors by 2050 (Danish Energy Agency, 2012) and, secondly, has defined targets of decreasing greenhouse gas emissions by 70 percent in 2030 and becoming fully decarbonized in 2050 (Danish Ministry of Climate, Energy and Utilities, 2020).

Moreover, Denmark started to produce oil and natural gas from the North Sea in the early 1980s and has been more than self-supplied with energy in the period 1997–2012. However, the Danish oil and gas resources are scarce and are likely to last for only a few decades. Can Denmark convert to 100 percent renewable energy and a fully decarbonized society within a few decades, or will it have to return once again to the former dependence on imported fossil fuels? This question is indeed relevant not only to Denmark but to Europe in general as well as the United States, China, and many other nations around the world.

The idea of this book is to unify the results and deduce the learning of a number of separate studies and thereby contribute to a coherent understanding of how society can implement renewable energy systems. The book is based on 25 years of involvement in a number of important and representative political decision-making processes in Denmark and other countries. As we will see, these processes reveal the lack of ability of organizations and institutions linked to existing technologies to produce and promote proposals and alternatives based on radical changes in technology.

On the other hand, the stabilization of the primary energy supply shown in Fig. 1.1 proves that the ability to act as a society has been possible, despite conflicts with representatives of the old technologies. In Denmark, during this period, official energy objectives and plans have been developed due to a constant interaction between Parliament and public participation. In this interaction, the description of new technologies and alternative energy plans has played an important role.

The theory of Choice Awareness seeks to understand and explain why the best alternatives are not described and developed per se and what can be done about it. Choice Awareness theory argues that public participation, and thus the awareness of choices, has been an important factor in successful decision-making processes and puts forward four strategies to help along these processes.

1 Book contents and structure

Fig. 1.2 shows the structure of this book. The Choice Awareness section (the gray area) includes a theoretical understanding and a framework for the development of renewable energy system analysis tools and methodologies (the white area). This chapter introduces both aspects and provides some important definitions.

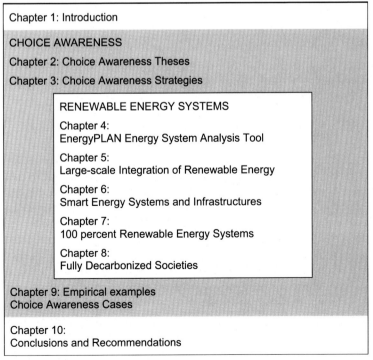

Fig. 1.2 Contents and overall structure of the book.

Chapter 2 introduces the Choice Awareness theory, which deals with how to implement radical technological changes such as renewable energy systems. This theory argues that the perception of reality and the interests of existing organizations will influence the societal perception of choices. Often these organizations seek to hinder radical institutional changes by which they expect to lose power and influence. Choice Awareness theory states that one key factor in this manifestation is the societal perception of having either a *choice* or *no choice*.

The Choice Awareness theory presents two theses. The first states that when society defines and wishes to implement objectives implying radical technological change, existing organizations will often seek to create the perception that the radical change in technologies is not an option and that society has *no choice* but to implement a solution involving the technologies that will save and constitute existing positions. The second thesis argues that, in such a situation, society will benefit from focusing on Choice Awareness, that is, raising the awareness that alternatives *do* exist and that it is possible to make a choice. Four key strategies are identified from which society will benefit when seeking to raise Choice Awareness.

Chapter 3 elaborates on the Choice Awareness strategies related to the second thesis: the design of concrete technical alternatives, feasibility studies based on institutional economic thinking, the design of public regulation measures, and the promotion of a democratic infrastructure based on new corporative regulation.

Chapter 4 describes the method of designing concrete technical alternatives based on renewable energy technologies. This method distinguishes between three implementation phases: introduction, large-scale integration, and 100 percent renewable energy systems. The need for simulation tools in the two latter phases is especially emphasized. Both methodology and tool development are discussed in relation to the theoretical framework of Choice Awareness, and the energy system analysis tool, EnergyPLAN, is described. The EnergyPLAN model is a freeware that can be accessed from the home page, www.EnergyPLAN.eu, together with documentation and a training program.

Chapter 5 refers to and deduces the essence of a wide range of studies of the Danish energy system. In these studies, the EnergyPLAN model has been applied to the analysis of large-scale integration of renewable energy. The Danish energy system is characterized by a high share of renewable energy and is therefore a suitable case for the analysis of further large-scale integration. The question in focus is how to design energy systems with a high capability of utilizing intermittent renewable energy sources. This chapter describes important methodology developments and compares the capability of different systems, including how they treat the fact that the fluctuations and intermittence of, for example, wind power differ from one year to another.

Chapter 6 adds to the previous chapter by focusing on infrastructures and introduces the term *Smart Energy Systems*. In recent years, a number of new names and definitions of subsystems have been promoted to define and describe new paradigms in the design of future energy systems, such as *smart grid* and *power-to-X*. These infrastructures are essential new components in the design of future renewable energy systems. However, each of them is also a subsystem that cannot be fully understood or analyzed if it is not properly put into the context of the overall energy system.

Chapter 6 illustrates how this context can be established by the use of the EnergyPLAN tool and presents the results of a number of recent studies including the role of the smart grid, future district heating technologies and systems, as well as smart transportation and power-to-X systems.

Chapter 7 proceeds with the topics of the previous chapters and presents a number of results achieved by applying the EnergyPLAN tool to the design of 100 percent renewable energy systems. The question in focus is how to compose and evaluate these systems. This chapter treats the principal changes in the methods of analysis and evaluation applied to these systems compared to systems based on fossil fuels with or without large-scale integration of renewable energy. Focus is on the application of concrete institutional economics. How can countries and regions afford to invest in renewable energy without further increasing public expenditures? And which methodologies should be applied to design such solutions?

Chapter 8 puts the previous chapters on renewable energy systems into the context of achieving a fully decarbonized society. To reach this goal, efforts within energy and transportation must be coordinated with other sectors contributing to greenhouse gas emissions, such as agriculture and industrial processes. The difficulties of decreasing greenhouse gasses in these sectors lead to the need for carbon sinks such as carbon capture and storage (CCS) and biochars. Therefore, the identification and inclusion of sustainable biomass become essential to meeting such political goals in a fast and affordable way.

Chapter 9 returns to the discussion of the theoretical framework. This chapter refers to a number of cases applying Choice Awareness strategies to specific decision-making processes of energy investments in the period since 1983. Typically, researchers have been involved in these processes through the design and introduction of concrete technical alternatives and/or the application of other Choice Awareness strategies. The cases refer to a large number of publications and documented cases. Chapter 9 seeks to deduce what can be learned from the cases with regard to the Choice Awareness theses and strategies formulated in Chapters 2 and 3.

Chapter 10 returns to the concerted discussions of the two aspects: Choice Awareness and renewable energy systems. This chapter presents reflections and conclusions drawn from this book in terms of the implementation of renewable energy systems at the societal level.

2 Definitions

Both the issue of Choice Awareness and the case of renewable energy systems involve some basic definitions, which are provided in the following section.

Choice Awareness

The theory of Choice Awareness addresses the societal level. It concerns collective decision making in a process involving many individuals and organizations representing different interests and discourses, as well as different levels of power to influence the decision-making process. The term *choice* obviously plays an important role in the definition of Choice Awareness. The Oxford English Dictionary (2008) defines choice as "the act of choosing; preferential determination between things proposed; selection, election." Choice involves the act of thinking and the process of judging the pros and cons of multiple options and selecting one of the options for action. This book distinguishes between a *true* choice and a *false* choice.

A *true* choice is a choice between two or more real options, while a *false* choice refers to a situation in which the choice is some sort of illusion. Some examples of false choices are a Catch-22 and Hobson's Choice—in other words, a free choice in which actually only one option is offered. These two types of false choices will be explained further in Chapter 2. More examples of false choices are blackmail and extortion, both involving the condition to either do what you are told or suffer unpleasant consequences.

The Oxford English Dictionary (2008) defines awareness as "the quality or state of being aware; consciousness." In biological psychology, awareness comprises a person's perception and cognitive reaction to a condition or event. In principle, awareness does not necessarily imply understanding; it is just an ability to be conscious of, feel, or perceive. However, here the term is combined with *choice*, which implies the acts of thinking and judging. Thus, Choice Awareness does involve an element of understanding. Choice Awareness is used here to describe the *collective perception* of having a *true choice*. Moreover, this situation involves a cognitive reaction in terms of judging the merits of relevant options and selecting one of them for action.

Collective perception is defined as a general perception in society. It does not include a few individuals who know better or different; the fact that a single person comes up with new ideas or invents new alternatives does not change the collective perception, as long as the person keeps these ideas to him- or herself. Only if individuals raise awareness by convincing or informing the public in general does this knowledge become part of the collective perception. In the same manner, the collective perception may be manipulated by individuals or organizations if they prove successful in convincing society in general that a certain alternative does not exist—meaning that it does not comply with technical requirements or other regulations.

Choice-eliminating mechanisms influence the collective perception in the direction of not having a choice at all or having a false choice, as just described. *Raising Choice Awareness* involves influencing the collective perception in the direction of having a true choice and identifying and understanding the pros and cons of relevant alternatives.

Radical technological change

Choice Awareness theory is concerned with the implementation of radical technological change. *Technology* is defined as one of the means by which mankind reproduces and expands its living conditions. The definition of *technology* embraces a combination of four elements—technique, knowledge, organization, and products—and is discussed further in Chapter 2.

Radical technological change is defined as a change of more than one of the four elements of technology. In Choice Awareness, special focus is placed on the change of existing organizations, and a distinction is made between organizations and institutions. *Organization* is defined as a social arrangement that pursues collective goals, controls its own performance, and has a boundary that separates it from its environment. Typical examples of organizations are companies, nongovernmental organizations, businesses, and administrative units. *Institutions* are structures and mechanisms of social order and cooperation. They govern the behavior of more than one individual and/or organization, and they include a formal regime for political rule making and enforcement. Thus, in short, one may say that institutions are organizations including all of the written laws and regulations and all of the unwritten codes of culture regulating them.

Applied and concrete economics

Choice Awareness theory involves four strategies in which concrete institutional economics, as opposed to applied neoclassical economics, plays an important role. *Applied neoclassical economics* is defined as neoclassical-based methods, such as cost-benefit analyses and equilibrium models, applied to existing real-life market economies. This is seen as opposed to theoretically correct methods applied to market economies that fulfill the theoretical assumptions of a free market. As discussed further in Chapter 3, the theory of neoclassical market economics is based on a number of assumptions that are not fulfilled in real-life market economies. The critique of neoclassical-based methods of this book is directed toward the real-life applications of the methods, and it is not decisive whether this critique is valid for theoretically correct applications or not.

Concrete institutional economics is defined as economics that deals with the concrete institutional conditions that form the development of a specific society. *Institutional economics* focuses on the understanding of the role of human-made institutions in the shaping of economic behavior. The concrete institutional conditions vary from one society to another, and the method linked to concrete institutional economics therefore deals with defining analytical aims, contexts, and aggregation levels for the analysis of the concrete societal institutions in a specific society (Hvelplund, 2005, pp. 91–95).

Renewable energy

Renewable energy is defined as energy that is produced by natural resources—such as sunlight, wind, rain, waves, tides, and geothermal heat—that are naturally replenished within a time span of a few years. Renewable energy includes the technologies that convert natural resources into useful energy services:

- Wind, wave, tidal, and hydropower (including micro- and river-off hydropower)
- Solar power (including photovoltaic), solar thermal, and geothermal
- Biomass and biofuel technologies (including biogas)
- Renewable fraction of waste (household and industrial waste)

Household and industrial waste is composed of different types of waste. Some parts are considered renewable energy sources—for example, potato peel—whereas other parts, such as plastic products, are not. Only the fraction of waste that is naturally replenished is usually included in the definition. In this book, however, for practical reasons, the whole waste fraction is included as part of the renewable energy sources identified in some analyses.

Renewable energy systems

Renewable energy systems are defined as complete energy supply and demand systems based on renewable energy as opposed to nuclear and fossil fuels. They include supply as well as demand. The transition from traditional nuclear and fossil fuel-based systems to renewable energy systems involves coordinated changes in the following:

- Demand technologies related to energy savings and conservation
- Efficiency improvements in the supply system, such as CHP
- Integration of fluctuating renewable energy sources, such as wind power

A distinction can be made between *end use* and *demand*. *Energy end use* is defined as the human call for energy services such as room temperature, light and transportation. *Energy demand* is defined as consumer demands for heat, electricity, and fuel. Consumers include households and industry as well as public and private service sectors. Fuel may be used for heating or transportation. Heat demand may be divided into different temperature levels such as district heating and process heating.

Within end use, one may distinguish further between, on the one hand, basic needs such as food, basic temperatures, and transportation from home to work, and on the other hand, specific requirements such as a certain number of square meters with a certain room temperature and a certain number of kilometers of driving. This distinction can be critical, for example, when analyzing the transportation infrastructure related to food production or to transportation between home and work. However, in the analyses presented in this book, it has not been necessary to make this distinction.

Changes such as insulation and efficiency improvements of electric devices leading to changes in the energy demand for heat, electricity, or fuel are defined as *changes in the demand system*. In addition to the preceding renewable energy technologies, renewable energy systems include both technologies which can convert from one form of energy into another—for example, electricity into hydrogen—as well as storage technologies that can save energy from one hour to another. Mathiesen and Lund (2009) and Blarke and Lund (2008) comprised these technologies under the designation relocation technologies. However, in the following, the difference between *energy conversion* and *energy storage* technologies is emphasized.

Energy conversion technologies are technologies that can convert from one demand (heat, electricity, or fuel) to another, such as the following:

- Conversion of fuel into heat and/or electricity by the use of power stations, boilers, and CHP (including steam turbines as well as fuel cells)
- Conversion of electricity into heat by the use of electric boilers and heat pumps

- Conversion of electricity and solid fuels into gas or liquid fuel by the use of electrolyzers and biomass conversion plants

Energy storage technologies are defined as technologies that can store various forms of energy from one hour to another, such as the following:

- Fuel, heat, and electricity storage technologies
- Compressed air energy storage
- Hydrogen storage technologies

The definition of *storage technologies* is broader than the concept of storage itself. For example, in the case of electricity, which is stored by converting it into hydrogen, the storage technology may include conversion technologies such as electrolyzers and fuel cells. The distinction between conversion and storage technologies is defined by the purpose of the technology in question. If the purpose is to convert electricity to hydrogen because a car needs hydrogen, then the electrolyzer is defined as a conversion technology. However, if the purpose is to store electricity, then the combination of electrolyzer, hydrogen storage, and fuel cell is defined as a storage technology.

In complex renewable energy systems, single components may be used for both purposes. For instance, the same electrolyzer may be used to supply cars with hydrogen and, at the same time, produce hydrogen for storage purposes. In this case, the electrolyzer is simply regarded as both a conversion and a storage technology.

The distinction between the two types of technologies is important when designing renewable energy systems, as will be elaborated on in Chapters 4–8. It is important to distinguish between, on the one hand, the need for balancing time, and on the other hand, the need for balancing the annual amounts of different types of energy demands.

Smart energy systems

The transformation toward future renewable energy systems poses a challenge as it involves substantial changes in the infrastructures to carry the energy, i.e., the electricity grid, the gas grid, and the district heating and cooling grids. Moreover, future energy systems may introduce new grid infrastructures to handle hydrogen and CO_2 and/or they may be part of reorganizing existing gas grids. All these grids face the common challenge of facilitating distributed activities that involve the interaction with consumers and bidirectional flows. To meet this challenge, all grids will benefit from the use of modern information and communication technologies as an integrated part of the grids at all levels. On this basis, as elaborated in Chapter 6, the following smart grid concepts have been defined:

Smart Electricity Grids are defined as electricity infrastructures that can intelligently integrate the actions of all users connected to them—generators, consumers, and those that do both—in order to efficiently deliver sustainable, economic, and secure electricity supplies. *Smart Thermal Grids* are defined as a network of pipes connecting the buildings in a neighborhood, town center, or whole city, so that they can be served from centralized plants as well as from a number of distributed heating or cooling production units including individual contributions from the connected buildings.

Smart Gas Grids are defined as gas infrastructures that can intelligently integrate the actions of all users connected to them—suppliers, consumers, and those that do both—in order to efficiently deliver sustainable, economic, and secure gas supplies and storage.

All three types of smart grids are important contributions to future renewable energy systems. However, each individual smart grid should not be seen as separate from the others or separate from the other parts of the overall energy system. First, it does not make much sense to convert one sector to renewable energy if this is not coordinated with a similar conversion of the other parts of the energy system. Second, this coordination makes it possible to identify additional and better solutions to the implementation of smart grid solutions within the individual sector, compared to the solutions identified with a sole focus on the sector in question. Consequently, this book promotes the concept of *Smart Energy Systems*.

Smart Energy Systems are defined as an approach in which smart electricity, thermal and gas grids are combined and coordinated to identify synergies between them to achieve an optimal solution for each individual sector as well as for the overall energy system.

3 Renewable versus sustainable

This book often uses the term *renewable* energy, but why not use *sustainable* energy instead? After all, in many situations, these two terms are used interchangeably. However, significant differences can be found between the two terms. Thus, renewable is used for the reasons described in the following.

Sustainable energy

Sustainable energy can be defined as energy sources that are not expected to be depleted in a time frame relevant to the human race; therefore they contribute to the sustainability of all species.

This definition of sustainable energy and the preceding definition of renewable energy represent typical definitions of both terms. They match rather closely the definitions given by the Internet encyclopedia Wikipedia. These definitions, however, reveal a difference in the significance of the two terms. Most important, Wikipedia (2008) includes the word *nuclear* in the sources defined as sustainable energy sources. However, as Wikipedia adds, for social and political reasons, there is a controversy as to whether nuclear sources should be regarded as sustainable. Nevertheless, at the present technological stage, nuclear is not sustainable, since it needs uranium, which is a scarce resource within the relevant time frame.

The same discussions seem to apply to carbon capture and storage (CCS) technologies, which have recently been promoted as an important solution in the debate on how to combat climate change. Very often this solution is proposed in relation to the use of fossil fuels, especially coal. Often the question is posed, why not continue to burn coal and even invest in new coal-fired power stations when this technology can be made sustainable by the use of CCS?

On the other hand, even though sustainable energy sources are most often considered to include all renewable sources, some renewable energy sources do not necessarily fulfill the requirements of sustainability. For instance, the production of biofuels such as ethanol from fermentation has in some life cycle analyses proven to be non-sustainable. Again, this is a controversy that has not yet found a consensus.

Nevertheless, the conclusion is that sustainable energy in some definitions *may* include nuclear and fossil fuels in combination with CCS, while these technologies and sources are not included in the definition of renewable energy. On the other hand, renewable energy may include some biomass resources that *may* prove not to be sustainable.

Political reasons for renewable energy

Besides the preceding difference, another important disparity exists between renewable and sustainable. This has to do with the reasons for wishing for technological change. Why does society *want* to implement renewable energy solutions? And why does society aim at implementing sustainable energy solutions? The reasons for introducing sustainable solutions are mainly, if not solely, related to an environmental motive. However, several reasons can be found for implementing renewable energy.

In the article "Choice Awareness" (Lund, 2000) and in Chapter 23 of the book *Tools for Sustainable Development* (Lund, 2007b), I described the recent history of Danish energy planning and policy since the first oil crisis in 1973. At least three main reasons can be defined for replacing fossil fuels by technologies related to renewable energy systems, including energy conservation and efficiency measures:

- *Energy security*, with an emphasis on oil dependence (and oil depletion). This reason played an all-important role in Danish society in the 1970s and has, in the beginning of the 21st century, experienced a revival caused by increasing oil prices in combination with the relations between the Western world and the governments in power of the remaining oil reserves. Moreover, it has once more become highly important as a result of the European energy crisis due to the Russian war in Ukraine.
- *Economics*, with an emphasis on job creation, industrial innovation, and the balance of payment. This reason took over and played a major role in Denmark in the 1980s. The main problem changed from being based on the issue of whether we could *get* the oil to whether we could *afford* it. This reason became the driving political force behind the industrial development of, among others, solar thermal and wind power in Denmark in the 1980s and 1990s.
- *Environment and development*, with an emphasis on climate change. This third reason became a key issue in the 1990s after the introduction of the Brundtland report (United Nations, 1987) and has since been of increasing social importance along with the rising discussions on global warming.

All three reasons have formed part of the political discussions and have been identified as political goals of the Danish Energy Policy during the entire period. However, the main focus has changed in such a way that each reason has been considered the most important one from one decade to another.

The concerns related to energy security are based on the underlying fact that fossil fuels constitute a limited resource. The United Nations' discussions on environment and development are based on the fact that energy consumptions are indeed not equally distributed between what is considered the rich and the poor countries of the world, respectively. This is described and discussed in relation to the rising global energy consumption in the paper "The Kyoto Mechanisms and Technological Change" (Lund, 2006a). In this paper, it is argued that the introduction of the so-called Kyoto mechanisms actually had the opposite effect of what the United Nations intended. The Kyoto mechanisms allow rich countries to implement climate change projects in other countries instead of decreasing their own emissions. Thereby, they increase the differences in energy consumption between rich and poor countries rather than decreasing them. Moreover, the mechanisms may slow down the needed technological development.

With regard to the discussion of the difference between *renewable* and *sustainable* energy, a main point is that if society accepts nuclear power and fossil fuels in combination with CCS as parts of the solution, society may be able to achieve parts of the environmental goals. However, society will not be able to solve the fundamental problems of scarce and limited resources of fossil fuels and uranium. Seen from the point of view of a Western country such as Denmark, society will not be able to meet the goals of energy security, and, with regard to economics, Denmark will still have to import fossil fuels and/or uranium.

Renewable energy and democracy

Another difference between renewable and sustainable energy is worth discussing. This difference has democracy as its main focus and is quite relevant to the issue of energy systems and the implementation of technological change such as those addressed by the Choice Awareness theory, as we will see in Chapter 2.

In the 1970s, an energy movement arose in Denmark as in many other Western countries. This movement was comprised of, among others, the antinuclear movement (OOA) and the Danish Organization for Renewable Energy (OVE). When the OOA was created and these energy problems were discussed, the issues of democracy and living conditions in local communities played major roles in the arguments against nuclear and in favor of renewable energy. With regard to nuclear, some were afraid of the consequences of implementing the technology in terms of security and ownership. The question was how to guard the plants and the transportation of radioactive waste without having to hire security staff and erecting fences. Who should own and operate these big power stations? If ownership was assigned to big companies, it would mean that local communities would lose influence. Also, how should space for nuclear power stations be allocated and radioactive waste be disposed of without impacting the quality of life of the communities involved? The antinuclear movement (OOA, 1980) discussed all of these concerns, as well as the relations between nuclear power and nuclear weapons.

The issue of local ownership also played an important role in the discussions on renewable energy. To have an influence on these decisions, the citizens in the local communities preferred to have their own renewable sources of energy instead of depending on nuclear or imported fossil fuels (OVE, 2000). As we will see in Chapter 2, these beliefs conform to the concept of Choice Awareness, which highlights the benefits of choice in creating a "life worth living" at both the individual and societal levels.

In this view, the difference between renewable and sustainable energy becomes important. If society accepts nuclear power and fossil fuels in combination with CCS as major parts of the solution, the technological change may not meet the local communities' wishes to improve their influence on decisions that are important to their lives. In short, one can say that the implementation of renewable energy systems helps to create what I, in Chapter 3, refer to as a *suitable democratic infrastructure* according to the Choice Awareness theory. This suitable democratic infrastructure may improve the awareness of choices and thereby, in general, create better living conditions. On the other hand, an improved democratic infrastructure will also improve the circumstances of making the choice of implementing renewable energy systems. More important, depending on the specific definition applied, this may not be the case with sustainable energy systems. Based on these considerations, the term *renewable* energy systems is used in this book.

Choice Awareness theses

2

This chapter introduces the two theses of the Choice Awareness theory. The theory deals with *how* to implement radical technological changes such as renewable energy systems. The step from nuclear and fossil fuel systems to renewable energy systems involves a radical technological change. As we will see, it cannot be implemented by existing organizations within existing institutions, but it will imply organizational and institutional changes. This means that someone will win and someone will lose as a result of these changes.

With a reference to discourse and power theories, the Choice Awareness theory emphasizes that different organizations see things differently. Existing organizational interests will thus seek to keep renewable energy proposals out of the agenda at many levels. The Choice Awareness theory is based on the observation that, in many cases, these conditions lead to a situation of *no choice*. As a society, we are subject to a collective perception that states, for example, that "We have no choice but to build another coal-fired power station." The Choice Awareness theory, however, maintains that this is not true: We *do* have a choice. The theory tells us how to be aware of this choice; thus, it enables us to debate our common future and make better decisions.

The theory addresses the societal level. It concerns collective decision making in a process that involves many individuals and organizations representing different interests and discourses, as well as different levels of power to influence the decision-making process. The theory is not comprehensive, but it emphasizes the key factor that existing organizational interests will often seek to eliminate choices from the political decision-making process.

The Choice Awareness theory advocates counterstrategies involving the design of technical alternatives, feasibility studies based on institutional economic thinking, and the design of public regulation measures seen in light of conflicting interests as well as changes in the democratic decision-making infrastructure. Those strategies are examined in more detail in Chapter 3.

1 Choice and change

The word *choice* obviously plays an important role in the definition of Choice Awareness. In this book, a distinction is made between a *true* choice and a *false* choice. As already defined in Chapter 1, a true choice is a choice between two or more real options, while a false choice refers to a situation in which choice is some sort of illusion. One example of a false choice is the concept referred to as Hobson's Choice, that is, a free choice in which only one option is offered. The "choice" is between deciding on the option or not. The phrase is said to originate from Thomas Hobson (1544–1630), who delivered mail between London and Cambridge by horse. When the horses were not needed for the mail delivery, they were rented out to students and academic

Renewable Energy Systems. https://doi.org/10.1016/B978-0-443-14137-9.00002-4

staff at the university. Hobson soon discovered that his best (and fastest) horses were the most popular ones and thus overworked. To prevent further exhaustion of his best horses, Hobson devised a strict rotation system, only allowing customers to rent the next horse in line. His policy, "This one or none," has come to be known as Hobson's Choice, when an apparent choice is in fact no choice at all (Smith, 1882).

Another example of a false choice is the prototypical Catch-22, as formulated by Joseph Heller in his novel of the same title (Heller, 1961). It considers the case of a bombardier who wishes to be excused from combat flight duty. To do so, he must submit an official medical certificate demonstrating that he is unfit because he is insane. However, according to Army regulations, any sane person would naturally not want to fly combat missions because they are so dangerous. By requesting permission not to fly combat missions on the grounds of insanity, the bombardier demonstrates that he is in fact sane and therefore is fit to fly. Conversely, any flyer who wanted to fly combat missions implicitly demonstrated that he was insane and was unfit to fly and should therefore be excused. To be excused, however, the unwilling individual had to submit a request, and, naturally, he never did. If the reluctant flyer did submit a request to be excused, the Catch-22 would assert itself, short circuiting any such attempt to escape from combat duty.

Choice/no choice at the individual level

As mentioned, the theory of Choice Awareness addresses the societal level, but the term itself is inspired by the activities that take place at the individual level. The term *Choice Awareness* is inspired by kinesiology, a method of treating emotional stress at the individual level. In 1984 and 1986, Stokes and Whiteside examined the emotional causes of physical problems, including dyslexia. They described a method of healing certain health problems by working with the emotional state of the individual. By using various psychological tools, Stokes and Whiteside believed that certain individual psychological problems could be cured.

An important tool is the behavioral barometer providing a systematic way of relating different feelings to one another. The idea of the barometer is that a person's feelings relate and respond to one another at the conscious, the subconscious, and the body levels, as defined by Stokes and Whiteside. For example, if a person feels "unappreciated" at the conscious level, according to the barometer, this feeling corresponds to other, related feelings at the subconscious and body levels. By understanding these relations between different feelings at different levels of consciousness and awareness, one can work with individual problems. In combination with various other methods, the behavioral barometer is a powerful tool for helping individual problems.

In kinesiology, the word *choice* plays an essential role. According to Stokes and Whiteside, every human being has from birth the feeling that "we have *no choice*; and without *choice*, we have no power."[a] The feeling of no choice is an emotional condition in which individuals may find themselves. This feeling may manifest itself

[a] Stokes and Whiteside (1986), Chapter 1, p. 8.

when a person experiences emotional problems. As expressed by Stokes and Whiteside, when "we buy in to no choice, we check out on our individuality, our self-worth, and the reality of spirit."[b]

According to Stokes and Whiteside, the feeling of no choice is both essential and fatal. If a person enters into a chronic state of feeling that he or she has no choice but to do something that he or she does not want to do, this condition is fatal. The cure is to make the person realize that he or she always has a choice. Even when experiencing the desperate situation of having no real alternative to begin with, one can still choose to say "no." The experience is that when a person accepts that he or she indeed has a choice to make—by saying no—he or she will be able to think of even better and real constructive alternatives. The point of Choice Awareness is that this *feeling*—or *perception*, as I prefer to call it—can also be observed at the collective and societal levels.

Choice/no choice at the societal level

During my participation in various decision-making processes (10 of which are discussed in Chapter 9), I found that the *perception* of no choice appeared many times and in several forms also at the *collective* and societal levels. By the term *collective perception*, I refer to the general perception in society. It does not include a few individuals who know better or differently. The fact that single persons get new ideas or come up with new alternatives does not change the collective perception, as long as they keep these ideas to themselves. Only if they raise awareness by convincing or informing the public in general does this knowledge become part of the collective perception. In the same manner, the collective perception may be manipulated by individuals or organizations if they prove successful in convincing society in general that a certain alternative does not exist, that is, it does not comply with certain requirements for technical or other reasons.

As we will see in Chapter 9, examples of the perception of no choice can be observed in the case of deciding for a new power station in my hometown of Aalborg in Denmark in the 1990s. At that time, the power company was owned by the municipality and electricity consumers, and a board of representatives and politicians were to make the decision. When the issue of electricity supply was put on the agenda, only one solution was presented: to build a coal-fired power station. One of the representatives was very frustrated by the decision-making process and really felt that he had no choice, since voting no would mean that, sooner or later, Aalborg would have no electricity supply at all. He wanted to consider an alternative in terms of a combined-cycle natural gas-fired power station, but this alternative was disregarded before the decision was to be made.

Later, the local county of Northern Jutland had to decide whether to approve the plant or not. In the public debate, the argument of job creation from the construction work played an important role. As described in detail in Chapter 9, the coal-fired power station was one of the least local job-creating alternatives one could imagine

[b] Stokes and Whiteside (1986), Chapter 1, p. 9.

when seen in relation to the lifetime of the plant. An investment in renewable energy in combination with fuel-saving technologies, such as conservation and distributed combined heat and power (CHP) plants, would save the imports of coal and leave more money for local job creation. However, the power companies' association in Western Denmark argued that if the coal-fired power station was not approved, they would spend the money investing in a power station somewhere else, outside the region of Northern Jutland. This led to a collective perception of no choice in the region. The alternative to the coal-fired power station and the jobs created, even though they were few, was nothing: "This power station or none!" A real Hobson's Choice.

At the regional level, no true choice existed. However, at a higher level, society did in principle have a choice. It would have been possible to implement an institutional change making CHP and renewable energy an option at the regional level.

I have also observed such collective perception of no choice in Eastern Germany, Thailand, and Vietnam. In all cases, the combat between discourses and the execution of power creates a situation in which the community ends up with the perception that "we have *no choice* but to build another coal-fired power station." Even though it is recognized that this solution is not beneficial to the environment, health, energy efficiency, security, job creation, or technological innovation, it seems to be the only option.

The general observation made here is that the construction of the collective perception of no choice plays an important role when making major societal decisions on energy planning. Choice Awareness, however, is crucial to the implementation of political aims and objectives when radical technological changes are needed.

While I was writing the first edition of this book, I saw a new example in the newspaper *Ingeniøren*, which was related to the discussion of climate change. In his opening speech to the Parliament in October 2006, Danish Prime Minister Anders Fogh Rasmussen announced the government's long-term objective for Danish energy policy: 100 percent independence from fossil fuels. When asked directly, the prime minister answered that nuclear sources will not form part of the solution. In other words, the long-term target is to convert to 100 percent renewable energy. This governmental objective has been repeated several times since then.

It is clear that coal does not form part of this long-term objective. However, the organizations that make a profit from burning coal have involved themselves in the promotion of coal in combination with carbon capture technology. The January 11, 2008, issue of Ingeniøren (2008a) dedicated two pages to a new report on how Denmark can decrease its greenhouse gas emissions by up to 80 percent before the year 2050. The article states that the report "is the first mapping of such large-scale reductions in Denmark."[c] It concludes with the statement

> *Electric vehicles, heat pumps, offshore wind power, and carbon capture and storage (CCS) are some of the technologies which we are simply forced to use.*[d]

---

[c] Translated from Danish: "Cowirapporten er den første kortlægning af så massive reduktioner i Danmark". *Ingeniøren,* January 11, 2008, p. 14.

[d] Translated from Danish: "Elbiler, varmepumper, havvindmøller og CCS er nogle af de teknologier, vi simpelthen bliver nødt til at benytte". *Ingeniøren,* January 11, 2008, p. 14.

This information, however, is not correct. Alternatives without coal and CCS do exist. At least three other surveys were conducted before the preceding one, one of them based on the joint effort of the Danish Society of Engineers (see Chapter 9) and all of them describing how 100 percent renewable energy systems without coal and CCS can be reached by the year 2050. However, the message of the article is clear: Denmark has no choice. We simply have to include coal and CCS in the future energy supply.

The following week, the front page of Ingeniøren (2008b) reported a statement from the development and research manager of DONG, a company that owns several coal-fired power stations:

> *The storage of CO_2 is absolutely necessary if we are to achieve a reduction of CO_2 emissions of such a scale as planned for the period after 2020. Europe cannot possibly do without coal. Therefore, we are forced to clean CO_2 from coal-fired power stations.*[e]

Again, the message is clear: Denmark and Europe have no choice but to burn coal and introduce CCS.

It is doubtful that the preceding statements deliberately ignored other studies or dismissed the fact that alternatives without coal do exist. Moreover, the individuals cited are very skilled and competent in the field. Nevertheless, the statements are clearly seeking to influence the collective perception of choice.

While I was writing the third edition of this book, another example arose in the autumn of 2022. Together with 15 colleagues from 5 different Danish Universities, I published a short paper called "Facts about Nuclear in Denmark" (Thellufsen et al., 2022). Among other points, we documented that electricity from a nuclear plant in Denmark was likely to have twice the cost of electricity from Danish wind and solar plants. Moreover, the construction time of the nuclear plant would likely be much longer than for wind and solar. Consequently, nuclear would not be a feasible solution in a situation in which Denmark wishes to speed up the green transition for reasons of urgency due to climate change as well as the energy crisis caused by the Russian war in Ukraine.

When our paper was referred to and discussed in a Danish news media,[f] a Norwegian scientist counterargued to our conclusion, stating that the high cost and long construction times were no surprise to him, but nevertheless society could not do without nuclear in Europe nor in Denmark. According to the Norwegian scientist, due to the green transition, society needs nuclear to balance the electricity grid and this cannot be done solely by wind and solar.

---

[e] Translated from Danish: "CO_2-lagring er bydende nødvendigt, hvis vi vil opnå CO_2-reduktioner i den størrelsesorden, der er på tale efter 2020. Europa kan umuligt klare sig uden kul. Derfor er vi nødt til at rense CO_2 fra kulkraftværkerne". *Ingeniøren,* January 18, 2008, p. 1.

[f] 27 October 2022 "Ny rapport: Atomkraft er for dyrt og for langsomt." https://videnskab.dk/teknologi/ny-rapport-atomkraft-er-for-dyrt-og-for-langsomt/.

Again the message is clear: We have no choice but to build nuclear, even if it costs more than wind and solar and has longer construction times; not to mention risk and radioactive waste.

Again, this statement is not correct. Societies can implement green transitions without nuclear, as also explained in detail in this book. In fact, the paper about nuclear in Denmark included such a scenario of green transition without nuclear. This is further elaborated in Chapter 8.

Again, it is doubtful that the preceding statements deliberately ignored other studies or dismissed the fact that alternatives without nuclear do exist. Nevertheless, the statements are clearly seeking to influence the collective perception of choice.

As we will see later, the preceding aspect is general for all of the cases discussed in Chapter 9. The reasons behind the resulting collective perceptions of no choice are much more fundamental. To understand the nature of the mechanisms leading to these perceptions, it is important to understand the term *radical technological change*.

Radical technological change

Not all technological changes are equally fundamental in the sense that they involve changes in existing organizations and institutions. Therefore, some technological changes challenge the political decision-making process more than others. Müller et al. (1984) defined technology as follows:

> *Technology embraces a combination of four constituents: Technique, Knowledge, Organization, and Products.*[g]

Hvelplund (2005) added *profit* as a fifth dimension of technology, which is considered useful when analyzing changes in the energy sector. The basic assumption of the technology theory of Müller, Remmen, and Christensen is that

> *a qualitative change in any of the components will eventually result in supplementary, compensatory, and/or retaliatory change in the others.*[h]

In other words, if one dimension is substantially changed, at least one of the others will follow. If they do not, the initial change will be abandoned over a period of time. Society cannot make a fundamental change in a technique without changing the knowledge and/or the organization and/or the product related to this technique. If one or more of the other dimensions are not changed, the new technique will not be implemented, and the traditional technique will be used once again.

After this definition of *technology* was introduced, Hvelplund (2005) defined the degree of radical change as increasing with the number of dimensions that must change. Hvelplund defined radical technological change as a change that affects more

---

[g] This definition is cited from *A Conceptual Framework for Technology Analysis* (Müller, 2003). The definition was first forwarded by Müller (1973) and was later expanded by Müller et al. (1984).
[h] Müller (2003), p. 30.

than one dimension.[i] The transition from nuclear and fossil fuel-based energy systems to renewable energy systems is to be considered a radical technological change. As we will see, this change in technique implies substantial changes in organization.

It should be emphasized that technological change must be placed in a historical and institutional context. Hvelplund (2005) pointed out that the existing institutional set-up may favor established technologies at many levels. Consequently, coherent and coordinated changes are needed at different levels in society. However, a technological change such as wind power replacing a coal-fired power station may result in radical changes for those who own and work in the coal mines, while at the same time this does not necessarily imply any change for the electricity consumer.

The technological change from nuclear and fossil fuel-based energy systems to renewable energy systems involves an economic redistribution, as investments in large power stations are replaced by investments in energy conservation and distributed CHP plants. Furthermore, for example, coal mining is replaced by the harvesting of biomass resources and investments in wind turbines and solar thermal power.

The description of the technological change from fossil fuel to renewable energy systems in this section is based on Hvelplund et al. (2007). The existing context is characterized by large supply companies on the one hand and many differentiated consumers divided into households and public and private enterprises on the other hand. Typically, the existing supply system is characterized by single-purpose companies, that is, enterprises that have the production and/or sale of energy services as their only purpose. They are often segmented into heat, electricity, or natural gas supply systems. Investments are capital intensive; they have a very long technical lifetime of 20–40 years and are almost 100 percent asset specific. Asset specificity means that the assets, such as district heating systems, supply stations, and power grids, can be used only for their present purposes. The organizations linked to the existing technologies are consolidated from an economic as well as a political point of view.

The existing consumers' system is characterized by many multipurpose organizations, which refers to the fact that households and private or public firms have other main purposes than investing in renewable energy system technologies. These organizations often lack capital for investing in renewable energy system technologies, including energy conservation activities, and they have no common organization of activities related to these technologies.

Unlike nuclear and fossil fuel technologies based on large power stations, renewable energy system technologies will typically benefit from a wide geographical distribution throughout their areas of consumption. The technological solutions differ from one place to another, and sometimes new, not well-proven technologies must be implemented. The maintenance of these new technologies is dependent on ownership and organization. Along with the implementation of new technologies, new types of organizations are therefore likely to develop.

Investments must be made by multipurpose organizations. Thus, electricity savings must be implemented by private households and industries with only a limited

---

[i] Hvelplund (2005), p. 12.

awareness of consumption and with main objectives quite unrelated to simply producing or consuming heat or electricity. This has to be compared with the former situation in which investments in supply technologies were carried out by single-purpose
organizations, such as utility companies, with energy production as their primary
objective.

The technologies must be implemented by many mutually independent organizations. Again, this has to be compared with the former situation with a limited number
of companies. The financial capital of these new organizations will often be scarce
compared with the financial capital of the existing supply companies. The political
capital of these new organizations will also be relatively scarce compared with that
of the existing companies.

All in all, this technological change can often be seen as a change from undifferentiated
solutions implemented by a few single-purpose organizations to differentiated solutions
implemented by many multipurpose organizations. Therefore, the change to renewable
energy systems is to be regarded as a radical technological change. The important point
is that this entails substantial changes in existing organizations and institutions, and these
changes will challenge these organizations. Moreover, it will influence the general perception of choice in society.

2   Choice perception and elimination

Since the radical technological change to renewable energy systems implies substantial challenges and poses a threat to existing organizations, these organizations will not
by themselves create and promote the alternatives required to implement this change.
Existing organizations can create certain alternatives, whereas other alternatives are
out of their perception. Even if they should wish to promote such alternatives, they
would often not be able to implement them within the existing institutional set-up.

Choice perception

The case of the power station Nordkraft (see Chapter 9) is an example from the early
1980s. In this case, the station was considering converting from oil to coal. The initial
proposal put forward by the power company was a *one—and only one—alternative*
proposal. The power company proposed that an oil boiler should be replaced by a coal
boiler, a technology fitting well into the existing organizations of the power companies. No other alternative was presented to the public when the proposal was to be
approved. City council members expressed preference for an alternative based on natural gas, but this alternative was not presented and did not form part of the basis for the
decision-making process.

The local citizens had to describe and promote a concrete technical alternative representing radical technological change—in this case, the insulation of houses and the
expansion of CHP outside the borders of the municipality. This alternative has subsequently proved to be the best solution in terms of economic feasibility when based
on actual historical fuel prices. However, the institutional set-up of existing power
companies could not identify and implement the best alternative by itself. The

alternative entailed a radical technological change; in other words, it could not be implemented without changes in institutions, including the existing organizations.

The discourse of the power companies referred to an optimization of the use of fuels within the existing technical and organizational set-up. The identification of radically and technologically different alternatives was not a part of their interest or perception of reality. And even if it was, the implementation of these alternatives would be out of their reach, since it would involve investments in the insulation of private houses as well as CHP units in district heating companies owned by others.

The main concern of the city council was to maintain low district heating consumer prices. Moreover, they also had to manage urban and environmental concerns in the physical planning. Again, the implementation of insulation and CHP outside the municipality was out of their reach. Natural gas was an option that was within the reach and perception of the city council. However, the city council did not have the power or the resources to ensure a proper analysis and description of this alternative when faced with the risk of substantial rises in district heating prices.

The proposal of technologically and radically different alternatives had to come from citizens outside the power companies and the city council. The existence of an alternative could raise the public awareness of the fact that, from a technoeconomic point of view, a choice did exist, and as a result, 700 citizens made claims for the description and inclusion of alternatives in the debate. However, given the institutional set-up, such technologically and radically different alternatives could not be implemented. Institutional changes were required at a higher level.

Discourse theory can be used to explain why existing organizations will design and promote some alternatives but not others. This theory is inspired by linguistic philosophy and it perceives social reality as a linguistic construction (Thomsen et al., 1996). Articulation and discourse are key concepts in this theory. Laclau and Mouffe (1985) defined articulation as "any practice establishing a relation among elements such that their identity is modified as a result of the articulation practice." Discourse is defined as the "structured totality resulting from the articulation practice."[j]

According to discourse theory, different organizations perceive and articulate things differently; they exercise different discourses. In the description and discussion of climate change, some politicians and environmental organizations may have one perception of reality, while other politicians and organizations have another. For example, industrial organizations do not seem to ascribe to the problems of climate change to the same extent as environmental organizations do. Moreover, industrial organizations do not draw the same conclusions with regard to the solution of the problems. They typically express the idea that an environmentally safe production should be available on existing markets, that is, managed by existing organizations and institutions (Thomsen et al., 1996).

A central element in discourse theory is the belief that reality is established through a combat between competing perceptions and articulations of reality and that these competing perceptions, to some extent, define one another. Mouffe (1993) argued that the collective identification of a "we" always raises the possibility that a "we"/"them"

---

[j] Laclau and Mouffe (1985), p. 105.

relationship is created and that the notion of "them" plays an important part in the definition of "we." Mouffe argued that, on the eve of the 21st century, the processes of redefining collective identities in our societies were "linked to the collapse of Communism and the disappearance of the democracy/totalitarianism opposition."[k]

Seen in relation to discourse theory, the implementation of renewable energy systems becomes a combat between different articulations and perceptions of reality. Choice Awareness theory argues that the perception of choice or no choice, including the collective perception of which alternatives to consider, is a core element. However, one perception cannot claim to be more real or true than the other. Reality is present in all of the different sets of articulations and perceptions. The main point in discourse theory is the idea that different perceptions of reality result in different mind constructions, that is, different ways of approaching the same real problems. Consequently, it is not solely a matter of different interests and views; the linguistic dimension also influences the construction of reality itself. According to Laclau and Mouffe,

Any discourse is constituted as an attempt to dominate the field of discursivity, to arrest the flow of differences, to construct a centre.[l]

Applied to the implementation of renewable energy systems, discourse theory implies the understanding that different organizations represent different perceptions of reality and therefore have different views on what should be done to solve the same problem. According to the theory, one could therefore expect that existing organizations linked to the burning of fossil fuels perceive climate change problems as a less severe threat that can be handled within the existing institutional framework, in other words, by technologies represented by existing organizations. If renewable sources are to be implemented according to this perception, those renewable technologies that fit into the framework of these organizations will be preferred.

One example of the advancement of technologies that fits well into existing organizations is the promotion of CCS, as described previously. Another example can be seen in the recent debate (2005–2015) on how to expand wind power in Denmark. The most cost-effective way is to increase the number of onshore wind turbines. From many years of experience, Danish society knows that this can be done if institutional frameworks are established in which neighbors can own shares of the wind turbines and make a profit. Danish society also knows that if neighbors are not involved, they are likely to protest against this solution. In Christensen and Lund (1998), an analysis is conducted of how society, by applying local ownership, has managed to implement wind power in a socially acceptable as well as environmentally benign way.

However, based on the argument that wind power should adjust to the market, the institutional framework for neighbor-owned wind turbines has gradually been abolished during the past 20 years and, instead, the government wishes to expand offshore wind farms. In the beginning, these wind farms were not economically feasible compared to onshore wind turbines, and they would increase the need for subsidy. However, offshore wind farms corresponded perfectly to the institutional framework

[k] Mouffe (1993), p. 3.
[l] Laclau and Mouffe (1985), p. 112.

of existing power companies. In short, the discourse that wind turbines should adjust to existing market institutions leads to the implementation of renewable technologies that are suitable for these institutions. This is the case even if these technologies are not economically feasible compared to alternatives that require the establishment of new organizations and involve neighbor ownership.

Different existing organizations reflect different perceptions of reality. Therefore, they can design project proposals and alternatives that are relevant to these perceptions, but they cannot be expected to design technologically and radically different alternatives. These alternatives have to come from someone else. To illustrate this point, one may refer to the preceding case of Nordkraft (also discussed further in Chapter 9), in which the local radio station asked Nordkraft's managing director if he could *imagine* that the project, which had met a lot of resistance, would not be implemented. The managing director answered that he could not imagine which other alternative one would suggest instead. Alternatives representing radical technological change simply do not form part of the imagination or perception of existing organizations.

However, the explanation for why and how certain choices do not become part of the collective perception is not only related to the discourses of existing power companies. When local citizens and environmental organizations raise Choice Awareness by introducing alternatives representing radical technological change, the organizations representing existing technologies may also respond by using different choice-eliminating mechanisms and strategies.

Choice-eliminating mechanisms

Radical technological changes pose a threat to those existing organizations that depend on the technologies to be replaced or diminished. These organizations respond to the threat. Choice Awareness theory argues that a core element in this response is the elimination of choice in the public debate and collective perception.

The existing organizations will execute power to protect their interests. However, in the investigation of the relationship between rationality and power, Flyvbjerg (1991) argued that the most important execution of power can be found and studied outside the scene of formal direct power. Flyvbjerg combined a theoretical approach with a comprehensive case study:

> In many cases, pivotal activities are not to be found in the design of objectives, policies, legislation, and plans nor in public participation and formal political decision making in relevant political assemblies. On the contrary, they take place before any objectives, policies, legislation, and plans have been formulated, in what one may call the genesis of planning and policies, and after the formal political decision has been made, during the implementation of plans and policies.[m]

[m] Translated from Danish: "De afgørende aktiviteter findes således mange gange ikke i udformningen af mål, politikker, lovgivning og planer eller i borgerdeltagelse og formel politisk behandling i relevante politiske forsamlinger. De findes derimod **før** der overhovedet er noget, som hedder mål, politikker, love og planer, i det man kunne kalde planlægningens og politikkens **genese**, og **efter** den formelle politiske vedtagelse, i planlægningens og politikkens implementering" (Flyvbjerg, 1991, p. 19).

Flyvbjerg distinguished between, on the one hand, *formal planning* and *formal politics*, and on the other hand, *real planning* and *real politics*. With regard to the relationship between implementation and formal and real politics, he pointed out that when important actors cannot implement their wishes through the *formal* political process, they will often seek to do so in the implementation phase by use of *real* politics, typically hidden from an immediate look. The only way to get access to such real politics is by conducting thorough studies of concrete planning and policy-making processes.

As we will see in Chapter 9, the study of concrete planning and policy-making processes reveals that choice-eliminating strategies and mechanisms executed outside the scene of formal power are manifold and take various forms. One form is the simple exclusion of alternatives from the agenda. The exclusion of a natural gas alternative—as in the case of Nordkraft power station in the early 1980s—has already been presented. Another example is the case of Aalborg heat planning in the mid-1980s, in which the municipality simply disregarded an alternative of small CHP plants. This solution did not fit well with the interests and perceptions of the city council and the municipality-owned district heating company. The solution was simply disregarded in the definition of optional alternatives in the public participation phase, initiated by the municipality in accordance with the legal procedures of heat planning.

Alternatives representing radical technological change had to come from the university and local citizens. The case of Aalborg heat planning revealed some interesting choice-eliminating mechanisms and strategies when these alternatives were designed and promoted. When addressing the public discussion phase, as already mentioned, the municipality simply left out certain alternatives. When these alternatives were proposed by the citizens, the municipality disregarded them in the comparative analyses. When comparative analyses could no longer be avoided and showed an inconvenient result, new analyses were made. Only the analyses that showed the most convenient result were put forward by the administration to the city council. And when citizens mailed "inconvenient" results to the city council, the content of the analysis was disregarded. Instead, a discussion of the letterhead used was initiated to incriminate the senders.

As just shown, choice-eliminating activities have been executed at many levels and in various forms. Power theory can be used to identify a systematic way of analyzing the different levels at which these activities are practiced. In their book *Silent Control—About Power and Participation*,[n] Christensen and Jensen (1986) described different levels of power and provided an understanding of how and where power is executed when existing organizations influence the decision-making process. The basic idea of the book is to understand how power is executed for the purpose of developing participation strategies. Consequently, their definition of power is quite broad: "Power is seen as the possibilities of actors to attend to their interests in relation to the allocation of goods and burdens in society (material as well as immaterial)."[o] With

[n] Translated from Danish.

[o] Translated from Danish, p. 12.

references to, among others, Dahl (1961), Bachrach and Baratz (1962), Lukes (1974), and March (1966), Christensen and Jensen categorized power into four levels: direct power, indirect power, mind-controlling power, and structural power.[P]

The three latter levels of power are of particular interest to the Choice Awareness theory. The execution of power leading to the collective perception of whether alternatives exist and how they should be evaluated often takes place before the scene is even set for the exercise of direct power.

Direct power is executed in a decision-making process, for example, as items on the agenda of a meeting in a board or a city council or in Parliament. As already mentioned, this execution of power is rare with regard to the elimination of choice. The choice-eliminating strategies and mechanisms are typically executed before the mere existence of alternatives leads to the perception of even having a choice. Thus, in the case of the Environmental Impact Assessment procedures (see Chapter 9) in the mid-1990s, the County Council of Northern Jutland was allowed to choose freely from different alternatives to the construction of a new coal-fired power station. According to the legal procedures, they had only to ensure the proper description (and existence) of alternatives, including those suggested in a public participation phase. They were not obligated to choose them. However, in general, the legal procedure failed to ensure a proper description of alternatives representing radical technological change (i.e., changes to the institutional set-up of the power companies and the regional authorities), even when such alternatives were promoted by local citizens in the public participation phase. By executing indirect power, the local county eliminated this choice before the item reached the agenda.

With *indirect power*, focus is placed on the execution of power both before and after the official meetings, for example, the decision of what is and is not put on the agenda. It is recognized that not all matters have equal access to the decision-making process. Consequently, powerful interest groups can seek to exclude certain items, or they can seek to influence the implementation in such a manner that the result has other consequences than expected. The execution of indirect power is indeed relevant to the identification of choice-eliminating mechanisms, and the examples are manifold, as we will see in Chapter 9. The examples involve disregarding alternatives and incriminating senders, as already mentioned, as well as claiming that alternatives are based on incorrect technical data, while withholding the "correct" data by referring to "national security." The latter was the situation in the case of the transmission line described in Chapter 9.

Indirect power also involves power being executed after decisions have been made, in the form of not implementing the decisions or implementing something else. The case of Nordjyllandsværket in the mid-1990s represents such a situation. The Danish Parliament had decided on an energy policy, called Energy 21, according to which no new coal-fired power station was needed. Instead, the Parliament decided to implement electricity savings and expand the number of small CHP plants. Still, the power

---

[P] Translated from Danish, pp. 13–14.

companies, supported by the minority government, succeeded in implementing another coal-fired power station.

At the *direct* and *indirect power* levels, choices are eliminated when it is decided how many and from which alternatives the decision body can choose. Time and resources are limited, and, consequently, someone must choose which alternatives to analyze and describe and which consequences to identify in which way. Moreover, the elimination of choice is executed when it is decided if matters should be discussed, where, and for how long. However, choices are not only eliminated by disregarding alternatives. Choice elimination also has to do with influencing the perception of those who imagine, design, and promote those alternatives. This phase is out of reach of the organizations that represent the existing technologies. Instead, this power is executed at the next two levels.

Mind-controlling power includes the execution of power in such a way that some actors influence other actors' perception of their interests and how these interests can be promoted in a legitimate way. Mind-controlling power and direct and indirect power have in common the fact that they are executed among actors. This is not the case with level four, *structural power*, which is executed through a collective unconscious acceptance of the societal framework constituted by habits, routines, and norms.

At the *mind-controlling* and *structural* levels, Choice Awareness is influenced, among other factors, by the perception and design of the democratic infrastructure, as explained in Chapter 3—for example, the perception of which and how well interests should be represented in key committee work. Another example of mind-controlling power is the discussion of *sustainable energy*, as explained in Chapter 1. Defining sustainable energy in such a way that it includes nuclear and/or coal in combination with CCS can be considered a way to make people who are in favor of renewable energy support and promote nuclear and coal instead.

Structural power can be seen as executed through the presence of the existing institutional set-up. For example, in the case of the Environmental Impact Assessments procedures of the mid-1990s, the choice-eliminating mechanisms were indeed related to the simple fact that the local county council wanted to stick to alternatives within the borderlines of their jurisdiction. And in the case of Nordkraft, the existing power company and the municipality could not perceive and design alternatives beyond the ownership of the power company or the borders of the municipality. Even if they had been able to do so, the implementation would be out of their hands in the given institutional set-up. In those cases, the structural power of the existing institutional set-up influenced the perception of potential alternatives.

In conclusion, it is important to be aware that choices are not only eliminated in the sense that they are erased from the collective perception, but are also eliminated due to the awareness among certain actors that certain alternatives cannot be implemented without introducing substantial changes to the institutional set-up. Therefore, my colleagues at Aalborg University and I have for many years executed the combined design and promotion of *technical alternatives* as well as *institutional alternatives*. We have used the promotion and public discussion of our technical alternatives to identify institutional barriers to be able to design institutional alternatives. In this way, we have for many years involved ourselves in raising Choice Awareness with regard to technical as well as institutional changes.

The first Choice Awareness thesis

The theory of Choice Awareness addresses the societal level. As mentioned, it concerns collective decision making in a process that involves many people and organizations representing different interests and discourses, as well as different levels of power to influence the decision-making process. The theory concerns the implementation of radical technological change in society, that is, changes that imply significant institutional reorganization.

With a reference to discourse and power theory, it is assumed that the perception of reality and the interests of existing organizations will often make them seek to hinder radical institutional changes by which they expect to lose power and influence. Choice Awareness theory advocates that one key factor in this manifestation is the societal perception of having a choice or having no choice.

The first thesis of the Choice Awareness theory states that when society defines and seeks to implement objectives implying radical technological change, the influence and discourse of existing institutions will affect the implementation. This impact will hinder the development of new solutions and eliminate certain alternatives and will seek to create a perception indicating that society has no choice but to implement technologies that will save and constitute existing positions. The results of this influence will take various forms, including the following

- The exclusion of technical alternatives from the debate and the decision-making arenas
- The technical evaluation of alternatives on the basis of methodologies that assess the radical new technology in question as not being relevant to or not complying with the requirements
- The design of feasibility studies in such a way that radical new technologies are assessed as not being economically feasible to society

These forms of influence will typically be based on the applied neoclassical perception that the existing institutional and technological set-up is defined by the market, which works in such a way that it will, by definition, identify and implement the best solutions

---

3   Raising Choice Awareness

The first Choice Awareness thesis is based on the observation that the collective perception of no choice is often constructed when major societal decisions on energy planning are discussed. As previously described, the mechanisms applied involve both the elimination of technical alternatives representing radical technological change as well as institutional barriers to the implementation. At the structural power level, this will again influence the collective perception of choice.

The theory argues that Choice Awareness is crucial when radical technological changes are at issue. But what can be done about it? The core element is to raise the awareness of the fact that society *does* have a choice. Radical technological change is possible. And since choice-eliminating mechanisms are executed at many levels, the second Choice Awareness thesis advocates that counterstrategies are introduced at the same variety of levels. Consequently, the rise of Choice Awareness involves the design and promotion of technical alternatives and the use of evaluation methods as well as the design of institutional alternatives. The latter includes both direct public regulation measures and the promotion of a suitable democratic infrastructure.

As already mentioned, I have observed the elimination of choice in many situations. However, I have also experienced how the introduction and description of concrete alternatives can raise the public awareness of the fact that we, as a society, *do* have a choice. Over a period of time, this awareness can lead to institutional changes in such a way that it actually becomes possible for society to choose among relevant alternatives. The recent history of Danish energy planning during four decades is an important example of such a transformation. Danish energy policy has been formed as a result of a process of conflicts. This process has led to the implementation of radical technological changes, and Denmark has been able to show remarkable results on the international stage. This ability to act as a society has been possible despite conflicts with representatives of the old technologies. Official energy objectives and plans have been developed as a result of constant interaction between Parliament and public participation, in which the description of new technologies and alternative energy plans has played an important role. Public participation, and thus the awareness of choices, has been an important factor in the ultimate decision-making process. The conflict-ridden debates should, therefore, be seen as necessary conditions for further improvements of energy initiatives and programs.

Some examples of the influence of this long-term institutional change are presented in Chapter 9. For example, in the aforementioned case of Aalborg heat planning, the promotion of concrete technical proposals led to the identification of specific institutional barriers in the energy taxation structure. As a follow-up on that case, a colleague and I promoted concrete proposals for how to change the Danish energy taxation system to create a situation in which the socioeconomically best solution would also generate the best consumer heat prices.

In the case of Nordjyllandsværket, the promotion of concrete technical alternatives played a role in the understanding of the contradiction between project proposal and the official parliamentary energy policy. Danish society was not powerful enough to avoid the central power stations, which fitted well into the organizations of existing power companies. However, the Parliament was powerful enough to promote small CHP plants more or less simultaneously with the approval of the central power stations. Such a decision brought substantial changes in the institutional set-up of market conditions for small CHP plants. Hereby, the Parliament opened up for investments in small CHP plants with a capacity of more than 1000 MW in the mid- and late-1990s.

The second Choice Awareness thesis advocates raising Choice Awareness at many levels. Very often step one is to design concrete technical alternatives. These ideas of the Choice Awareness theory are much in line with the recommendations of O'Brian (2000) on replacing risk assessment with alternatives assessment:

> *Risk assessment is one of the major methods by which parts (corporations such as Monsanto or Hyundai, "private landowners", industrial nations) can act on their wants at the expenses of wholes (e.g., whole communities and countries, or seventh-generation from now) without appearing to be doing so. Risk assessment lets them appear simply "scientific" or "rational" as they numerically estimate whether or how many deaths or what birth defects will be caused, and ignore other regions of human experiences that also matter to people.[q]*

Thus, risk assessment becomes one of the ways to justify "one alternative decision."

O'Brian advocated a replacement of risk assessment with alternatives assessment, defined as the evaluation of the pros and cons of a wide range of options. The goal is to replace the assessment of a narrow range of options with the public assessment of a wider range. O'Brian pointed out that this methodology does not reflect the interests of existing institutions. Alternatives assessment is

> *a simple and sensible alternative, but it is resisted mightily because the consideration of options is a threat to business as usual and business as planned—and often, to established power arrangements.[r]*

Like Choice Awareness, the idea of alternatives assessment has also been inspired by processes at the individual level. O'Brian referred to her sister, a psychiatric social worker, who told her that one sign that a client might be suicidal is when the person is convinced that he or she has only one or two options, and both of them are terrible. O'Brian explained as follows:

> *Replacing risk assessment with alternatives assessment involves the same simple principle. Instead of allowing ourselves to be limited to one or two options that are terrible, we can insist on public consideration of a range of alternatives that seem good for different reasons. We can evaluate these alternatives and choose the one that seems best.[s]*

The ideas of the Choice Awareness theory are also in accordance with Groupthink theory. In his books *Victims of Groupthink* (1972) and *Groupthink* (1982), Janis introduced the term *groupthink*, which is defined as the psychological drive for consensus at any cost that suppresses disagreement and prevents the appraisal of alternatives in cohesive decision-making groups. Janis showed how this phenomenon contributed to some of the major US foreign policy fiascos, such as the Korean War stalemate, the

---

[q] O'Brian (2000), p. xviii.

[r] O'Brian (2000), p. xiii.

[s] O'Brian (2000), p. 129.

escalation of the Vietnam War, the failure to prepare for the attack on Pearl Harbor, and the Bay of Pigs blunder. Janis used the term *groupthink* as

> *a quick and easy way to refer to a mode of thinking that people engage in when they are deeply involved in a cohesive ingroup, when the members' strivings for unanimity override their motivation to realistically appraise alternative courses of action.*[t]

He described seven symptoms and consequences of groupthink in which the "incomplete survey of alternatives" comes in as number one on the list.[u] Thus, group-think is very much in line with Choice Awareness theory in general and, in particular, with the Choice Awareness strategy emphasizing the importance of promoting a new corporative democratic infrastructure, which can produce relevant alternatives for the decision-making process (see Chapter 3).

Janis (1982) recommended a number of strategies to prevent groupthink. One of these is the suggestion that

> *organizations should routinely follow the administrative practice of setting up several independent policy-planning and evaluation groups to work on the same policy question, each carrying out its deliberations under a different leader.*[v]

This recommendation is also very much in line with the recommendations of Choice Awareness to promote a better democratic infrastructure, which is explained further in Chapter 3.

In his thesis "Alternatives, Nature and Farming," Christensen (1998) discussed the human relationship with nature and the perception of alternatives in agriculture. Christensen engaged in the discussion of what is a "real" alternative. For example, is organic farming a real alternative to conventional agriculture? The environmental and nature protection problems of our time require innovative thinking regarding our perception of nature as well as our agricultural practice. According to Christensen, it is not solely a matter of getting new ideas. Alternative visions have to be combined with level-headed and complex analyses. The key issue is to raise the question of how to inspire to a fruitful change in such a way that alternatives are not isolated or end up being another part of the existing systems.

Christensen related to the term *radical change*, which corresponds well to the concept of radical technological change as used when formulating the two theses of the Choice Awareness theory. Christensen discussed radical change regarding alternatives of organic farming: "The proposals intend to unify two goals: social viability and environmental sustainability."[w] However, Christensen pointed out, "It seems that a focus on social viability very easily may reduce the extent to which environmental problems are taken into account."[w] Thus, Christensen emphasized the problem of how

[t] Janis (1982), p. 9.

[u] Janis (1982), p. 175.

[v] Janis (1982), p. 264.

[w] Christensen (1998), p. 446.

to implement radical technological changes without adjusting technology to social viability to such an extent that it is integrated into the existing systems and radical changes are not implemented after all. This issue is a core element in Choice Awareness. The four Choice Awareness strategies elaborated in Chapter 3 address exactly such a challenge.

The second Choice Awareness thesis

Choice Awareness theory addresses a situation in which society defines and seeks to implement objectives that will imply radical technological changes.

The second thesis of the Choice Awareness theory argues that in this situation, society will benefit from focusing on Choice Awareness, that is, raising the awareness that alternatives do exist and that it is possible to make a choice. The awareness can be promoted by various means, including the following

- Promoting the description of concrete technological alternatives in various debates and decisions on new plans and projects at all levels
- Promoting feasibility study methodologies that include relevant political objectives in the analyses
- Promoting the concrete description of public regulation measures to advance new technologies

The advancement of these three promotions can in general be helped by changes in the democratic decision-making infrastructure, which advance the representatives of new technologies

---

The concept of Choice Awareness, including the two theses, emphasizes the words *conflict* and *process*. The decision-making procedures, including the definition of alternatives and how these should be assessed, are to be seen as a *conflict*. This conflict is a fight between different interests, influences, and discourses, in which well-established organizations seek not to lose power and influence. Furthermore, it is a *process* over time. Developing societal procedures that allow the design of alternatives, proper assessment methodologies, and good public regulation measures takes time. It is a process, and in this process, it is not important to win every battle; it is important to win the war.

Choice Awareness strategies

3

The basic idea of Choice Awareness is to understand that existing institutional perceptions and organizational interests will often seek to eliminate certain choices from the political decision-making process when the introduction of radical technological change is discussed. The counterstrategy is to raise public awareness of the fact that alternatives do exist and that it is possible to make a choice. Becoming aware of choice-eliminating mechanisms is in itself an important part of raising this awareness. Furthermore, Choice Awareness can be promoted by the strategies mentioned in Chapter 2 and illustrated in Fig. 3.1. Each of the strategies is elaborated on in this chapter.

1 Technical alternatives

The description and promotion of concrete alternatives is a core strategy in Choice Awareness. It is the essential first step that must be taken to change the focus of a public discussion. Typically, "the one and only" solution presented, such as another coal-fired power station without combined heat and power (CHP) production, is reckoned to have a negative impact on pollution and energy efficiencies. However, the general perception seems to be "If this is the only alternative, so what?" Society has no choice but to choose it. Or as mentioned in the previous chapter, even if nuclear has high costs and long construction times, it will still be needed. We have no choice but to build nuclear plants. However, if someone succeeds in putting forward and promoting a concrete alternative, two changes take place in this process. First, it becomes obvious that society indeed does have a choice. Second, the focus of the discussion changes from "Yes, it is bad, but so what?" to "Which of the alternatives is the best solution?"

When a second alternative is presented, the risk arises that even though this alternative is considered the best solution, it cannot be implemented for various institutional reasons, such as organizational and economic barriers. In this case, the concrete identification of specific barriers becomes important in relation to further discussions. Again, one should remember that the implementation of renewable energy systems is a long-term process. It is not important to win every battle, but it *is* important to win the war.

But how should good, concrete renewable energy system alternatives be designed? And which counterarguments should be considered? In the book *Public Regulation and Technological Change*,[a] Lund and Hvelplund (1994) described some of the arguments that the introduction of an alternative may face. Our analysis is based on the

---

[a] *Offentlig Regulering og Teknologisk Kursændring. Sagen om Nordjyllandsværket.*

Renewable Energy Systems. https://doi.org/10.1016/B978-0-443-14137-9.00003-6

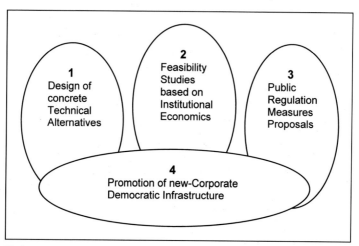

Fig. 3.1 Choice Awareness strategies.

cases of Nordjyllandsværket and the transmission line (see Chapter 9), and these descriptions are called "the anatomy of *half-true* statements."

Half-true statements can be grouped into the following categories:

The half-true statement of an incorrect context: This is a statement concerning one element put into a context that is not in accordance with the context assumed by the addressee. One example from the Nordjyllandsværket case is when the chairman of the board of the power company ELSAM informed a number of local politicians that the planned power station would be feasible and would improve the environment. This statement was half-true because it did not make clear that it was valid for only certain special cases, which were not in accordance with the implementation of the governmental energy policy. This type of statement will typically be seen as a comparison of the present solution and the new alternative but without mentioning other possible—and better—alternatives. For example, it might be reported that a new coal-fired power station is better for the environment than an old one, but other alternatives that may be even better for the environment are not mentioned.

The half-true statement of the incorrect time dimension: This statement disregards dynamics in economic cost structures and technological innovation. For example, in the Nordjyllandsværket case, the power companies wanted to build two power stations instead of one to achieve discount advantage in relation to the investment costs. In this case, the energy authorities informed the politicians that surplus capacity would have no influence on the electricity demand without considering the change in cost structure that was likely to be produced by such overcapacity.

The half-true statement of non-equalized evaluation: This statement promotes only advantages and disregards disadvantages. The Nordjyllandsværket case contained several examples of this. For example, the discount of 300 million DKK achieved by buying two power stations was intensively promoted, while the disadvantage of losing technological innovation (better efficiencies) by building the second plant now instead of later was disregarded. We calculated this disadvantage to correspond to approximately 700 million DKK (see Chapter 9).

A half-true statement may be used by "manipulators" to promote a solution. From their point of view, a good half-true statement is characterized as a "part" of the truth that can be communicated and understood easier than a comprehensive view of the truth itself.

It is easy to understand that a new power station is better than an old one, so the environment will benefit from a replacement of the old plant. It is much more complex to explain that, for example, a new power station was not even necessary in the case of Nordjyllandsværket. This explanation involves detailed discussions on the forecasting of demands and small CHP plants and the technological advantages of waiting until the need arises. It is also easy to understand that surplus capacity does not influence the electricity demand. It is much more complex to explain that the dynamics of politics and economics in a situation of surplus capacity typically result in economic barriers to the implementation of new conservation and electricity-saving technologies. It seems to be a good idea to get a discount by buying two power stations. It is much more complicated to explain that, some years into the future, this technology will be old compared to the ones accessible when the need really arises.

The construction of a good half-true statement is to some extent built into the institutional and organizational set-up of existing interests. Typically, one can observe a certain division of labor between the administrative and the managerial and political levels. For example, in the case of Nordjyllandsværket, the planning department of the power company produced technical analyses that were correct on the given premises. Thus, the department made an assessment of the economic feasibility of replacing old power stations with new ones compared to prolonging the lifetime of the old ones. These analyses led to the statement that based on certain assumptions, it would be a good idea to invest in two new power stations. Subsequently, the communication department and the chairman of the board generalized the statement in such a way that the public understood that the investment in two power stations was good for the environment in all situations and was implemented in line with the governmental energy policy. Politicians then used the same argument in favor of the power station. The energy authorities and the power company's planning department remained passive and let this generalization pass without reacting.

The preceding story forms a good introduction to why and how good alternatives should be designed. A major purpose is to avoid the dominance of half-true statements. By introducing a concrete alternative, a statement such as "new coal-fired power stations are better than old ones" will have to argue against even better alternatives. By introducing concrete alternatives, focus shifts from accepting bad solutions, because society has no choice, to discussing various options, for example, which choice to make. However, the introduction of alternatives should expect resistance from existing organizational interests. Consequently, the specific design of alternatives does matter.

Based on the experience from the many cases discussed in Chapter 9, the following guidelines can be defined:

1. *Alternatives must be designed in such a way that they are equally comparable in terms of the central parameters, such as capacity and energy production.* Otherwise, they can easily be disregarded. Energy savings may be included if saved capacity is calculated. If the main

proposal may lead to overcapacity, which is actually often the case regarding large power stations, the alternative should still be designed with the same capacity. However, the costs of creating overcapacity in the main proposal should be illustrated by distributing the investment in capacity over a period of time in the alternative. The calculation of investment should also include the benefit achieved by cost reduction and maybe even the benefit achieved by expected technological innovation in terms of better efficiencies or lower costs.

2. *Elements from all three aspects of renewable energy systems should be involved.* This includes savings in demand, efficiency improvements in supply such as CHP, and renewable energy sources. Thereby, the alternative is typically not an option that exists only once. By including all aspects, it becomes a generalized example of the radical technological change represented by the transition to renewable energy systems. If the alternative was only to replace coal by biomass, one could easily argue that this is not a long-term solution. The resources of biomass are not sufficient, and soon society would have to build a new coal-fired plant anyway. However, by combining all of the elements of renewable energy systems, the alternative will, in principle, be able to replace all future coal-fired plants.

3. *The alternative should be designed in such a way that the direct costs correspond to those of the main proposal.* This aspect actually complies very well with the idea of including all three elements of renewable energy systems, since savings in demand are typically very cost-effective. Moreover, cost reduction may be achieved by postponing some of the investments instead of establishing overcapacity.

By designing the alternative in such a way that the capacity and energy production (including savings) and the direct costs of the two solutions are the same, it may be possible to focus on a discussion of the issues in which the main proposal and the alternative differ. This involves comparing the alternatives with the main proposal in terms of all the central parameters: environment, local jobs, balance of payment, rural development, technological innovation, and industrial development. How to design feasibility studies regarding such parameters is discussed in the next section.

Typical examples of the design of alternatives are cases like Nordjyllandsværket (Denmark), Lausitz (Germany), or Prachuap Khiri Khan (Thailand). In all of these cases, the main proposal was to build a new or to expand an existing coal-fired power station without considering the benefits that could be achieved by increasing CHP production. Moreover, the proposed power stations were typically very large, and overcapacity was a general result.

As long as the main proposal is the only alternative presented, the problem of not integrating CHP is unclear. However, in all three examples, substantial heat production was left to heat-alone boilers. When designing an alternative of small CHP plants, the heat-alone boilers were included in the discussion and the fuel efficiency improvements of CHP were shown. Consequently, in all cases, it was possible to design better alternatives that were still comparable in terms of direct costs if fuel efficiency improvements were included.

2 Economic feasibility studies

Choice Awareness theory is based on the basic assumption that in societal decision-making processes involving radical technological change, existing institutional interests will try to influence the process in the direction of no choice. This influence involves the elimination of technical alternatives from the agenda, as well as the use of feasibility studies based on methodologies and assumptions supporting existing organizational interests. Consequently, Choice Awareness includes the awareness of how feasibility studies are and should be carried out.

This section is based on the book *Feasibility Studies and Public Regulation in a Market Economy* by Hvelplund and Lund (1998a) and two chapters from the book *Tools for Sustainable Development* (Hvelplund et al., 2007). As will be discussed further in the next section, the theory of neoclassical market economics is based on a number of assumptions that are not fulfilled in real-life market economics. However, the critique of neoclassical-based methods of this book is directed toward their real-life application. It is not of decisive importance whether this critique is valid for theoretically correct applications. Consequently, the term *applied neoclassical economics* is used and defined as neoclassical-based methods, such as cost-benefit analyses and equilibrium models, when applied to existing market economies. This is seen as opposed to theoretically correct methods when applied to market economies that fulfill the theoretical assumptions of a free market. Every market economy is formed by a number of market institutions. In applied neoclassical economics, the analysis of those institutions is usually outside the scope of the analytical model, and these institutions are therefore, in practice, treated as static or unchangeable. But, as explained in Chapter 2, in situations of radical technological change, it is very important to address the issue of changes in the institutions in which the market is embedded.

Neoclassical economics is based on the concept of a *free market*. The theoretical *free market* requires a number of institutional preconditions, such as many mutually independent suppliers of a product, many mutually independent buyers of a product, full information regarding quality and prices of available products, agents on the market acting with rational behavior, sellers who maximize profits, and buyers who maximize utility.

When these and other conditions are fulfilled, it can be argued that if all consumers and producers act to optimize their individual profits, the market will define what is best for society. The market becomes a democratic place, where free and rational buyers, who are well informed about their options, buy the goods they want. Therefore, when the previously mentioned institutional preconditions are present, any interference with this free-market process can be regarded as non-democratic.

It is important to emphasize that a *free market* is not a market without public intervention but a market where public regulation in a decisive way acts to establish and maintain the institutional preconditions of the free market. Most market economies are mixed economies consisting of a private and a public sector. In the private sector, an exchange takes place on the market, based on supply and demand, between sellers and buyers of goods, labor, and capital. The public sector redistributes incomes and produces goods and services, mainly outside the market.

In applied neoclassical economics, the market is usually considered a *free market*. Activities on the market are the result of well-informed, free, and rational agents, each optimizing its utility function, and no private regulation influences the allocation process. In neoclassical economics, the relationship between the public sector and the market sector is usually one in which the state autonomously constructs the laws and institutional framework of the market via public regulation measures; taxes, goods, and services from the public sector are defined as neutral from an allocation point of view.

Consequently, in applied neoclassical economics, the public sector is defined as neutral with regard to the effects of regulation on the market processes. The combination of the two premises—the private and the public sectors do not distort the allocation process and the market process is governed by free, rational, and well-informed actors—means that the production at any given time can be regarded as optimal. This premise of optimality is a precondition in most econometric models inspired by neoclassical economic thinking when these models are used for planning purposes by macroeconomists.

From the premise "We are living in the best of all worlds," we can deduce that any change away from this optimum represents socioeconomic losses to society. For instance, all costs related to policies of reducing greenhouse gas emissions or increasing the share of renewable energy are considered extra societal costs in all computations. In the case of the IDA Energy Plan 2030 (see Chapter 7), an example is given of the discussion of such computations. In these econometric models, systematic institutional mistakes do not exist in the economic process. Meanwhile, in real life, this premise is not fulfilled. Therefore, it is necessary to see the economy of a country or region as an institutional economy, in which the present situation may very well not be optimal at all.

In the following, some guidelines are presented for performing feasibility studies in a situation of radical technological change. The guidelines refer to issues involving environmental concerns and have technological innovation and institutional change as their main objectives. Therefore, the scope and focus of these feasibility studies are much broader than in the conventional cost-effectiveness and cost-benefit analysis approach.

The feasibility study should include the design of feasible technological alternatives; an evaluation of the social, environmental, and economic costs; an overview of the innovative potential of these alternatives; and an analysis of the institutional conditions that influence the implementation of the alternatives. Since these feasibility studies can be applied to both public and private decision making, one should distinguish between socioeconomic and business economic evaluations. In socioeconomic feasibility studies, the question is whether a project is profitable to society as a whole, whereas in business economic feasibility studies, the goal is only to determine whether a project is profitable from a business perspective. However, companies can also apply the criteria of socioeconomic feasibility studies to achieve a better understanding of the impact of their decisions on society as a whole.

Thus, any feasibility study or public regulation activity should begin with a series of specific analyses of the institutional and political contexts of the given project in the country or region in question. Therefore, feasibility studies should not be conducted on the basis of the neoclassical assumption of a society placed in an economic optimum. It is not an easy task to perform these institutional analyses. It is, however, an interesting task that may result in the identification of projects that have not been implemented under present institutional conditions, despite being both economically and environmentally feasible from a socioeconomic point of view.

In situations of radical technological change, new technologies must be developed, and investments in these technologies must be made while they compete with and are compared to well-established existing technologies. More importantly, these competitions or comparisons never occur in the ideal free-market situation but are mainly embedded in certain political and institutional frameworks. These frameworks are usually created over time and thus favor old technological schemes. The introduction of a new technology is, therefore, far more than an issue of cost-benefit calculations. It requires a thorough examination of the total construction of systems; that these systems are connected to the broader objectives and conditions of society; and, finally, that the idea of just maintaining old technological systems instead of investing in newer and cleaner ones is challenged.

Therefore, in such situations, feasibility studies are not simply cost-benefit calculations that assume that the present institutional conditions will lead to the best of all worlds. When designing feasibility studies, one must understand that society often does not find itself in an economically optimal situation. Furthermore, it is important to recognize that other possibilities may exist that will benefit both the economy and the environment. Feasibility studies should, therefore, be designed to identify such possibilities as well as the institutional policies that will make it possible for visionary politicians to implement them. It should also be understood that because of their significant impacts on the development of different technologies, feasibility studies are mostly subject to influences or pressures from different political and economic interests. It is important to be aware of this and to prepare an integrative, communicative, innovative, and transparent study process to ensure that the study groups and society in question will not be trapped in the dominant technological schemes and interests. In short, feasibility studies aim to serve as an effective social learning tool that can be used for overcoming these institutional and political barriers and pave the way for radical technological change.

The preceding conclusions regarding the present technical and institutional situation lead to the following guidelines on the design of feasibility studies:

- Start by making a systematic analysis of the decision-making context: Who will use the feasibility study and for which purpose? Which relevant political objectives should be included in the study? Make sure to design the concrete study in such a way that it will provide relevant information in relation to the actual context.
- Seek to have an open political process in the discussion of methodologies and parameters. Raise public awareness of the fact that feasibility study methodologies form part of the study itself.
- Perform analyses with a very long time horizon to find the best solutions independent of existing technological systems.
- Analyze the bindings of existing technological systems. This is particularly important in case overcapacity is found in the existing system. For example, a power system with overcapacity tends to either result in energy prices close to the short-term marginal costs or lead to pressure from the energy companies on the political process, urging the politicians to protect these companies from the competition of newer technologies.

- Analyze the links between the economics of a project and future technological changes. For example, if we deal with renewable energy, what happens to the economics of solar heating systems in a system with a high share of nuclear power or a system with a high share of CHP? This may be designated a *technical sensibility analysis*.
- Analyze the links between the economics of the project and the legislation needed to make it feasible. For example, what happens if the rules ensuring the right to sell electricity to the public grid are abolished? What happens if economic policy results in increased interest rates? This may be designated an *institutional sensibility analysis*.
- Analyze the links between the institutional sensibility analysis and the political process. For example, which agents on the energy market have the financial and political motivation to "kill" newcomer technologies? Can any counterforces be defined that can support newcomer technologies? How can the political balance of power in the sector in question be described? Which political scenarios can be developed and what effects will they have on the particular project? This is designated a *political sensibility analysis*.

3 Public regulation

Successful public regulation measures to implement radical technological change cannot be designed on the basis of the aforementioned preconditions of applied neoclassical economic theory. The main problem is that the necessary technical solutions often require new organizations and new institutions. In general, the applied neoclassical model considers the institutional conditions as given and does not consider them to be modifiable via public regulation. It is essential to distinguish between the *free market*, as previously described, and the *real market* with its institutions at a given place and a given time.

The *real market* is the market with its specific institutions as they exist in reality and as they relate to private market power, public regulation, infrastructure, accessibility of information, business structure, and so forth. This market very often has a considerable level of private regulation. In particular, the traditional energy supplies are usually monopolistic or oligopolistic, with one or only a few suppliers of given goods. Even when many suppliers are on the market, they are often interconnected through ownership relations and are therefore not mutually independent. Information is frequently kept secret by referring to commercial interests, despite the fact that full information and openness are preconditions for well-functioning free markets.

All in all, the *real market* does not fulfill the institutional preconditions of the *free market* described in economic textbooks. The interplay between the *real market* and the *free market* is often one of ideology, where the strongest actors on an oligopolistic *real market* use the ideology of the *free market* to argue for no public regulation without removing their own private regulation of the market. In the world of reality, the argument "let the free market decide" is synonymous with the sentence "let us decide." "Us" means the strongest actors on an oligopolistic market with a few dominating companies.

Furthermore, the *free-market* premise of the public sector is neutral in the sense that market allocation processes are not fulfilled. It seems obvious that the use of public funds for education, roads, harbors, defense, and medical care has different effects on the direction of market processes. A high level of infrastructure investments in roads will support the economic sectors linked to the car industry. Military expenses support economic sectors linked to military projects. The examples are infinite and, consequently, the premise that the activity of the public sector does not influence the direction of the market processes is not valid.

Thus, even if the market sector was "free" in the sense described in economics textbooks, the market would still not be free from the interference of the public sector. Therefore, it can be concluded that in the real-market world, some form of either public or private hierarchal regulation will always take place. Regulation is defined as any organized purpose directed to influence the framework of the market and the organization of cooperation on the market.

When designing public regulation measures, it therefore becomes important to distinguish between socioeconomic feasibility studies, where the purpose is to examine whether a given project is profitable from a societal point of view, and business economic feasibility studies, where the purpose is to examine whether a given project is profitable from the point of view of a specific company. Fig. 3.2 shows the relationship between business economy, socioeconomy, and public regulation on a given market in the case of capital-intensive technologies with a long technical lifetime.

Fig. 3.2 shows that in the present situation ("Situation I"), specific market conditions and a specific legislation can be identified ("Market Economy I" and "Public Regulation I"). Typically, under current conditions, existing technologies are favored and will turn up as the most profitable alternative in a business economic evaluation, while the economy of alternatives of radical technological change may be bad from a business economic point of view ("Business Economy I"). Nevertheless, socioeconomic feasibility studies ("Socioeconomy I") may show that the development and investment in new technologies are profitable from a societal point of view.

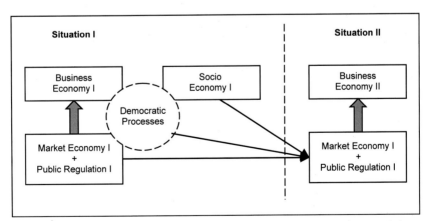

Fig. 3.2 The relationship between business economy, socioeconomy, and public regulation on a given market.
Source: Hvelplund, F., Lund, H., Sukkumnoed, D., 2007. Feasibility studies and technological innovation. In: Kørnøv, L., Thrane, M., Remmen, A., Lund, H. (Eds.), Tools for Sustainable Development. Aalborg University Press, Copenhagen, pp. 593–618.

In this case, what is good for society is not good for business, and the situation should then be discussed by the democratic institutions (the dotted circle). Eventually, these discussions may lead to the development and implementation of a new regulation strategy ("Public Regulation II"). This should ensure that the best solution from the point of view of society also becomes best for business, as a new business economic strategy ("Business Economy II") is established with the same priority as "Socioeconomy I." Now an ideal situation is identified in which the market companies will act in accordance with what is considered best for society.

In "Situation I," a company that wants to evaluate alternatives of radical technological change should carry out business economic feasibility studies ("Evaluate Business Economy I") to estimate the economic consequences under current institutional and technical conditions ("Market Economy I" and "Public Regulation I"). But it is also recommended that the same company should perform socioeconomic feasibility studies ("Analyze Socioeconomy I") to evaluate the socioeconomic consequences of potential future actions implemented by the government. In this way, a company may be able to identify in advance the types of changes in public regulation ("Public Regulation II") that the government could introduce.

The government should carry out socioeconomic feasibility studies ("Analyze Socioeconomy I") to develop an environmental policy that pursues the goals of society. However, it is also recommended that the government conducts business economic feasibility studies ("Analyze Business Economy I") to understand the calculations made by companies under the given market conditions. Consequently, both business and socioeconomic feasibility studies are important and should be conducted by both governmental institutions and private organizations. Several successful examples can be reviewed from Hvelplund and Lund (1998a), especially the cases of wind energy development and CHP district heating systems in Denmark. The feasibility studies of these cases clearly show the overall benefits that can be achieved by Danish society. They encourage the Danish government to introduce various public regulation measures that will lead to the emergence and expansion of new technologies and industries and provide several socioeconomic benefits.

Similar studies have been made of the three Baltic countries regarding the replacement of, among others, nuclear and central power stations based on oil shale by CHP generation (Lund et al., 1999a, 2000, 2005; Rasburskis et al., 2007). In all cases, it has been shown how the countries, from a socioeconomic point of view including job creation and the influence on the balance of payment, would benefit from CHP, whereas from a business economic point of view, it would not pay for companies to invest in CHP under the present public regulation conditions.

A similar analysis was made in the case of the European Communities as a whole in the article "Energy, Employment and the Environment: Towards an Integrated Approach" (Lund and Hvelplund, 1998). The study uses the European Union (EU) figures of supply and demand of the late 1990s as a reference point. As a result of increasing demands and decreasing reserves in the EU, the imports of fossil fuels from other countries are expected to increase.

Following this, the article proposes an alternative composed by a number of technically feasible measures. The aim is to improve energy efficiency and assist in the realization of the EU's CO_2 emissions reduction target of the late 1990s and to improve energy security. The technically feasible measures include 20 percent savings in demand, a 50 percent increase in CHP, and a 10 percent increase in wind power. If adopted, these measures would result in a 25 percent reduction of CO_2 emissions in the EU, compared with the level of 1990. It would also lead to fossil fuel imports 50 percent lower than those predicted at that time.

Moreover, through these measures, the EU would be able to improve its possibilities of job creation without a significant impact on the EU balance of payments. The application of the proposed measures would create around 1.5 million jobs in Europe from 2000 to 2010, because of the construction work required. The balance of payments would initially be affected by this phase of investment in new energy technologies, although this effect could be minimized by subsidizing EU suppliers of these technologies and associated services. Furthermore, a positive net effect would develop as the imports of fuel would decrease.

This article further points out that the relative labor force in Europe was expected to decrease, so it would be easier to explore these opportunities at that time than in the future. The aim of reducing CO_2 emissions and improving energy security measures, such as those proposed in the article, will have to be explored at some point. The article concludes by posing the question "Why wait until 2010 to solve the problems when the EU at that time is likely to have higher CO_2 emissions, higher fuel imports, and possibly a smaller labor force?"

Another example of what can be gained by public regulation when making both socioeconomic and business economic feasibility studies is the case of electric heating conversion in Denmark, as described in the article "Implementation of energy-conservation policies: the case of electric heating conversion in Denmark" (Lund, 1999a). This article analyzes which kinds of public planning, regulation, and initiatives are suitable for the implementation of energy conservation policies. The case of electric heating conversion in Denmark illustrates a number of general problems and demonstrates which radical technological changes are needed to implement the political objectives of CO_2 reduction. The case also provides an example of how public regulation can deal with those problems.

First, it must be understood that the implementation of CO_2 reduction policies is characterized by a change in technology. This change requires not only minor technological modifications but also large organizational changes that may include the establishment of completely new organizations. Second, the existing institutional set-up is strongly influenced by existing organizations, and these are closely linked to the old technologies that will no longer be needed.

Therefore, the implementation of new technologies represents a challenge to public regulation. On the one hand, it must be expected that it will meet resistance from representatives of the old technologies; on the other hand, it must initiate the establishment of new institutional set-ups. In the case of electric heating conversion in Denmark, these barriers have so far been overcome partly by introducing a mixture of numerous, differentiated, and multipurpose public regulation instruments and partly by changing strategy in terms of implementing the "least-feasible" conversions first, thereby avoiding conflicts with the old technologies. The specific details of this case are described in Lund's (1999a) article.

The design of public regulation measures based on concrete institutional economics is not only relevant when radical technological change is in question. The California energy crisis in the summer of 2000 and spring of 2001, during which the electricity supply was not capable of meeting the demands and electricity prices increased dramatically, raised the question of whether deregulated markets can ensure security of supply (Clark and Lund, 2001). California had to confront the basic issue of whether a public good like electricity supply could be left to the *free market* or if the government should play an active role in certain infrastructure sectors, such as the energy sector. Clark and Bradshaw (2004) presented a strategy to avoid such crises in the future, emphasizing the point that both government and industry need clear, concise, long-term-oriented, and consistent market rules, standards, codes, and operating protocols to achieve the goals of sustainable society and business. Also, in the identification and design of market rules, one can benefit from analyzing the concrete institutional market conditions.

4   Democratic infrastructure

The three previously mentioned strategies deal with the issues of describing concrete technical alternatives, using suitable feasibility studies, and designing proper public regulation measures. However, *who* should do all this? It is clear that society cannot expect such initiatives to come from existing organizations that depend on existing institutional set-ups. Someone else has to do it, but *who?* The principal answer is the representatives of future societal interests and the representatives of potential new technologies. These representatives must be found among citizens, non-governmental organizations, small emerging companies, and politicians who are involved in public decision making.

However, it is important to realize that public decision making does not occur in a political vacuum. The decision-making process is shaped by various political and economic interest groups in society who strive to protect their profits or pursue their values. When seeking the implementation of radical technological change, it is therefore important to be aware that, typically, existing technologies are well represented in the democratic decision-making infrastructure, whereas potential future technologies are weakly represented, if represented at all.

In 1995, Hvelplund et al. (1995) described and analyzed the democratic infrastructure of Danish energy planning at that time. Based on these descriptions, we proposed certain changes to achieve a better fulfillment of political renewable energy objectives in a book that defines the terms *old-corporative* and *new-corporative* regulation. *Corporative* indicates that in any mixed-market economy, the authorities typically cooperate with the representatives of different technologies.

The analysis revealed that the Danish regulation was "old-corporative" in the sense that the authorities at all levels favored the old technologies, while the representatives of new technologies experienced difficulties in being heard. This old-corporative regulation was particularly visible in the committees who conducted the technical and economic analyses and provided information to the political decision making. These committees always seemed to include many representatives from power stations and natural gas and district heating companies. None or very few, however, were recruited from the renewable energy industry or from independent environmental and energy efficiency organizations.

Unfortunately, meetings behind closed doors were also characteristic features of the old-corporative regulation. The authorities and the representatives of the old technologies would not allow the public to know what was going on. Indeed, the public was not informed until after the decision was made. As part of the analysis, we asked to see documents from the work of three very important committees of the early 1990s, namely, the Electric Heating Conversion Committee, the Electricity Strategy Committee, and the High Voltage Transmission Line Committee. In all three cases, the result was the same. The doors remained closed until the committees' work was finalized. The public was not allowed to interfere but was informed when the decision had been made.

In accordance with Danish law, the public had the right to be informed of a dialog between different bodies of authorities and private companies, such as utilities. Consequently, the public had a legal right to see documents, which were passed on from one authority to another or to private companies. However, to be able to make this dialog in private, the authorities defined new "autonomous authorities," including representatives from both the authorities and the utility companies and "personal members," who kept their "personal" papers in the archive of these authorities. These "personal" papers, the authorities argued, were not part of the legal act. This practice was approved by the appeal authority: the Danish ombudsman institution. In some cases, the authorities could not manage to distinguish between all of the "autonomous authorities" and "personal members." In these cases, the authorities had to apologize, and their apology was accepted by the ombudsman. However, no measures were taken to hinder such a procedure from ever happening again.

The representation of the committees could be seen in the results. One example is the Combined Heat and Power Committee from 1985, which is described in detail in Hvelplund et al. (1995) and Hvelplund (2005). The job of the committee was to identify the potential for small CHP plants in Denmark. The committee concluded that the potential was so small that it was of no interest or importance, that is, technical potential of 450 MW. However, time has shown that the potential was indeed much larger, since already a few years later, more than 2000 MW was implemented. Other important examples are documented in Hvelplund et al. (1995).

Old-corporative regulation makes both the authorities and society blind to the potential of new technologies. Practice makes it difficult to implement radical technological changes such as renewable energy systems. As a consequence, we advocated the replacement of old-corporative regulation with new-corporative regulation in Denmark in 1995. Among other suggestions, we proposed that representatives of new technologies should be involved in committee work at all levels.

The preceding example illustrates how important it is to pay attention to the democratic infrastructure when society seeks to implement radical technological change. Consequently, a focus on old-corporative versus new-corporative regulation becomes a very important part of the Choice Awareness strategy. Changes in the democratic infrastructure may be the key to initiating the fulfillment of the three other strategies, namely, to promote the description of alternatives, to conduct suitable feasibility studies, and to present concrete proposals for public regulation measures.

5 Research methodology

According to the Choice Awareness theory, as described in Chapter 2, one cannot expect organizations linked to existing technologies to initiate and promote radical technological changes. It does not form part of their perception and interests. Proposals representing such changes have to come from someone else. Accepting this fact has an impact on the development of strategies to raise Choice Awareness, as described in the previous sections of this chapter, as well as an impact on the research method one has to apply.

As a researcher, one cannot expect to be able to observe all the many facets of choice-eliminating mechanisms, if no one puts forward any alternatives representing radical technological change. Furthermore, one cannot expect such alternatives to emerge by themselves. Therefore, the Energy Planning research group at Aalborg University has developed a research method, in which we, on the basis of our expertise in the energy field, design and promote concrete technical alternatives as well as coherent institutional alternatives. For many years, the group has involved researchers with technical as well as socioeconomic expertise. This enables our group to construct, design, and promote alternative energy plans, which are described both technically and in terms of their socioeconomic impacts.

Furthermore, we often contribute to the public debate with suggestions regarding public regulation at the socioeconomic level. Our energy plans and suggestions are published and discussed in public with the public administration at the national and municipal levels as well as with different energy companies. Members of the group have developed and used this research method since 1975 and have made alternative energy projects and plans for Denmark and its regions, as well as applied the method to other countries.

The aim of this method is twofold. First, we wish to increase the number of technical and societal possibilities in general, which are sometimes blocked by economic and politically vested interests. In this way, we seek to fulfill our "public service" obligation as employees of a public university. Second, this research method can generate new information about the dynamics of society because we have established a socioeconomic experiment by introducing discussions that may otherwise not have taken place.

This research method has been applied to most of the cases described in Chapter 9. Without applying this method, we would probably not have been able to do our research. For example, in the case of European Environmental Impact Assessment (EIA) procedures in Denmark, our aim was to examine whether the Danish implementation of EIA would ensure a thorough examination of alternatives representing radical technological change. However, the authorities had no legal power to require it, and the power company had no motivation for proposing such alternatives. The situation made it unlikely that cleaner technology alternatives would even be put on the EIA agenda. Consequently, our research method included a participatory component, by which we introduced renewable energy technology alternatives to the EIA authorities (the county). Furthermore, we filed complaints to the Nature Protection Appeal Board when we felt that our alternatives were not examined properly by the council authorities.

One could describe our method as "questioning" by introducing technical alternatives and institutional alternatives followed by subsequent complaints. In the preceding example, the different EIA authorities reacted and, in this way, provided answers to our "questioning." By this method, society received important information on how EIA authorities in practice react to the proposal of alternatives. Moreover, awareness was raised of the fact that cleaner technology alternatives did exist and it was questioned why they were not implemented.

When applying the research method to our involvement in different cases, we have used the steps shown in Fig. 3.3.

Step 1: To identify and design relevant technical alternatives. These alternatives have typically been related to the fulfillment of certain political goals. The identification of relevant goals very often includes a combination of energy policy objectives and economic objectives. Energy policy objectives often comprise energy security, environmental protection, and the promotion of renewable energy, while economic objectives typically comprise economic growth and, depending on the economic situation, job creation, balance of payment, innovation, and industrial development. One may refer to well-defined parliamentary goals or introduce other goals into the discussion.

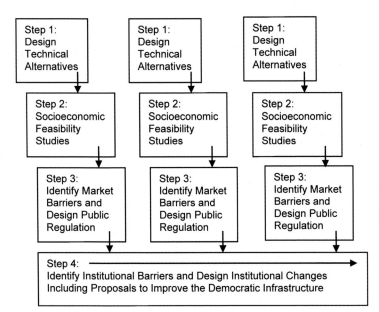

Fig. 3.3 Step-by-step research method. Technical alternatives and socioeconomic evaluations lead to the identification of, first, market barriers, and second, other institutional barriers.

Step 2: To conduct socioeconomic feasibility studies that can provide relevant information in terms of defining which of the alternatives will be able to fulfill the political goals in the best way. Such information is typically hard to achieve by applying methods such as cost-benefit analyses and macroeconomic equilibrium models based on applied neoclassical economics. Instead, the methods based on what has here been defined as concrete institutional economics are recommended.

Step 3: To identify market economic institutional barriers to the implementation of socio-economic least-cost solutions (i.e., the alternatives that can best fulfill the political energy policy and economic objectives). These barriers can be distinguished from others by making business economic feasibility studies and comparing the results of these to the results of the socioeconomic studies. Part of this step is also to make concrete proposals for short-term public regulation measures (changes in taxes, subsidies, financing options, energy sales, and connection rules and agreements).

Step 4: To identify further institutional barriers of a more general nature, such as the lack of proper organizations, the lack of knowledge, or the lack of institutions providing relevant information for the decision-making process, and design proposals for the long-term institutional changes in organizations and in the democratic infrastructure. In our case, step 4 has been carried out by using inputs from several cases.

The EnergyPLAN energy system analysis model

4

This chapter deals with the development of energy system analysis tools and methodologies that are suitable for the design and evaluation of renewable as well as fully decarbonized energy system alternatives. The specific purpose of this chapter is to present the energy system analysis model EnergyPLAN and describe how to use the model for the design of relevant alternatives. In Chapters 5–8, this model is used for the analysis of renewable energy systems and alternatives.

The EnergyPLAN model is an input/output energy systems analysis model that has been developed and expanded on a continuous basis since 1999. It is deterministic and aims to identify suitable energy system designs and operation strategies using hourly simulations over a 1-year time period. The model analyzes national energy systems on an aggregated basis and emphasizes the evaluation of potential synergies between the different subsectors. Thus, the model involves hourly balances of district heating and cooling as well as electricity and gas grids. It also includes a wide range of cross-sector technologies such as heat pumps, combined heat and power (CHP) plants, electrolyzers, and electric vehicles as well as gasification, hydrogenation, and co-electrolyzer units. Moreover, the model includes a wide range of biomass and carbon conversion technologies. Focus is on the integration of energy solutions with other greenhouse gas emitting sectors and the purpose is to identify suitable pathways to a fully decarbonized society.

The model is a freeware that has been widely used in different countries across the world. In this chapter, EnergyPLAN is described and compared to various other energy system analysis models.

First, this chapter presents some overall considerations for the use and construction of models for the purpose of designing alternative energy systems. The overall understanding of how, when, and why such specific descriptions of alternatives are needed was discussed in Chapters 2 and 3 with regard to Choice Awareness. In this chapter, these issues are examined in the context of the design of energy system analysis models and concrete technical alternatives based on renewable energy technologies. This methodology distinguishes between four implementation phases: introduction, large-scale integration, 100 percent renewable energy systems, and fully decarbonized societies.

1 Overall considerations

In accordance with the idea of Choice Awareness, the overall aim of the EnergyPLAN model is to analyze energy systems for the purpose of assisting the design of alternatives based on renewable energy system technologies. Based on the description of Choice Awareness in Chapter 3, the following key considerations can be highlighted.

Renewable Energy Systems. https://doi.org/10.1016/B978-0-443-14137-9.00004-8

The model should be able to make a consistent and comparative analysis of all alternatives in question as well as a reference. It is important that all alternatives including the reference are calculated and analyzed equally to create the basis for a consistent comparison. The reference may be an existing proposal that one may wish to challenge by introducing alternatives, or it may be an official plan made by authorities that can relate the discussion of the alternatives in question to other proposals.

The model should be able to analyze radical technological changes. Consequently, it should be able to analyze both the existing system as well as other systems that are radically different both technically and institutionally. This means that the model should not be too influenced by the technical design of the existing system. Nor should the model focus solely on the existing institutional set-up, such as the existing electricity market design. A balance must be created for the model to conduct an analysis on the basis of existing technical and institutional set-ups without depending on these conditions to such an extent that the analysis of radical changes cannot be made properly. For example, if the existing version of a specific electricity market such as Nord Pool is an integrated part of the model, it may become impossible to analyze radically different alternatives. Or if the existing power station structure is an integrated part of the model, it may become impossible to analyze radically different technical options.

The model should be able to provide suitable information for feasibility studies and the design of public regulation measures based on concrete institutional economics. Therefore, it should be able to contribute to feasibility studies with an analysis of relevant parameters such as external costs, job creation, and industrial innovation. In the case of investment optimization based on market prices, the model should be able to distinguish between business and socioeconomic feasibility studies.

The model should have a transparent and consistent methodology that produces understandable results. This means that it should have a consistent documentation and be publicly available, user friendly, and easily accessed. Moreover, good references or other forms of public acceptance will improve the model.

In addition to the preceding four important criteria, one may add that the model, in some cases, should be able to help identify and design proper alternatives for future systems in which the number of alternative combinations is almost infinite. Therefore, the model should be able to explore a wide range of future options. It should be fast and capable of managing changes in various inputs in a straightforward and systematic way.

Moreover, if used to identify 100 percent renewable energy systems and fully decarbonized societies, the model should be able to analyze the two major challenges of these systems as elaborated in the following section.

The two major challenges of 100 percent renewable energy systems and fully decarbonized societies

The implementation of 100 percent renewable energy systems and fully decarbonized societies involves several substantial challenges. However, from the viewpoint of energy system analysis models, two may be considered the most important: (1) the amount of biomass resources that can be used for energy is limited and substantially lower than the present level of fossil fuels used and (2) the remaining sources, mostly wind and solar, are fluctuating and intermittent.

To include the first challenge, it becomes essential for the energy system analysis model to include methodologies and technologies that optimize the use of the limited biomass resources, while also building a bridge between the biomass resource and the need for gas or liquid fuels to supplement the direct use of electricity in the transportation sector. Moreover, to identify solutions to a fully decarbonized society, the model should be able to align with other greenhouse gas emitting sectors. Therefore, it becomes important to include technologies to provide carbon sinks, such as carbon capture and biochar, in the modeling.

To meet the second challenge, it becomes essential to include temporal distributions and the intermittency of renewable energy sources in the analysis. The time steps may be hourly or similar. As a decisive factor, the model must be able to include the impacts of fluctuations in renewable energy sources in the analysis in a suitable way. This is normally done on an hourly basis as opposed to annual or monthly time steps. However, the need for such accuracy depends on the degree of implementation of renewable energy in the system in question. In the following section, three implementation phases are defined.

Three implementation phases

The need for energy systems analysis tools depends on the share of renewable energy in the system. The following three phases of implementing renewable energy technologies can be defined.

The introduction phase: This phase represents a situation in which no or only a small share of renewable energy is present in the existing energy system. The phase is characterized by marginal proposals for the introduction of renewable energy, for example, wind turbines integrated into a system without or with only a small share of wind power. The system will respond in the same way during all hours of the year, and the technical influence of the integration on the system is easy to identify in terms of annual fuel savings.

The large-scale integration phase: This phase represents a situation in which a large share of renewable energy already exists in the system; for example, when more wind turbines are added to a system that already has a large share of wind power. In this phase, further increases in renewable energy will have an influence on the system, which will vary from one hour to another, depending, for example, on whether a heat storage is full or whether the electricity demand is high or low during the given hour. The influence of wind power integration on the system, and thereby the calculation of annual fuel savings, becomes complex and requires a detailed calculation with hourly simulation models.

The 100 percent renewable energy phase: This phase represents a situation in which the energy system has been or is being transformed into a system based 100 percent on renewable energy. In this type of system, new investments in renewable energy must be compared not to nuclear or fossil fuels, but to other sorts of renewable energy system technologies. These technologies include conservation, efficiency improvements, and storage and conversion technologies, as well as the use of smart grids (electricity, district heating, and gas). The influence on the system is complex, not only in terms of differences from one hour to another, but also regarding the identification of adequate conversion and storage technologies as well as the smart operation of grid infrastructures.

The fully decarbonized society phase: This phase represents a situation in which the energy system is seen as part of achieving a net-zero greenhouse gas emission solution. In this system, it is seen that some sectors outside the energy sector, such as agriculture and industrial processes beyond the burning of fossil fuels in itself, may not be reasonable to bring to a net-zero within the sector in question. Therefore, the best solution may be found when the energy sector provides compensation in the form of sinks from technologies such as carbon capture and biochar.

The definition of these four implementation phases can be used in the selection and design of proper tools for the technical analyses. In the first phase, the technical calculations are rather simple and do not require complex models. Typically, annual fuel savings can be calculated without models or by using simple models based on duration curves or similar data. However, in the next phase, it becomes essential to make hour-by-hour calculations due to the fluctuations in most renewable energy sources. In the third phase, it also becomes essential to include proper analyses of advanced conversion and storage technologies as well as smart grid infrastructures in the system. In the fourth phase, it becomes essential to consider various biomass conversion technologies as well as their relation to carbon capture and other potentials for creating sinks.

Different types of energy system analysis models

On a global scale, a large number of different computer models exist that can all be called energy system analysis models because they make calculations related to the analysis of energy systems. Based on a list of 68 models, Connolly et al. (2010) provided a detailed description and review of 37 different models. Table 4.1 lists some of the models. In general, all models address the implementation of renewable energy sources or other technologies related to renewable energy systems such as CHP. For practical reasons, only a few of the many existing tools and models have been

Table 4.1 Energy system analysis models.

Name	Description
BALMOREL	The purpose of the BALMOREL project is to support modeling and analyses of the energy sector with emphasis on the electricity and combined heat and power sectors. These analyses typically cover a number of countries and include aspects of energy, environment, and economy. The project maintains and develops the BALMOREL model, a tool that can be used by energy system experts, energy companies, authorities, transmission system operators, researchers, and others for the analyses of future developments of a regional energy sector. This model is developed and distributed under open source ideals. Hosted by the BALMOREL project, Denmark

Table 4.1 Continued

Name	Description
CHPSizer	A tool for conducting preliminary evaluations of CHP for hospitals and hotels in the United Kingdom. The software enables the user to make a preliminary evaluation of CHP for a building. This will guide the user in deciding whether to proceed with a more detailed examination of CHP for the building in question. Rather than being based on theoretical calculations, the software has been developed using actual energy profile data collected from buildings in the United Kingdom
EnergyBALANCE	This model is a simple energy balance spreadsheet that provides a good comprehensive view of a regional or national energy system. It is part of the Energy Planning Tool. The energy balance methodology is intended to be very simple and very easy to implement. Basically, the energy balance of a country or region can be calculated on one page in a spreadsheet. Hosted by the Danish Organization for Renewable Energy
EnergyPLAN	Computer model for hour-by-hour simulations of complete regional or national energy systems, including electricity, individual and district heating, cooling, industry, and transportation. Focuses on the design and evaluation of renewable energy systems with high penetration of fluctuating renewable energy sources, CHP, and different energy storage options. Hosted by Aalborg University, Denmark. Can be accessed from www.EnergyPLAN.eu
energyPRO	A complete modeling software package for combined technoeconomic design, analysis, and optimization of both fossil and bio-fueled cogeneration and trigeneration projects, as well as other types of complex energy projects. Simulates and optimizes energy production in fixed and fluctuating electricity tariff systems by active use of thermal and fuel store. Hosted by EMD International A/S, Denmark. Can be accessed from www.emd.dk
ENPEP	Suite of models for integrated energy/environment analysis. ENPEP was developed by Argonne National Laboratory and is distributed for use in over 70 countries. This model provides state-of-the-art capabilities for use in energy policy evaluation, energy pricing studies, assessing energy efficiency and renewable resource potential, assessing overall energy sector development strategies, and analyzing environmental burdens and greenhouse gas mitigation options. Hosted by Argonne National Laboratory for the International Atomic Energy Agency
H2RES	A model designed for balancing between hourly time series of water, electricity, and hydrogen demand, appropriate storage, and supply (wind, solar, hydro, diesel, or mainland grid). The main purpose is energy planning of islands and isolated regions that operate as standalone systems, but it may also serve other purposes. Hosted by Zagreb University

Continued

Table 4.1 Continued

Name	Description
HOMER	This model is made particularly for small, isolated power systems, although it enables grid connection. Optimization and sensitivity analysis algorithms provide the basis for an evaluation of the economic and technical feasibility of a large number of technologies. Models both conventional and renewable energy technologies. Hosted by National Energy Laboratory, the United States
HYDROGEMS	Library of computer models for simulation of integrated hydrogen systems based on renewable energy. The objective is to provide a set of modeling tools that can be used to optimize the design and control of RE/H2 systems. Hosted by Institute for Energy Technology, Norway
LEAP	Scenario-based energy-environment modeling tool. Its scenarios are based on comprehensive accounting of how energy is consumed, converted, and produced in a given region or economy under a range of alternative assumptions on population, economic development, technology, price, and so on. Scenarios can be built and then compared to assess their energy requirements, social costs and benefits, and environmental impacts. Hosted by Stockholm Environment Institute, Boston, the United States
MARKAL	Integrated energy/environment analysis. MARKAL is a generic model tailored by the input data to represent the evolution of a specific energy system over a period of usually 40–50 years at the national, regional, state, province, or community level. Hosted by the International Energy Agency's Energy Technology Systems Analysis Program
MESAP	MESAP is an energy systems toolbox for application-oriented system solutions in many areas: market analysis in electricity trade, database for power plant operation control, data pool for technical reporting, management of control data in grid companies, CO_2 monitoring, emission inventories for air pollutants, and database for energy models, as well as systems for common statistics administration. MESAP is the only software for all of these applications. It is hosted by Seven2one Informationssysteme GmbH, Karlsruhe, Germany
PRIMES	Modeling system that simulates a market equilibrium solution for energy supply and demand in the European Union Member States. This model determines the equilibrium by finding the prices of each energy form. Thereby, the quantity producers find the best solution that matches the quantity demanded by consumers. The equilibrium is static (within each time period) but repeated in a time-forward path under dynamic relationships. Hosted by National Technical University of Athens, Greece
RAMSES	Simulation/planning model for electricity and district heating supply. Semilinear hour simulation of Nordic electricity and district heating system. INPUT: Plant database (existing and new plants),

Table 4.1 Continued

Name	Description
	transmission lines, prices and taxes, electricity and district heating demand, and load curve sets. OUTPUT: Electricity price, fuel consumption, emissions, cash flows, loss of load probability, and so on. Hosted by the Danish Energy Agency
Ready Reckoner	Model to assist users with a "first pass" technical and financial analysis of cogeneration at their site. The program is a Ready Reckoner intended for quick preliminary evaluations. The Ready Reckoner conducts a simple technical and financial analysis of a cogeneration opportunity. Should the cogeneration opportunity appear attractive in this evaluation, then the user is recommended to conduct more detailed analyses or engage suitable advisers to consider the project evaluation to the extent necessary to commit funds. Hosted by Department of Industry Science and Resources and the Australian EcoGeneration Association, Australia
RETScreen	A model that can be used to evaluate the energy production and savings, life cycle costs, emission reductions, financial viability, and risk of various types of energy-efficient and renewable energy technologies. The software also includes product, cost, and climate databases. Hosted by RETScreen International Clean Energy Decision Support Centre
SIVAEL	SIVAEL is a simulation program for a thermal power system with related CHP areas. The program makes a simulation with start/stop and load distribution on an hourly basis. The simulation period is from 1 day to 1 year. The program can handle condensing plants, CHP plants—both back pressure and extraction—and also wind power, electricity storage (battery or pumping power), and trade with foreign countries. Hosted by the Danish TSO https://energinet.dk
WASP	Long-term electricity generation planning including environment analysis. This model determines the least cost-generating system expansion plan that adequately meets the demand for electrical power while respecting user-specified constraints on system reliability. WASP uses probabilistic simulation to calculate the production costs of a large number of possible future system configurations and dynamic programming. It also determines the optimal expansion plan for the electric power system considered. Hosted by the International Atomic Energy Agency

included here. The following tables should only be considered examples to illustrate some important points. For a more complete survey, please consult review papers such as the above-mentioned or Chang et al. (2021).

While completing the review in Connolly et al. (2010), it became apparent from discussions with the different tool developers that the developers did not share a common language to classify different types of energy tools. Consequently, to ensure that the tools were described correctly, seven general definitions were created and sent to the developers to distinguish between the different types of energy tools. One or more of these definitions can be used to describe an energy tool. The energy tool types include:

1. A *simulation tool* simulates the operation of a given energy system to supply a given set of energy demands. Typically, a simulation tool is operated in hourly time steps over a 1-year time period.
2. A *scenario tool* usually combines a series of years into a long-term scenario. Typically, scenario tools function in time steps of 1 year and combine the annual results into a scenario of typically 20–50 years.
3. An *equilibrium tool* seeks to explain the behavior of supply, demand, and prices in a whole economy or part of an economy (general or partial) with several or many markets. It is often assumed that agents are price takers and that equilibrium can be identified.
4. A *top-down tool* is a macroeconomic tool using general macroeconomic data to determine growth in energy prices and demands. Typically, top-down tools are also equilibrium tools.
5. A *bottom-up tool* identifies and analyzes the specific energy technologies and thereby identifies investment options and alternatives.
6. *Operation optimization tools* optimize the operation of a given energy system. Typically, operation optimization tools are also simulation tools.
7. *Investment optimization tools* optimize the investments in an energy system. Typically, optimization tools are also scenario tools optimizing investments in new energy stations and technologies.

Chang et al. (2021) used the same definitions when making a review of trends in the field of energy systems modeling including 54 analysis tools. The study identifies three main trends, i.e. the increasing modeling of cross-sectoral synergies, the growing focus on open access, and the improved temporal detail to deal with planning future scenarios with high levels of variable renewable energy sources. However, Chang et al. (2021) conclude that key challenges remain in terms of representing high resolution energy demand in all sectors; in terms of understanding how tools are coupled together, their different levels of openness and accessibility as well as the level of engagement between tool developers and policy/decision makers.

Regarding the idea of Choice Awareness and the introduction of the four implementation phases just described, important differences can be found among the models. One important difference is whether the model makes a detailed hour-by-hour simulation or is based on aggregated annual calculations, possibly made by using duration curves or similar data. Another important difference is whether the model addresses the national or regional system level or the project or single station level. In Table 4.2, some chosen models are grouped according to these essential differences.

Not all models are designed in such a way that they fit perfectly into these groupings. Consequently, one model is located in two groups, and other models may have aspects across more than one group. The MESAP model is not shown in Table 4.2 because it is really a database system that combines the use of other models. Moreover, some models are being further developed on a continuous basis. Typically, models based on aggregated annual calculations also contain some hourly simulations

Table 4.2 Grouping of energy system analysis models.

	Aggregated annual calculations	Detailed hour-by-hour simulations
Regional/national system level	EnergyBALANCE TransportPLAN IndustryPLAN339 LEAP MARKAL PRIMES ENPEP	EnergyPLAN LEAP RAMSES BALMOREL SIVAEL WASP H2RES HOMER
Project/station system level	Ready Reckoner CHPSizer RETScreen	energyPRO HYDROGEMS

in different parts of the calculation process. Nevertheless, Table 4.2 shows two important differences.

As described regarding the three implementation phases, models based on aggregated data are typically suitable for the *introduction phase*. They can provide an adequate level of detail, since in this phase it is not necessary to run detailed hourly simulations, which is often a complicated and dataconsuming process. Moreover, models based on aggregated data typically have the advantage of being easy to document and communicate compared to hourly simulation models. However, for the analysis of large-scale integration or 100 percent renewable energy systems, hour-by-hour simulation models are essential.

In the division concerning areas usually considered, the models at the project/station level typically cannot evaluate the influence of, for example, fluctuating renewable energy sources on the overall regional and/or national system. On the other hand, they are typically better equipped for making detailed analyses of the business economic operation and design of single stations. Consequently, in the groupings in Table 4.2, only detailed hour-by-hour simulation models at the regional/national level can fulfill the requirements for energy system analyses of renewable energy systems on the large-scale and 100 percent renewable energy implementation phases.

It should be emphasized, however, that the design of alternatives by use of these models in many cases may benefit from being combined with either models based on aggregated data or models at the project/station level. For example, analyses made with the EnergyPLAN model have, in some cases, been combined with analyses conducted by one of the "sister models": energyPRO and EnergyBALANCE (Lund et al., 2004). The energyPRO model has been used for additional analyses of how single stations will respond to changes in the energy system design defined by use of the EnergyPLAN model. The EnergyBALANCE model has been used to make fast and easy first estimates of different large-scale options, before involving the

EnergyPLAN model. A review of trends in the development and coupling of energy system analysis models can be found in Chang et al. (2021).

Hourly simulation models at the national level

Historically, energy system analysis models based on hourly simulations at the regional/national level have typically been developed with two purposes in mind. Some models have been developed to optimize the load dispatch between single power stations in utility systems, whereas others have been made for planning purposes, that is, identifying proper investment strategies.

The overall aim of the load dispatch models is to design suitable operation strategies on a day-to-day basis. These models have, prior to the introduction of national and international electricity markets, been used and designed by electricity system operators to plan least-cost production strategies of electricity supply systems with several production units. As these models must be able to calculate exact operational costs and emissions, they are typically very comprehensive and detailed in their description of each individual power station.

The overall aim of planning models is to identify suitable future investment strategies. These models have been used and designed by public planning authorities, utility companies, and different nongovernmental organizations, including universities and research institutions. Sometimes load dispatch models have been used for planning purposes, or planning models have been developed on the basis of load dispatch models. In practice, these models have sometimes proven to be very conservative in the sense that they have mainly been able to analyze small, short-term adjustments to the existing system rather than radical changes in the overall system design and regulation. Moreover, the use of these models is typically very time-consuming because of the need for detailed data.

Both load dispatch models and planning models have been influenced by the introduction of international electricity markets, which has taken place in many countries around the world since the late 1990s. Operation strategies for power production units are now determined by the market, and the models are used to identify optimal market behavior. Planning models have begun to include the modeling of international electricity markets in the analyses. Some models have chosen to convert to solely making simulations of the results of market operations based on the present institutional set-up of the international electricity market, while, for example, the EnergyPLAN model has been equipped with the ability to make both market analyses as well as simulations on the basis of pure technical optimizations.

Consequently, hour-by-hour simulation models at the regional/national level have important differences. One is whether the present electricity market structure forms the only institutional basis for the simulation and/or operational optimization procedure of the model or if this simulation is also based on technical and/or economic optimization. In the latter case, it is also important to define whether this optimization is made on the basis of a business economic market strategy or some sort of socioeconomic least-cost strategy.

Another important difference is whether the model combines all sectors of a regional/national energy system or includes only parts of this system, such as the

Table 4.3 National level hour-by-hour simulation models.

Detailed hour-by-hour simulation models at the regional/national level	Operational optimization based on technoeconomic optimization	Operational optimization based on electricity market simulations
Include all sectors: electricity, district heating, individual heating, industry, transportation Include mainly the electricity sector	EnergyPLAN H2RES HOMER	EnergyPLAN RAMSES (BALMOREL) BALMOREL SIVAEL WASP

electricity supply. This difference is important to the analysis of large-scale integration and especially 100 percent renewable energy systems. These systems will benefit from efficiency measures such as CHP production, which makes an integrated analysis of the electricity and the heating sector relevant. They will also benefit from the use of electricity for transportation purposes, which makes the combined analysis of the electricity and transport sectors essential.

In Table 4.3, the hour-by-hour simulation models applied to the regional/national level have been grouped according to their optimization and their operation levels. Again, this grouping is not 100 percent accurate for all models. The LEAP model is left out, which is characterized as a simulation model rather than an optimization model because other models optimize the operation of the system.

In Table 4.3, the EnergyPLAN model is placed in both the technoeconomic optimization and the electricity market optimization model groups. This model is designed to calculate the consequences of both types of optimization strategies. Moreover, the calculation of the market-economic strategy is based on a business economic optimization of individual stations in which different taxes and subsidies can be specified. The aim is to be able to provide information for the discussion on suitable public regulation measures.

It must be emphasized that the models have differences other than those presented in Tables 4.2 and 4.3. Models, such as LEAP and RAMSES, emphasize the analyses of scenarios and include the calculation for a series of years. However, these characteristics illustrate examples of the variety of models currently available. As already mentioned, many more models and tools exist and can be found, e.g., in Chang et al. (2021).

2 The EnergyPLAN model

EnergyPLAN is a computer model designed for energy systems analysis. This model has been developed and expanded on a continuous basis since 1999. It is a user-friendly tool designed in a series of Windows tab sheets and programmed in Delphi

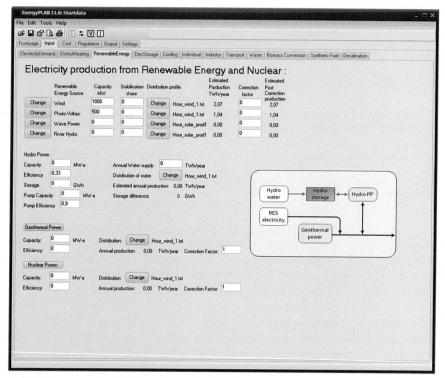

Fig. 4.1 Example of a Windows input tab sheet in the EnergyPLAN model.

Pascal. The inputs are defined by the user in a number of technical Windows input tab sheets and a few cost specification input tab sheets (see Fig. 4.1).

The following section provides a brief description of the EnergyPLAN model. A full documentation of it can be found on www.EnergyPLAN.eu and a summary can be found in Lund et al. (2021c). Moreover, it is described and compared to other models in Lund and Münster (2003a,b), Lund et al. (2007b), Connolly et al. (2010), and Chang et al. (2021).

Purpose and application

The main purpose of this model is to assist the design of national or regional energy planning strategies on the basis of technical and economic analyses of the consequences of implementing different energy systems and investments. It encompasses the whole national or regional energy system, including heat, gas, and electricity supplies as well as the transportation and industrial sectors. Moreover, in recent years, a wide range of biomass conversion technologies and related energy storage options have been added to the model. Regarding electricity supply, this model emphasizes the analysis of different regulation strategies and focuses on the interaction between

CHP and fluctuating renewable energy sources. Moreover, it includes various biomass conversion and power-to-X options.

The EnergyPLAN model is a deterministic input/output model. General inputs are demands, renewable energy sources, energy station capacities, costs, and a number of optional different regulation strategies emphasizing import/export and excess electricity production. Outputs are energy balances and resulting annual production, fuel consumption, import/exports of electricity, and total costs, including income from the exchange of electricity (see Fig. 4.2).

Compared to other similar models, the following characteristics of EnergyPLAN can be highlighted:

- EnergyPLAN is a *deterministic* model as opposed to a stochastic model or models using Monte Carlo methods. With the same input, it will always come to the same results. However, as we will see in Chapter 5, it can perform a calculation on the basis of RES data of a stochastic and intermittent nature and still provide system results that are valid for future RES data inputs.
- EnergyPLAN is an *hour-simulation* model as opposed to a model based on aggregated annual demands and production. Consequently, the model can analyze the influence of fluctuating RES on the system as well as weekly and seasonal differences in electricity, gas, and heat demands and water inputs to large hydropower systems.
- EnergyPLAN is *aggregated in its system description* as opposed to models in which each individual station and component is described. For example, in EnergyPLAN, the district heating systems are aggregated and defined as three principal groups.
- EnergyPLAN *optimizes the operation* of a given system as opposed to models that optimize investments in the system. However, by analyzing different systems (investments), this model can be used for identifying feasible investments, as we will see in Chapters 5–8.
- EnergyPLAN provides a choice between *different regulation strategies* for a given system as opposed to models into which a specific institutional framework (such as the Nord Pool electricity market) is incorporated.
- EnergyPLAN *analyzes 1 year* in steps of 1 hour as opposed to scenario models analyzing a series of years. However, several analyses each covering 1 year may be combined into scenarios.
- EnergyPLAN is based on *analytical programming* as opposed to dynamic programming or advanced mathematical tools. This makes the calculations direct and the model very fast when performing calculations. In the programming, any procedures that would increase the calculation time have been avoided, and the computation of 1 year requires only a few seconds on a normal computer, even in the case of complicated national energy systems.
- EnergyPLAN includes hourly analyses of the complete *smart energy system*, i.e., district heating and cooling as well as electricity and gas grids and infrastructures, as opposed to models that have a sole focus on, for instance, the electricity sector.
- EnergyPLAN includes a wide range of biomass conversion and carbon capture technologies and thus enables the analysis of relations to other greenhouse gas emitting sectors.

Energy systems analysis structure

The EnergyPLAN model can be used to calculate the consequences of operating a given energy system in such a way that it meets the set of energy demands of a given year. Different operation strategies can be analyzed. Basically, the model

EnergyPLAN

INPUT

Demands
Electricity
Cooling
District Heating
Individual Heating
Fuel for Industry
Fuel for Transport

RES
Wind
Solar Thermal
Photovoltaic
Geothermal
Hydro Power
Wave

Capacities & efficiencies
Power Plant
Boilers
CHP
Heat Pumps
Electric Boilers
Micro CHP

Storage
Heat Storage
Hydrogen Storage
Electricity Storage
CAES

Transport
Petrol/Diesel Vehicles
Gas Vehicles
Electric Vehicles
V2G Electric Vehicles
Hydrogen Vehicles
Biofuel Vehicles

Regulation
Technical Limitations
Choice of Strategy
CEEP Strategies
Transmission Cap.
External Electricity
Market

Fuel Cost
Types of fuel
CO2 Emission Factor
CO2 Emission Costs
Fuel Prices

Cost
Variable Operation
Fixed Operation
Investment
Interest Rate

Distribution data

Electricity Demand | District Heating | Wind | Hydro | Wave | Waste
Solar thermal | Photovoltaic | Geothermal | | Individual Heating
Industrial CHP | Transportation | | Market Prices

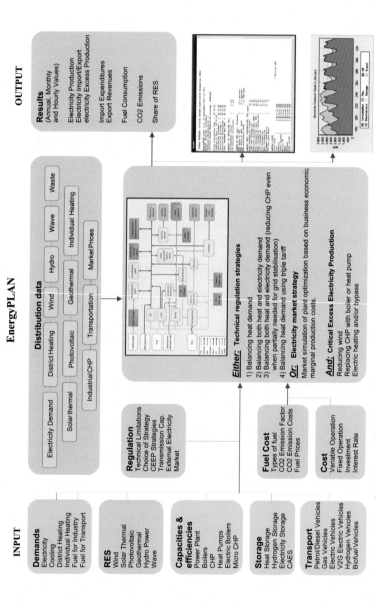

Either: Technical regulation strategies

1) Balancing heat demand
2) Balancing both heat and electricity demand
3) Balancing both heat and electricity demand (reducing CHP even when partially needed for grid stabilisation)
4) Balancing heat demand using triple tariff

Or: Electricity market strategy
Market simulation of plant optimization based on business economic marginal production costs.

And: Critical Excess Electricity Production
Reducing wind
Replacing CHP with boiler or heat pump
Electric heating and/or bypass

OUTPUT

Results
(Annual, Monthly and Hourly Values)

Electricity Production
Electricity Import/Export
electricity Excess Production

Import Expenditures
Export Revenues

Fuel Consumption

CO2 Emissions

Share of RES

Fig. 4.2 Input-output structure of the EnergyPLAN model.

distinguishes between technical regulation (i.e., identifying the least fuel-consuming solution) and market-economic regulation (i.e., identifying the consequences of operating each station on the electricity market with the aim of optimizing the business economic profit). In both situations, most technologies can be actively involved in the regulation, and in both situations, the total costs of the systems can be calculated. In the documentation of the model, a list of energy demands is presented as well as an overview of all components of the model. A short description of how they are operated in relation to the two different regulation strategies is also presented together with a list of the main inputs for each component.

The model includes a large number of traditional technologies, such as power stations, CHP, and boilers, as well as energy conversion and technologies used in renewable energy systems, such as heat pumps, electrolyzers, and heat, electricity, and hydrogen storage technologies, including Compressed Air Energy Storage (CAES). It can also include a number of alternative vehicles, such as sophisticated technologies like V2G (vehicle to grid), in which vehicles supply the electric grid, and synthetic fuel vehicles, which use methanol and/or methane. Moreover, the model includes various renewable energy sources, such as solar thermal and photovoltaic (PV), wind, wave, and hydropower.

The EnergyPLAN model is further expanded and improved on an ongoing basis in a dialog with the users of the model. Since the first edition of this book was published, the model has been expanded in terms of hourly analyses of the gas grid, including gas storage and a number of biomass and/or power to gas conversion units such as gasification and hydrogenation. In Chapters 6 and 7, examples are presented showing the use of these facilities to study fuel pathways for transportation and to make hourly analyses of the complete energy system including the use of various smart grids and infrastructures.

Fig. 4.3 shows the procedure of the energy system analysis. As a first step, calculations are based on a small computation, which is made simultaneously with the typing of input data in the input and cost tab sheets. The next step consists of a series of initial calculations that do not involve electricity balancing. Then the procedure is divided into *either* a technical *or* a market-economic optimization. The user chooses which one to apply. However, each calculation lasts only a few seconds and, consequently, it is possible to make both calculations, one after another. The technical optimization minimizes the import/export of electricity and seeks to identify the least fuel-consuming solution. On the other hand, the market-economic optimization identifies the least-cost solution on the basis of the business economic costs for each production unit. In both situations, the model can calculate the socioeconomic consequences that provide important information for the design of different public regulation measures.

The principle of the energy system of the EnergyPLAN model is shown in Fig. 4.4. In the EnergyPLAN model, an aggregated analysis is made of the many individual

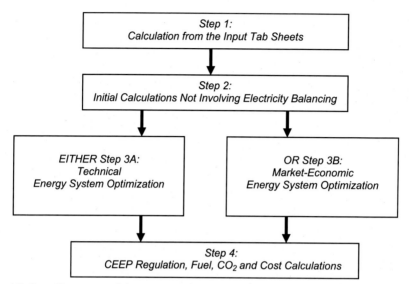

Fig. 4.3 Overall structure of the energy system analysis procedures.

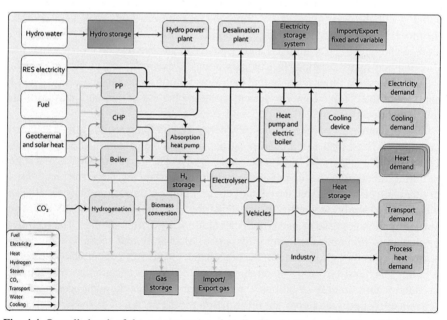

Fig. 4.4 Overall sketch of the energy system described in the EnergyPLAN model.

stations that together form a regional or national energy system. Thus, all boilers, CHP stations, and so forth that produce heat for district heating are grouped into three district heating systems. Moreover, the EnergyPLAN model includes a series of optional renewable energy sources as well as a large number of conversion and storage technologies. In this way, the model can make comprehensive analyses of rather complex

100 percent renewable energy systems without the need for an enormous quantity of detailed data.

Validation of model

In Lund and Mathiesen (2012), the validation of a model like EnergyPLAN is described. The validation of this kind of model is complicated, since the models are typically huge and involve a substantial number of assumptions and formulas of which all cannot easily be described. Kleindorfer et al. (1998) argued that the validation of such models may be compared to the validation of miniature scientific theories. The principle of validation is discussed in relation to different philosophical positions, including Rationalism, Empiricism, and Positive Economics, emphasizing a discussion of the objectivist approach versus the relativist approach: foundationalism versus anti-foundationalism. On the one side, an extreme objectivist believes that model validation can be separated from the model builder and its context and that validation is an algorithmic process that is not open to interpretation or debate. In contrast, an extreme relativist believes that the model and model builder are inseparable and that validation is a matter of opinion.

Kleindorfer et al. (1998) argued that most practitioners have instinctively adopted a middle ground in this debate and they compare the validation of simulation models to the situation in a courthouse. The prosecutor does not have to prove the guilt in any foundationalist sense, but rather "beyond reasonable doubt." Extending the courthouse metaphor, the authors argue that

> The model builder would be free to establish and increase the credibility of the model through any reasonable means. This process would also involve other model stakeholders, such as model users and referees of journal articles.

In the process of defining reasonable, one may refer to papers such as Pidd (2010) and Qudrat-Ullah and Seong (2010) in which the purpose of the model is highlighted as essential, i.e., if the model is acceptable for its intended use. Following these guidelines, it should initially be highlighted that the purpose of the EnergyPLAN model is to assist in the design of complete renewable energy systems seen in the light of a wish to transform the present energy system and support the implementation of a fully decarbonized society, which from a Choice Awareness perspective requires a radical technological change.

Seen in this light, the following items are highlighted in relation to the validation of the EnergyPLAN model. First, the EnergyPLAN energy system analysis model has a complete documentation of which an updated version can always be downloaded from www.EnergyPLAN.eu. Next, EnergyPLAN has proven its ability to form the basis for the modeling of complete national energy systems in various studies. Thus, the EnergyPLAN model has been used in a number of recent studies in different countries including Denmark, Romania, China, and Ireland. These studies typically involve the analysis of a reference that has been compared to official statistics or similar. Furthermore, the model has been used to analyze the role of a number of different technologies in future sustainable energy systems including wind, wave and PV, CHP, heat pumps, waste to energy, CAES, and electricity and biofuels for transportation including V2G. All these studies have been published in refereed journal papers and a list of most of them can be found on www.EnergyPLAN.eu. Third, the EnergyPLAN model is a freeware that can be downloaded from the home page already mentioned. This means that anyone can access the model and then repeat and/or evaluate the studies completed by others.

Østergaard et al. (2022a) synthesize EnergyPLAN applications through an analysis of its use, both from a bibliometric and a case-geographical point of view, and through a review of the evolution in the issues addressed and the results obtained using EnergyPLAN. This synthesis is provided with a view to addressing the validity and contribution of EnergyPLAN-based research. As of July 1st, 2022, EnergyPLAN had been applied in 315 peer-reviewed articles. In addition, the paper shows how the complexity of energy systems analyses has increased over time, with early studies focusing on the role of wind power and the cogeneration of heat and power and later studies addressing contemporarily novel issues like the sector integration offered by using power-to-X in fully integrated renewable energy systems.

Energy system analysis methodology

A short outline of how to use the model in the design of energy alternatives at the national level (Lund et al., 2007a) is presented here. On the home page of the model, www.EnergyPLAN.eu, a whole set of exercises and assignments can be downloaded, including detailed answers, that together constitute a comprehensive set of user guidelines. As already shown in Table 4.3, the model can be used for different kinds of energy systems analyses.

> **Technical analysis**: Design and analysis of complex energy systems at the national or regional level and according to different technical regulation strategies. In this analysis, input is a description of energy demands, production capacities and efficiencies, and energy sources. Output consists of annual energy balances, fuel consumptions, and CO_2 emissions.
>
> **Market exchange analysis**: Further analysis of trade and exchange on international electricity markets. In this case, the model needs further input to identify market prices and to determine the response of these prices to changes in import and export. Input is also needed to determine marginal production costs of the individual electricity-producing unit. The modeling is based on the fundamental assumption that each station optimizes according to business economic profits, including any taxes and CO_2 emissions costs.
>
> **Feasibility studies**: Calculation of feasibility in terms of total annual costs of the system according to different designs and regulation strategies. In this case, inputs such as investment costs and fixed operation and maintenance costs must be added together with lifetime periods and an interest rate. The model determines the socioeconomic consequences (taxes and subsidies are not included) for the energy system. The costs are divided into fuel costs, variable operation costs, investment costs, fixed operation costs, electricity exchange costs and benefits, and possible CO_2 payments.

A step-by-step approach to national energy systems analysis

The approach to energy systems analysis used in the EnergyPLAN model can be divided into four steps:

Step 1: Defining reference energy demands
Step 2: Defining a reference energy supply system
Step 3: Defining the regulation of the energy supply system
Step 4: Defining alternatives

Step 1: Defining reference energy demands

The first step is to define a reference energy demand. The electricity demand is simply defined by identifying an annual demand (TWh/year) and choosing an hourly distribution. The distribution data can be picked from the database of the model, or one can create a new distribution (for more information, please consult the model documentation). Energy conservation is included by modifying the energy demands. The model has been designed in such a way that it helps to change the hourly distribution where necessary; for example, the electricity demand is altered when electric heating or cooling demands decrease as a result of energy-saving measures.

Though the annual demand is sufficient, the model also offers the possibility of defining two additional demands with separate distribution datasets. One is meant to be used for adding electricity to transportation and another for fixed import/exports, but they can be used for any purpose; the model simply adds up the three demands.

The model can also include flexible electricity demands: these are demands that are included in the regulation of securing a balance between supply and demand. One must choose among flexibility within 1 day, 1 week, or 1 month (4 weeks). For each group, a flexible demand can be identified by two values: annual demand (TWh/year) and maximum capacity (MW).

District heating demand is defined in the same way as the electricity demand: it uses an annual demand (TWh/year) and a selected distribution dataset. The model divides district heating supply into three groups, and the demand of each group must be defined. The first group comprises traditional district heating stations with boilers; the second group consists of small CHP stations, and the third group includes large CHP stations based on thermal extraction stations. However, the three groups can also be used for other types of separation depending on the specific study and context.

With regard to energy systems analysis, the model takes a smart energy systems approach to the identification of suitable solutions. The main focus is put on cross-sectoral synergies in order to find the best and most affordable solutions. The production is specified in the same way as the demand, by using an annual production and a selected distribution dataset.

With regard to the transportation and individual heating sectors, various options exist in terms of several different electric, hydrogen, and biofuel vehicles and various solar thermal and micro-CHP systems, including conversion and storage technologies such as electrolyzers. If these technologies involve the district heating or electricity sectors, they are included in the hourly balancing during the computation.

With regard to biomass, a wide range of biomass technologies are included in the model as well as the interaction between different technologies in terms of utilizing by-products from one technology as input to another.

Step 2: Defining a reference energy supply system

Step 2 is to define the reference energy supply system, divided into renewable energy sources, capacities, and efficiencies of energy production units, and the division of fuel based on annual average consumption. Renewable energy sources for electricity production, such as wind power, PV power, and wave power, are defined by an installed capacity and by a selected distribution dataset. Again, the distribution data can be chosen from the existing database, or new distributions can be made. The model offers the possibility of adding a factor (between 0 and 1) to modify the distribution curve. The factor adjusts the distribution curve to higher annual productions if, for instance, new wind power capacities are being built on locations with better wind potentials.

Renewable energy sources for district heating production, such as solar thermal, can be specified for each of the three district heating groups in the same way as heat production from industry: in other words, by an annual production (TWh/year) and a selected distribution dataset. Moreover, heat, electricity, and hydrogen storage and energy conversion technologies, such as electrolyzers, can be specified.

Capacities and efficiencies of the energy production units are defined as average values for each type of station in each of the three district heating groups. For group one (district heating boilers), only the efficiency needs to be stated, since the capacity of the boiler must always be sufficient. For groups two and three (CHP), capacities (MW_e and MW_{th}) and efficiencies are given for CHP units and boilers. Moreover, a heat pump and a heat storage capacity (GWh) can be defined for each of the two groups. The heat pump is defined by the capacity (MW_e) and the coefficient of performance (COP) factor (heat output divided by electric input). Moreover, the maximum share of heat production from the heat pump can be specified to achieve the specified COP.

Finally, the capacity (MW) and the efficiency of a condensing power station are given. The model distinguishes between the CHP stations in group three and the condensing power stations. However, in practice, these stations may be the same units (extraction stations). Consequently, the model makes the calculation assuming that the capacity input of the condensing power stations constitutes the total maximum capacity of both the condensing stations and the CHP stations in group three. Thus, if the CHP capacity at a certain stage of the analysis is not used for CHP production, the same capacity might be used for condensing power production (but with a different efficiency).

The fuel consumption of the stations can be calculated by the model on the basis of efficiencies. To track fuel use and CO_2 emissions, the model needs inputs in terms of the share of fuel types at the different stations. The shares are given by relative numbers, and all types of fuels are increased or decreased accordingly. However, the model enables an adjustment of the amount of one or more types of fuels.

Step 3: Defining the regulation of the energy supply system

The regulation strategy is defined by choosing one of the predefined general strategies and then specifying some limitations and additional options. Basically, the technical analyses distinguish between a technical optimization and an electricity market optimization. In the market-economic optimization, electricity production is determined on the basis of business economic marginal production costs of the different types of electricity-producing units. Moreover, electricity-consuming units, such as heat pumps and electrolyzers, are also included. One can specify various taxes on different fuels and types of production and thereby conduct analyses of the consequences of changing taxes and/or introducing new ones.

With regard to the technical optimization, one must choose one of the two following strategies:

Technical regulation strategy 1: Meeting heat demand. In this strategy, all units produce solely according to the heat demand. In district heating systems without CHP, the boiler simply supplies the difference between the district heating demand and the production from solar thermal and industrial CHP. For district heating with CHP, the units are prioritized according to the sequence (1) solar thermal, (2) industrial CHP, (3) CHP units, (4) heat pumps, and (5) peak load boilers. The model offers the option of operating the small CHP units according to a triple tariff, giving an incentive to allocate electricity production during hours of high and peak demand.

Technical regulation strategy 2: Meeting both heat and electricity demands. When choosing Strategy 2, the export of electricity is minimized mainly by replacing CHP heat production with boilers or heat pumps. This strategy simultaneously increases electricity consumption and decreases electricity production, as the CHP units must lower their heat production. Similarly, when there is extra capacity available at the CHP stations and space in the heat storage, the production at the condensation stations is replaced with CHP production, thus increasing the overall efficiency of the energy system.

The model also considers the ancillary services required to secure grid stability in the electricity system. Limitations in identifying optimal operation strategies can be specified in terms of the minimum share of electricity production required from a unit for it to supply ancillary services. The condensing power stations and the CHP stations in group three are always assumed to have these abilities. Any share of small CHP stations and renewable energy sources with ancillary service abilities can be specified as an input value.

As part of the regulation strategy, one can specify system limitations on the export/import of electricity represented by transmission line capacities (MW). Depending on the situation and the chosen regulation strategy, bottlenecks may occur that require higher exports of electricity than the amount allowed by the transmission lines. This is called critical excess electricity production. Consequently, one can specify strategies to avoid this problem. As shown in Chapters 5–8, the description and analysis of the reference system can be used to establish a common point of departure when promoting and discussing alternative strategies.

Step 4: Defining alternatives

When a reference is described, the analysis of alternatives is relatively easy. The computation of the whole system takes only a few seconds on a normal personal computer.

Analyzing different regulation strategies is, in many cases, a simple matter of pressing a button to change regulation and run the computation once again. Changing technologies is a matter of choosing other technologies. Of course, this change depends on a proper definition of inputs in terms of efficiencies and costs, which may be time-consuming to find for new technologies.

Sister models to EnergyPLAN

The EnergyPLAN computer model has a series of sister models (sister models in the sense that they originate from Aalborg University and have been made to supplement and support one another): EnergyBALANCE, energyPRO, TransportPLAN and IndustryPLAN. The EnergyBALANCE model is a simple spreadsheet model based on aggregated annual calculations of energy balances. It is designed for an easy integration of typical inputs from national statistics. This model adds different data of efficiencies, and so forth, to make overall analyses of changes in demand and supply technologies. For a long time, it was available from the Danish Organization for Renewable Energy (OVE) via its home page. However, now it is mostly used by consultancies in their own more advanced versions. www.orgve.dk.

The energyPRO model excels in modeling and optimizing the operation of a single station. It has the ability to evaluate many different types of technologies and performance criteria for generating units, particularly CHP stations, and enables the user to add detailed definitions of parameters, such as heat production, electricity production, fuel costs, power curves, and control strategies. It can also conduct a sophisticated economic analysis that considers varying values for revenues (such as heat prices and spot market electricity prices) and costs (such as fuel costs, taxes, and other operational expenses). However, this level of detail may also require a significant amount of research and data input to properly initialize the model, and it implies a high level of understanding of the specific performance characteristics of the station.

The model focuses primarily on production aspects and, apart from a few exceptions such as heat distribution losses, it does not consider how the station fits into the broader energy system. It is an advanced computer tool for the design and operation of CHP stations, and it has been used to design most of the existing small CHP stations in Denmark. The initial version of energyPRO was designed in the late 1980s. Shortly after that, the program was made commercially available by the software company Energy and environMental Data (EMD). Based on an ongoing dialog with users, EMD has refined and added new facilities and features to the model on a continuous basis. It has become a widely used software package for the analysis of local energy stations based on gas engines, gas turbines, and steam turbines burning both waste and wood chips, as well as stations based on boilers only. EnergyPRO has been used in Lund and Andersen (2005) and Andersen and Lund (2007).

TransportPLAN is a back-casting modeling tool, which allows for detailed scenario analysis of transportation systems (Kany et al., 2022). The detailed resolution of input data provides the possibility of adjusting the development of the transportation demand precisely and in-depth. The tool requires inputs regarding annual transportation demand, vehicle fleet composition, utilization rates and fuel distribution in the first modeling year. Transportation demand is divided between passenger and freight transportation, measured in passenger-kilometers for passenger transportation and tons-kilometers for freight transportation. Transportation demand data in combination with capacity utilization factors enable the calculation of total kilometers traveled for all modes of transportation (in the following referred to as either traffic work or vehicle-kilometers). Additionally, an average energy consumption for the fleet of vehicles for all modes of transportation is necessary to calculate the final energy consumption.

TransportPLAN is built to enable the development of transportation system strategies and scenarios. The tool provides options to increase/decrease annual growth in transportation demand and annual modal shifts across all modes of transportation. Additionally, the tool can create pathways to implement alternative fuels and propulsion technologies as well as introduce different trajectories for the development of energy efficiency improvements in existing and new engine technologies and improvements in capacity utilization rates. The model results consist of annual transportation demand and traffic work for each modeled year, annual final energy consumption, greenhouse gas emissions and transportation system costs. TransportPLAN can provide detailed input regarding the transportation sector for further analysis of systems integration by use of the EnergyPLAN model. IndustryPLAN is a tool for analyzing the industrial energy demands of European countries (Johannsen et al., 2023). The tool is developed as a Microsoft Excel spreadsheet using a combination of Excel functions and VBA coding, making the tool accessible to a wide audience. The tool provides a bottom-up top-down approach in which first the "black box" of industry is opened with country-based data and technology data from the bottom-up assessment of each industrial sector. In a top-down approach, measures are implemented on the sub-sectors, aggregated, and connected to GDP development and saturation rates of new technologies. The IndustryPLAN tool is established based on the following overarching ambitions:

1. To provide a platform for implementing the guiding principles for energy efficiency in industry.
2. Enabling the establishment of tangible future scenarios for the industry sector as a part of smart energy systems.

In IndustryPLAN, users can design scenarios with varying implementation of energy efficiency improvements and fossil fuel replacement measures. I.e., a user may choose to implement all the available energy efficiency improvements, none of the improvements, or anything in between. Measures to be implemented are selected based on a least-cost principle, prioritizing measures based on their cost per energy saved.

IndustryPLAN provides many different results aimed at evaluating and quantifying future industrial energy demands. Combined with the included scenario design functionality, these outputs can aid in the investigation of a wide array of research questions, such as analyses on the importance of energy savings, and the impact of extensive electrification or conversion of fossil fuel-based processes to biomass and hydrogen-based processes. While such analyses and results are interesting on their own, specifically for the industry sector, an important capacity of IndustryPLAN is that it can provide inputs for holistic energy system models such as EnergyPLAN, encompassing complete national energy systems, including the heat, electricity, industry, and transportation sectors. Thus, the disaggregated and detailed industry assessment from IndustryPLAN can provide a more thorough representation of the industry sector in the modeling of integrated energy systems.

3 Reflections

Based on the formulation of the Choice Awareness theory in Chapter 3, this chapter discussed some overall key issues to consider when designing tools for the analysis and assessment of renewable energy alternatives, representing radical technological change. The following reflections can be made regarding these key considerations and the EnergyPLAN model:

- The EnergyPLAN model can make a consistent and comparative analysis of different energy systems based on fossil fuels, nuclear energy, and renewable energy. When the reference energy system is described, EnergyPLAN makes it possible to conduct a fast and easy analysis of radically different alternatives without losing coherence and consistency in the technical assessment of even complex renewable energy systems.
- The EnergyPLAN model seeks to enable the analysis of radical technological changes. The model describes existing fossil fuel systems in aggregated technical terms, which relatively easily can be changed into radically different systems, such as systems based on 100 percent renewable energy sources. This model divides the input to market-economic analyses into taxes and fuel costs, making it possible to analyze different institutional frameworks in the form of different taxes. Moreover, if more radical institutional structures are to be analyzed, it can provide purely technical optimizations. This makes it possible to separate the discussion of institutional frameworks, such as specific electricity market designs, from the analysis of fuel and/or CO_2 emissions alternatives. Compared to many other models, EnergyPLAN has not incorporated the institutional set-up of the electricity market of today as the only institutional framework.
- The model can calculate the costs of the total system divided into investment costs, operation costs, and taxes such as CO_2 emissions trading costs. Thereby, the model can create data for further analysis in socioeconomic feasibility studies, including balance of payment, job creation, industrial innovation, and so on.
- The model has a coherent documentation and seeks to provide a user-friendly communication in input/output tab sheets. Moreover, it is very fast. On a normal PC, the complete hour-by-hour simulations of even very complex national energy systems take only a few seconds. Consequently, the model can be used interactively to test different input combinations in the design of references as well as to make several different calculations of many options without taking very much time. This is further helped by the library of distribution data incorporated into the model, which makes it rather fast and easy to implement comprehensive changes in the input.
- Regarding the four different implementation phases, this model includes a very high number of different technologies that are relevant to renewable energy systems. Consequently, it serves as a good tool for making detailed and comprehensive analyses of a very wide spectrum of large-scale integration possibilities, as well as 100 percent renewable energy systems and energy systems suitable for a fully decarbonized society.

Large-scale integration of renewable energy

5

The large-scale integration of renewable energy sources into existing energy systems must meet the challenge of coordinating variable renewable energy production with the rest of the energy system. Meeting this challenge is essential, especially with regard to electricity production, since electricity systems depend on an exact balance between demand and supply at any time. Given the nature of photovoltaic (PV), wind, wave, and tidal power, little can be gained by regulating the renewable source itself. Large hydropower producers are an exception, since these units are typically well suited for electricity balancing. In general, however, the possibilities of achieving a suitable integration are to be found in the surrounding supply system, that is, in power and CHP stations. The regulation in supply may be facilitated by flexible demands, for example, heat pumps, consumers' demand, and electricity for transportation such as Power-to-X and the like. Moreover, the integration can be helped by different energy storage technologies. However, not all measures are equally efficient and effective.

This chapter examines and deduces the essence from a series of studies in which the EnergyPLAN model has been applied to the analysis of large-scale integration of renewable energy sources (RES) into the Danish energy system. When these studies were conducted in 2000 and onwards, the Danish energy system already had a relatively high share of renewable energy and was therefore suitable for the analysis of further large-scale integration. The studies conducted address the integration of RES into future energy systems and seek to identify the best suitable means. The analyses are based on official projections of the Danish energy system made by the Danish Energy Agency in 2001. The projections are presented in the beginning of this chapter.

In addition to these studies, this chapter presents a method for comparing different energy systems in terms of their ability to integrate RES on a large scale. The question in focus is how to design energy systems with a high capability of utilizing variable RES, also considering the problem that the variations in production of, for example, wind power differ from one year to another. This challenge is met by analyzing and illustrating different energy systems in so-called excess electricity diagrams. In these diagrams, a curve represents the system in all of the years, regardless of the fact that the variations of RES differ from one year to another.

A number of studies of large-scale integration of RES are presented and, finally, some reflections and conclusions sum up the chapter with regard to the methodologies and principles as well as the technical measures involved. This leads to a series of recommendations concerning the most feasible technical measures, how to combine the measures, and when to use them considering the share of RES in the system.

Renewable Energy Systems. https://doi.org/10.1016/B978-0-443-14137-9.00005-X

1 The Danish reference energy system

The different analyses of large-scale integration of renewable energy presented in this chapter are all based on a projection of the future Danish energy supply by the year 2020. In 2001, at the request of the Danish Parliament, the Danish Energy Agency formed an expert group for the purpose of investigating and analyzing possible means and strategies for managing the problem of excess electricity production from combined heat and power (CHP) and RES (Danish Energy Agency, 2001). According to the official Danish Energy Policy of that time, as expressed in the government's energy plan, Energy 21 (Danish Ministry of Environment and Energy, 1996), the share of CHP and especially the share of wind power were expected to increase.

The expert group defined the two terms *exportable excess electricity production* (EEEP) and *critical excess electricity production* (CEEP).[a] EEEP can be exported, while CEEP refers to a situation in which the electricity produced exceeds both the demand and the export capacity of transmission lines out of the system (Denmark). This situation must be avoided so the electricity system will not collapse. Based on these definitions, the expert group defined a reference scenario showing the resulting development in both CEEP and EEEP if expansions in CHP, wind power, and demand were to be implemented according to the official energy policy. Three years—2005, 2010, and 2020—were chosen for the analysis.

At that time, the Danish electricity system was divided into two separate geographical areas: East Denmark and West Denmark. Excess electricity production can arise in one area without being present in the other. Subsequently, the Danish government decided to connect the two systems by a DC connection, but this decision was not final at the time the projection was done. Consequently, it was decided to analyze each area separately. The reference system is characterized by the following development:

- The Danish electricity demand is expected to rise from 35.3 TWh in 2001 to 41.1 TWh in 2020, equal to an annual rise of approximately 0.8 percent.
- The installed capacity of wind power is expected to rise from 570 to 1850 MW in East Denmark and from 1870 to 3860 MW in West Denmark from 2001 to 2020. The increase is primarily due to the expected implementation of one 150 MW offshore wind farm each year.
- Existing large coal-fired CHP steam turbines are to be replaced by new natural gas-fired combined cycle CHP units when the lifetime of each of the old CHP stations runs out. Additionally, distributed CHP stations and industrial CHPs are due for a small expansion.

Denmark had quite good transmission line capacities to its neighboring countries. Thus, East Denmark was connected to Sweden (1700 MW AC) and East Germany (600 MW DC), and West Denmark was connected to North Germany (1200 MW AC), Sweden (600 MW DC), and Norway (1000 MW DC). When defining CEEP, the capacities of all the existing transmission lines were included apart from the AC connection to North Germany, as this area had a very high wind power production and had similar excess production problems during the same hours as West Denmark.

---

[a] Translated from Danish: Kritisk og Eksporterbart Eloverløb.

Based on the preceding assumptions, the expert group evaluated the magnitude of the expected excess electricity production problem divided into EEEP and CEEP. The result of the analysis is shown in Table 5.1. In the reference scenario, excess electricity production was expected to increase considerably in the period toward the year 2020. The expected excess production of 1680 GWh in East Denmark equaled 11 percent of the demand in 2020. In West Denmark, excess production equaled 28 percent of the demand in 2020. The expectations of high excess production illustrated in Table 5.1 can be explained mainly by two assumptions. First, in the reference scenario, small- and medium-scale CHP stations were not expected to regulate according to variations in wind power but solely according to heat demands. In Denmark, CHP stations have been paid through a triple-tariff system with high payments in the morning and the afternoon, reflecting a high electricity demand during these periods, and low payments during night hours, weekends, and holidays.

Consequently, Danish CHP has been designed with relatively high CHP capacities and heat storage, making it possible to produce mainly during high-tariff periods. When electricity sales prices are high, the CHP unit operates at full capacity and stores the heat. When prices are low, the CHP unit stops, and heat for district heating is supplied from the storage. By 2001, this regulation ability had not been used to integrate variations in renewable energy. It had only been used to adjust to changes in electricity demand by applying the so-called triple tariff. This means that production is given a low, medium, or high price, depending on production conditions, in other words, whether or not production takes place during peak load.

The second assumption behind the resulting high excess production is that the task of securing grid (voltage and frequency) stability has been managed solely by large power stations. Consequently, distributed production from small CHP units and wind turbines was considered a burden to the fulfillment of this task. As part of the study, the EnergyPLAN model was used to conduct analyses of how to avoid excess production problems (Lund and Münster, 2001, 2003b).

Most of the analyses in the coming sections of this chapter apply to West Denmark. However, because it was later decided to connect the two Danish systems, some

Table 5.1 Expected excess electricity production in the Danish reference scenario defined in 2001.

Reference scenario (GWh)	2000	2005	2010	2020
East Denmark				
EEEP	2	190	460	1680
CEEP	0	0	0	0
Total	2	190	460	1680
West Denmark				
EEEP	520	3130	3360	5070
CEEP	0	170	290	1330
Total	520	3300	3650	6400

Table 5.2 Reference energy system: Denmark 2020.

	TWh/year
Key data	
Electricity demand	41.1
District heating demand	30.0
Excess electricity production (CEEP + EEEP)	8.4
Primary energy supply	
Wind power	17.7
Fuel for CHP and power stations	92.3
Fuel for households	19.7
Fuel for industry	20.2
Fuel for transport	50.7
Fuel for refinery and so forth	17.4
Total	218.0

analyses were based on a joint reference system including all of Denmark. Moreover, the work of the expert group only included analyses of the electricity system. Consequently, data for the remaining sectors, including the transportation sector, have been added on the basis of the official Danish energy plan, Energy 21 (Danish Ministry of Environment and Energy, 1996). The main data of the joint reference scenario are given in Table 5.2.

Electrification of transportation scenario

Several of the studies presented in this chapter include the conversion of parts of the transportation fleet into electric vehicles in combination with hydrogen fuel cell vehicles. All of the studies are based on a scenario described by Risø National Laboratory, Electric Vehicles and Renewable Energy in the Transport Sector—Energy System Consequences (Nielsen and Jørgensen, 2000). This report concludes that the technical performance, particularly the range, of battery cars and hydrogen fuel cell cars will gradually improve in the coming decades, making it feasible to replace a substantial part of the transportation task of passenger cars and small delivery vans below 2 tons by these types of cars. By the year 2030, 80 percent of the Danish vehicles weighing less than 2 tons are to be replaced by a combination of battery electric vehicles (BEVs) and hydrogen fuel cell vehicles (HFCVs). According to the study, this transformation will lead to a rise in the electricity consumption by 7.3 TWh/year and fuel savings by 20.8 TWh/year. When applied to West Denmark, an alternative system has been defined in which 12.6 TWh of gasoline is replaced by 4.4 TWh of electricity, equal to the share of West Denmark.

2 Excess electricity diagrams[b]

This section is based on Lund's (2003a) article "Excess Electricity Diagrams and the Integration of Renewable Energy," which presents a method for demonstrating the ability of a given energy system to use specific RES in the electricity supply. In the article, the method is applied to the large-scale integration of wind, PV, and wave power into a future Danish reference energy system. The RES integration potential is expressed in terms of the ability of the system to avoid excess electricity production. The different energy sources are analyzed according to an electricity production ranging from 0 to 100 percent of the electricity demand. The analyses have taken into account the fact that certain ancillary services are needed to secure the grid stabilization (voltage and frequency) of the electricity supply system. As a conclusion, excess electricity diagrams show the different patterns of each of the RES. As we will see, these diagrams are capable of showing the general characteristics of a given system even though the variations of RES differ from one year to another.

The analyses that follow were made for West Denmark in 2020, as described in the previous section. For such reference energy systems, the ability to integrate a variable RES can be illustrated in diagrams, as shown in Fig. 5.1. The diagram shows the resulting annual excess production of the system as a function of the share of wind power, assuming that wind power has the exact same hourly distribution as it did in West Denmark in 2001.

Fig. 5.1 is based on a series of 1-year complete energy system analyses of the system made in the EnergyPLAN model. Each analysis includes hour-by-hour calculations of all electricity production and demands given the specified production units

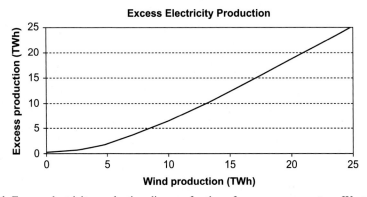

Fig. 5.1 Excess electricity production diagram for the reference energy system, West Denmark, 2020.

---

[b] Excerpts reprinted from *International Journal of Sustainable Energy,* 23/4, Henrik Lund, "Excess electricity diagrams and the integration of renewable energy", pp. 149–156 (2003), with permission from Taylor and Francis.

and regulation strategies. Based on these calculations, annual electricity production is identified, including excess electricity production defined as the difference between the total electricity production and the demand. In the case of Fig. 5.1, the system was first analyzed with a wind power input of 0 TWh/year. Then the input was raised in steps of 5 TWh/year up to 25 TWh/year.

The x-axis shows the wind power production between 0 and 25 TWh, equal to a variation from 0 to 100 percent of the demand (24.87 TWh). In addition, the y-axis shows excess production in TWh. The lower the excess production is, the better the integration of RES. In the article and in Fig. 5.1, the analysis has been made with the following restrictions in ancillary services to achieve grid stability (voltage and frequency): At least 30 percent of the power (at any hour) must come from power production units that are capable of supplying ancillary services. At least a 350 MW running capacity of large power stations must be available at any moment. Distributed generation from CHP and RES is not capable of supplying ancillary services. As can be seen, the excess production in such a system is substantial.

In Lund's article, it was analyzed how the resulting excess electricity curve varies with the wind power production from one year to another. The same kind of analysis is made for PV and wave power. The hour-by-hour distributions of the different RES have been based on actual measurements whenever possible. In the case of PV, data based on actual measurements have been available. The distribution of electricity production derives from the Danish Sol300 project. The project involves 267 PV systems installed in typical one-family houses at eight locations in Denmark beginning in 2000. Distribution data have been provided for two years, as illustrated in Fig. 5.2.

For onshore wind power, which has existed in Denmark for many years, the distribution is based on the actual production of wind turbines located in the reference area: West Denmark. These data have been provided by the transmission system operator (TSO) company of the region. Three years have been analyzed, and the data are shown in Fig. 5.3.

Data based on actual measurements were not yet available for Danish wave power. So far, wave power stations in Denmark had existed only as small test facilities. The distribution of wave power is therefore made on the basis of wave measurements in the North Sea off the west coast of Denmark. Distribution data have been provided for two years, as illustrated in Fig. 5.4. Lund's article describes in more detail the sources of the data for all three types of RES.

When you look at Figs. 5.2–5.4, two obvious observations can be made. First, the electricity production from all three types of variable RES evidently differs from one year to another. When comparing a specific hour on the same specific date of each of the two years, one may find that wind power production is high in 2000 and low in 2001. The same applies to the other two types of sources. However, even though the production differs, some main characteristics can be found and an overall picture can be identified. Thus, looking at these figures, it is quite easy to distinguish between wave and wind power.

Excess electricity production diagrams have been made for the three types of sources and the years shown in Figs. 5.2–5.4. Moreover, a "synthetic" PV year is

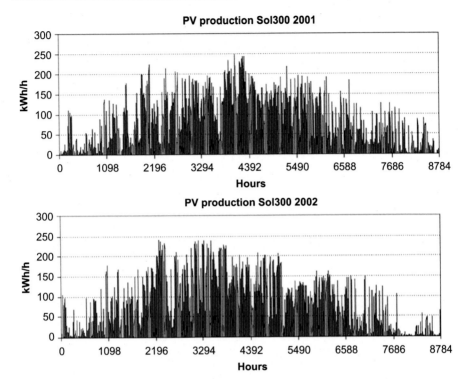

Fig. 5.2 Hourly distributions of PV electricity production in the Danish Sol300 project.

analyzed on the basis of statistically typical numbers and distributions of solar power in Denmark—the so-called Test Reference Year. The results of the different RES are shown in Fig. 5.5. In general, PV is the RES that generates the highest excess production, followed by wave power and onshore wind power in the analyzed system.

The analysis revealed one important fact: It was discovered that even though significant variations can be seen from one year to another, the results in terms of excess production curves are almost exactly the same for each of the individual RES. This discovery is important because the excess electricity diagrams may then serve as an illustration of the ability of a system to integrate variable RES, which does not depend on the difference in variations between the different years. Consequently, the distribution of wind power of all years can be shown by the same curve. This makes it possible to compare different energy systems in terms of their ability to integrate RES on a large scale by comparing two curves in the same diagram. Depending on whether a year is good or bad in terms of wind, one may have to go a bit up or down the curve from one year to another, but it is still the same curve.

The only curve that differs in Fig. 5.5 is the curve of the *synthetic* PV, which shows an excess production slightly higher than the two curves based on actual measurements. This is due to the fact that the actual measurements include an element of correlation between the many locations, whereas the synthetically generated distribution,

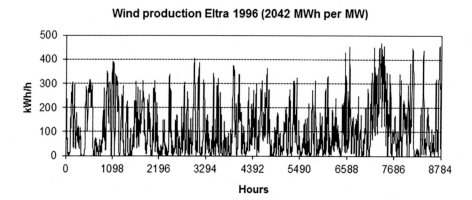

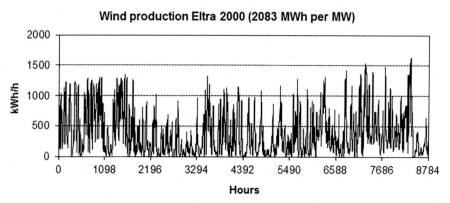

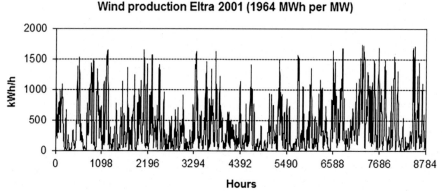

Fig. 5.3 Hourly distributions of electricity production from onshore wind power in West Denmark (actual measurements of electricity production).

in principle, assumes all installations to be located on the same spot. This emphasizes the importance of using measurements and not synthetic data. Actual data that reflect the dispatched locations of distributed RES are better than both actual measurements of one location and synthetic data. This also indicates that the result of the synthetic wave power data may provide a minor overestimation of the excess production.

SUSTAINABLE ENERGY

SUSTAINABLE ENERGY

Towards a Zero-Carbon Economy using Chemistry, Electrochemistry and Catalysis

JULIAN R.H. ROSS

Emeritus Professor, University of Limerick, Limerick, Ireland;
Member of the Royal Irish Academy (MRIA); Fellow of the
Royal Society of Chemistry (FRSC)

ELSEVIER

Elsevier
Radarweg 29, PO Box 211, 1000 AE Amsterdam, Netherlands
The Boulevard, Langford Lane, Kidlington, Oxford OX5 1GB, United Kingdom
50 Hampshire Street, 5th Floor, Cambridge, MA 02139, United States

Library of Congress Cataloging-in-Publication Data
A catalog record for this book is available from the Library of Congress

British Library Cataloguing-in-Publication Data
A catalogue record for this book is available from the British Library

ISBN: 978-0-12-823375-7

For information on all Elsevier publications
visit our website at https://www.elsevier.com/books-and-journals

Publisher: Susan Dennis
Acquisitions Editor: Anita Koch
Editorial Project Manager: Charlotte Kent
Production Project Manager: Bharatwaj Varatharajan
Cover Designer: Christian J. Bilbow

Typeset by STRAIVE, India

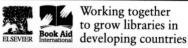

Working together
to grow libraries in
developing countries

www.elsevier.com • www.bookaid.org

Contents

Preface

It is not possible to open a newspaper or magazine without reading of some aspect of the global problem of climate change and of the measures that are necessary to combat it so that we can achieve 'zero carbon' before the year 2050. There has been a steady increase in the emission of greenhouse gases since the Industrial Revolution and the aim of all those countries that have signed up to the Paris Accord is to bring back the resultant temperature rise to no more than 2°C (and even to 1.5°C) within fewer than 30 years.

This book considers many aspects of the potential uses of 'sustainable energy'. In this context, this is the energy that can be obtained by using renewable resources such as wind power, hydroelectric power or solar radiation, and the book discusses how this energy can be used in place of conventionally derived energy from fossil reserves: coal, oil and natural gas. In order to set the scene, the book also discusses in some detail the many ways in which conventional energy is currently used.

The first chapter sets the scene by considering some aspects of the greenhouse effect and outlines the objectives of the Paris Accord that is aimed at reducing the emissions responsible for the effect. It then traces the origins of the greenhouse effect, discussing some human activities (many of which are discussed later in the book) that have taken place since the Industrial Revolution and have contributed to the increased emissions.

The book then considers some important existing industrial activities, all related to the use of energy created from the use of fossil fuels, coal oil and natural gas, each of which results in the emission of greenhouse gases. Some of these emissions can be reduced by methods such as carbon collection and storage, but an alternative is to produce some of the chemicals and fuels on which we rely by using biomass–derived materials. Hence, the use of biomass as a source of energy and chemicals is then considered.

Transport, in one form or the other, is responsible for a significant share of our greenhouse gas emissions. The developments that have occurred since the Industrial Revolution of various forms of transport are outlined and modern developments such as the use of hybrid engines, battery power and fuel cells are then considered. This leads to a detailed discussion of various types of batteries and fuel cells followed by a section considering the potential importance of electrolysis brought about using renewable energy as a means of producing hydrogen and syngas.

The final chapter considers how green hydrogen or syngas produced using electrolytic methods fuelled by renewable electricity can be used in industrial applications such as ammonia and methanol synthesis, the production of steel and cement manufacture. It also considers the importance of achieving reductions in emissions from commercial, domestic and agricultural sources.

The reductions required to allow us reach the targets set in the Paris Accord are enormous and the progress towards achieving these aims has been disappointingly slow until now. Governments and responsible agencies must therefore pay significantly greater attention to ways in which objectives can be achieved and can only manage that by applying the 'carrot and stick approach': offering incentives to all energy users that encourage energy-saving initiatives and the introduction of new methods while at the same time penalising inactivity.

Acknowledgements

As I did in my previous two books, I first thank the very many people with whom I have worked over the years for their efforts and enthusiasm, especially the students and postdocs from my various research groups, too many to name individually, who have helped me build up my knowledge of catalysis and related fields. Thanks are also due to the many scientists and engineers with whom my different research groups have collaborated and from whom I have learnt much about the applications and exploitation of fundamental research in the field of heterogeneous catalysis. This collaborative work was carried out with funding provided by many sources, particularly by various EU research programmes.

I thank Elsevier and the many people from that company with whom I have collaborated during my editorial work for *Applied Catalysis* and *Catalysis Today* and in the production of the three books that I have now written and published with them. In particular, I thank Kostas Marinakis who not only guided me through the process involved in the planning of this book but with whom I have had many previous interactions during my work as an editor. I wish him well in his retirement. Thanks are also due to Kostas's successor, Anita Koch, for her more recent involvement with the production of this book; to Narmatha Mohan for her assistance in ensuring that the necessary permission had been obtained to reproduce copyright material; and to Bharatwaj Varatharajan for his careful and helpful work on the final production and during the proofreading stage. I particularly thank Alice Grant who, as the most recent Elsevier desk editor involved, has cheerfully and helpfully worked with me for most of the writing process.

My thanks are due to two good friends who, each in particular way, helped me during the writing phase: firstly, my colleague and long-standing collaborator, Michael Hayes, who very kindly read through the first draft of Chapter 5 (Biomass as a Source of Energy and Chemicals) and not only provided me with useful comments but also gave me invaluable information on soil organic matter; and secondly, Tony Hilley, a retired offshore oil and gas engineer, who encouraged me throughout the writing phase by providing me with a large number of important web links to recent developments in the field of energy. I also thank Miguel Bañares for his comments on the contents of the completed manuscript and for suggesting the term 'Mount Sustainable'.

Finally, I must once more express my sincere thanks to my wife, Anne, who has encouraged and supported me during the writing of yet another book. This support was even more important for the current volume as she has patiently tolerated my involvement in the task during a period when COVID-19 intruded on our existence and forced long periods of self-isolation.

Julian R.H. Ross

CHAPTER 1

Introduction

Energy production and the greenhouse effect
Solar activity and global warming

For centuries, we have relied on our natural resources for the provision of energy. Early man relied on the combustion of biomass (predominantly wood) to provide heat and fuel for cooking. Very much later, roughly at the time of the Industrial Revolution, he discovered coal, oil and natural gas and these discoveries led to our current almost total dependence on fossil fuels for the provision of energy.[a] Until the Industrial Revolution, the earth's population was predominantly agrarian and any fluctuations in climate that occurred were related only to variations in solar activity. Since then, however, there has been a steady increase in the average global temperature and it is now generally recognised that this change of temperature is related to increased emissions of the so-called greenhouse gases.

Fig. 1.1 shows the values of the solar irradiance and also the global temperature that have been measured over the period since 1880; although there have been some significant changes in the solar activity (and there was a marked maximum value around 1960), the measured values have remained relatively steady over the last 50 years. However, there has been a very significant increase in global temperature during the same period. It is now generally accepted (see Fig. 1.2) that human activities have been responsible for this increase in temperature.[b]

---

[a] We also rely on petroleum derivatives for the manufacture of many of the other resources that we now take for granted: polymers, dyestuffs, pharmaceuticals, detergents, etc. However, our fossil fuel reserves are gradually diminishing and they must therefore be used much more strategically.

[b] A useful summary of some aspects of climate change are to be found in the publication "Vital Climate Change Graphics" published by UNEP/GRID-Arendal; this is available as a free pdf from https://www.grida.no/publications/254/.

Sustainable Energy
https://doi.org/10.1016/B978-0-12-823375-7.00006-8

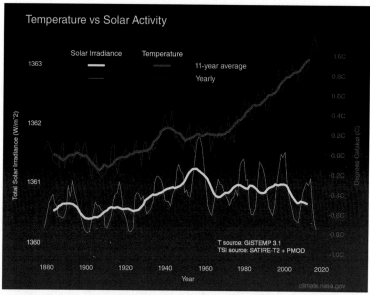

Fig. 1.1 Global temperature and solar activity since 1880. The yearly variations of both these parameters are shown by lighter curves and these have been averaged to give the more distinct curves. *(Source: https://climate.nasa.gov/.)*

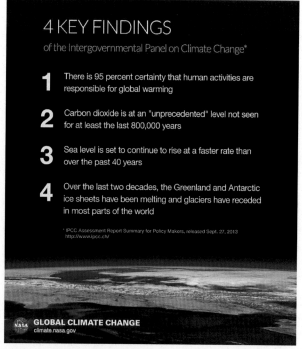

Fig. 1.2 IPPC key findings. Predicted major changes due to global warming. *(Source: https://climate.nasa.gov/.)*

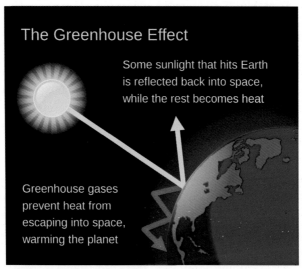

Fig. 1.3 Schematic representation of the greenhouse effect. *(Source: From Wikipedia (https://en.wikipedia.org/wiki/Greenhouse_effect/).)*

The greenhouse effect

Much life on earth as we know it depends on the light radiation from the sun that penetrates through the atmosphere to warm the earth's surface. Without the atmosphere, much of the incident radiation would be re-emitted from the surface and would be totally lost in space. Fortunately however, the atmosphere acts in the same way as does the glass in a greenhouse,[c] absorbing and reflecting back much of the re-emitted radiation and ensuring that the temperature of the atmosphere is increased. This process is shown schematically in Fig. 1.3. The resultant temperature on earth is a delicate balance of the levels of incoming and reflected radiation and is thus very susceptible to changes in the composition of the atmosphere; if too much of the reflected radiation is retained by the atmosphere, the temperature of the earth will rise.

---

[c] With a greenhouse, almost all the incident light passes through the glass and is absorbed by the soil within the structure; some of the energy is then re-emitted at a different wavelength but this is now absorbed by the glass, ensuring that the increased temperature in the greenhouse is maintained.

Greenhouse gases

Table 1.1 lists the main greenhouse gases associated with global warming, giving for each the chemical formula, the global warming potential relative to that for CO_2 over a 100-year lifespan and the atmospheric lifetime in years.

Table 1.2 shows the main sources of these greenhouse gases and also gives the pre-industrial atmospheric concentrations and the current atmospheric concentrations. The first three gases all existed in the pre-industrial era, although the concentrations have all increased since, while the last entries all refer to man-made gases introduced over the last century. We obtain an approximation to the relative contributions of the relevant gases to global warming if we multiply the current concentrations of each gas by the global warming potential from Table 1.1. The resultant figures show that the main culprits are CO_2, methane and nitrous oxide: not taking into account the small contributions of the fluorine-containing molecules, CO_2 contributes 73.3% of the total global warming potential of these gases while methane contributes 8.5% and N_2O contributes 18.2%. Although the contributions of the various fluorinated molecules are relatively low, it needs to be recognised that the lifetimes of these species are significantly above those of the other greenhouse gases and it is for this reason that they are no longer manufactured. As we will see below, there are a number of other greenhouse gases, some of which contribute to global warming while others do not. Water vapour is one example of a gas which does not contribute directly to global warming and ozone is

Table 1.1 Global warming potential and atmospheric lifetime for the most important greenhouse gases.

Greenhouse gas	Chemical formula	Global warming potential, 100-year time-span	Atmospheric lifetime/years
Carbon dioxide	CO_2	1	100
Methane	CH_4	25	12
Nitrous oxide	N_2O	298	114
Chlorofluorocarbon-12 (CFC-12)	CCl_2F_2	10,900	100
Hydofluorocarbon-23 (HFC-23)	CHF_3	1,48,800	270
Sulfur hexafluoride	SF_6	22,800	3200
Nitrogen trifluoride	NF_3	17,200	740

Reproduced from the Fourth Assessment Report (Intergovernmental Panel on Climate Change, IPCC, 2007).

Table 1.2 The most important sources of the major greenhouse gases and their preindustrial and recent (2011) concentrations.

Greenhouse gas	Major sources	Pre-industrial concentration/ ppb	2011 concentration/ ppb
Carbon dioxide	Fossil fuel combustion Deforestation Cement production	278,000	390,000
Methane	Fossil fuel production Agriculture Landfills	722	1803
Nitrous oxide	Fertilizer application Fossil fuel and biomass combustion Industrial processes	271	324
Chlorofluorocarbon-12 (CFC-12)	Refrigerants	0	0.0527
Hydofluorocarbon-23 (HFC-23)	Refrigerants	0	0.024
Sulfur hexafluoride	Electricity transmission	0	0.0073
Nitrogen trifluoride	Semiconductor manufacturing	0	0.00086

another. We will now consider each greenhouse gas in turn, starting with water vapour.

Water vapour

The most important greenhouse gases are water vapour and carbon dioxide. Both of these result from the combustion of fossil fuels but may also arise from other sources. Water-vapour, which results predominantly from the evaporation of surface water, has a feedback effect: it forms clouds in the atmosphere and these lead to precipitation, this having the consequence that the level of water-vapour in the atmosphere is well controlled. The clouds also reflect some of the radiation (UV, visible and infra-red) reaching the atmosphere from the sun, this also restricting the temperature rise. One consequence of the presence of increased partial pressures of carbon dioxide in the atmosphere (see below) is that the resulting temperature rise also causes an increase in the partial pressure of the water in the atmosphere, thus giving rise to a further increase in the temperature. Hence, water vapour has an indirect effect on global warming.

Carbon dioxide

Even though the concentration of carbon dioxide in the atmosphere is much lower than that of water, its effect is much greater since there is no equivalent feedback mechanism to that with water: once the carbon dioxide reaches the atmosphere, its residence time there is very much greater than that of water. The double bonds of the $C{=}O$ linkages of the CO_2 absorb much of the infrared radiation emitted from the earth and prevent this radiation from leaving the atmosphere. The result is an increase in atmospheric temperature. It should be recognised that the CO_2 reaching the atmosphere can come from many sources apart from combustion, for example, respiration and volcanic eruptions. It can also arise from deforestation and changes in land use. As discussed above, the increase in atmospheric temperature caused by the CO_2 also has an effect on the level of water vapour in the atmosphere since the saturation vapour pressure of the water increases with increasing temperature and hence this magnifies the effect of the increase in CO_2 concentration. Atmospheric CO_2 is essential for the growth of plants and all types of vegetation. Hence, we rely on a steady partial pressure of CO_2 to enable agricultural activities. We will return to the subject of CO_2 utilisation in subsequent chapters. As shown in Fig. 1.4 of Box 1.1, there has been a dramatic increase in the concentration of CO_2 in the atmosphere over the last 70 years.

Methane

Methane (CH_4), the simplest hydrocarbon molecule, may arise from a number of sources, both natural and man-made. It is produced by the decomposition of wastes in landfills, from agricultural sources such as rice paddies, from digestive processes of ruminants (e.g. cattle and sheep) and manure management from domestic livestock. It was also commonly emitted as waste from oil well operations and as leakages from chemical processing; however, both of these sources are now much more carefully controlled. (See Box 1.2 for an example of methane emission.) Methane is a much more active greenhouse gas than is CO_2, its 'global warming potential' being much higher (see Table 1.1). The atmosphere also contains yet lower concentrations of other hydrocarbons such as the vapours of petroleum and diesel components and these too are greenhouse gases. Methane and the other hydrocarbons have much longer lifetimes than does CO_2 in the atmosphere; while CO_2 is removed by natural processes, the

BOX 1.1 Variation of global CO_2 concentrations as a function of time

Fig. 1.4 shows the concentration of CO_2 in the atmosphere as a function of time over many centuries. These data have been compiled from the analysis of air bubbles trapped in ice over the last 400,000 years. During ice ages, the levels were about 200 ppm (ppm) and they rose to around 280 ppm in the warmer interglacial periods. The rise after about 1950 is attributable to a rapid increase in the use of fossil fuels as will be discussed further in later sections.

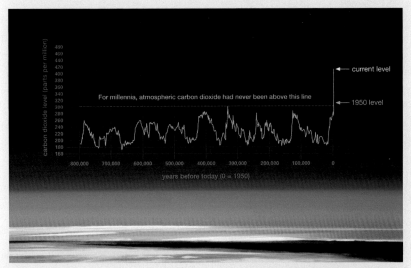

Fig. 1.4 The variation in carbon dioxide concentration as a function of time. It is clear that there has been a dramatic increase in the level of carbon dioxide since 1950 that is well outside the normal temporal variations. (*Source: https://climate. nasa.gov/resources/*)

BOX 1.2 Methane emissions from the production of bitumen from oil sands

There is a significant industry based on the extraction of bitumen from underground reservoirs containing oil sands. The bitumen is heated using the injection of steam to decrease its viscosity and to make it flow more easily. The steam is generated by the combustion of natural gas (methane) and this process gives rise to significant CO_2 emissions, these contributing to global warming. Canada's Oil Sands Initiative Alliance (COSIA) is attempting to find ways of reducing these emissions and has announced that it will assist innovators in developing new routes to reduce the emissions formed during the steam generation step, preferably producing a sequestration-ready product (e.g. concentrated CO_2) or a saleable product (e.g. carbon black).

https://cosia.ca/blog/helping-clear-air-oil-sands-emissions-natural-gas-decarbonization/

hydrocarbons are relatively stable. As a result, they all have higher global warming potentials (Table 1.1).[d]

Nitrous oxide

Nitrous oxide (N_2O) is a very powerful greenhouse gas (see Table 1.1) that is formed by soil cultivation practices, especially by the use of nitrogenous fertilisers; it is also formed by fossil fuel combustion, nitric acid production and biomass burning. It should be recognised that N_2O is only a minor constituent of so–called NO_x, a mixture of the oxides of nitrogen (N_2O, NO and NO_2), formed in high-temperature combustion processes[e] such as those involved in electricity generation and internal combustion engines. NO_x is considered to be an atmospheric pollutant and its emission is associated with the formation of 'acid rain'; the NO_x emissions from these sources are generally controlled by catalytic reduction processes.[f]

---

[d] It should be noted that the global warming potential of methane is time dependent as it is gradually destroyed by oxidation processes in the atmosphere; over periods less than 100 years, the value of the global warming potential is much larger.

[e] At high temperatures and in excess oxygen, thermodynamics favours the formation of NO_2.

[f] For a description of the control of NOx emissions from power stations and automobiles, see Contemporary Catalysis – Fundamentals and Current Applications, Julian R.H. Ross, www.elsevier.com/books/contemporary-catalysis/ross/978-0-444-634740-0. See also Chapter 2.

Ozone and chlorofluorocarbons

Ozone is also a greenhouse gas. It is formed in the troposphere by the inter-action of sunlight with other emissions such as carbon monoxide or methane and also by interaction with hydrocarbons and NOx from automobile emis-sions. The lifetime of ozone is relatively very short (days to weeks) and its dis-tribution is very variable. It absorbs harmful UV radiation and we are therefore dependent on the presence of the ozone layer. The creation of an ozone hole over the Antarctic is ascribed to the emission of chlorofluorocarbons (CFCs), another class of powerful greenhouse gas, and this has led to the banning of the production of these molecules; the production of other hydrofluorocarbons (HFCs) and perfluorocarbons (PFCs) is also being phased out.[g]

Consequences of the greenhouse effect

It is generally recognised that it is difficult to predict the consequences of changing the composition of the naturally occurring atmospheric green-house that surrounds the Earth. However, it is extremely likely that the aver-age temperature of the Earth will continue to rise; even though some areas will become cooler, others will become warmer. Warmer conditions will probably give rise to more evaporation and precipitation although some regions will become wetter and others will become drier. A stronger green-house effect will warm the world's oceans and these will expand and increase sea levels; additionally, glaciers and other ice will melt, thus further increas-ing the sea level. The increased CO_2 concentration in the atmosphere will encourage some crops and other plants to grow more rapidly and to use water more efficiently; however, at the same time, higher temperatures and change in the climate patterns may cause changes in the distribution of the areas where crops grow best. Although climate change has been the subject of great concern for quite some time, it was only about 30 years ago that scientists became particularly concerned about the changes which were occurring;[h] see Box 1.3. Arrhenius discussed in 1886 the importance of the increases in emissions of carbon dioxide resulting from coal-burning; he argued that this would lead to improved agricultural practices and better

[g] This phasing out is part of the Kyoto Protocol (2005); the US has not ratified this international agreement.
[h] An excellent article by Andrew Revkin outlining some of the history of awareness of the problems of climate change is to be found in the National Geographic Magazine of July 2018 (https://www.national geographic.com/magazine/2018/07/embark-essay-climate-change-pollution-revkin/). However, there are many other such articles available on the web.

growth of crops. An article by Waldemar Kaempffert in the New York Times as early as 1956 (October 28)[i] predicted that the increased emissions from energy production would lead to long-lasting environmental changes. This article pointed out very clearly that an impediment to counteracting these changes was the abundance of coal and oil in many parts of the world and that these fossil fuels would continue to feature in industrial use as long as it was financially beneficial to use them. However, it was not until 1988 that the World Meteorological Organisation (WMO) established the Intergovernmental Panel on Climate Change (IPCC). The IPCC summarises the scientific developments in countering climate change in the IPCC Assessment Reports that are published every five to six years, these being compiled in association with a number of other related reports from the panel. (The Sixth Synthesis Report is due in 2022.)[j] In parallel to these activities, there have been a large number of reports by other agencies, both international and national, some of which will be quoted below. There have also been two important international agreements on how climate change should be counteracted, the Kyoto Agreement of 1992 and the Paris Agreement of 2015, both established under the auspices of the United Nations Framework Convention on Climate Change (UNFCCC). Just fewer than 200 countries were signatories to these agreements, the aims of which being to reduce the emission of greenhouse gases.[k] These agreements will be discussed further below.

BOX 1.3 The importance of the existence of a greenhouse effect on earth

If the earth did not have an effective shielding greenhouse layer containing high levels of water vapour as well as CO_2 and the other greenhouse gases, much of the light of all wavelengths reaching the surface of the earth from the sun would be reflected back into space without warming the planet. The planet Mars has a relatively thin atmosphere consisting largely of carbon dioxide but little or no water vapour and this results in a weak greenhouse effect. As a

Continued

---

[i] See New York Times December 8, 2015: https://www.nytimes.com/interactive/projects/cp/climate/2015-paris-climate-talks/from-the-archives-1956-the-rising-threat-of-carbon-dioxide.

[j] The Montreal Protocol to reduce the emission of compounds that affect the ozone layer such as the chlorofluorocarbons mentioned above was agreed in 1987.

[k] The US under President Trump had announced that it was going to leave the Paris Agreement but that decision has now been reversed by President Biden; of the countries with over 1% share of the global emissions, only Iran and Turkey are not parties to the agreement.

BOX 1.3 The importance of the existence of a greenhouse effect on earth—cont'd

result, the surface of Mars is largely frozen and there is no evidence of any life form. In contrast, Venus has a much higher concentration of CO_2 in its atmosphere (150,000 times as much as on Earth and 19,000 as much as on Mars) and the surface temperature is +460°C. Again, this atmosphere would not be amenable to life as we know it. (See https://agreenerfutureblog.wordpress.com/1-the-natural-greenhouse-effect/1-4-greenhouse-effect-on-other-planets/; https://earthsky.org/space/venus-mars-atmosphere-teach-us-about-earth/)

The global emission of all greenhouse gases in 2010 was 48 gigatonnes of CO_2 equivalent and it was estimated that this figure would increase to 53.5 gigatonnes in 2020. What is much more alarming is that the amount will increase to 70 gigatonnes by 2050 unless action is taken to reduce greenhouse emissions. These unfettered increases will give rise to totally unacceptable global temperature increases, these in turn giving rise to myriad problems for the world's inhabitants. The Kyoto Protocol and the Paris Agreement have led to two targets relating to temperature rise: the first, a limit of 2.0°C compared to the emissions in the pre-industrial era; and the second, pursuing at the same time means to limit the temperature increase to 1.5°C.

Fig. 1.5 illustrates the predicted changes occurring for different scenarios envisaged in the run-up to the Paris Accord, ranging from the absence of any policy (resulting in a very high chance that the global temperature rise will be well above 4°C) to a very strict policy (when the chance of the temperature approaching pre-industrial levels will be much greater). The solid curves of the diagram show the predicted emissions of CO_2 for different scenarios, from no action (top curve) to the most ambitious series of actions with higher rates of decarbonisation (1.5°C warming, bottom curve). It will be seen that only the latter approach gives any significant reduction in the emission of greenhouse gases. The slightly lower set of ambitions, with constant rates of decarbonisation (2°C warming) gives a levelling off of the levels of CO_2 after about 2030.

The sources of greenhouse gas emissions

There have been a number of very detailed national and international reports that give information on emissions of greenhouse gases and list the main

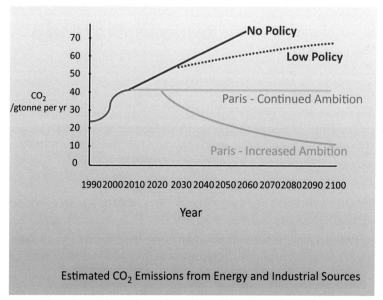

Estimated CO_2 Emissions from Energy and Industrial Sources

Fig. 1.5 Changes in carbon dioxide emissions envisioned in Paris Accord. The effects on CO_2 emissions of either no change or adopting various strategies (see text). *(Adapted from Climate Science Special Report (US Global Change Program), Fourth National Climate Assessment (NCA4) Vol. 1, Chapter 14.2: https://science2017.globalchange.gov/.)*

sources. Some of these reports will be discussed in more detail in later sections. One of the most relevant of these reports for the present purposes is one by the European Environment Agency[1] and the following sections will summarise some of its most relevant material.

Table 1.3 shows data for the total emissions of greenhouse gases from European countries in 2017 and also shows the changes in emissions that have occurred in the period between 1990 and 2017, the greenhouse gas emissions per capita and the change in the total energy intensity of each country in the period 1990–2017. The majority of these countries are members of the EU but the data for Norway and Turkey have also been added for completeness. It should be recognised that the United Kingdom is also included as an EU country as the data relate to a period prior to BREXIT. (It should be noted that the data for a number of smaller EU countries have been omitted for clarity; full details are available in the source report which also gives some more detail for each country.) The figures for the whole EU are given in the last row; the EU figures for the % changes in GHG

[1] The European Environment – State and Outlook 2020, European Environment Agency, (2019), doi: https://doi.org/10.2800/96749, downloadable from https://eea.europa.eu/.

Table 1.3 Greenhouse gas emissions from EU countries.

Country	Total GHG emissions 2017/ MtCO$_2$ equiv.	Change in GHG emissions 1990–2017/%	GHG emissions per capita in 2017/ tCO$_2$ equiv. per person	Change in the total energy intensity of the economy 1990–2017/%
Austria	84.5	+6.2	9.6	−18.3
Belgium	119.4	−20.3	10.5	−27.1
Bulgaria	62.1	−39.5	8.8	−54.0
Czech Republic	130.5	−34.7	12.3	−48.4
Denmark	50.8	−29.5	8.8	−35.5
Finland	57.5	−20.5	10.4	−24.5
France	482.0	−13.4	7.2	−25.5
Germany	936.0	−25.9	11.3	−40.1
Greece	98.9	−6.4	9.2	−13.0
Hungary	64.5	−31.5	6.6	−38.5
Ireland	63.8	+12.9	13.3	−66.1
Italy	439.0	−15.9	7.3	−10.8
Netherlands	205.8	−9.1	12.0	−34.2
Poland	416.3	−12.4	11.0	−61.7
Portugal	74.6	+22.8	7.2	−4.0
Romania	114.8	−53.9	5.9	−69.6
Slovakia	43.5	−40.8	8.0	−63.6
Spain	357.3	+21.8	7.7	−14.3
Sweden	55.5	−23.7	5.5	−39.8
UK	505.4	−37.6	7.7	−49.3
(Norway	*54.4*	*+4.9*	*10.3*	*−22.4)*
(Turkey	*537.4*	*+144.5*	*6.7*	*−12.8)*
EU-28	**4483.1**	**−21.7**	**8.8**	**−36.3**

The data for two non-EU countries, Norway and Turkey, are also included in italics and the total emissions from the EU (EU-28) are shown in bold figures.
Some data concerning greenhouse gas emissions from some European countries.

emissions, the GHG emissions per capita and the change in the energy intensity of the economy allow one to see how each country has performed compared with the average result for the whole EU. It can be seen that Bulgaria, the Czech Republic, Germany, Ireland, Poland, Romania, Slovakia and the UK have performed better than the average.

Fig. 1.6 shows the emissions of the principal greenhouse gases per source type (given as CO$_2$ equivalents in millions of tons per year) for all the EU states for the period 1990 to 2017 and Table 1.3 gives some additional information. Most of the categories included in the figure and table have shown

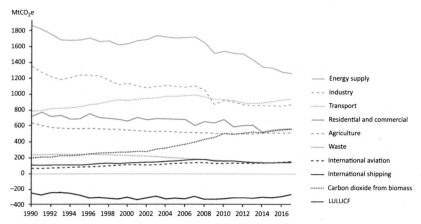

Fig. 1.6 EU greenhouse emissions per sector. Equivalent CO_2 emissions from different European sectors over the period 1990 to 2017. *(Source: Europe environment state and Outlook 2020 (https://www.eea.europa.eu).)*

significant decreases in emissions. A particularly large decrease occurred in energy supply, this being the result of increased use of renewable sources of energy, particularly wind and solar power.[m] The introduction of these technologies has been accompanied by a decrease in the use of coal combustion for electricity production. There has also been a significant drop in industrial energy consumption, due largely to improved efficiencies in industrial processes and to a switch away from the use of coal and oil to natural gas for the supply of energy, this change also being associated with improvements of technologies (Chapter 3). Some EU countries have been at the forefront in decreasing greenhouse gas emissions. For example, as can be seen from Table 1.3, Germany, originally one of the greatest emitters of greenhouse gases in Europe, had reduced its emissions by 25.9% over the period to a level of 936 million tons of CO_2 equivalent. (This has been achieved partly by a significant decrease in the use of lignite as a fuel.) Unfortunately, however, some countries had increased their contributions; for example, Ireland's contribution increased by 12.9% to 63.8 million tons of CO_2 equivalent and that of Cyprus increased by 55.7% to 10.0 million tons of CO_2 equivalent. These rather different results appear to be related to differences in the structures of the different economies. For example, Ireland has very little heavy industry but has a strong agricultural economy as well as

---

[m] Nuclear energy is also considered to be a renewable source of energy. France is particularly reliant on nuclear power. Germany, on the other hand, has decided to close down all of its nuclear reactors and is well on its way to doing so. Other countries such as Ireland have never had nuclear facilities.

large numbers of new high-technology companies consuming relatively large quantities of energy.

Table 1.4 gives a breakdown of the changes that have occurred in the emissions of greenhouse gases (both CO_2 and others) for the period 1990 to 2017 from all European countries, the data being given in million tons of CO_2 equivalents. It can be seen that road transportation continues to give increased emissions as do refrigeration and air conditioning. In all other sectors, there have been significant reductions; particularly important reductions have been found in residential heating, iron and steel production, the manufacturing industries and public electricity and heat production. Many of these improvements will be discussed in subsequent sections and chapters.

Fig. 1.6 shows that there has been a slightly decreasing contribution to greenhouse emissions from agriculture since 1990. These emissions, mainly from ruminants (cattle and sheep) are still significant and are therefore a continuing cause for concern, particularly in Ireland. A similar set of figures will apply globally although there will be national differences arising from different

Table 1.4 EU emissions per source.

Emission source	MtCO$_2$ e
Road transportation	170
Refrigeration and air conditioning	93
Aluminium production	−21
Agricultural soils: direct emissions of N$_2$O from managed soils	−22
Cement production	−26
Fluorochemical production	−29
Fugitive production from natural gas	−37
Commercial/institutional	−38
Enteric fermentation—cattle	−43
Nitric acid production	−46
Adipic acid production	−56
Manufacture of solid fuels and other energy industries	−60
Coal mining and handling	−66
Managed waste disposal sites	−73
Residential—fuels	−115
Iron and steel production	−116
Manufacturing industries	−253
Public electricity and heat production	−433
International aviation	*+89*
International navigation	*+35*

The emissions arising from international aviation and navigation outside the EU are also shown (in bold italics) for completeness.
Trends in EU emissions from the predominant sources in the period 1990 to 2017.

Million people

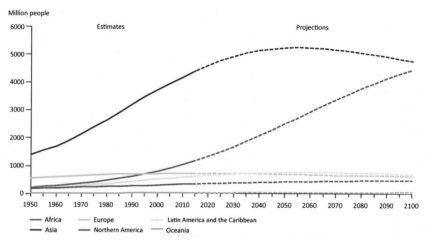

Fig. 1.7 World population trends. Estimated and predicted world population changes from 1950 to 2100. *(Source: Europe environment state and Outlook 2020 (https://www.eea. europa.eu).)*

methods of agriculture in each country. The global contribution to green-house gas emissions from agriculture is related to the very significant growth of population that has occurred over a couple of centuries since the start of the industrial revolution (Fig. 1.7).

Prior to the industrial revolution, the world population was only about 700 million. However, with improvements in food production and also in medical standards, the population grew rapidly to 1.6 billion people in 1900 and thereafter even more rapidly to reach 6 billion before the end of the twentieth century. The world population is already above 7 billion and it is projected to reach 8 billion by 2030. A significant proportion of the increased emission of greenhouse gases in the period since the industrial rev-olution has emanated from changes in land use that have been needed to provide food for the increasing population, particularly in relation to the use of fertilisers and the production of meat for human consumption. Hence, one aspect of the control of greenhouse gas emission in the future will have to be related to improvements in agricultural practice to reduce emissions such as those of methane from ruminants and N_2O from fertilizer applica-tions. Of particular relevance to the present text is the conversion of biomass to valuable products: biofuels, industrially relevant chemicals and hydrogen. These topics will be considered in more detail in Chapter 5. The only cat-egory with negative emissions shown in Fig. 1.6 is LULUCF (land use, land-use change and forestry) in which CO_2 is consumed by the growth of

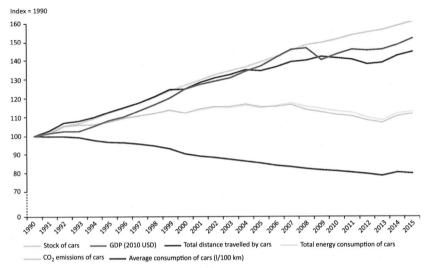

Index = 1990

Legend:
Stock of cars — GDP (2010 USD) — Total distance travelled by cars — Total energy consumption of cars
CO_2 emissions of cars — Average consumption of cars (l/100 km)

Fig. 1.8 European car usage. Changes (relative to data from 1990) in European car usage and performance over the period 1990 to 2017. *(Source: Europe environment state and Outlook 2020 (https://www.eea.europa.eu).)*

vegetation. It is clearly important that deforestation is very strictly controlled internationally and those significant efforts are made to increase the area of land devoted to forestry and other crops with much more efficient and careful use of fertilisers. The subject of the bio-economy is one which will receive little further attention in this text but the interested reader is referred to an open-access book edited by Lewandowski and published by Springer which gives excellent coverage of the topic.[n]

It is interesting to note that the emissions from transport shown in Fig. 1.6 increased steadily up until about 2006 but that they have decreased significantly since then. This is further illustrated by the data of Fig. 1.8 which shows the emissions arising from the use of private vehicles over the period 1990 to 2017. Although there has been a marked increase in the numbers of private cars and in the total distances travelled over this period (these figures being closely related to the parallel improvement in GDP), the total energy consumption and the emission of CO_2 have not increased in the same way; both of these parameters increased somewhat until about 2006 but then decreased significantly. This can be explained by the very significant improvement in engine efficiency and the consequent

[n] "Bioeconomy – Shaping the Transition to a Sustainable, Biobased Economy", Edited by
I. Lewandowski, ISBN 978-3-319-68152-8, https://doi.org/10.1007/978-3-319-68152-8.

decrease in the average consumption per vehicle that occurred over the period considered.

International aviation and international shipping (see Fig. 1.6) have both shown steady increases and it is related to these figures (which are duplicated in statistics for non–EU nations) that the carbon footprint of international travel has recently attracted so much concern. This is highlighted further in Fig. 1.9 which compares the energy usage for different transport modes; only the usage by international travel continues to increase while the other modes of transport have shown significant decreases since about 2005.

The funding of research and development work related to energy supply in Europe has shown some significant changes over the last 40 years. Fig. 1.10 shows the trends in European spending over that period for various different technologies. Up until the mid–1980s, the predominant area of research that was funded related to nuclear energy while research in fossil fuels also attracted significant research effort. Hydrogen and fuel cells attracted some added interest around 1990 but all research activities decreased somewhat from 1990 until the early years of the new century. Since about 2005, there have been marked increases in research activities, particularly significant research expenditures having occurred in the areas of renewables and energy efficiency. The funding

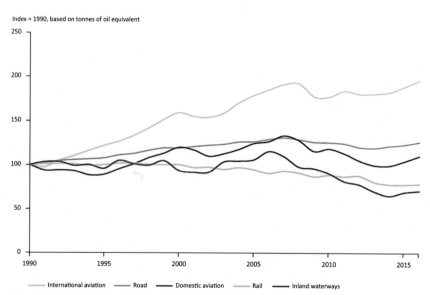

Fig. 1.9 Energy use per transport mode. The variation in energy use in the EU over the period 1990 to 2017 per transport mode relative to the usage in 1990. *(Source: Europe environment state and Outlook 2020 (https://www.eea.europa.eu).)*

Million USD (2017 prices, PPP)

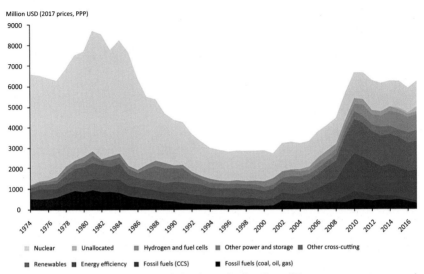

Fig. 1.10 European energy research funding. The funding of European energy research in various categories during the period 1974 to 2017. *(Source: Europe environment state and Outlook 2020. (https://www.eea.europa.eu/legal/copyright).)*

level for nuclear research (see Chapter 3), although much lower than in the period around 1980, is still significant. This is related to the fact that nuclear fusion has the potential, if ever achievable in a safe and controlled fashion, of supplying vast amounts of renewable energy without any undesirable by-products. Until fusion technology is available, however, we must seek other solutions to our energy requirements which will have minimal effect on our greenhouse gas emissions. We must also seek ways in which the emissions from traditional industries related to products such as iron and steel or cement can be reduced (Chapter 2).

This book, written from the point of view of a physical chemist, considers a number of important uses of energy, discussing in turn a number of chemical and other processes from which significant quantities of greenhouse gases are currently emitted. It then outlines strategies currently under development, or already in place, aimed at reducing these emissions. Particular emphasis is placed on the very considerable merits of a 'hydrogen economy' and hence special attention is given to the production, storage, transport and uses of hydrogen. Much of the material included is available in one form or another from the web and similar sources but often without an adequate discussion of the chemistry involved. Hence, one of the main aims of the book is to give a chemist's view of the technologies which are currently being explored and developed.

CHAPTER 2

Traditional methods of producing, transmitting and using energy

Introduction

The rapid growth of the highly industrialised world as we know it stemmed from the developments of new technologies introduced following the Industrial Revolution. Many of these technologies depended on the availability of sources of energy. In the early days of mechanisation associated with the Industrial Revolution, the major source of the energy required for the operation of the new machinery was coal, used to power steam engines; there was also some use of wind and water power. More recently, the world has come to depend more and more on other sources of energy, particularly oil and natural gas. Now, with the recognition of the problems of global warming, emphasis is being placed more and more on renewable energies: biomass, nuclear, solar, tidal, etc. As noted in Contemporary Catalysis—Fundamentals and Current Applications (2019), the use of these energy sources have followed a series of Kondratiev cycles (see Fig. 2.1).[a] Some aspects of each of these cycles will now be discussed briefly in turn.

Coal

All the sources the fossil fuels that we use derive originally from solar radiation since they were all formed in a series of geological processes from biomass, the growth of which depended on energy arriving from the sun. Coal was generated from layers of peat (formed in a process involving biomass degeneration) as a combustible sedimentary rock by the effect of pressure and heat over a period of millions of years. The world has very large reserves of coal. As shown in Table 2.1, the total global estimated reserves are of the order of 1055 billion tonnes and these are distributed reasonably equally

---

[a] It should be noted that the grouping labelled Nuclear, Solar, Tidal should also include other energy sources such as wind, hydroelectric and geothermal. The various sources of energy are discussed in more detail in this chapter.

Sustainable Energy
https://doi.org/10.1016/B978-0-12-823375-7.00007-X

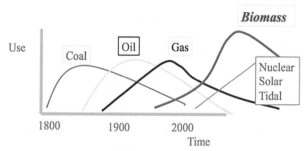

Fig. 2.1 Kondratiev cycles associated with the use of energy sources. Periodic cycles in the use of coal, oil, gas, biomass and renewable energy sources. *(Reproduced with the kind permission of Elsevier from Contemporary Catalysis—Fundamentals and Current Applications, J.R.H. Ross, 2018.)*

Table 2.1 Global coal reserves and expected lifetimes of these reserves.

Region	Total reserves/ million tonnes	Expected lifetime based on 2018 usage	Share of world total
North America	258,012	342	24.5%
US	250,219	365	23.7%
Central and South America	14,016	158	1.3%
Europe	134,593	215	12.8%
Germany	36,103	214	3.4%
Poland	26,479	216	2.5%
Ukraine	34,375	>500	3.3%
UK	29	11	<0.05%
CIS	188,853	329	17.9%
Kazakstan	25,605	217	2.4%
Russian Federation	160,364	364	15.2%
Middle East and Africa	14,420	53	1.4%
Asia Pacific	444,888	79	42.2%
Australia	147,435	304	14%
China	138,819	38	13.2
India	101,363	132	9.6
Indonesia	37,000	67	3.5
Total World	**1,054,782**	**132**	**100%**

Source: https://www.bp.com/content/dam/bp/business-sites/en/global/corporate/pdfs/energy-economics/statistical-review/bp-stats-review-2019-full-report.pdf.

throughout the world; as we will see later, the regions which have lower reserves (e.g. the Middle east and Africa) have larger reserves of oil and gas. It is interesting to note that the UK, where the Industrial Revolution commenced, has very low reserves. The known reserves of coal are made up of both 'hard coal' (anthracite and bituminous coal) and soft coal (sub-bituminous coal and lignite) and the breakdown of these for the major coal-producing countries are shown in Table 2.2.

As we will see, coal is used largely as a source of energy. The Kondratiev cycle of Fig. 2.1 associated with the use of coal is shown as having a maximum in the mid-1800s, coinciding roughly with its important contribution to the emergence of the Industrial Revolution. However, the use of coal is still very significant because of the large global reserves and hence it is still very important to the world economy; it is used very largely for energy generation although it can also be used as a source of chemicals.

Coal combustion for heating purposes

Coal is composed mainly of carbon but also contains some oxygen, hydrogen and nitrogen as well as sulphur. The combustion of coal therefore gives predominantly CO_2 from the reaction of the carbon content with oxygen of the air:

$$C + O_2 \rightarrow CO_2 \; \Delta H° = -393.5 \, kJ/mol$$

Table 2.2 Hard and soft coal reserves of the major coal-producing countries.

Country	Hard coal/million tonnes	Soft coal/million tonnes	Total/million tonnes
USA	111,338 (23.3%)	135,305 (31.4)	246,643 (27%)
Russia	49,088 (10.3%)	107,922 (25.1%)	157,010 (17%)
China	62,200 (13%)	52,300 (12.2%)	114,500 (13%)
India	48,787 (10.2%)	45,660 (10.6%)	94,447 (10%)
Australia	38,600 (8.1%)	39,900 (9.3%)	78,500 (9%)
South Africa	48,750 (10.2%)	0 (0%)	48,750 (5%)
Ukraine	16,274 (3.4%)	17,879 (4.2%)	34,153 (4%)
Kazakhstan	28,151 (5.9%)	3128 (0.7%)	31,279 (3%)
Poland	14,000 (2.9%)	0 (0%)	14,000 (2%)

The table shows the reserves in million metric tonnes of hard coal (anthracite and bituminous coal) and soft coal (sub-bituminous coal and lignite) for each country; the percentages of total world reserves of each are shown in brackets.
Source: https://www.bp.com/content/dam/bp/business-sites/en/global/corporate/pdfs/energy-economics/statistical-review/bp-stats-review-2019-full-report.pdf.

In principle, the CO_2 produced can be trapped and stored (using carbon capture and storage, CCS, a topic to be discussed further in later chapters) but this technology is not yet sufficiently developed to permit it to be adopted for general use. Hence, the CO_2 emitted is one of the main sources of greenhouse gas (25% of global greenhouse gas emissions, see Preface). Another serious problem associated with the combustion of coal is that the nitrogen and sulphur contents give rise to the emission of NO_x (N_2O, NO and NO_2) and SO_2; these gases are generally accompanied by the formation of particulates and trace metals. All these emissions have serious effects on human health by contributing to the formation of acid rain and smog (see Fig. 2.2).

The NO_x arises predominantly from the combustion of nitrogen compounds in the coal such as pyridine (so-called 'fuel-NO_x') but some are also formed by the chemical combination of the nitrogen and oxygen of the flue gases at the high combustion temperatures used (so-called 'prompt NO_x'). While the technologies exist for the removal of SO_2 and NO_x from the flue gases of power stations operating with coal combustion (SO_2 scrubbing and selective catalytic reduction (SCR) respectively), these are not always

Fig. 2.2 Nelson's Column, London, during the great smog of December 1952. *(Photo by N.T. Stobbs.) (https://commons.wikimedia.org/wiki/File:Nelson%27s_Column_during_the_Great_Smog_of_1952.jpg)*

practiced as they add significantly to the selling price of the electricity produced. Further, these methods cannot be applied when coal combustion is used for domestic heating purposes.

The hydrogen, nitrogen and sulphur contents of coal are due to the presence of a wide variety of chemical compounds arising from the original biomass, many of them aromatic in character and including a complex mixture of heterocyclic compounds related to pyridine and thiophene; coal also has a significant oxygen content, this being largely associated with compounds such as pyrroles and phenols. Because of these components, coal was used, prior to the increased availability of crude oil, as the source of many important chemicals (see Fig. 2.3). Coal was also used until relatively recently for the production (by pyrolysis) of Towns' Gas used for heating and lighting purposes in many parts of the world. It was also used as the source of the hydrogen for ammonia production and of the syngas used in the

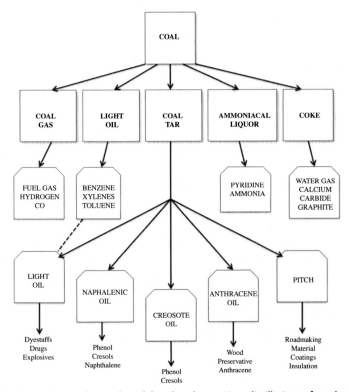

Fig. 2.3 Some chemicals produced by the destructive distillation of coal at about 1100°C.

Fischer–Tropsch Process operated in South Africa by Sasol during Apartheid; this entailed the operation of Lurgi gasifiers of the type developed in Germany in the period leading up to the Second World War. Ammonia synthesis and Fischer Tropsch Synthesis will be discussed further in a later section.

Coal for power generation and the steam engine

Coal was for many centuries used for heating purposes. However, its use in providing energy resulted from the development of the steam engine, the introduction of which as a source of power for machinery being responsible for the industrial revolution discussed in Preface. The earliest use of these steam engines was in the UK's cotton and woollen industries of Lancashire and Yorkshire; this development led to the first significant increases in greenhouse gas emissions discussed in Preface. Steam engines were soon thereafter used in railway engines (Fig. 2.4).

Coal for electricity generation

With the introduction of electrical power in the early 1800s, the steam engine was also used for the generation of electricity. The later development of the more efficient steam turbine led to the type of power station to which

Fig. 2.4 A typical steam engine (North Yorkshire Moors Railway, UK).

Fig. 2.5 A modern coal-burning power station: Moneypoint Power Station, Co. Clare, Ireland. *(With kind permission of ESB Archives.)*

we are all now accustomed. Fig. 2.5 is an aerial view of a typical modern power station used largely for coal combustion, showing the large area used for coal storage.

Fig. 2.6 is a schematic depiction of a typical cogeneration system that uses a primary gas turbine system coupled with a secondary steam turbine, this arrangement giving much higher all-over efficiency than one or other of these technologies alone. As we will see in subsequent sections, the use of coal as a fuel in such power stations has now almost completely been replaced

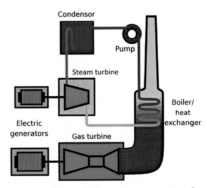

Fig. 2.6 Schematic representation of the arrangement of a typical cogeneration power plant. *(Source: Wikipedia.)*

by the combustion of natural gas or oil but the steam turbine is still the basis for electricity generation. Natural gas is now the fuel of choice as the emission of CO_2 is much lower than with the use of either coal or oil, as long as there are no serious emissions of methane during its transportation to the power station.

A turbine operating on its own would liberate a significant amount of energy as hot flue gases. With cogeneration, the hot flue gases are used to generate steam that is fed to a steam turbine that also produces electricity, adding to the all-over efficiency of the system. Some of the heat energy produced can be used for local heating systems.

It was not until the mid-1900s that it was generally recognised that coal combustion is associated with serious emissions of SO_2 and NO_x (mixtures of the oxides of nitrogen, predominantly NO and NO_2) and that these emissions cause serious health problems as well as bringing about the formation of 'acid rain', responsible for very significant damage to forestry. In consequence, clean air legislation has gradually been introduced to reduce these emissions and also those from domestic heating sources. With such legislation, the general use of smokeless fuel has become mandatory.

However, the greatest problem now associated with coal-burning power stations is the emission of greenhouse gas CO_2. For this reason, many coal-fired power stations have been converted to the use of either oil or natural gas since the amount of CO_2 emitted for a given amount of energy is thereby significantly reduced (see Box 2.1).

The proportion of coal used in the provision of energy varies significantly from country to country, depending very markedly on the availability of alternative sources of energy. As an example, Fig. 2.7 shows the use of coal and of various other sources of energy for electricity production in the US for the period 1950 to 2019. It can be seen clearly that the use of coal in the US since about 2000 has decreased steadily relative to that of other energy sources. The US Energy Information Administration expects that the use of coal in the US for electricity generation will decrease from 24% currently to about 13% in 2050. Similar decreasing usage patterns are expected for many other countries. However, China, India and Australia are still using increasing quantities of coal since these countries do not have a significant supply of alternative fuels. China, India and Australia are therefore currently very major contributors to greenhouse gas emissions.

It is of interest to examine the current use of electricity in some of the world's major economies. Considering first the United States, Fig. 2.8 shows the breakdown of electricity use for 2019. It can be seen that domestic use

BOX 2.1 Relative enthalpies of combustion of coal, n-octane and natural gas

The use of coal, oil and natural gas combustion for power stations give rise to different levels of the emission of CO_2. If we approximate the case of coal using the data for pure carbon and oil by a hydrocarbon with the data for n-octane ($n = 10$), we obtain the following data:

$$C + O_2 \rightarrow CO_2; \Delta H^\circ = -393.7 \, kJ \, (C \, atom)^{-1}$$

$$C_8H_{18} + 12.5O_2 \rightarrow 8CO_2 + 9H_2O; \Delta H^\circ = -5104 \, kJ \, mol^{-1}$$

$$or \, \Delta H^\circ = -638 \, kJ \, (C \, atom)^{-1}$$

$$CH_4 + 2O_2 \rightarrow CO_2 + 2H_2O; \Delta H^\circ = -800 \, kJ \, (C \, atom)^{-1}$$

It can be seen that the quantity of heat liberated per C atom combusted increases in the order: $C < C_8H_{18} < CH_4$ and that methane is the most effective fuel in terms of CO_2 emission. When the actual fuels used are considered rather than C and n-octane and the figures are now given as the emissions in kg CO_2 per kWh or per GJ energy produced (see Table 2.3), it again can be seen that natural gas combustion is significantly preferable to the combustion of coal or of oil as long as methane losses on the way to the plant are low.[b]

The data in Table 2.3 also show that the combustion of wood is only justifiable if sustainable sources are used. It can also be seen that peat and lignite are particularly undesirable energy sources and this is reflected in the fact that Ireland has closed its peat-burning stations and that Germany has completely ceased the use of lignite as a fuel.

Table 2.3 Emissions of carbon dioxide from various fuels as a function of the power generated.

Fuel	Emissions in kg CO_2/kWh	Emissions in kg CO_2/GJ
Wood	0.39	109.6
Peat	0.38	106.0
Lignite (Rhineland)	0.41	114.0
Hard Coal	0.34	94.6
Fuel Oil	0.28	77.4
Gasoline	0.25	69.3
Diesel	0.27	74.1
Liquid Petroleum Gas	0.23	63.1
Natural Gas	0.20	56.1

[b]Data from www.volker-quaschning.de.

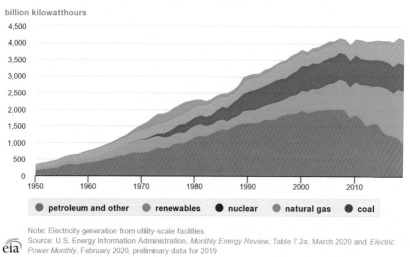

Fig. 2.7 Fuels used for electricity generation in the US from 1950 to 2019. *(www.epa. gov.)*

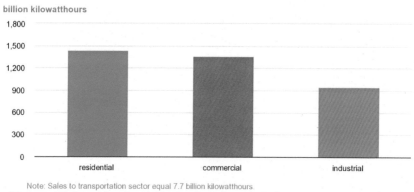

Fig. 2.8 US electricity usage. *(www.epa.gov.)*

was the most significant sector (about 40%) and that this was closely followed by the use by commercial enterprises, the third most significant user being the industrial sector.

The use of electricity for domestic purposes in the US is further broken down in Fig. 2.9. Approximately 50% of the usage can be attributed to

U.S. residential sector electricity consumption by major end uses, 2018

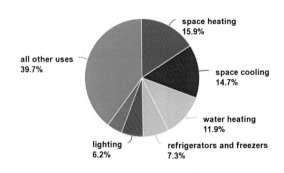

Note: Space heating includes consumption for heat and operating furnace fans and boiler pumps.
All other uses includes miscellaneous appliances, clothes washers and dryers, computers and related equipment, stoves,
dishwashers, heating elements, and motors not included in other uses.
Source: U.S. Energy Information Administration, *Annual Energy Outlook, 2019*, Table 4, January 2019

Fig. 2.9 US domestic electricity usage. *(www.epa.gov.)*

heating, cooling, water heating and refrigeration. While such usage will vary significantly from region to region, depending particularly on its geographical location, it can be seen that a very significant proportion of the US national energy usage is therefore devoted to maintaining comfortable living conditions.

Equivalent data for the European Union for 2017 show that 24.7% of the energy used in the residential sector comes from electricity and that less than 30% of this is used for space and water heating and cooling purposes (Fig. 2.10); however, when the European data on the use of energy includes the use of other resources such as natural gas, the categories space and water heating account for 75% of the usage. It is clear that a significant difference between Europe and the US is that the use of electricity for cooling purposes in Europe is much less and that natural gas is much more generally used for domestic heating.

Ireland is rather an exception in the European scene as domestic heating there is fuelled largely by oil. This is related to the fact that there is a large rural population for whom natural gas is not available. Table 2.4 shows the energy sources used for Irish electricity generation for the period 2005 to 2018. The use of coal has remained very steady, reflecting the use of the large Moneypoint plant shown in Fig. 2.5. (The much lower figure for 2018 reflects a problem that year with the turbines of the plant.) A large proportion of the coal burnt in Moneypoint has come from an open-cast mine in Columbia.

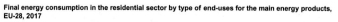

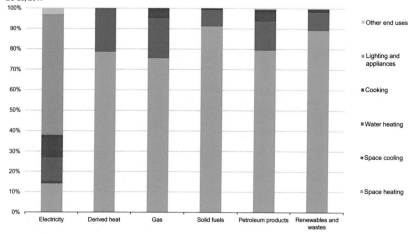

Final energy consumption in the residential sector by type of end-uses for the main energy products, EU-28, 2017

Source: Eurostat

eurostat

Fig. 2.10 EU domestic energy usage. *(https://ec.europa.eu/eurostat.)*

Table 2.4 Irish energy sources used for electricity generation.

Year	Coal	Peat	Oil	Wastes	Gas	Renewables	Imports
2005	549	211	287	0	995	161	176
2006	506	184	244	0	1186	213	153
2007	473	187	165	0	1330	240	114
2008	442	237	147	0	1438	309	39
2009	344	226	79	0	1402	353	66
2010	306	187	52	0	1558	321	40
2011	339	183	20	0	1327	466	42
2012	432	210	20	4	1216	452	36
2013	368	196	16	6	1129	484	193
2014	340	215	22	6	1087	550	185
2015	419	217	35	6	1064	676	58
2016	404	199	25	6	1318	646	0
2017	313	186	12	14	1348	764	0
2018	185	180	12	26	1377	877	0

Energy generated per fuel type; figures are in kton equivalents (ktoe).
Source: https://www.seai.ie/publications/Energy-in-Ireland-2018.pdf.

Natural gas is the major source of energy for Irish electricity generation and oil usage has almost completely stopped.[c] The use of renewable methods for electricity generation has also increased steadily; this is a topic to which we will return in Chapter 3.

There have also been a variety of other important uses of coal and a number of these still continue. The use of pyrolysis products for the production of important chemicals now produced from oil has already been mentioned; see Fig. 2.3. Town Gas production was also a very significant use and the gas produced was for many decades used for heating purposes. Prior to the widespread use of electricity, gas was also used for lighting. The use of Town Gas has now largely been replaced by natural gas. Some aspects of the use of natural gas will be discussed in a later section.

Coal use in cement production

Historically, another very significant consumer of coal has been the cement industry. Cement production (see Box 2.2) gives rise to the emission of large

BOX 2.2 Cement production

Cement has a long history. The Egyptians used cement made by calcining gypsum while the Greeks and Romans made their cement by calcining limestone to give lime and then adding either sand, to make mortar, or a mixture of sand and gravel, to give concrete. While the modern cement industry also depends on the decomposition of limestone (calcium carbonate), the chemistry is somewhat more complex. For example, the formation of an early form of Portland Cement was patented in 1824 by Joseph Aspdin, who produced it by firing a finely-ground clay together with limestone until the limestone had decomposed; the name Portland Cement was given since the resultant concrete looked like Portland Stone. Isaac Johnson made the first version of the modern Portland Cement by calcining a mixture of chalk and clay at high temperatures, 1400–1500°C, this giving a 'clinker' that has to be pulverised before the addition of other additives and subsequent use. Further developments followed such as the use of rotary kilns instead of the earlier vertical shaft kilns and the addition of gypsum to control the setting properties of the cement. The interested reader should further examine the fascinating history of the development of modern cements by carrying out a suitable Google search.

[c] Moneypoint now uses natural gas for most of the time.

quantities of CO_2, both from the burning of the fuel used in this high-temperature process and from the decomposition of the calcium carbonate component of the cement formulation involved in the all-over chemical process. It has been estimated that approximately 8% of the world's carbon dioxide emissions result from cement manufacture. A brief outline of the history of cement production is given in Box 2.2; the development of Port-land cement was closely associated with the emergence of the Industrial Revolution since a high-strength cement resistant to aging in water was required to build the light-houses essential to a growing shipping trade.

If we consider only the decomposition of calcium carbonate:

$$CaCO_{3(s)} \rightarrow CaO_{(s)} + CO_{2(g)}$$

the standard enthalpy change, $\Delta H°$, is $178.3\,kJ\,mol^{-1}$. If coal is used as the fuel and we consider it to be entirely composed of carbon, the enthalpy of its combustion is $393.7\,kJ\,mol^{-1}$ (Box 2.1). Hence, if the combustion process is 100% efficient, an additional 0.45 molecules of CO_2 are formed by the heat-ing process in addition to that formed in the decomposition of the $CaCO_3$. The efficiency of the heating process is far from 100% and so the ratio obtained in practice is well above 1.5. It is clear that using either oil or natural gas will improve the efficiency of the process. However, even if natural gas is used, cement production is one of the most serious contributors to the Greenhouse Effect. It has been estimated that anything between 8 and 10% of the world's annual emissions of CO_2 is emitted as a consequence of the production of various types of cement. The annual world production of cement exceeds 3000 million tons per year, the main producers being China (ca. 1800 million tons per year), India (ca. 220 million tons per year) and the US (ca. 63.5 million tons per year). A total of almost 900 kg of CO_2 are emitted for every 1000 kg of Portland Cement produced. For this reason, much research is currently being carried out on improved methods for pro-ducing cement: the use of different raw materials and additives; and modi-fications of heating methods used for the calcination process. We will return in Chapter 8 to a discussion of some developments in the methods used for cement production aimed at decreasing its carbon footprint.

Coal usage in iron and steel production

A further significant contributor to global CO_2 emissions is the iron and steel industry, this contributing between 4% and 7% of annual emissions. Table 2.5 shows the production figures for 2018 for the principal manufacturing countries and also includes, for comparison purposes, the data for the UK. The relatively low values for the UK are included because of its very significant contribution of developments in the iron and steel industries to the Industrial Revolution of the 18th Century; it is clear that while the UK was once at the forefront of iron and steel production, the major producers are now China, India and Japan, followed by the United States, South Korea and Russia.

A very clear description of the history of the production of iron and steel is to be found in an article entitled 'The Entire History of Steel' by Schifman in the magazine Popular Mechanics.[d] Schifman traces the discovery of metallic iron alloyed in meteorites and goes on to describe the emergence of the iron age, with cast and wrought iron and the importance of the incorporation of small amounts of carbon in the iron during the reduction process. Around 400 BC, Indian metal workers developed the use of a crucible to smelt a mixture of iron bars and charcoal pieces, heating this in a furnace using bellows to increase the temperature and producing the first steel; Indian steel was exported internationally. Early Japanese smiths also contributed to the development of modern steel by manufacturing intricate samurai swords. The use of coal addition to the iron ore during the smelting process was developed in the UK in the early 1700s and Sheffield became the centre of the steel production industry, using crucibles to smelt the iron ore–coal mixture. The blast furnace was invented by Henry Bessemer in the UK in 1856. The iron and steel industry in the US expanded

Table 2.5 Major crude steel production countries.

Country	Tonnage/million tonnes
China	928.3
India	106.5
Japan	104.3
United States	86.6
South Korea	72.5
Russia	71.7
Germany	42.4
Turkey	37.3
Brazil	34.9
UK	7.3

Steel production figures for 2018.
Data from www.worldsteel.org.

[d] https://www.popularmechanics.com/technology/infrastructure/a20722505/history-of-steel/.

Fig. 2.11 A Bessemer furnace in operation in Youngstown, Ohio, in 1941. *(Photo: Alfred T. Palmer Credit: Library of Congress, Prints & Photographs Division, Farm Security Administration/Office of War Information Color Photographs.)*

significantly after the Civil War using the Bessemer process (see Fig. 2.11). One of the pioneers in the new industry was Andrew Carnegie who helped establish the US iron and steel industries as the most important in the world. By 1900 the US was producing more than 11 million tonnes of steel a year, more than the UK and German industries combined. With the development of methods for producing stainless steel, US steel production was gradually overtaken by Japan and China.

As discussed further in Chapter 8, two main methods are currently used for the production of iron and steel: production from the ores (the so-called 'Integrated Route') and the reuse of scrap iron and steel ('Recycling Route'). The integrated route involves: (a) production of coke from coal by pyrolysis at 1000–1200°C in the absence of oxygen to drive off volatile components; (b) production of pig iron in a blast furnace at about 1200°C in which air is fed to a mix of iron ore and coke (plus 'fluxes' such as limestone to collect impurities), the coke forming the CO needed to reduce the ore; and (c) mixing of the pig iron with coke and up to 30% scrap, together with a small amount of flux, and heating using an oxygen lance to about 1700°C to produce molten steel. About 70% of the global production of steel involves

Fig. 2.12 Blast furnaces at Koninklijke Hoogovens plant at Ijmuiden, NL. *(Photo by J. Schoen, http://members.lycos.nl/fotoarchiefvon/hoogovens.JPG.)*

this integrated route. An example of a blast furnace facility is shown in Fig. 2.12. Such a plant gives rise to very significant emissions of CO_2 (Box 2.3).

The production of steel by recycling routes utilising scrap metal generally uses an electric arc furnace. As metallic iron is the raw material (rusted metal

BOX 2.3 CO_2 emission from steel production

Around 770 kg of coal is used to produce one tonne of steel from iron ore. Iron ore is commonly found as magnetite (Fe_3O_4), hematite (Fe_2O_3), goethite ($FeO(OH)$), limonite ($Fe(OH)\cdot nH_2O$) or siderite ($FeCO_3$). For magnetite, the reduction by coke or coal can be represented by the following all-over reaction:

$$Fe_3O_4 + 2\,C \rightarrow 3\,Fe + 2\,CO_2; \quad \Delta H^\circ = +330\,kJmol^{-1}$$

As indicated in the main text, in this reaction the carbon is first converted to CO and this CO is used as reductant but the all-over chemical reaction is as shown. It is clear that not only is a significant quantity of CO_2 formed in the reduction process but that CO_2 is also formed by the combustion of additional carbon in order to provide the all-over enthalpy change required for the reduction and also to provide energy to heat the system to the reaction temperature.

can also be used), much less energy is required (about 80% less) and the process also therefore produces much less CO_2. The production of steel by recycling is limited by the availability and purity of the scrap metal supply. About 20% of all steel was produced by recycling in the 1970s and that amount has increased to about 40% today, this figure including the scrap put into the feed for the integrated route described above.

Crude oil

Just as previous generations relied to a very great extent on the use of coal, much of our current way of life is dependent on the use of crude oil and its derivatives. This section starts by discussing the global reserves of crude oil, showing that it is a finite resource, and it then goes on to consider its current importance.

Table 2.6 shows the estimated oil reserves (in hundreds of million tons), the predicted lifetime of these reserves based on the production at the end of 2018, and the percentage of the total world reserves of each region of the world (figures in bold); it also shows the equivalent data for each significant oil-producing country in each region. The Middle East region has the largest estimated reserves, just short of 50% of the worldwide reserves, and also the largest rate of production. These figures are divided over a number of very significant oil-producing countries in that region. It is interesting to note that Canada is shown as having almost 10% of the world's reserves but it should be recognised that the majority of these are in the form of shale oil which is not currently exploited to any significant extent. A similar situation is found for Venezuela for which a significant proportion of its shale oil reserves, located in the Orinoco Belt, are also as yet unexploited; as a result, the predicted lifetime of the Venezuelan reserves is shown as being greater than 500 years. Of the Asian Pacific region, only China has any significant reserves but these are much less than the country's needs. What is particularly significant for the oil-producing countries of the world (with the exception of Venezuela, as noted above) is that the predicted lifetime of the known reserves in almost all cases is significantly less than 100 years. Even before the current concern regarding global warming, there had been a realisation that our current dependence on oil as a fuel cannot continue. Indeed, the Association for the Study of Peak Oil has estimated that the peak in the recovery of all liquid hydrocarbons is already past and that the reserves will be very much lower by 2050. As will be discussed further below, crude

Table 2.6 Global crude oil reserves.

Region/ Country	Reserves/100 million tons	Reserves/ Production (R/P)	% share of World Reserves
North	**35.4**	**28.7**	**13.7**
America	27.1	88.3	9.7
Canada	1.1	10.2	0.4
Mexico	7.3	28.7	3.5
United States			
South	**51.1**	**136.2**	**18.8**
America	48.0	>500	17.5
Venezuela			
Europe	**1.9**	**11.1**	**0.8**
Norway	1.1	12.8	0.5
CIS	**19.6**	**27.4**	**8.4**
Kazakstan	3.9	42.7	1.7
Russian	14.6	25.4	6.1
Federation			
Middle East	**113.2**	**72.1**	**48.3**
Iran	21.4	90.4	9.0
Iraq	19.9	87.4	8.5
Kuwait	14.0	91.2	5.9
Saudi Arabia	40.9	66.4	17.2
United Arab	13.0	68.0	5.7
Emirates			
Africa	**16.6**	**41.9**	**7.2**
Algeria	1.5	22.1	0.7
Libya	6.3	131.3	2.8
Nigeria	5.1	50.0	2.2
Asia-Pacific	**6.3**	**17.1**	**2.8**
China	3.5	18.7	1.5
Total World	*244.1*	*50.0*	*100*

Oil Reserves (R) and Reserves/Production (R/P) at the end of 2018 for the major oil-producing countries. R/P is equivalent to the predicted lifetime of these reserves in years based on the 2018 usage. Data extracted from BP Statistical Review of World Energy, 2019.

oil is the source of many of the important chemical products that we now rely on and it therefore appears totally irresponsible to squander our reserves by simply burning them for energy production.

Crude oil and its various derivatives contribute about 8% of the global greenhouse emissions associated with fossil fuels. The process of extraction and the transportation of the crude oil to the refineries, generally located remotely from the oil sources, contributes very significantly to this figure. The latter emissions have many different origins, from the venting or flaring

Table 2.7 Average carbon intensity for upstream crude oil production.

Country	Average carbon intensity—Well to refinery/gCO$_2$eq/MJ
Algeria	2.05
Venezuela	2.0
Canada	1.76
Iran	1.68
Iraq	1.68
USA	1.09
Russia	0.96
UAE	0.7
Qatar	0.64
Norway	0.52
Bahrain	0.48
Saudi Arabia	0.45

The average volume-weighted global carbon intensity is approximately 10.3 g CO$_2$eq/MJ.

of associated gases to the energy inputs associated with heavy oil extraction and upgrading and they include emissions generated during transportation. In an important article published in Science in 2018, a large team of American scientists having as its lead author Adam R. Brandt, based at Stanford University, have given results of their estimates of 'well-to-refinery carbon intensity' values for all the major oil fields currently active globally.[e] They found that the well-to-refinery greenhouse gas emissions for 2015 were approximately 1.7 Gt CO$_2$ equivalent, this figure corresponding to about 5% of the total global fuel combustion emissions for that year. Table 2.7 summarises their data, showing the average carbon intensity for oil production by some of the more significant countries included in the survey. The global average value of the carbon intensity is ca. 10.3 g CO$_2$ eq/MJ.

It can be seen that the carbon intensity values given vary quite considerably from one oil-producing country to another. It is important to recognise that the emissions considered are not all of CO$_2$ but include emissions of methane (34% of the total CO$_2$ equivalent emissions); as was discussed in Chapter 1 the greenhouse effect of methane is approximately 25 times that of CO$_2$. Oil wells that vent natural gas, therefore, contribute more CO$_2$ equivalent than do wells

[e] https://www.osti.gov/pages/servlets/purl/1485127.

that flare the natural gas. Flaring contributes about 22% of the global volume-weighted upstream carbon intensity listed in Table 2.7, the counties contributing most to flaring include Iran and the US. It can be seen from the data of Table 2.7 however that the well-established oil producers of the Middle East are well below the global average. The authors of the Stanford study point out that a producer such as Saudi Arabia has a small number of extremely large and productive oil reservoirs and has low flaring rates; these wells also have low rates of water production and this helps minimise their emissions as the separation of any water requires significant amounts of energy. At the other end of the spectrum are Canada and Venezuela, both of which are producers of unconventional heavy oils the extraction of which is energy intensive and gives rise to significant CO_2 emissions.

Fig. 2.13 shows some of the steps involved in processing crude oil to usable products. The crude oil is first fed to a distillation column which separates the components into products ranging from gases to heavy residues. Each component is then further treated to give products ranging from fuel gas, used as a refinery fuel, to various components used for diesel and petroleum formulation and also to petroleum coke (also burnt to provide heat) and asphalt. Many of the units involved require significant amounts of energy: total emissions of CO_2 range from 1 million tonnes per year for a typical refinery to 3.5 million tonnes per year for a complex facility. (These high values fall only just below those for standard power plants.) According to the Greenhouse Gas Reporting Program (GHGRP) of the US EPA, the largest proportion of the CO_2 emission in US refineries comes from the various energy-producing combustion processes used (68%) while the majority of the remainder comes from the catalytic cracking and reforming units (27%); the remaining 5% comes from flaring and other sources. Very similar figures apply to the refineries in other parts of the world.

The majority of the products from an oil refinery are used for either transportation or heating purposes but a proportion is used as feedstocks for the petrochemicals industry. Many of the uses of the products of the oil refinery require the removal of all traces of sulphur and other potentially contaminated products such as aromatic components. Hence, a large proportion of the units in a typical refinery are devoted to purification steps such as hydrotreating. Other units are devoted to processes such as reforming, fluid catalytic cracking and alkylation. Further information on these processes can be obtained from the references given in the caption to Fig. 2.13.[f]

---

[f] It should be noted that the greenhouse gas emissions from the various products from an oil refinery are generally given separately from figures for transportation or energy production; however, the total well-to-wheel figures that are given in the literature relating to the transport sector include them.

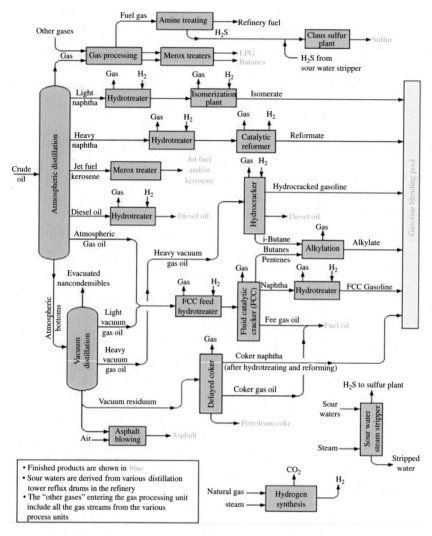

Fig. 2.13 Processes in a typical petroleum refinery. *(http://en.wikipedia.org/wiki/File: RefineryFlow.png. See also Contemporary Catalysis, Julian R. H. Ross, Elsevier 2018, ISBN: 9780444634740.)*

Natural gas

Natural gas is another important contributor to our current existence and it too has finite reserves. This section starts by outlining the global reserves of natural gas and then goes on to discuss some of the processes that depend on its widespread availability.

Table 2.8 Natural gas reserves.

Region/Country	Reserves as at end of 2018 /trillion m^3	Reserves/Production (Expected lifetime/yrs)	% Global share
North America	**13.9**	**13.2**	**7.1**
Canada	1.9	10.0	0.9
U.S.	10.0	14.3	6.0
South and Central America	**8.2**	**46.3**	**4.2**
Venezuela	6.3	190.7	3.2
Europe	**3.9**	**15.5**	**2.0**
Netherlands[a]	0.6	18.2	0.3
Norway	1.6	13.3	0.8
Ukriane	1.1	54.9	0.6
U.K.[a]	0.2	4.6	0.1
CIS	**62.8**	**75.6**	**31.9**
Azerbaijan	2.1	113.6	1.1
Russian Federation	38.9	58.2	19.8
Turkmenistan	19.5	316.8	9.9
Uzbekistan	1.2	21.4	0.6
Middle East	**75.5**	**109.9**	**38.4**
Iran	31.9	133.3	16.2
Iraq	3.6	273.8	1.8
Qatar	24.7	140.7	12.5
Saudi Arabia	5.9	52.6	3.0
United Arab Emirates	5.9	91.8	3.0
Africa	**14.4**	**61.0**	**7.3**
Algeria	4.3	47.0	2.2
Nigeria	5.3	108.6	2.7
Asia Pacific	**18.1**	**50.9**	**9.2**
Australia	2.4	18.4	1.2
China	6.1	37.6	3.1
Indonesia	2.8	37.7	1.4
Malaysia	2.4	33.0	1.2

[a]The Netherlands and the UK have relatively small reserves but are included because they are currently significant producers of natural gas.
Data for the countries with the predominant reserves in each region. (Figures for the end of 2018.)

The data given in Table 2.8 give the estimated natural gas reserves for some of the major gas-producing regions of the world. The global reserves are distributed throughout all the geographical regions but the majority of the individual national reserves are concentrated in countries of the Middle

Table 2.9 Natural gas composition—before and after processing.

Component	Before processing/%	Pipeline—after processing/%
CH_4	78.3	92.8
C^{2+}	17.8	5.44
N_2	1.77	0.55
CO_2	1.15	0.47
H_2S	0.5	0.01
H_2O	0.12	0.01

Data from Bradbury et al. (Footnote 11).

East and the CIS. The values of these reserves are traditionally quoted in cubic metres. The total yearly world production rate at the end of 2018 was 196.9 trillion cubic metres (final row of Table 2.8), this being the equivalent of the energy content of 3326 million tonnes of oil. As indicated earlier in this chapter (see Box 2.1), if the combustion of natural gas is used as a source of energy, the emission of CO_2 is significantly lower than that resulting from the combustion of coal or even of oil. It should be recognised that, in common with the burning of coal or oil, the combustion of natural gas can give rise to significant proportions of NO_x due to the combination of nitrogen and oxygen of the air at the high temperatures of the combustion process (so-called 'prompt NO_x', as discussed under coal combustion). Natural gas as recovered is a mixture of methane and lower volatile hydrocarbons as well as N_2, CO_2, H_2S and H_2O; the quantities of these components vary considerably with source. Before the gas is piped to the consumer, a significant proportion of the impurities are removed. Table 2.9 gives data from a US. Department of Energy report that shows the average compositions of natural gas as recovered and after purification, these being compiled from data for 2012. As natural gas is odourless, small quantities of mercaptan (methanethiol) are added to the gas (at levels up to 10 ppm) before it is distributed for use in heating applications since the harmless but pungent odour of this component is easily detected at concentrations as low as 1.6 ppb; the combustion of the added mercaptan gives rise to the emission of only small quantities of SO_2.

As shown in Table 2.8, the United States currently possesses about 6% of the world's reserves of natural gas, this corresponding to about 13.9 trillion m^3 (500 trillion ft^3). This value is a relatively recent development, the known

U.S. natural gas production (gross withdrawals)

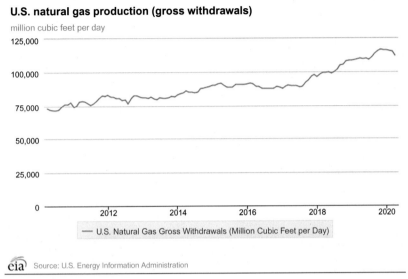

Fig. 2.14 The increase in US natural gas production since 2010. *(https://www.eia.gov/naturalgas/crudeoilreserves/.)*

reserves having more than doubled from a value of about 5.8 trillion m^3 (210 trillion ft^3) in 1978; the data of Fig. 2.14 show that this growth in production is still increasing continuously. This very large increase in production is due to the fact that much of the natural gas now produced in the United States is recovered by hydraulic fracturing, more commonly known as 'fracking' (and sometimes called 'fraccing'). Fracking has been practiced for some decades as a means of improving flows during the extraction of oil and gas from reservoir rocks. It entails the injection under pressure of a mixture of water, proppants (commonly sand) and chemical additives. The injection process fractures the rocks and the injected proppant maintains the size of the fissures, this enabling outwards flow of the oil or gas. The method was first used in 1947 to improve the flow of crude oil from oil wells. Most natural gas is now produced in the US using so-called 'horizontal slickwater fracturing' from deposits of shale using more water and higher pressures than used previously. The first of the wells using this method was put into production in 1998 at the Barnett Shale Field in North Texas. Since 1998, the number of such wells has increased very rapidly so that natural gas is now produced in most regions of the US. The fracking technique is also now applied in other countries such as Canada, Germany and the Netherlands and in the UK's gas fields in the North Sea.

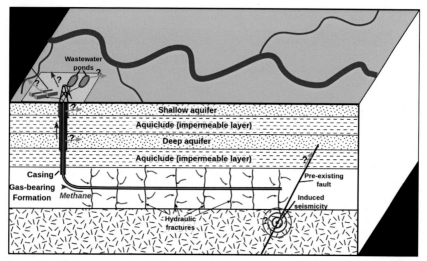

Fig. 2.15 Schematic representation of hydraulic fracturing ('fracking'). The figure also shows the routes for possible leakage of natural gas and fracking liquids into the environment. https://en.wikipedia.org/wiki/Hydraulic_fracturing.

As shown schematically in Fig. 2.15, the modern application of fracking involves the drilling of horizontal bore-holes in the gas-bearing shale bed. This shale layer may be at a depth of several thousand metres. The hydraulic fracturing process, brought about by the injection of the fracking fluid at high pressure, gives rise to a series of fissures that are connected to the bore-hole through which the gas passes to the surface. Although significant precautions are taken to prevent leakage of gases from the borehole, there is evidence of significant losses of methane to the atmosphere. There can also be appreciable leakage of the fracking fluid into the surrounding rock structures and also into aquifers as shown by arrows with a question mark of Fig. 2.15. There is in consequence a significant resistance to the introduction of the fracking technique in many countries, not only as a result of environmental concerns caused by this potential for leakage of natural gas but also by fears that fracking can cause earthquakes leading to the creation of additional leakage paths and structural damage at the earth's surface. Fracking is currently banned in several US states and also in a number of European countries including Ireland.

It is interesting to note that while the production and transportation of natural gas by more traditional means has a relatively small carbon footprint, this is not the case for natural gas produced by fracking as a proportion of the gas is used at each stage of the recovery, processing, transmission, storage and distribution stages. A relatively recent report [11] shows that the proportion of

Table 2.10 Proportions of US natural gas used prior to delivery.

Stage of production	Natural gas volume/ Billion cubic feet	% of Total production
Total Production	26,826	100
Production Stages	1283	4.7
Processing	1224	4.6
Transmission and Storage	842	3.1
Distribution	66	0.24
Total Losses	*3415*	*12.7*
Quantity to Consumers	**23,411**	**87**

the natural gas produced in the US that reaches the final consumer is only 87%; see Table 2.10. The total CO_2 emissions for 2012 associated with the 13% of the natural gas used in production and distribution were 163.7 million metric tonnes while the associated emissions of unburnt methane were 154.7 million metric tonnes CO_2 equivalent; the total CO_2 emissions associated with the end use were 1234 million metric tonnes.

Concluding remarks

This chapter has given a brief introduction to the widespread use of fossil fuels – coal, oil and natural gas – for energy-related purposes, all of these having been introduced before the emergence of the current state of awareness of the problems of greenhouse gas emissions. Each of these traditional uses gives rise to large quantities of emitted CO_2. The advantage of using natural gas as opposed to oil or coal as an energy source is very significant due to the lower emissions of CO_2 for a given amount of energy produced. Nevertheless, unless the CO_2 produced can be collected and used or safely stored (Chapter 4), the emissions from natural gas combustion are still very substantial. Chapter 3 now examines some of the ways in which the energy-related CO_2 emissions discussed in this chapter can be reduced by the introduction of new 'renewable' technologies. Chapter 4 then examines the production of 'grey' and 'blue' hydrogen from natural gas and of 'green' hydrogen by the electrolysis of water using 'renewable' electricity, a topic considered in more detail in Chapter 8. Chapter 4 also discusses the important role that the hydrogen produced plays in many different processes, including the synthesis of ammonia and of methanol.

CHAPTER 3

Less conventional energy sources

Introduction

Having considered in Chapter 2 the production of energy from unsustainable fossil fuels as well as some of the uses of these fossil fuels as feedstocks for the chemical industry, we will now consider the production of energy from more sustainable sources. In this context, the use of the word 'sustainable' is used to indicate that these sources are either totally renewable or that the method of supply makes significant use of renewable sources.

The sun emits light and other radiation at a rate of 3.846×10^{26} watts. This energy is formed as a result of nuclear reactions occurring within the sun's mantle and so the use of this energy has a negligible consequence on subsequent supplies. As shown schematically in Fig. 3.1, the intensity of the radiation reaching the earth's atmosphere is 1368 watts per square metre (W/m^2); of this, some is absorbed by the atmosphere and the average density at the earth's surface on a surface perpendicular to the rays is approximately $1000\,W/m^2$. This quantity is far in excess of that required for the world's energy needs.[a]

Fig. 3.2 shows the relative contributions of CO_2 equivalent emissions of each of the most important forms of 'renewable energy' relative to those of coal and gas (Chapter 2), these data having been calculated in 2014; the data take into account the emission of any greenhouse gases formed during the production of the energy, for example by the operation of pumps and other equipment.

Most of the routes that will be discussed below depend in some way or another on the radiation reaching the earth from the sun, the only exceptions being the use of nuclear energy and of those that make use of energy stored in the earth's crust (geothermal energy) or formed

---

[a] It has been shown that radiation density is such that if all the energy falling on the US state of Texas at noon could be converted to electricity, the amount produced would be approximately 300 times the total amount of electricity generated by all the earth's power plants.

Sustainable Energy
https://doi.org/10.1016/B978-0-12-823375-7.00001-9
49

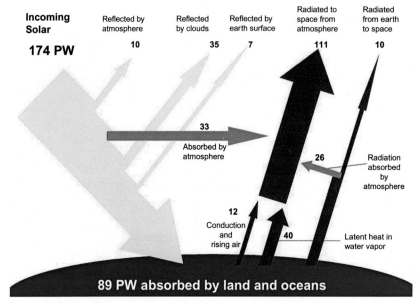

Fig. 3.1 Uptake and re-emission of solar power by the earth's surface. *(From https://en. wikipedia.org/wiki/Solar_energy)*

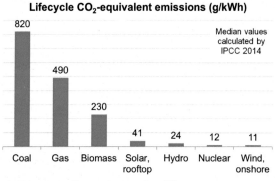

Fig. 3.2 Life-cycle CO_2-equivalent emissions by different energy sources. *(From https:// www.ipcc.ch/report/ar5/syr/)*

as a consequence of gravitational forces between the earth and the moon (tidal energy).

Nuclear energy

There are three types of nuclear reactions, all of which create energy: fission, radioactive decay and fusion. Only fission is currently used for the generation of energy; this is described further below after a brief outline of each of the processes.

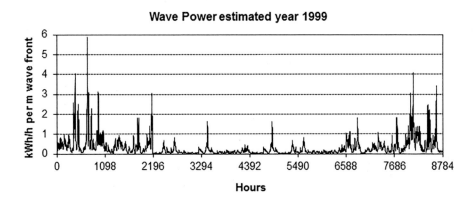

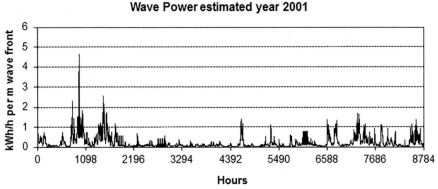

Fig. 5.4 Estimated hourly distributions of electricity production from wave power based on wave measurements in the Danish North Sea.

In Fig. 5.6, a curve representing 2001 is shown for the three different types of RES and compared to one another. As can be seen, the curves representing PV are a little higher than the two others.

The methodology of excess electricity diagrams does not only apply to the preceding reference system but to radically different systems as well. In Fig. 5.7, the same diagrams are shown for two systems (which will be elaborated on in Chapter 7): a Danish business-as-usual reference system year 2030 (BAU2030) and a radical change in design prepared by the Danish Society of Engineers (IDA2030). Both systems have been analyzed in the EnergyPLAN model for three different wind years: Wind 1996, Wind 2000, and Wind 2001. As can be seen, the differences in wind years only result in marginal changes in the excess electricity diagrams compared to the impacts of changing the system from BAU2030 to IDA2030. Thus, Fig. 5.7 illustrates how these excess electricity diagrams can compare different energy systems in terms of their ability to use variable sources such as wind. More important, this illustration does not change from one year to another even though the variations in wind production do change.

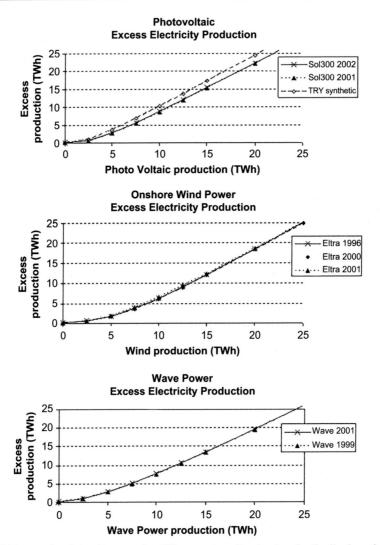

Fig. 5.5 Excess electricity diagrams for different years, showing hourly distribution of electricity production from wind, wave, and PV power.

3 Optimal combinations of RES[c]

This section is based on Lund's (2006b) article "Large-Scale Integration of Optimal Combinations of PV, Wind, and Wave Power into the Electricity Supply," which presents the results of a series of analyses of large-scale integration of wind, PV, and

[c] Excerpts reprinted from *Renewable Energy*, 31/4, Henrik Lund, "Large-Scale Integration of Optimal Combinations of PV, Wind, and Wave Power into the Electricity Supply", pp. 503–515 (2006), with permission from Elsevier.

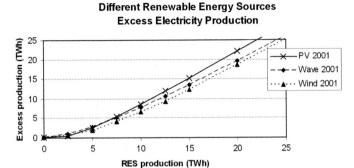

Fig. 5.6 Excess electricity diagram comparing wind, wave, and PV power fed into the same energy system.

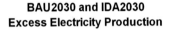

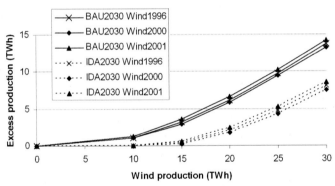

Fig. 5.7 Excess electricity diagrams for two different energy systems (BAU2030 and IDA2030). The systems have been analyzed for three different years of hourly distributions of electricity production from wind power: Wind 1996, Wind 2000, and Wind 2001.

wave power into the Danish energy system. It is based on the same data as already presented in Figs. 5.2–5.4, and, again, the reference used is the projection of the future energy system in West Denmark by 2020. The idea is to use the different patterns in the variations of different renewable sources, and the purpose is to identify optimal combinations of RES from a technical point of view.

The analysis is made on the basis of hourly distributions beginning in 2001, since data are available for all three types of RES in that year. By basing the analysis on the same year, any correlation between, for example, wind and waves is included in the study. The results are shown in curves of excess electricity production generated by increasing RES inputs. Results have been generated for each of the different RES and for relevant combinations to identify an optimal mixture. It should be added that electricity production from a single RES that equals 100 percent of the demand is hard to imagine in practice, especially for technologies such as PV and wave power. For example, an annual production of 25 TWh from PV requires approximately

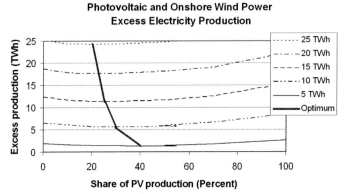

Fig. 5.8 The analysis identifying optimal combinations of PV and onshore wind power.

25,000 MWp installed capacity, which is not financially or practically realistic. The same applies to wave power. Moreover, this amount of RES is not likely to be added to the reference system without improving the integration ability of the system and thereby decreasing excess production. Such improved systems are discussed in more detail later in this chapter. The analysis constitutes a valid illustration of the differences among the three different RES and serves as an important basis for the identification of optimal combinations.

The optimal combinations of two types of RES technologies are shown in Fig. 5.8, along with the annual excess electricity production for the combination of onshore wind and PV power. The excess production generated has been calculated with different inputs of total electricity from RES. This is illustrated by the curves starting with a total of 5 TWh and rising in steps of 5 TWh to a total of 25 TWh. Each curve shows the resulting excess production of different shares of PV and wind power.

On the left side of Fig. 5.8, the share of PV is zero, and on the right side, the share of PV is 100 percent. All curves show an optimal combination in which excess production is minimal. An optimal combination is achieved with a PV share of 20 percent when the total RES electricity production is high, and 40 percent when the total production is low. In the same way, optimal combinations of PV, onshore wind, and wave power have been identified. Fig. 5.9 shows the optimal combination of the three RES, that is, the combination that generates minimum excess electricity production.

In Fig. 5.10, the excess electricity diagram of the optimal combination (Optimix) is compared to the results of each of the different RES technologies, assuming that the same production is to come from only one type of RES. In general, PV is the RES with the highest excess production, followed by wave power and onshore wind power. The optimal combination of the three types of RES results in less excess production.

Fig. 5.10 illustrates how the combination of different sources reduces the integration problem. However, the excess production is still considerable. It should be emphasized that a very high percentage of RES can be integrated into the electricity supply without any excess production if other measures are implemented, such as flexible energy systems (see the following sections). Fig. 5.10 shows how a certain combination of RES units will enhance the effect of these implementation strategies.

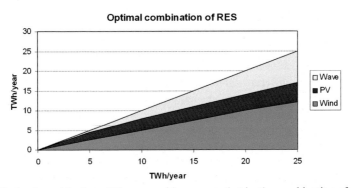

Fig. 5.9 Optimal combination of the renewable sources, that is, the combination of minimum excess production.

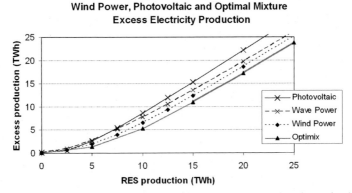

Fig. 5.10 The integration of individual renewable sources compared to the optimal mixture.

The result illustrates how excess production increases when the RES input in terms of wind, PV, and wave power is raised. Meanwhile, combinations of different RES can slow down the increase in excess production. For example, an optimal combination of 20–40 percent PV and consequently 60–80 percent wind power has been found to generate less excess production than 100 percent of either PV or wind power alone.

As can be seen from Fig. 5.9, the optimal mixture seems to be reached by an onshore wind power production of approximately 50 percent of the total renewable electricity production. Meanwhile, the optimal mixture between PV and wave power seems to depend on the total amount of electricity produced by RES. When the total RES input is below 20 percent of the demand, PV should cover 40 percent and wave power only 10 percent. When the total input is above 80 percent of the demand, PV should cover 20 percent and wave power 30 percent. The combination of different RES as the only measure is far from a solution to the RES integration problem. Other measures such as the investment in flexible energy supply and demand systems and the integration of the transportation sector have much more potential for solving the problem.

4 Flexible energy systems[d]

This section is based on Lund's (2003b) article "Flexible Energy Systems: Integration of Electricity Production from CHP and Fluctuating Renewable Energy," which discusses and analyzes different national strategies for large-scale integration of renewable energy. It points out key changes required in the energy system to benefit from a high percentage of wind and CHP without generating excess electricity production. Here, the altered system is referred to as a *flexible energy system*.

This study was made prior to the work of the Energy Agency's expert group mentioned previously. Thus, the study is not based on exactly the same reference scenario. However, the study used the expected development of the official Danish energy policy expressed in the government's energy plan, Energy 21, as a reference, and this reference is very similar to the previous one. According to Energy 21, the rate of wind power was expected to increase to approximately 50 percent in 2030, and excess electricity production was expected to increase accordingly and create serious problems for the regulation of the electricity supply. The magnitude of the problem is illustrated in Fig. 5.11.

Even though electricity consumption was expected to decrease slightly, the production had to increase substantially. By 2030, the excess production was expected to constitute 80 percent of the electricity production from wind and other renewable resources. The implementation of the Danish energy policy according to Energy 21 was based on three initiatives: energy conservation resulting in slightly decreasing electricity consumption, the integration of more renewable energy sources, and the efficient use of fuels by CHP. The variations in demand and production resulting in excess production are, in principle, the result of applying these three means at the same time. The fact that a high level of excess electricity production was expected, as illustrated in Fig. 5.11, can be specifically explained by the assumptions that (1) the

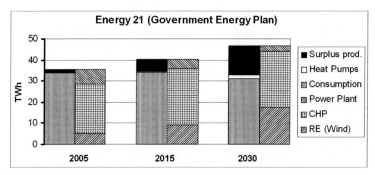

Fig. 5.11 Electricity balance according to the Danish government's energy plan, Energy 21.

[d] Excerpts reprinted from *International Journal of Energy Technology and Policy,* 1/3, Henrik Lund, "Flexible Energy Systems: Integration of Electricity Production from CHP and Fluctuating Renewable Energy", pp. 250–261 (2003), with permission from Inderscience.

small- and medium-sized CHP stations were not expected to regulate according to variations in wind power but solely according to heat demands and (2) the task of securing grid stability (voltage and frequency) was managed solely by large power stations.

Fig. 5.11 illustrates an important condition; namely, that the excess production problem changes radically during the period. In 2005, the excess production was much lower than the production from condensing power stations. This means that, in principle, the problem can be solved solely by moving and/or storing electricity, in other words, by using *energy storage technologies*, as defined in Chapter 1. In 2030, on the other hand, excess production is expected to be much higher than the production from condensing power stations, which means that moving and/or storing the production cannot solve the problem. In that case, other sorts of changes are needed, such as *energy conversion technologies*, as defined in Chapter 1. As will be elaborated further in Chapter 6, one might say that, in the 2005 situation, the problem may be solved solely within the electricity sector, i.e., within the concept of *smart grid*; while in 2030, the problem can only be solved by involving the other sectors, i.e., within the concept of *smart energy systems*.

The reference was modeled in the EnergyPLAN model, and the results are shown in Fig. 5.12. The diagrams show the implementation of the Energy 21 reference from Fig. 5.11 in four weeks representing winter, spring, summer, and fall. Consumption is shown to the left and production to the right. "Export" represents excess production. Fig. 5.12 illustrates how excess electricity production is mainly the result of combinations of wind power and CHP. Meanwhile, the constraints of maintaining grid stability sometimes require that power stations without CHP produce in periods with excess production and thus add to the problem. Fig. 5.12 shows that the excess electricity production problem is substantial all year round.

Flexible energy system

Based on the Energy 21 scenario, the EnergyPLAN model was used to analyze various investment and regulation means to find suitable designs of flexible energy systems. The following initiatives seemed to be the most important ones:

Regulation of CHP: The fact that CHP stations are not expected to regulate according to variations in wind power, but solely according to heat demand, is significant to the size of the excess electricity production problem shown in Figs. 5.11 and 5.12. Consequently, an alternative regulation strategy was analyzed in which CHP stations are partly replaced by heat production from boilers in case of excess electricity production. This change in strategy can solve the problem in the beginning of the period, when the problem is small. Meanwhile, the benefit of using CHP decreases, and the result is lower fuel efficiency (the rate of fuel per unit of heat and electricity produced increases).

Investments in heat pumps and heat storage capacity: Adding heat pumps to the systems means that they can be used instead of boilers to restore fuel efficiency. Additionally, the energy system becomes much more flexible in more than one way. First, using heat pumps can decrease excess electricity production. Second, by replacing CHP heat production by heat pumps, the flexibility of the CHP stations is increased as long as the capacity is maintained. Third, by adding heat storage capacity to the system, the flexibility is further

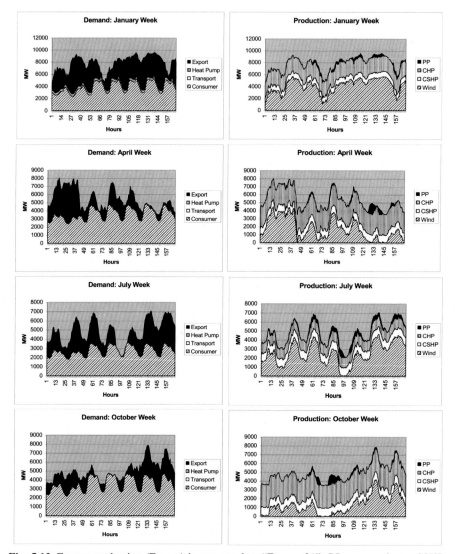

Fig. 5.12 Excess production (Export) in energy plan, "Energy 21". *PP*, power plants; *CSHP*, industrial combined steam and heat power.

increased. This flexibility can solve most of the excess electricity production problems. Meanwhile, situations in which the production from large power stations is needed to maintain grid stability will arise more often and limit the decrease in excess electricity production. *Grid-stabilizing CHP and wind power*: The task of securing grid stability (voltage and frequency) had so far been managed only by large power stations. Consequently, distributed production from small- and medium-sized CHP units and wind power may become a burden and may set a limit to the fulfillment of this task. This limit can be overcome by involving distributed production units in securing grid stability.

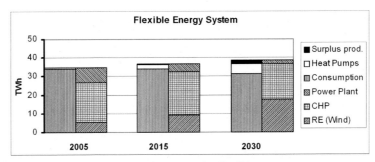

Fig. 5.13 Electricity balance when implemented by flexible energy systems.

Therefore, the following flexible energy system has been analyzed using the EnergyPLAN model:

- The CHP units in the energy system are supplemented by heat pumps equal to ~1000 MW of electric power in 2030 and heat storage capacity equal to the heat consumption of approximately 1 day.
- CHP units and heat pumps are operated according to a strategy of meeting the difference between demand and wind power production.
- All CHP units and wind turbines built after 2005 are involved in securing grid stability.

The result of implementing this system is shown in Fig. 5.13. Fuel efficiency is maintained, while most of the excess electricity production has been avoided.

Flexible energy systems including electricity for transportation

As the percentage of wind power increases, it becomes more and more difficult for flexible energy systems, based on CHP, heat pumps, and heat storage, to manage excess electricity production. Sooner or later, further initiatives will have to be taken. The electrification of the transportation sector results in better flexibility and, at the same time, improves the fuel efficiency. The analyses used an early version of the EnergyPLAN model in which the modeling of transportation was not a specific subject. However, it was possible to do a modeling of the flexible electricity demands arising from the use of electricity for transportation (batteries and/or hydrogen). These demands were assumed to be evenly distributed over the year but could be made flexible within shorter periods of time.

The EnergyPLAN model has evaluated the possibilities of integrating the transportation sector into the energy system. By 2030, 80 percent of the Danish vehicles weighing less than 2 tons will be replaced by a combination of BEVs and HFCVs, leading to a rise in the electricity consumption by 7.30 TWh/year and fuel savings by 20.83 TWh/year.

The electricity transportation scenario has been analyzed along with the preceding flexible energy systems based on CHP, heat pumps, and heat storage. The results of the analysis are shown in Figs. 5.14 and 5.15. In the analysis shown, electricity for transportation has been made flexible within a period of 1 day. Fig. 5.14 illustrates

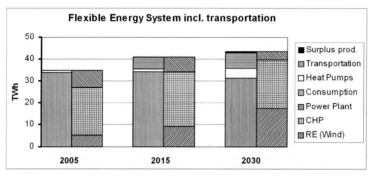

Fig. 5.14 Electricity balance when implemented by flexible energy systems, including electricity for transportation.

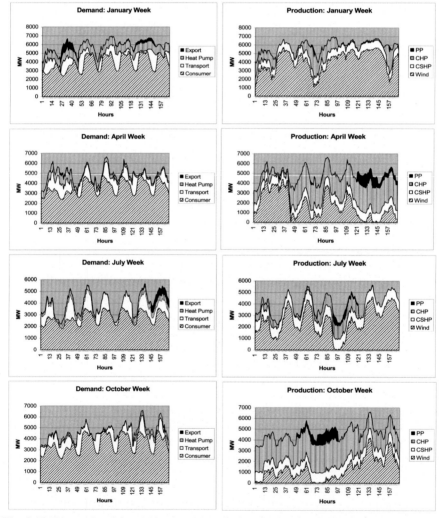

Fig. 5.15 Excess electricity production (export) when implemented with flexible energy systems, including transportation. *PP*, power plants; *CSHP*, industrial combined steam and heat power.

how excess electricity production is nearly removed, even in 2030. Fig. 5.15 shows how consumption and production are balanced by regulating CHP and heat pumps together with the use of electricity for transportation for a period of 4 weeks.

When the official Danish energy plan, Energy 21, was launched in 1995, Denmark's main policy was to export the problem simply by selling future excess production on the European electricity market. However, Europe cannot solve the total problem if every individual country adopts the same policy. Furthermore, Denmark has to face problems of low revenues and high investments in transmission lines and meet agreements of reducing CO_2 emissions. The introduction of flexible energy systems can solve the problem of integrating variations from a high percentage of renewable energy and distributed CHP production. The key factors of this solution seem to be the following:

- Making CHP stations operate according to variations in RES
- Investing in heat pumps and heat storage capacity
- Integrating small CHP stations and wind power when securing grid stability
- Integrating the electricity sector and the transportation sector by introducing electric vehicles (battery and hydrogen)

By introducing such flexible energy systems to reduce the export of excess electricity production, fuel consumption and CO_2 emissions in Denmark are reduced accordingly. In the case of integrating the transportation sector by converting to battery and hydrogen vehicles, the total reduction in fuel consumption is very high because of the low efficiencies of vehicle combustion engines.

5 Different energy systems[e]

This section is based on Lund's (2005) article "Large-Scale Integration of Wind Power into Different Energy Systems." Earlier in this chapter, excess electricity diagrams were presented for the integration of renewable energy into a reference energy system of West Denmark 2020. Here, the same reference system is compared to other energy systems. Excess electricity diagrams are supplemented by CO_2 emission reduction diagrams, and the impacts of adding flexible technologies are presented. Thus, the article discusses the abilities of different energy systems and regulation strategies to integrate wind power.

In the article, the reference system has been defined as the existing regulation adjusted by a number of measures that may be introduced to avoid critical excess production. Thus, the reference regulation can be described as follows:

- All wind turbines produce according to variations in the wind.
- All CHP stations produce according to heat demand (or triple tariff).
- Only large power stations participate in the task of balancing supply and demand and securing grid stability.

[e] Excerpts reprinted from *Energy,* 30/13, Henrik Lund, "Large-Scale Integration of Wind Power into Different Energy Systems", pp. 2402–2412 (2005), with permission from Elsevier.

- Minimum 300 MW and minimum 30 percent of the production must come from grid-stabilizing power stations.
- CEEP is avoided by applying the following priorities: (1) replacing CHP with boilers; (2) using electric heating; and (3), if necessary, stopping the wind turbines.

This reference system has been compared to the following three alternative energy systems:

- *50 percent more CHP*: In the reference system, 21.21 TWh equaling approximately 50 percent of the total Danish heating demand is produced by CHP. An alternative system has been defined in which the share of CHP is increased by 50 percent to 31.82 TWh.
- *Fuel cell technology*: Improvements of electric efficiencies in CHP units and power stations (as, for example, fuel cells) increase the efficiency and consequently decrease the fuel consumption. An alternative system has been defined by raising CHP electricity efficiencies from the average of 38 to 55 percent and power station efficiencies from 50 to 60 percent.
- *Electrification of cars*: Based on the study of the electrification of cars (partly BEVs and HFCVs), an alternative system has been defined in which 12.6 TWh of gasoline can be replaced by 4.4 TWh of electricity.

The results of the analyses of the alternative systems are shown in Fig. 5.16. The diagram illustrates how improvements in terms of more CHP (50 percent CHP) and better efficiencies (fuel cell) accelerate the excess production problem, while the electrification of cars (transport) decreases this problem. At the starting point, without any wind power, all improvements decrease CO_2 emissions compared to the reference energy system. However, along with the increase of wind input, only the electrification of cars maintains a good CO_2 reduction ability.

If all three alternative improvements in Fig. 5.16 are combined, the excess production becomes severe even in the case of no wind power. The ability of such a system to integrate wind power has been compared with a number of alternative regulations. These are based on the principle that small CHP units are involved in the grid stabilization task and that these CHP units operate to integrate wind power by reducing their electricity production at hours of excess production. Three variants replacing heat production by other devices have been analyzed:

- *CHPregB*: Boilers replace CHP heat production.
- *CHPregEH*: Electric heaters (boilers) replace CHP heat production.
- *CHPregHP*: Heat pumps replace CHP heat production.

Fig. 5.17 shows the results of adding alternative regulation systems to the reference system shown in Fig. 5.16. Fig. 5.17 shows how important it is to involve the CHP units in the regulation. This measure alone decreases excess production radically. Meanwhile, if the CHP units are replaced by boilers (CHPregB), the fuel efficiency is decreased and the potential for reducing CO_2 emissions is not fully exploited. Adding electric heating to the system (CHPregEH) does not solve the problem, but the introduction of heat pumps (CHPregHP) makes it possible to decrease the excess production and, at the same time, maintain fuel efficiencies. In the next chapter, an example is given of how the solution of including existing small CHP units in the regulation was implemented in the Danish system, which is a part of the Nord Pool electricity market.

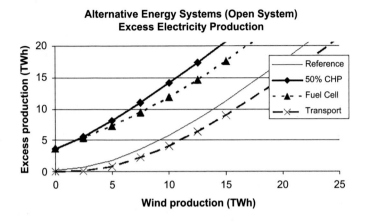

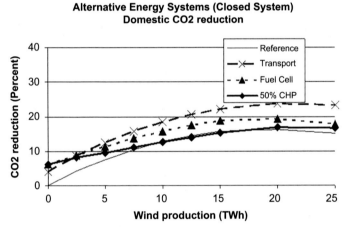

Fig. 5.16 Excess electricity production and domestic CO_2 reduction in the three alternative energy systems compared to the reference.

Fig. 5.18 shows the results of adding the previously mentioned alternative regulations to a system in which all three system changes presented in Fig. 5.16 are implemented, that is, increasing CHP and introducing fuel cells and electricity for transport. Again, the diagram shows how important it is to involve the CHP units in the regulation.

Together, Figs. 5.17 and 5.18 confirm the results of the analysis of flexible energy systems and show that this conclusion is also valid for the three other systems. Moreover, the article illustrates how different energy systems can be compared to one another in the same diagram in terms of their ability to integrate RES on a large scale.

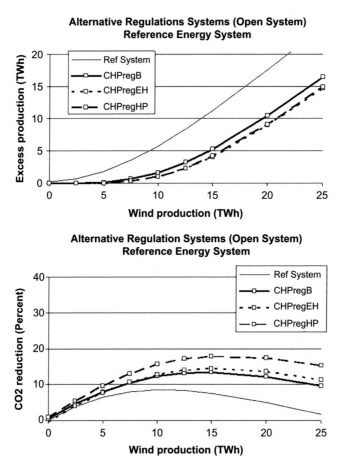

Fig. 5.17 Excess electricity production and domestic CO_2 reduction of the three alternative regulation systems applied to the reference energy system.

6 Grid stability[f]

This section is based on Lund's (2004) article "Electric Grid Stability and the Design of Sustainable Energy Systems," which analyzes the significance of including future distributed CHP and renewable power production units in the task of grid stabilization, that is, securing voltage and frequency stability of the electricity supply. At that time, in most countries, electricity was produced either from hydropower or large steam turbines on the basis of fossil fuels or nuclear power. Electricity from distributed generation constituted only a small part of the production. Until then, the tasks of balancing

[f] Excerpts reprinted from *International Journal of Sustainable Energy*, 24/1, Henrik Lund, "Electric Grid Stability and the Design of Sustainable Energy Systems", pp. 45–54 (2004), with permission from Taylor and Francis.

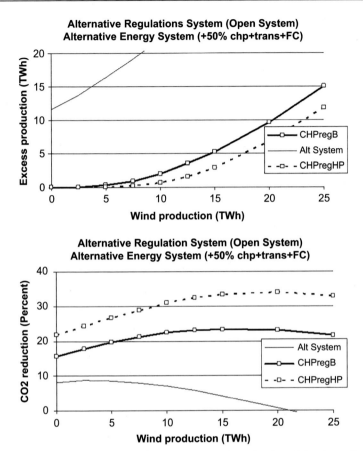

Fig. 5.18 Excess electricity production and domestic CO_2 reduction of the three alternative regulation systems applied to the alternative energy system including an increase in CHP as well as the introduction of fuel cells and electricity for transportation.

supply and demand and securing frequency and voltage on the grid were managed only by these large production units.

However, the implementation of cleaner technologies, such as renewable energy, CHP, and energy conservation, is necessary to secure future renewable energy systems. Consequently, such distributed production units sooner or later need to contribute to the task of securing a balance between electricity production and consumer demands. The article presented technical designs of potential future flexible energy systems, which will be able both to balance production and demand and to fulfill voltage and frequency stability requirements to the grid. Again, the analysis is based on the reference scenario for West Denmark by the year 2020, and again, an example is given in Chapter 6 of how the solution of including existing small CHP units in the grid stabilization has been implemented in the Danish system, which is part of the Nord Pool electricity market.

Electricity Balance and Grid Stability

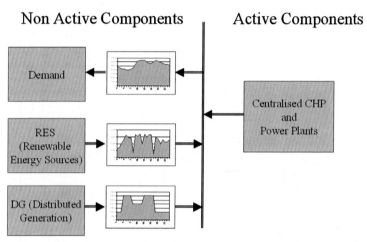

Non Active Components | Active Components

Fig. 5.19 The current electricity system in Denmark.

Fig. 5.19 shows the starting point of the analysis. The task of securing a balance between electricity production and consumer demand had so far only been managed by large power stations. However, small-scale CHP units have the technical potential for solving some of the balancing problems. When the analysis was made, small- and medium-sized CHP stations did not participate directly in balancing wind power in Denmark. These stations, however, did contribute to the balancing of variations in the demand. CHP stations had been paid through a triple-tariff system, with high payment between morning and late afternoon, reflecting a high electricity demand during this period, and low payment during night hours, weekends, and holidays.

Consequently, the Danish CHP units were designed with relatively high production and heat storage capacities, making it possible to produce mainly during the high-tariff period. When electricity sales prices were high, the CHP unit operated at full capacity and stored excess heat in the heat storage. When prices were low, the CHP unit stopped, and heat for district heating was supplied from the storage. Until 2004, this regulation ability was not used to integrate variations in renewable energy.

Small CHP units can be used to balance the variable output of wind power. The heat storage facilities of the CHP station are important features of this technique. Provided that the excess heat production can be stored for future use, the CHP station is able to increase the electricity production when and if required for balancing activities, without any economic penalty.

In the current Danish system, both tasks are solved primarily by large power stations. In some countries, large hydropower stations participate in fulfilling the task as well. When the share of distributed generation is small, balance requirements will not be compromised. However, when the share increases, the balance may be at stake. Consequently, the system design has defined limits to the integration of CHP and RES. In Fig. 5.19, small- and medium-sized CHP stations are illustrated as distributed generation components operating in accordance with a fixed triple tariff.

To identify limits and possible solutions to increasing the shares of RES and CHP, three potential future systems were analyzed and compared with the existing system. Fig. 5.20 compares a reference (system 0) with three potential future alternatives:

System 0 (reference) is characterized by the following:

- Distributed generation units are operating according to variations in heat demand during the season in question and to a fixed triple tariff during the period of 1 week.
- RES (i.e., wind turbines) are operated according to variations in the wind.
- Centralized power stations, including large CHP stations, are operated to secure a balance between electricity demands and production while still meeting seasonal variations in the district heating demand.
- The task of securing frequency and voltage stability is left solely to the centralized units, under the restriction that the production of these units must always correspond to at least 30 percent of the total electricity production, and the production must always be at least 350 MW (in order to have the necessary units operating).

This system represented the system in Denmark at that time.

System 1 (activating small- and medium-sized CHP stations): The system is the same as the reference apart from the fact that all CHP stations are operated to balance both heat and electricity production. If the electricity production exceeds the demand, parts of the CHP units are replaced by boilers. The heat storage capacities are used to minimize such replacements. The system has been analyzed both in a situation in which small- and medium-sized CHP stations do not participate in the grid-stabilizing task (System 1A) and in a situation in which they do participate (System 1B).

System 2 (adding heat pumps): In System 1, CHP production is replaced by heat production from boilers in periods of excess electricity production, and, consequently, the fuel efficiency is decreased. The idea of System 2 is to compensate for the increase in fuel consumption by adding heat pumps to the system. Furthermore, heat pumps increase the flexibility of the system because they can consume electricity at hours of excess production and, at the same time, replace the heat production of CHP units. Again, the system has been analyzed both in a situation in which the small stations do not participate in the grid-stabilizing task (System 2A) and in a situation in which they do participate (System 2B).

System 3 (including electricity for transport): In System 3, the electricity consumption for transportation is added to the regulation system according to the scenario described earlier in this chapter. Vehicles that weigh less than 2 tons are replaced by BEVs and HFCVs, respectively. In 2030, 20.8 TWh of oil will be replaced by 7.3 TWh of electricity. On the basis of this national scenario, it was chosen to analyze a scenario for West Denmark in 2020 in which 3.2 TWh of electricity substitutes 9.8 TWh of oil.

In Lund (2004), both the reference and the three alternative regulation systems have been analyzed in terms of their ability to balance supply and demand in the given scenario with different wind power inputs. The results of the three regulation systems are shown in Figs. 5.21 and 5.22. In System B, small CHP stations are involved in the task

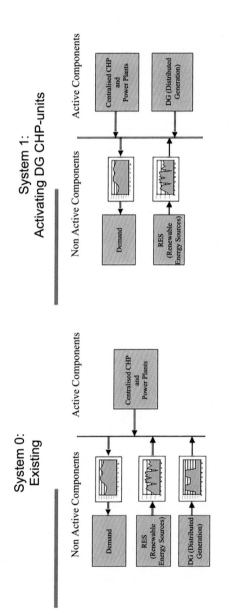

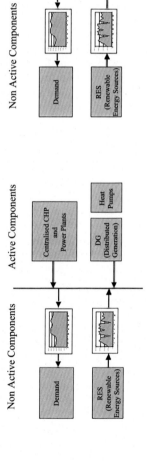

Fig. 5.20 Different energy system designs with regard to grid stabilization.

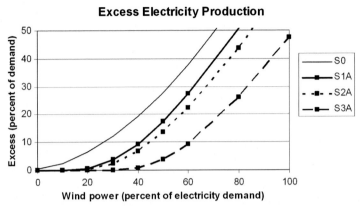

Fig. 5.21 Excess electricity production in percentage of demand if small CHP units do *not* participate in the task of grid stabilization.

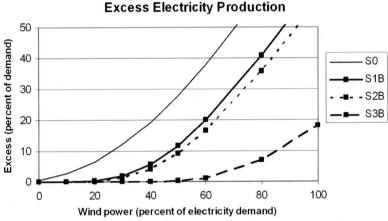

Fig. 5.22 Excess electricity production in percentage of demand if small CHP units *do* participate in the task of grid stabilization.

of grid stabilization, while in System A, they are not. In both diagrams, the results are compared with the reference (System 0). In Figs. 5.21 and 5.22, the regulation ability is illustrated in terms of excess electricity produced as a function of wind power input. Both values are given as percentages of the electricity demand.

The diagram should be read as follows: for a given wind power input of 40 percent, the S0 curve representing the reference shows an excess electricity production of 20 percent. Consequently, only half of the wind power input can be used directly in the system. For the same input, the S3 curve (both S3A and S3B) shows an excess production of approximately zero. Consequently, all wind power produced can be utilized directly if this regulation system is implemented.

In the West Denmark reference scenario, the wind power input in 2000 was 20 percent. Figs. 5.21 and 5.22 illustrate how this wind power input creates a small excess production, which was actually the case in that year. Meanwhile, the excess production problem can easily be avoided by involving small- and medium-sized CHP units in the balancing task. In the Danish case, this step was partly implemented starting from 2004 and did decrease the problem significantly. In the reference scenario, the wind power input was planned to increase to 25 percent in 2005 and 35 percent in 2010. Consequently, further steps, such as involving CHP units in the grid stabilization task and/or investing in heat pumps, should be considered. By 2020, wind power had increased to almost 50 percent in the reference scenario. Thus, the inclusion of electricity for transportation should be considered, if excess electricity production is to be avoided. It might be added that even though the implementation of wind power in Denmark was slowed down in the period after 2004, the implementation was not far behind the plans referred to earlier. As a result, the wind power share in 2012 was nearly 30 percent and actually increased to 50 percent by 2020.

7 Local energy markets[g]

This section is based on Lund and Münster's (2006a) article "Integrated Energy Systems and Local Energy Markets," which takes its point of departure in the two research reports "Local Energy Markets" (Lund et al., 2004) and "MOSAIK" (Østergaard et al., 2004). The article adds to the analyses in the previous sections by making an economic feasibility study of flexible energy technologies. Moreover, the analysis includes the modeling of the Nordic electricity market Nord Pool. With the technical analyses of the previous sections of this chapter as the starting point, the study focused on how Denmark could benefit from international electricity trade while integrating RES.

The conclusion is that significant benefits can be achieved by increasing the flexibility of the Danish energy system. On the one hand, the flexible energy system makes it possible to benefit from trading electricity with neighboring countries, and on the other, Denmark will be able to make better use of wind power and other types of renewable energy in the future. The article analyzes different ways of increasing the flexibility in the Danish energy system by using the same flexible technologies as described in the previous sections of this chapter. The strategy is compared with the opposite extreme, that is, trying to solve all balancing problems via electricity trade on the international market. The analysis concludes that it is feasible for Danish society to involve CHP stations in the balancing of variable wind power. Moreover, major advantages can be derived from equipping both small and large CHP stations with heat pumps. By doing so, it will be possible to increase the share of wind power from the 2004 level of 20–40 percent without causing significant problems of imbalance between electricity consumption and production, as already shown in the previous sections. As we will see in the following, this investment is also economically feasible to Danish society. Furthermore, it will have the positive side effect that the feasibility of large-scale wind power production is improved.

[g] Excerpts reprinted from Energy Policy, 34/10, Henrik Lund & Ebbe Münster, "Integrated Energy Systems and Local Energy Markets", pp. 1152–1160 (2006), with permission from Elsevier.

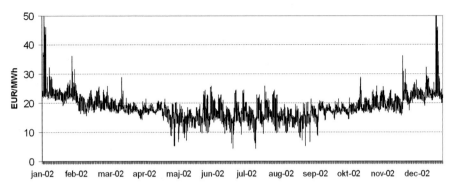

Fig. 5.23 Typical spot market price variations on the Nordic Nord Pool electricity market used as a basis for the analysis.

This article describes how the Nordic electricity spot market Nord Pool is modeled in the EnergyPLAN model and which data are used to make an economic feasibility study of the integration of wind power into different energy systems. The model is based on typical historical variations in the Nord Pool prices, as shown in Fig. 5.23.

Based on this price variation, the model includes a price elasticity function that evaluates the impacts of electricity export and import on the Nord Pool spot market prices. Export comprises excess electricity produced from CHP and wind power, while import covers electricity produced from hydropower in Norway and Sweden. The model includes the influence from a CO_2 emission trading market. Moreover, the model includes a combination of "wet," "dry," and "normal" years with regard to differences in the water content in the hydro reservoirs in Norway and Sweden. The analysis is based on a statistical 7-year cycle and includes the average of a 7-year period. The modeling as well as the data behind the analyses are described in detail in the article.

The model described above has been used to analyze the ability of flexible energy systems to integrate RES, as well as the consequences of this integration. The main purpose of the analysis has been to identify the changes required in the energy system to increase the flexibility and thereby the ability to integrate and utilize more wind power in the system. The EnergyPLAN model has been used to calculate the annual costs of the energy supply of West Denmark, including the exchange of electricity on the Nord Pool market and the cost of investing in flexibility, if any. The input of wind power has been analyzed in relation to an annual wind electricity production ranging from 0 to 25 TWh, equal to 100 percent of the electricity demand in West Denmark.

First, the value of wind power has been calculated as integrated into the reference energy system with no investments in flexibility. Then, the value has been calculated together with different kinds of flexibility investments. In Fig. 5.24, the annual income of wind power including CO_2 payment has been calculated for each of the three different years: wet year, normal year, and dry year. Also, the average of a 7-year period is shown in the diagram. The result is the net income of Danish society of different levels of annual wind production assuming that the system seeks to optimize exchange

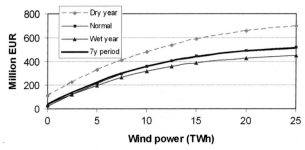

Fig. 5.24 Annual income (including CO_2 payment) from trading wind power on Nord Pool.

on Nord Pool. The net income is shown as the extra income when compared to the situation with no wind and "no trading," in other words, running the system with as little import/export as possible. No investment costs are included in Fig. 5.24. The diagram illustrates how market exchange in combination with wind power is an advantage for Denmark in all situations, but the net profit achieved by a marginal expansion of wind power decreases with the increase of annual wind power production.

In Fig. 5.25 (top), the curve of the average of the 7-year period is shown including investment costs of wind power (20-year lifetime, 5 percent net interest rate). The optimal wind power investment is given for the maximum value of the curve. The location of this point is better illustrated in Fig. 5.25 (bottom), in which marginal net profit (without investments) is compared with marginal production costs, including investments in wind power expansion. Fig. 5.25 is based on the reference assumption of an

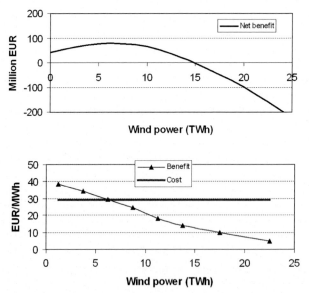

Fig. 5.25 Net profit (top) and marginal cost benefit (bottom) of wind power (per year).

international CO_2 price of 13 EUR/ton and, as illustrated, a wind power production cost of 29 EUR/MWh.

It has been analyzed whether investments in new large steam turbine power stations are feasible in such a system. The analysis concludes that the possibilities of increasing the profit of trading are very limited, and the additional income is far from sufficient to repay the investment. As illustrated, the value of wind power in the reference energy system decreases rapidly as wind power investment increases. The marginal profit gained will soon be lower than the marginal costs of installing new wind power capacities, even when a CO_2 payment of 13 EUR/ton is included. The rapid decrease illustrates the problems of integrating variable electricity production from renewable sources when the reference energy system is not regulated for the purpose of integration. Consequently, the following alternative systems have been analyzed in which small additional investments have been made to improve the flexibility of the system to which the wind turbines are connected:

- *regCHP*: CHP production is replaced by boilers in situations of excess production and in periods when Nord Pool electricity prices are low.
- *regCHP + HP*: Corresponds to "regCHP" plus 350 MWe of heat pump capacity used in combination with heat storage capacity to replace boilers, whenever feasible due to low Nord Pool electricity prices.

Fig. 5.26 shows the results, including both investments in wind power and heat pumps. The investments in flexible energy systems seem to be very profitable. In particular, the alternative including heat pumps raises the net income of the Danish energy supply. Thus, the annual net profit (after investment costs are paid) is raised by approximately 5 million EUR/year for a wind power production of 5 TWh/year and increasing to more than 80 million EUR/year for a wind power production of 15 TWh/year. This rise in profits must be compared with the annual costs of integrating heat pumps into the system of approximately 16 million EUR/year including capital costs. Consequently, the internal rate of return of these measures is as high as several hundred percent for a wide range of wind power capacities.

As illustrated, investments in flexibility measures, such as adding heat pumps to the energy system, are feasible when wind power production exceeds approximately

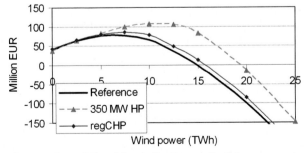

Fig. 5.26 Annual net earnings of electricity trading for two flexible energy systems compared to a reference. Better regulation of small CHP stations (regCHP) and investments in 350-MWe heat pump capacity (350 MW HP) increase the annual earnings of wind power production.

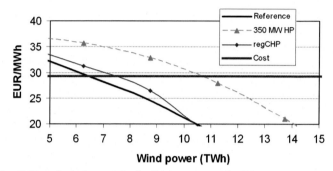

Fig. 5.27 Feasibility of wind power in flexible versus nonflexible energy systems. The net earnings of increasing wind power are higher for flexible energy systems than in the reference. The optimum in which the marginal net earnings cannot exceed the marginal costs is changed from approximately 6 to more than 10 TWh/year.

20 percent of the demand. Moreover, such investments influence the feasibility of wind power, as shown in Fig. 5.27. This figure is comparable to Fig. 5.25 (bottom) and identifies the optimal Danish wind power investment when optimizing the exchange on Nord Pool and including CO_2 payment in the income. Fig. 5.27 illustrates how the optimal wind power investment is higher for the two flexible energy systems than for the reference.

Given a production price for new wind power capacities of 29 EUR/MWh and a CO_2 payment of 13 EUR/ton, the optimal investment in the reference system is 6–7 TWh, equal to 25 percent of the demand. Meanwhile, the same optimum is 30 percent if CHP units are included in the regulation (regCHP) and above 40 percent if heat pumps are included.

A comprehensive sensitivity analysis of the preceding feasibility study was conducted including the following parameters:

- A 50 percent increase in the investment costs of heat pumps
- Changes in CO_2 payment (between 0 and 33 EUR/ton)
- Changes in wind power production costs (between 23 and 37 EUR/MWh)
- Changes in fuel costs
- Changes in marginal CO_2 savings on the Nord Pool market
- Changes in the influence of CO_2 reductions on Nord Pool spot prices
- Changes in future average price on Nord Pool from 32 to 40 EUR/MWh
- Change in import/export to Germany
- Change in the range of Nord Pool price variations (more volatile prices)

From this sensitivity analysis, it can be seen that the feasibility of new wind power investments was very sensitive to, especially, CO_2 payment and wind power production costs, as illustrated in Fig. 5.28 for the reference system.

The same influence was found for the flexible energy systems; however, the feasibility of these systems is better in general. Two factors have proven to be resistant to any changes in the assumptions: the high feasibility of investing in flexible energy systems (such as heat pumps) whenever wind power exceeds 20 percent of annual

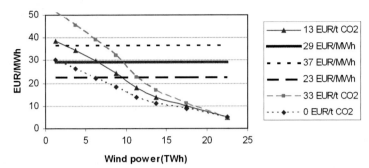

Fig. 5.28 Sensitivity analysis. The feasibility of wind power in flexible versus nonflexible energy systems depends mainly on CO_2 emission costs and wind power investment costs.

electricity production and the fact that flexible energy systems also improve the feasibility of wind power.

8   Integration of transportation[h]

This section is based on Lund and Münster's (2006b) article "Integrated Transportation and Energy Sector CO_2 Emission Control Strategies," which illustrates and quantifies the mutual benefits of integrating the transportation and energy sectors in the analysis of the Danish energy system. As shown earlier in this chapter, this issue is very relevant in relation to the large-scale integration of renewable energy. In short, the energy sector can help the transportation sector replace oil by renewable energy and CHP, while the transportation sector can assist the energy sector in integrating a higher share of variable RES and CHP.

Again, the reference energy system applied is the system of West Denmark in the year 2020. To investigate the impact that a potential electrification of part of the transportation system would have on the existing electrical system, two scenarios were defined for the year 2020. One has already been analyzed in the previous sections: the study that concluded that the technical performance, particularly the range, of BEVs and HFCVs will gradually improve in the coming decades, making it feasible to replace a substantial part of the transportation task of passenger cars and small delivery vans below 2 tons by these types of cars. Fuel cell cars powered by synthetic fuels like methanol were left out because the study expected these to have a poorer overall efficiency.

The scenario shown in Fig. 5.29 is scaled down to correspond to West Denmark only. The scenario assumed that by 2020, 27 percent of passenger cars and small vans based on an internal combustion engine (ICE) would be replaced by battery cars, while 14 percent would be substituted by HFCVs. The batteries of the cars were assumed to be large enough to level out consumption on a 24-hour basis (loading during the

[h] Excerpts reprinted from *Transport Policy*, 13/5, Henrik Lund & Ebbe Münster, "Integrated Transportation and Energy Sector CO_2 Emission Control Strategies", pp. 426–433 (2006), with permission from Elsevier.

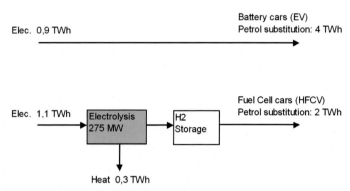

Fig. 5.29 Transport scenario #1: BEVs in combination with HFCVs.

night), while the combined hydrogen storage of the electrolyzer stations and cars was assumed to level out consumption on a 4-week basis. The electrolyzers were dimensioned to operate approximately 4000 hours/year. The heat produced by the electrolyzers was not considered in the model. If the electrolyzers were placed close to CHP stations, the heat produced in periods when the system needed to increase electricity consumption would have a positive effect on the balance of the grid. If the produced heat was used by the district heating network, the CHP station would have a lower heat and thus electricity production. However, this effect was not included in the analysis.

An alternative scenario based on the liquid fuels (biofuels and synthetic fuels) that ICE cars use is presented in Fig. 5.30. This scenario was based on the REtrol-vision[i] of the Danish power company ELSAM, now DONG Energy (ELSAM, 2005). It was

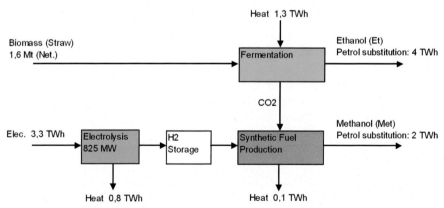

Fig. 5.30 Transport scenario #2: Ethanol and methanol in vehicles using internal combustion engines.

[i] Translated from Danish: *VEnzin*.

scaled to provide the same gasoline substitution as scenario #1. As can be seen, this scenario had a lower overall efficiency, but it could not be directly compared to scenario #1 because it assumed the use of ICEbased cars, which were either standard cars (low percentage mix of ethanol or methanol with gasoline) or slightly converted cars (higher percentage mix). Hence, the total costs of the system, including conversion of the fleet, were much lower. In this case, the heat balance was negative because the heat produced by the ICE of the cars was not considered. The consumed heat was provided as waste heat from condensing power stations. An important asset of this scenario is the fact that ethanol fermenters produced the carbon needed for the production of methanol. In this way, the total system, including the cars, could be regarded as CO_2 neutral. Apart from ethanol, the fermenter produced a solid biofuel. This fuel was subtracted from the biomass input. As in scenario #1, the electricity consumption of the electrolyzers was assumed to be flexible to the extent that it was leveled out on a 4-week basis.

In this article, the different alternatives are first compared in an excess electricity production diagram, as shown in Fig. 5.31. The "Ref" curve shows how most of the wind power electricity in 2020 would have to be exported from West Denmark, if the reference regulation method described in the previous sections was used. Please note that 25 TWh of wind power corresponded to 100 percent of the electricity demand of West Denmark. If 350 MW electric heat pumps were established at the CHP stations and the alternative regulation method was used (HP 350 MW), the situation would improve considerably.

If transportation scenario #1 (EV/HFCV) was introduced instead, it would have more or less the same effect. Transportation scenario #2 (Et/Met) had a larger impact because it used more electricity. A combination of heat pumps and scenario #2 was only marginally better, because the minimum fraction needed by the power stations for stabilizing purposes put a limit to the regulation possibilities.

If this constraint was eased by assuming that 50 percent of the wind turbines were supplied with advanced high-voltage semiconductor regulation equipment and were

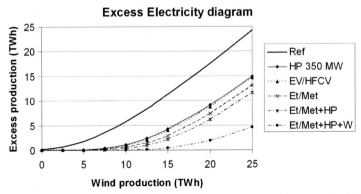

Fig. 5.31 Excess electricity diagram comparing the two transportation scenarios (EV/HFCV and Et/Met) with and without heat pumps (HP) to the reference system (Ref).

thereby able to perform phase and frequency regulation, the situation would again improve considerably (Et/Met+HP+W). This type of equipment is available today and is considered economically feasible for the very big off-shore turbines that are to be established in the future. It is particularly relevant to the combination of wind turbine and electrolyzer because this combination can perform both upward and downward regulations when both parts are active (Østergaard et al., 2004). The economic impacts of converting the entire fleet have not been calculated, but the positive effects on the economy of such a conversion of the energy system as a whole have been evaluated.

Fig. 5.32 shows the influence of the alternatives on the feasibility of wind power and the marginal benefits of adding extra wind power to the system. In the diagram, it is seen how the optimal share of wind power production increased with the flexibility of the system. Optimal wind power production moved from about the previous level of the reference system to 40 and even 50 percent of the demand when heat pumps and transportation electrification were assumed. Scenario #2 (Et/Met+HP) had the highest optimal value because it used more electricity than scenario #1 (EV/HFCV + HP).

Thus, we can see how the use of electricity for transportation increased the optimal amount of wind turbines in West Denmark. While the establishment of 350 MWe heat pumps at the CHP stations led to an increase of this optimum from approximately 25 percent to approximately 40 percent in 2020, the additional electrification of the transportation fleet further increased the optimum to approximately 50 percent.

The article includes calculations of CO_2 balances, which show that the two scenarios result in savings of approximately 1 Mt CO_2 per year for West Denmark. If the indirect CO_2 savings achieved in the neighboring countries by exporting electricity are considered, even larger savings (~ 2 Mt) can be gained with the previously mentioned facilitation of an increase in the number of wind turbines.

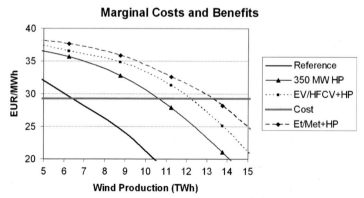

Fig. 5.32 Marginal costs and benefits of wind power for West Denmark 2020. Compared to the reference, the net earnings of increasing wind power are higher if one of the two transportation scenarios is implemented. The optimum in which the marginal net earnings cannot exceed the marginal costs of wind power is changed from approximately 6 to between 10 and 14 TWh/year.

9 Electric vehicles and V2G[j]

This section is based on Lund and Kempton's (2008) article "Integration of Renewable Energy into the Transport and Electricity Sectors through V2G," which adds to the preceding analyses of the transportation sector by including vehicles to grid (V2Gs), that is, vehicles supplying electricity to the grid. Plug-in electric vehicles (EVs) can reduce or eliminate the use of oil for the light vehicle fleet. Adding V2G technology to EVs can provide storage possibilities, matching the generation and loading time.

In the article, two national energy reference systems were selected: the projections of the energy system in West Denmark by the year 2020 and a joint system that included all of Denmark. The projections used in the previous sections of this chapter included only analyses of the electricity system. Here, for a more integrated energy system view, data for the rest of the energy sectors, including the transportation sector, have been added on the basis of the before mentioned Danish governmental energy plan, Energy 21.

The Danish reference case, with its high share of CHP, is not typical for most countries. Therefore, a non-CHP reference was defined simply by replacing all CHP in the Danish system by heat production from district heating thermal boilers and electricity production from condensing power stations. The second national reference system was set at the same total size as the Danish energy system, for comparison purposes. The modeling of transportation demands was based on Danish statistics from 2001, in which the vehicle fleet consisted of 1.9 million combustion cars driving an average of 20,000 km/year and in total consuming 2700 million liters of gasoline, equal to 25.5 TWh/year. The reference combustion vehicle fleet (REF) was compared to four electric vehicle alternatives:

- BEV: battery electric vehicles, with night charge
- InBEV: intelligent battery electric vehicles
- V2G: vehicle to grid cars
- V2G+: vehicle to grid cars with a battery three times larger than normal

Except for the combustion case, all are referred to as EVs. All EVs, except V2G + (discussed later), were assumed to have a battery capacity of 30 kWh and a grid connection of 10 kW. The EVs had an efficiency of 6 km/kWh and consumed 3333 kWh/year to drive 20,000 km. Based on these statistics and assumptions, the reference fleet and three alternative vehicle fleets were defined, as shown in Table 5.3.

The night charge BEV was assumed to charge during the night, starting after 4 p.m., when it was plugged in, and continuing slowly until the battery was fully charged. Unlike the night charge, the InBEV and V2G charging was based on signals from the electric system, as described in detail in Lund and Kempton's (2008) article. The InBEV recharged as much as possible when excess power was available. The

[j] Excerpts reprinted from Energy Policy, 36/9, Henrik Lund & Willett Kempton, "Integration of Renewable Energy into the Transport and Electricity Sectors through V2G", pp. 3578–3587 (2008), with permission from Elsevier.

Table 5.3 Input parameters of transportation reference case and three alternatives.

	REF reference	BEV night charge	InBEV intelligent charge	V2G
Number of vehicles	1.9 million	1.9 million	1.9 million	1.9 million
Average use	20,000 km/year	20,000 km/year	20,000 km/year	20,000 km/year
Vehicle efficiency	14 km/liter	6 km/kWh	6 km/kWh	6 km/kWh
Gasoline consumption	25.5 TWh/year	–	–	–
Electricity consumption	–	6.33 TWh/year	6.33 TWh/year	6.33 TWh/year
Charging capacity	–	19 GW	19 GW	19 GW
Battery storage	–	57 GWh	57 GWh	57 GWh
Discharging capacity	–	0	0	19 GW

V2G also did this, which additionally supplied the grid with power when the production from power stations, wind turbines, or running CHP stations was low. The aggregated national demand for transportation was based on time-specific driving data from the United States.

In both the CHP and the non-CHP system, the impacts of EVs and V2Gs were calculated in the case of wind power ranging from 0 to 45 TWh/year in a national system of the size of Denmark. Wind power of 45 TWh/year would be approximately 100 percent of the foreseen Danish national electricity demand in 2020, including the electric vehicles. This is equivalent to an average power output of 5.2 GW. Figs. 5.33–5.36 show the results of the modeling for the entire energy system.

At the top of Fig. 5.33, the excess electricity production in the CHP system is illustrated. As the fraction of wind power increased beyond 5 TWh, the excess production of electricity increased. Following the dark line for the REF case, at 10 percent wind power (about 5 TWh), there was a little excess production, whereas at 50 percent (22.5 TWh), a substantial fraction (approximately 50 percent) of the wind power produced was excess production. The other lines show that the excess production was reduced successively by BEV with night charge, BEV with intelligent charging, and V2G. In short, the excess production decreased partly because cars were added as load, themselves increasing the electricity demand from 41 TWh/year to approximately 47 TWh/year. Additionally, each refinement of the vehicle fleet successively reduced excess production. The combined reductions were significant. For example, in the 50 percent wind scenario, the change from reference combustion fleet to V2G reduced excess electricity production by 50 percent.

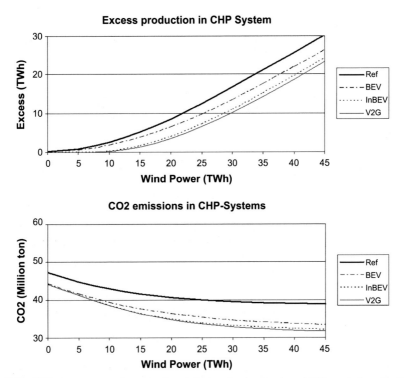

Fig. 5.33 CHP system, annual excess electricity production (top), and CO_2 emissions (bottom) as electricity from wind power increases.

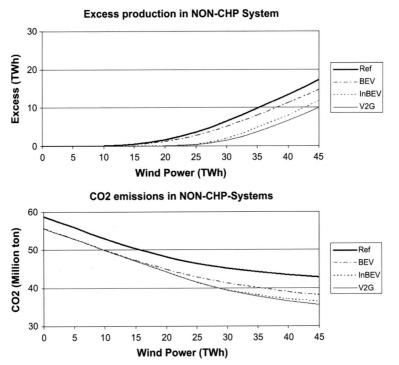

Fig. 5.34 Non-CHP system, annual excess electricity production (top), and CO_2 emissions (bottom) as electricity from wind power increases.

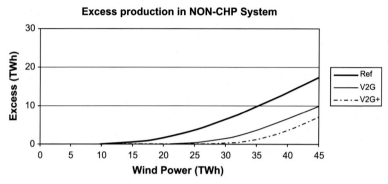

Fig. 5.35 V2G impact on annual excess electricity production in the non-CHP system compared to a system with three times higher battery storage capacity (V2G+).

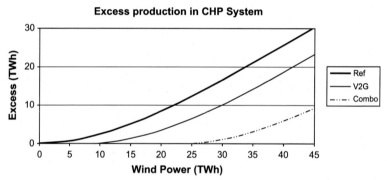

Fig. 5.36 V2G impact on annual excess electricity production in the CHP system compared to a system in which V2G cars are combined with an active regulation of CHP stations, including heat pumps and heat storage (Combo).

The BEV with night charge significantly reduced both excess production and CO_2 emissions (Fig. 5.33, bottom). From the BEV night charge line, the incremental benefit to the InBEV intelligent charging and to V2G was small. These results suggest that the ability of EVs to absorb excess power from wind may be at least as important as their ability to return bulk power at times of need. However, the small additional decrement via V2G was also due to the model assumptions at this point. The "night charge" was actually more intelligent than plug-in vehicles, as the model assumed charging only during the night. Also, the small incremental benefit achieved by using intelligent charging to V2G was partly because the study did not include the ability of V2G to provide regulating power.

The bottom of Fig. 5.33 shows the CO_2 emissions of the CHP system. The solid dark line representing the reference (combustion) vehicle shows that with no electrical vehicles, increasing wind generation reduced CO_2 emissions. However, the slope of CO_2 reduction began to level off at about 10–15 TWh of wind power production and was almost flat at three-quarters of wind power (33 TWh). Again, the addition of night charge BEV and other types of intelligent charging and V2G substantially reduced CO_2 emissions.

The leftmost edge of Fig. 5.33 (bottom), with 0 TWh of wind, depicts the CO_2 impact from the replacement of gasoline-fueled cars by electric cars. The zero shows the reduction of CO_2 emissions achieved by displacing gasoline. The reduction in CO_2 emissions achieved at a level of zero wind power was substantial, even though the power stations were fueled by fossil fuels, due to the much better vehicle efficiencies (one electric vehicle displaces 13,000 kWh of gasoline with 3333 kWh of electricity). This connection between REF and V2G becomes of growing importance as the proportion of wind power increases. As more electricity is produced from wind, an electric fleet will have an increasingly beneficial CO_2 impact.

The right side of Fig. 5.33 demonstrates the REF-V2G values if wind power is 45 TWh/year. These values are almost twice the size of the values shown on the left side, where wind power equals zero. This means that the size of the direct CO_2 reduction achieved by completely eliminating motor fuels was smaller than the indirect effect created by reducing CO_2 from electric generators by using EV and V2G.

Fig. 5.34 shows the results for the non-CHP system, which represents the system of a typical industrialized country, with heat provided by independent devices (in this case district heating) rather than combined with power stations. CHP is more efficient, which means that when wind power equaled zero, the CHP system had considerably lower CO_2 emissions. This effect continued at all wind power levels. On the other hand, excess electricity production began earlier in the CHP system, before reaching 5 TWh of wind (for combustion cars), and the excess production was considerably higher when wind power production was high. This is because electricity production from CHP adds to the excess wind power production.

The addition of EVs had a larger effect, proportionally, in the non-CHP system. With V2G, even at high wind fractions—for instance, three-quarters of the electricity or 34 TWh produced from wind—the CHP system still had 12 TWh of excess production, while only 4 TWh of excess production was found in the non-CHP system. The introduction of all types of EVs led to a reduction in both excess production and CO_2 emissions for high shares of wind. However, the influence did not completely erase excess power or CO_2. The lines phasing out toward the right mean that the beneficial effects of the EVs were reduced as the fraction of wind became higher.

One important factor is the limitation of the capacity in the battery storage. In Fig. 5.35, the impact of a V2G fleet with base characteristics is compared to the impact of a V2G fleet with a storage capacity three times higher (here called V2G +), in other words, 90 kWh/vehicle or 171 GWh altogether. A 90 kWh vehicle later on became normal. As seen in Fig. 5.35, this increase in the storage capacity significantly reduced excess production (and CO_2 emissions, not shown here).

In the preceding results, CHP stations are not included in the regulation of wind power, and small CHP stations do not contribute to the fulfillment of maintaining grid stability. To examine the effect of including CHP stations, the same analysis was conducted for alternative scenarios, in which CHP stations participated in the regulation and heat pumps and heat storage capacities were added to the system. As shown in the previous sections, heat pumps in combination with heat storage and CHP have proven to be a very efficient technology in terms of integrating wind power. Consequently, these investments decreased excess production as well as CO_2 emissions in general. Fig. 5.36 shows how the combination of V2G with an active regulation of CHP stations, including heat pumps and heat storage together (here called "Combo"), significantly reduced excess electricity production with even very high shares of wind power.

Altogether, these analyses show that EVs with night charging and, to a larger extent, V2G improve the efficiency of national energy systems, reduce CO_2 emissions, and improve the ability of the systems to integrate large shares of wind power. Moreover, V2G can be combined with other measures, such as heat pumps and active regulation of CHP stations, in such a way that they together form a coherent solution to the large-scale integration of wind power into renewable energy systems.

It should be noted that the assumptions in this first application of the national energy model to V2G are conservative in the sense that they probably underestimate the value of V2G. First, the most important advantage of V2G over simple EVs is its potential to completely replace regulating power stations and thereby provide grid stability (both voltage and frequency). This benefit was not included in the above analysis. However, as part of the CEESA study (see Chapter 7), Pillai and Bak-Jensen (2011) subsequently showed how V2G could provide a significant contribution in terms of providing grid stabilization. Second, it was assumed that the V2G controllers did not relate to their drivers' operating schedules and were thus required to fully charge the battery each morning. In this way, more power station operation was required during nights with little excess electricity, and less battery capacity was available during the day to absorb this electricity.

10   Electricity storage options[k]

This section is based on Lund and Salgi's (2009) article "The Role of Compressed Air Energy Storage (CAES) in Future Sustainable Energy Systems," which focuses on CAES technologies as one example of introducing electricity storage technologies in the energy system with the aim of improving the large-scale integration of RES. Moreover, the article includes a comparative study of other relevant electricity storage options and thereby serves as a general study on electricity storage technologies.

The primary purpose of the study was to examine the feasibility of CAES stations when applying load level to improve the integration of wind power into regional or national energy systems. Part of the analysis is to compare CAES stations to other technologies with the same aim. In its nature, this analysis, of course, depends on the energy system in question and especially on the share of wind power and other

[k] Excerpts reprinted from *Energy Conversion and Management,* 50/5, Henrik Lund & Georges Salgi, "The Role of Compressed Air Energy Storage (CAES) in Future Sustainable Energy Systems", pp. 1172–1179 (2009), with permission from Elsevier.

production units with limitations or restrictions in the system, such as, for example, distributed CHP stations. In this case, the starting point for the evaluation was a business-as-usual extension of the present Danish energy system into 2030, as conducted by the Danish Energy Agency. By using the EnergyPLAN model, the system was analyzed, applying different shares of wind power and leading to low or high amounts of excess electricity production.

First, an infinite CAES station was added to the system. The analysis showed that when the wind power share was up to 59 percent (wind production compared to electricity demand), the infinite CAES station was able to remove all excess electricity production and use all stored energy to replace the power production of non-CHP power stations. With a wind power share above 59 percent, even the infinite CAES station was not able to use all of its excess production for the simple reason that the power stations did not produce enough power that could be replaced by the wind turbine production. Similarly, the infinite CAES station can replace all non-CHP power production in the system when the wind power share is down to 59 percent. Below 59 percent, the lack of excess electricity production will set the limit on filling the storage. In other words, in systems with low wind power shares, the lack of excess power sets the limitations, and in systems with high wind power shares, the lack of non-CHP power production sets the limitations on a full utilization of CAES stations. In the Danish system as expected for the year 2030, the optimal use of CAES is found to be a wind power share of 59 percent.

In the next analysis, the infinite CAES station was replaced by a 360 MW station described in detail in the article and analyzed in the optimal situation of 59 percent wind power. The results showed that the limited capacity of the specific station significantly reduced the influence of the CAES station. The reason for this is illustrated in Fig. 5.37.

Fig. 5.37 shows the duration curves of excess electricity (to the right) and non-CHP power production (to the left) and compares these to the resulting annual turbine

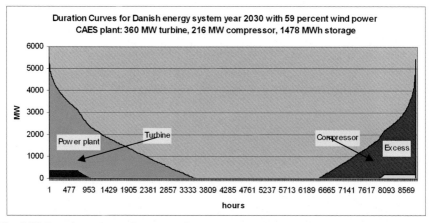

Fig. 5.37 Duration curves of power station and excess electricity production compared to CAES turbine and compressor contributions.

and compressor consumption and production, respectively, when modeled in EnergyPLAN with the aim of reducing excess and power station electricity production. As seen from the diagram, the influence of the CAES station is marginal. The capacity of the share of excess electricity produced is much higher than that of the compressor, and, consequently, most of the excess production cannot be absorbed by the compressor. The problem is the same with regard to power production and turbine capacity. Also, the storage capacity sets a limitation, which can be identified by conducting the same analysis of a CAES station with infinite storage. However, even in this situation, the influence of the CAES station is marginal.

It is important to understand that the reason for this marginal influence from CAES—and for that matter from any other electricity storage technology—is based on the nature of the variations of RES productions. The nature dictates relatively high power outputs from the RES during relatively few hours, as shown by the duration curves in Fig. 5.37. Thus, the situation requires high compressor and turbine capacities and, consequently, high investments that can only be utilized in a few hours. Moreover, the combination of the amounts of energy and the time span in which the energy needs to be stored requires a huge storage capacity that again will be filled and emptied relatively few times.

However, even though the influence of CAES with regard to using excess production is marginal, the CAES station itself may be economically feasible for the system. The feasibility was assessed by identifying the value of the fuel saved in the system. The results are shown in Fig. 5.38 in terms of annual net operational income of the CAES station. The net income was calculated as the difference between, on the one hand, the values of variable operation and excess electricity costs and, on the other, the fuel and variable operation costs saved in the system. Three different sets of fuel prices were used, as expressed by the three different oil prices of 40, 68, and 96 USD/barrel. If the excess electricity produced is not used in CAES systems, one alternative is to sell it on the Nordic electricity market. In Fig. 5.38, three different selling prices have been analyzed: 0, 13, and 27 EUR/MWh, respectively.

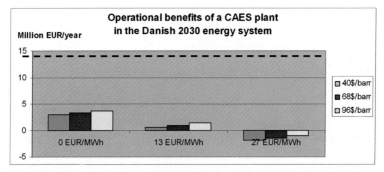

Fig. 5.38 Net operational income of a CAES station in the Danish year 2030 energy system compared to the annual investment and fixed operation cost of 14 million EUR/year (the dotted line).

In Fig. 5.38, the annual net operational income is compared to the investment and fixed operational cost per year of 14 million EUR (marked by a dotted line). As shown, variations in the fuel price mean little to the results because CAES stations burn natural gas, and, consequently, fuel savings at the power stations are counterbalanced by fuel costs at the CAES station. The price of excess electricity production is important. However, even if the excess power produced is free, annual net earnings are far from the level of annual investment and fixed operation costs.

The system economic feasibility of CAES stations was compared to other technologies that may also use excess electricity production and contribute to a better system integration of wind power. All alternative technologies were designed in such a way that they all had the same annual investment and fixed operation costs as explained previously, namely, 14 million EUR/year. The cost components were based on Mathiesen and Lund (2009) and explained in Lund and Salgi's (2009) article, and the results are shown in Fig. 5.39. The results in Fig. 5.39 apply an oil price of 68 USD/barrel and an excess electricity price of 13 EUR/MWh. However, only minor changes were achieved by applying other price levels, apart from the electric boiler, which was especially sensitive to the price of excess electricity.

It can be seen from Fig. 5.39 that CAES stations are not feasible, and other options are significantly more attractive. It must, however, be emphasized that this analysis was solely concerned with the benefits of better operation, including the integration of wind power. In one important aspect, the technologies are not comparable; the CAES and the hydrogen/fuel cell technology add production capacities to the system, which the other options do not.

Consequently, CAES may reduce the need for backup capacity and thereby save investment costs in the system. A sensitivity analysis was made in which the saved investment costs were included, resulting in up to 1 million EUR/MW of steam turbines with a lifetime of 30 years. In this analysis, the CAES station became

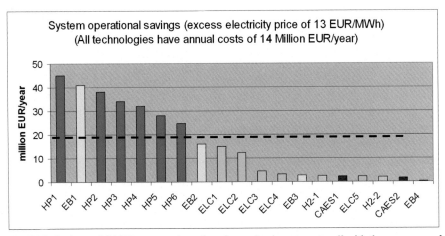

Fig. 5.39 Savings of CAES compared to other alternative investments, all with the same annual investment and fixed operation costs of 14 million EUR/year (the dotted line).

economically feasible, even though it still could not compete with some of the other technologies.

With regard to the CAES station, it is concluded that these stations cannot alone solve the problems of excess electricity production, and that other options are significantly more attractive. However, if CAES stations can save investments in power station capacities in the system, the CAES technology has economically feasible potential.

With regard to electricity storage technologies in general, it is concluded that these technologies have only a minor influence on the large-scale integration of renewable energy and that the economic feasibility is in general very low. In particular, when compared to conversion technologies, such as heat pumps in combination with heat storage capacities, which have a major impact, electricity storage technologies in general cannot compete. If these technologies are to be competitive, they must provide other benefits such as saving power station capacity and/or supporting grid stability.

11   Reflections

The reflections and conclusions on the different analyses of large-scale integration of renewable energy presented in this chapter are drawn with regard to principles and methodologies as well as the practical implementation of renewable energy systems in Denmark and in other countries.

Principles and methodologies

As mentioned at the beginning of this chapter, the discipline of analyzing the large-scale integration of renewable energy into existing systems must address the challenge of redesigning the systems based on the characteristics of the variable renewable sources. The systems must be designed in such a way that they are able to cope with the variable nature of renewable energy sources, especially with regard to the electricity supply.

From a methodological point of view, this raises the problem of how to deal with variations in production of the same installed capacity of RES, such as PV, wind, wave, and tidal power, varying from one year to another. One may be able to design a suitable energy system to cope with the variations of last year, but how can one be sure that such a system will also be able to meet the variations of the present and coming years in the same efficient way? When comparing two alternative investments in flexible energy systems, how can one be sure that the best option considering last year's variations will also be the best way to cope with the variations of coming years?

To deal with these problems, excess electricity production diagrams have proven to be a suitable and workable methodology. In these diagrams, each energy system is represented by only one curve that shows the ability of the system to integrate variable productions of RES, representing several years of different hour-by-hour variations. The diagrams have proven usable for PV, wind, and wave power.

Moreover, the analyses show that in the design of systems suitable for the large-scale integration of RES, one should distinguish between two different issues. The first issue is concerned with *annual amounts* of energy; that is, on an annual basis, the supply of each of the amounts of electricity, heating, and fuel must meet the corresponding demands. The other issue is concerned with *time*; that the supply of the same amounts of energy must meet the timely duration of the corresponding demands. The latter is of special importance to the electricity supply.

This distinction between amounts and time is important, since one cannot solve the problem of time if the amounts produced do not match the demand. One cannot use a storage capacity efficiently if the electricity production on an annual basis significantly exceeds the demand. Under this condition, one may be able to fill the storage but not empty it again. Consequently, when designing energy systems that are suitable for the large-scale integration of RES, one should first design the system in such a way that a balance of the annual amounts can be created and *then* look into the time problem; that is, design the system in such a way that the hourly distribution of especially the electricity demands can be met.

As a result, it is important to distinguish between *conversion* technologies and *storage* technologies, as described in the definitions in Chapter 1. With regard to recommendations, a general understanding can be drawn from the analyses. Technologies converting electricity into thermal and transport needs, such as heat pumps and electric vehicles, typically improve the efficiency of the system, at the same time as they entail the possibilities of cost-effective and efficient storage options. On the other hand, "pure" electricity storage technologies, such as CAES and hydrogen/fuel cell systems, only contribute marginally to the integration of variable RES and also have a low feasibility.

12   Conclusions and recommendations

The large-scale integration of renewable energy should be considered a way to approach renewable energy systems. The integration of RES must be coordinated with energy conservation and efficiency improvements, including the use of CHP and the introduction of fuel cells. All these measures each improve the fuel efficiency of the system. However, they also add to the electricity balancing problem and contribute to the excess electricity production.

The point is that RES should not be regarded as the only measure when conducting analyses of large-scale integration. The long-term relevant systems are those in which these measures are combined with energy conservation and system efficiency improvements. In that respect, the Danish energy system with a high share of CHP can be regarded a front runner and a system well suited for the analysis of the large-scale integration of renewable energy. In systems with a high share of CHP (in this chapter represented by the Danish energy system), excess electricity production can best be dealt with by giving priority to the following technologies:

1. CHP stations should be operated in such a way that they produce less when the RES input is high and more when the RES input is low. When including heat storage capacity, these

measures are likely to integrate variable RES up to 10–20 percent of the demand without losing fuel efficiency in the overall system. After this point, the system will begin to lose efficiency as heat production from CHP units is replaced by thermal or electric boilers.

2. Heat pumps, and maybe additional heat storage capacity, should be added to the CHP stations and operated in such a way that further RES can be efficiently integrated. These measures will allow the integration of up to 40 percent of variable RES into the electricity supply without losing overall system efficiency. The economic feasibility of the investments in heat pumps proves very high for Danish society. Moreover, the investment in wind power is substantially improved.

3. Electricity should be used in the transportation sector, preferably in electric vehicles. This measure will serve as an efficient improvement of the integration of variable RES.

4. In general, it is not beneficial to include electricity storage capacity in the preceding steps. This storage capacity is both inefficient and expensive compared to the benefits that may be achieved. Moreover, the nature of variable RES dictates the need for high capacities of both conversion units and storage in combination with a low number of full load hours. Thus, the electricity storage technologies call for high investments in combination with low utilization. If these technologies are to be competitive, they should provide further benefits, such as saving power station capacity and/or securing grid stability.

5. It is not necessary to include flexible consumer demands of the "classic" electricity demands in the regulation. The use of such a measure raises the same problems as for electricity storage technologies. The nature of variable RES calls for high energy amounts and longtime spans to such an extent that a realistic flexible consumer demand cannot really do the job. However, this statement only goes for the classic electricity demand and not the future flexible electricity demands of heat pumps, electric vehicles and Power-to-X technologies.

6. It is much more important to involve the new flexible technologies, such as CHP, heat pumps, and the electrification of transportation (batteries and electrolyzers), in the grid stabilization tasks, in other words, to secure and maintain voltage and frequency in the electricity supply. Such involvement becomes increasingly important along with the acceleration of the share of RES.

7. With regard to the relation between technoeconomic analyses in a closed system and economic analyses of the benefits of international electricity exchange, studies show that the flexible technologies that can efficiently integrate variable RES at the same time prove feasible in the sense that they make an economic profit from trading on international markets.

The main focus in this chapter has been on the challenge of integrating variable renewable energy into the electricity supply. However, the implementation of a renewable energy system has at least one additional major challenge, namely, the limitations to biomass resources available for energy and the complexity of the transportation needs. The influence of this challenge and how to analyze it will be one of the main topics of the next three chapters.

Smart energy systems and infrastructures

6

In recent years, a number of new terms and definitions of subenergy systems and infrastructures have been promoted to define and describe new paradigms in the design of future energy systems such as *smart grid, fourth generation district heating, CCUS* and *power-to-X*. All these infrastructures are essential new contributions and represent an important shift in paradigm in the design of future renewable energy strategies. However, they are also all subsystems and subinfrastructures which cannot be fully understood or analyzed if not properly placed in the context of the overall energy system. Moreover, they are not always well defined and/or are defined differently by different institutions.

This chapter introduces the concept of *smart energy systems*. As opposed to, for instance, the *smart grid* concept, which puts the sole focus on the electricity sector, smart energy systems include the entire energy system in its approach to identifying suitable energy infrastructure designs and operation strategies. The typical smart grid sole focus on the electricity sector often leads to the definition of transmission lines, flexible electricity demands, and electricity storage as the primary means to deal with the integration of variable renewable sources. However, as already highlighted in Chapter 5, because of the nature of wind power and other renewable energy sources, they are neither very effective nor cost-efficient. The most effective and least-cost solutions are to be found when the electricity sector is combined with the other sectors, such as the heating sector and/or the transportation sector. Moreover, as will be explained in this chapter, the combination of electricity and gas infrastructures may play an important role in the design of future renewable energy systems.

This chapter begins by discussing the challenges as well as the concepts and definitions of various smart grids and energy systems. Then it presents the results of a list of studies relevant to the understanding of the challenges of the different energy infrastructures and how to meet these. One main point is that these analyses are contextual and, in order to do a proper analysis, one has to define the overall energy system in which the infrastructure should operate. Another main point is that different subsectors influence one another and one has to take such an influence into consideration if the best solutions are to be identified. Moreover, this chapter gives an example of how to apply concrete institutional economics in times of economic crisis. This focus is of particular relevance to investment-intensive technologies with long lifetimes such as infrastructures.

Renewable Energy Systems. https://doi.org/10.1016/B978-0-443-14137-9.00006-1

1   Theory and definitions[a]

In this section, definitions of smart infrastructures are discussed in the light of the subject of this book, i.e., the design and implementation of future renewable energy systems.

Smart electricity grid

As highlighted in Chapter 5, the large-scale integration of renewable energy sources into existing energy systems as well as the implementation of 100 percent renewable energy systems involve the challenge of coordinating variable renewable energy production with the rest of the energy system. Especially with regard to electricity production, meeting this challenge is essential since electricity systems depend on an exact balance between demand and supply at any time. The need for change in the current electricity grid and power design and operation to meet this challenge has been recognized and discussed for several years under different labels.

Rosager and Lund (1986) and Lund (1990) published on the subject as early as in 1986, by which time the idea of a regulation hierarchy was introduced to manage distributed generation without causing feedback in the system. Later, the subject was discussed under the label "Distributed generation" (Lund, 2003a,b), and was also a part of the discussion of individual innovative technological concepts such as vehicle to grid (V2G) as described in Chapter 5. Parallel to the previously mentioned discussion regarding the large-scale electricity grid, for many years similar discussions have been part of the debate on the design of microgrids as well as local, regional, and national energy systems.

In 2005, Amin and Wollenberg wrote a paper called "Towards a Smart Grid." The paper points out that the key elements and principles of operating interconnected power systems were established before the 1960s, i.e., before the emergence of extensive computer and communication networks. Today, computation is used at all levels of the power network. However, coordination across networks is still not being used to its full potential. As Amin and Wollenberg emphasized, practical methods, tools, and technologies are allowing "power grids and other infrastructures to locally self-regulate, including automatic reconfiguration in the event of failures, threats or disturbances." They did not include a definition of smart grid in the paper; however, it can be understood from the context that a *smart grid* is a power network using modern computer and communication technology to achieve a network that can better deal with potential failures.

Later, the discussion of the need for changes in future power infrastructures were related to the "smart grid" concept in a large number of reports and papers. Many of them, such as Crossley and Beviz (2009) and Orecchini and Santiangeli (2011),

[a] The definitions of thermal grids and fourth generation district heating in this chapter are based on valuable inputs and discussions with Professor Sven Werner, Halmstad University in Sweden; Professor Svend Svendsen, Technical University of Denmark; and Robin Wiltshire, Building Research Establishment in Watford, United Kingdom.

argued for the need for smart grids to facilitate better integration of variable renewable energy. Thus, the concept of smart grids is widely used, but even so there is no consensus with regard to the definition. As the promotion of smart grids has been included in several political strategies and research programs, the concept has been defined. However, even though the definitions have many similarities, there are important differences between them. The following are four such definitions:

A smart grid *is an electricity grid that uses information and communications technology to gather and act on information, such as information about the behaviors of suppliers and consumers, in an automated fashion to improve the efficiency, reliability, economics, and sustainability of the production and distribution of electricity.*

US Department of Energy (2012)

Smart Grids ... *[concern] an electricity network that can intelligently integrate the actions of all users connected to it—generators, consumers and those that do both—in order to efficiently deliver sustainable, economic and secure electricity supplies.*

SmartGrids European Technology Platform (2006)

A Smart Grid *is an electricity network that can cost efficiently integrate the behaviour and actions of all users connected to it—generators, consumers and those that do both—in order to ensure economically efficient, sustainable power system with low losses and high levels of quality and security of supply and safety.*

European Commission (2011a)

Smart grids *are networks that monitor and manage the transport of electricity from all generation sources to meet the varying electricity demands of end users .... The widespread deployment of smart grids is crucial to achieving a more secure and sustainable energy future.*

International Energy Agency (2013)

As can be seen, the general essence seems to be the use of information technology on electricity grids. However, it varies slightly if this use concerns only suppliers or both suppliers and consumers and if the ability of the grid has a focus on intelligent or cost-efficient integration, and if the purpose is to raise security and safety or secure a more sustainable energy future or both. However, it is evident that all definitions take a sole focus on the electricity grid.

One important aspect that appears in some definitions and discussions is the bidirectional power flow, i.e., the consumers also produce to the grid. This is different from the traditional grid in which there is a clear separation between producers on the one side and consumers on the other side, resulting in a unidirectional power flow. Consequently, concepts mentioned earlier such as regulation hierarchies, distributed generation, and V2G concepts as well as many microgrids all become smart grids or part of the smart grid concept.

These definitions as well as papers and approaches regarding smart grids all seem to predominantly focus on the electricity sector. Only Orecchini and Santiangeli (2011) emphasized the need for the intelligent management of a complete set of energy forms including electricity, heat, hydrogen, and biofuels.

This chapter emphasizes why smart electricity grids should not be seen as separate from the other energy subsectors and what the integration of the other subsectors means in terms of identifying proper solutions to the integration problem. Two main points can be emphasized: First, it does not make much sense to convert the electricity supply to renewable energy if this is not coordinated with a similar conversion of the other parts of the energy system. Second, this coordination makes it possible to identify additional and better solutions to the implementation of, e.g., smart electricity grids, compared to the solutions identified with a sole focus on the subsector in question.

The subject of this book is the design and implementation of future renewable energy systems and, in this context, the following definition has been chosen:

Smart electricity grids are defined as electricity infrastructures that can intelligently integrate the actions of all users connected to them—generators, consumers, and those that do both—in order to efficiently deliver sustainable, economic, and secure electricity supplies.

Smart thermal grids (district heating and cooling)

As we will see in Chapter 7, the design of future renewable energy systems is typically based on a combination of variable renewable energy sources such as wind and solar power on the one hand and residual resources such as waste and biomass on the other hand. In relation to waste and biomass, a pressure on the resources is expected due to the environmental impact and future alternative demands for food and material. To ease the pressure on the biomass resources and the investments in renewable energy, feasible solutions to future renewable energy systems involve substantial elements of energy conservation and energy efficiency measures.

District heating has an important role to play in the task of making these scarce resources meet the demands. District heating comprises a network of pipes connecting the buildings in a neighborhood, town center, or whole city, so that they can be served from a centralized plant or a number of distributed heat producing units. This approach allows the use of any available source of heat. Compared with a scenario without district heating, the inclusion of district heating in future renewable energy systems allows the use of combined heat and power (CHP) together with the utilization of heat from waste-to-energy and various industrial surplus heat sources as well as the inclusion of geothermal and solar thermal heat. In the future, these industrial processes may involve various procedures of converting solid biomass fractions into bio(syn)gas and/or different types of liquid biofuels for transportation fuel purposes, among other things.

To be able to fulfill its role in future renewable energy systems, district heating will have to be able to

1. Supply low-energy buildings with low-temperature district heating
2. Distribute heat in networks with low grid losses
3. Recycle heat from low- as well as high-temperature sources and integrate renewable heat sources such as solar and geothermal heat
4. Be an integrated part of the operation of smart energy systems (i.e., integrated smart electricity, gas, and thermal grids)

Consequently, the present district heating system must undergo a radical change into low-temperature district heating networks interacting with low-energy buildings as well as smart electricity grids. In the European Commission's (2011b) Energy 2020—A Strategy for a Competitive, Sustainable, and Secure Energy, the need for "high efficiency cogeneration, district heating, and cooling" is highlighted (page 8). This paper launches projects to promote, among other things, "smart electricity grids" along with "smart heating and cooling grids" (page 16). In previous state-of-the-art papers (Werner, 2005; Persson and Werner, 2011) and discussions (Wiltshire and Williams, 2008; Wiltshire, 2011), these future district heating technologies have been defined as fourth generation district heating technologies and systems (4GDH). Later, the concept was elaborated and further discussed in Lund et al. (2014, 2018) leading to the illustration shown in Fig. 6.1. Werner (2004) defined the first three generations in the following way:

> The *first generation* of district heating systems using steam as a heat carrier was introduced in the United States in the 1880s, and almost all district heating systems established until 1930 used this technology. Today, steam-based district heating is an outdated technology because of high losses and safety reasons. The technology is still used in, among other places, Manhattan and Paris, but replacement programs have been successful in Hamburg, Copenhagen and Munich.
>
> The *second generation* used pressurized hot water as a heat carrier, with temperatures typically above 100 °C. These systems emerged in the 1930s and dominated all new systems until the 1970s. Remains of this technology can still be found in the older parts of the current water-based district heating systems.
>
> The *third generation* was introduced in the 1970s and gained a major share of all extensions in the 1980s. Pressurized water is still the heat carrier, but supply temperatures are usually below 100 °C. This technology is used for all replacements in Central and Eastern Europe and the former USSR. All extensions and all new systems in China, Korea, Europe, the United States, and Canada use this third generation technology.

The development of these three generations has been aimed at lower distribution temperatures, the creation of components requiring less material, and prefabrication, resulting in lower manpower involvement at the construction sites. Following these identified directions, a future fourth generation of district heating technologies should embrace lower distribution temperatures, assembly-oriented components, and more flexible materials.

For obvious reasons, district heating has higher potential in countries with cold climates than countries with warm climates. However, in warm climates, district cooling may be an option and, in some countries, both types of networks and/or a combination

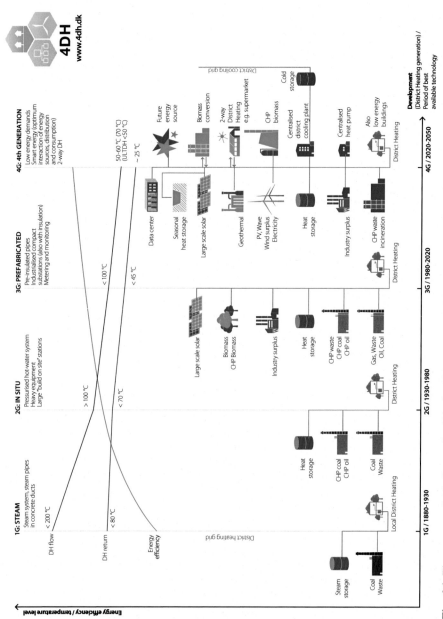

Fig. 6.1 The concept of 4th Generation District Heating (4GDH) compared to previous generations.

would be desirable. In principle, district cooling can be applied in two different ways. One solution is to use a district heating network to distribute heat, which then by individual absorption units is turned into cooling in the individual building. This option is well suited for locations at which both heating and cooling of buildings are required during different seasons of the year. Moreover, the network can be used to supply both cooling and hot water to the building at the same time. The other solution is to produce central cooling and distribute the cold water. This option has the advantage of being able to include "natural" cooling, such as cold water from rivers or harbors. With regard to smart cooling, the challenge in principle is the same as for heating; i.e., to optimize temperature levels and thereby decrease losses in the grid as well as in production.

On this basis, the following definition has been derived.

Smart thermal grids are defined as a network of pipes connecting the buildings in a neighborhood, town center, or whole city, so that they can be served from centralized plants as well as from a number of distributed heating and/or cooling production units including individual contributions from the connected buildings.

The concept of smart thermal grids can be regarded as a parallel to smart electricity grids. Both concepts focus on the integration and efficient use of potential future renewable energy sources as well as the operation of a grid structure allowing distributed generation, which may involve interaction with consumers. Further, both concepts involve the use of modern information and communication technologies such as smart meters.

The two concepts, however, differ in the sense that smart thermal grids face their major challenges in, e.g., the utilization of low-temperature heat sources and the interaction with low-energy buildings, while smart electricity grids face their major challenges in the integration of variable renewable electricity production as well as in securing the reliability and safety of the grid.

4GDH systems are consequently defined here as a coherent technological and institutional concept, which by means of smart thermal grids assists the appropriate development of sustainable energy systems by providing heat supply to low-energy buildings with low grid losses in a way in which the use of low-temperature heat sources can be integrated with the operation of smart energy systems. The concept involves the development of an institutional and organizational framework to facilitate suitable planning, cost and motivation structures.

Since it was defined in 2014, the concept of 4GDH created a common framework for research and industry alike, and pointed to potential futures for district heating which could benefit from low-temperature heating in buildings. The fully developed fourth generation district heating includes the cross-sectoral integration into the smart energy system. Later, Østergaard et al. (2022b) defined four generations of district cooling to make a similar useful framework for district cooling. An illustration of the concept is shown in Fig. 6.2. The first generation with pipelines for brine or compressed refrigerants was introduced in the late 19th century. The second generation was mainly based on large compression chillers and cold water as distribution fluid. The third generation has a more diversified cold supply such as natural cooling, and the fourth generation combines cooling with other energy sectors, sometimes into a renewable energy-based smart energy systems context, including combined heating and cooling.

After its definition, the 4GDH label also led to the publication of several papers on a concept called fifth generation district heating and cooling (5GDHC). This concept and its relation to 4GDH was discussed in Lund et al. (2021b) identifying differences and similarities between 4GDH and 5GDHC regarding aims and abilities. The analysis shows that the two concepts have in common not only the overarching aim of decarbonization, but also to some extent, the five essential abilities first defined for 4GDH. The main driver for 5GDHC has been a strong focus on combined heating and cooling, using a collective network close to ambient temperature levels as common heat source or sink for building-level heat pumps. It is found that 5GDHC can be regarded as a promising technology with its own merits, yet a complementary technology that may coexist in parallel with other 4GDH technologies. However, the term "generation" implies a chronological succession, and the label 5GDHC does not seem compatible with the established labels 1GDH to 4GDH.

Smart gas grids

In terms of implementing future renewable energy systems, existing natural gas grids face similar challenges as the other grids. Moreover, the future energy system may benefit from new gas grids, such as a hydrogen grid, or hydrogen may be integrated into the use of existing gas grids. To understand the specific challenge of such gas grids, two characteristics of the implementation of 100 percent renewable energy systems must be emphasized. One is that the biomass resources available for energy purposes are limited due to demands for food and materials as well as biodiversity. Furthermore, they are limited to such a degree that it is hard to see how biomass alone could cover current energy demands in the transportation sector. The other characteristic is that a transportation system based solely on renewable energy—as we will see later in this and the following chapter—requires some sort of biomass-based gas and/or liquid fuel to supplement the direct use of electricity. The point is that for the sake of transportation, some biomass needs to be converted into either gas (such as methane) or liquid fuel (such as e.g., methanol). Moreover, biomass in the form of gas contributes to achieving better flexibility and efficiencies in future CHP and power plants.

However, not only biomass is relevant to gas production, but electricity in "power to gas" systems may also be highly relevant to boost and supplement the limited biomass resources. Such technologies may have substantial synergies if they are combined with the production of gas from biomass in technologies such as fermentation, gasification, and hydrogenation. These will be elaborated on later in this chapter. Moreover, the whole biomass question—including the important issue of sustainability—also has to be coordinated with the options of achieving negative CO_2 emission—or sinks—using technologies such as CCS and Biochar in order to

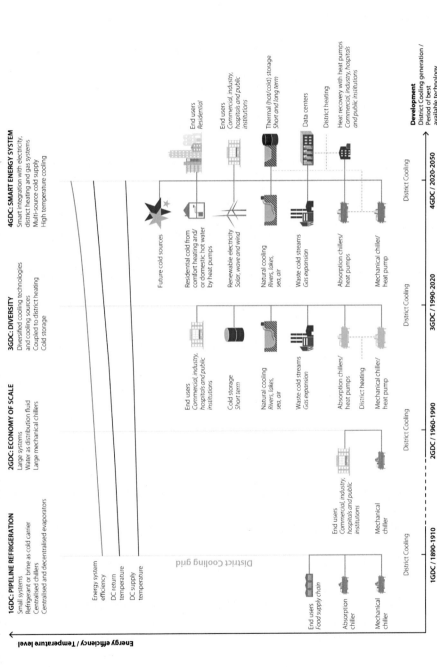

Fig. 6.2 The concept of 4th Generation of District Cooling (4GDC). The major driver for the 4GDC is the cross-sectoral integration into the RES-based smart energy system including the exploitation of the CHC synergy wherever appropriate. Depending on the context, a 4GDC system may not include all the shown components. E.g., in climatic zones with no heat demand, current 2GDC systems may upgrade to 4GDC systems simply by integrating with other sectors and providing flexibility to the smart energy systems using solar, wind, and wave power.

reach the goals of a carbon-neutral or even a carbon-negative society. That will be elaborated in Chapter 8.

The two major challenges of smart gas grids compared to existing natural gas grids are, first, that the smart grid has to deal with a bidirectional flow as opposed to the existing unidirectional flow, and second, that the smart gas grid needs to handle different types of gas with different characteristics including different heating values, which may also call for an additional gas grid, e.g., for hydrogen

On this basis, the following definition has been derived.

Smart gas grids are defined as gas infrastructures that can intelligently integrate the actions of all users connected to it—suppliers, consumers, and those that do both—in order to efficiently deliver sustainable, economic, and secure gas supplies and storage.

As mentioned earlier, the concept of smart gas grids is similar to other smart grids. All concepts focus on the integration and efficient use of potential future renewable energy sources as well as the operation of a grid structure allowing distributed generation, which may involve the interaction with consumers. Further, all concepts involve the use of modern information and communication technologies such as smart meters.

The three concepts, however, differ in terms of their major challenges. *Smart thermal grids* face their major challenge in the temperature level and the interaction with low-energy buildings. *Smart electricity grids* face their major challenge in the integration of variable renewable electricity production. *Smart gas grids* face their major challenge in bi-directional flows and in the efficient use of limited biomass resources. It should be emphasized that the three concepts supplement one another and all of them are to be regarded as necessary in the implementation of renewable energy systems.

Smart energy systems

As illustrated previously, all smart grids are important contributors to future renewable energy systems. However, each individual smart grid should not be seen as separate from the others or separate from the other parts of the overall energy system. First, it does not make much sense to convert, e.g., the electricity supply to renewable energy if this is not coordinated with a similar conversion of the other parts of the energy system. Second, better solutions arise for the implementation of the smart energy system and the individual sectors if their implementation is coordinated.

In other words, there are several synergies connected to taking a coherent approach to the complete smart energy system compared to looking at only one sector. This does not only apply to finding the best solution for the total system, but also to finding the best role to play for each individual subsector. As already highlighted in Chapter 5, one can find better and cheaper solutions to the electricity balancing problem if including, e.g., the heating sector in the analysis compared to only looking at the electricity sector. Such synergies include the following:

- Electricity for heating purposes makes it possible to use heat storage instead of electricity storage, which is both more efficient and cost-effective. Moreover, it provides a more flexible CHP production.
- Electricity for heating may be used for providing electricity system balancing, e.g., in different types of balancing markets.
- Biomass conversion to gas and liquid fuel needs steam, which may be produced in CHP plants, and produces low-temperature heat, which may be utilized by district heating and cooling grids.
- Biogas production needs low-temperature heat, which may be supplied more efficiently by district heating compared to being produced at the plant.
- Electricity for gas such as hydrogenation makes it possible to use gas storage instead of electricity storage, which is cheaper and more efficient.
- Energy savings in the space heating of buildings make it possible to use low-temperature district heating which, in addition, makes it possible to utilize better low-temperature sources from industrial surplus heat and CHP.
- Electricity for vehicles can be used to replace fuel and provide for electricity balancing.

Based on these considerations, the following definition has been made.

Smart energy systems are defined as an approach in which smart electricity, thermal, and gas grids are combined and coordinated to identify synergies between them in order to achieve an optimal solution for each individual sector as well as for the overall energy system.

Smart energy systems theory and tools

The concept of Smart Energy Systems is not only a set of definitions as expressed above; it is also an approach to address the design of an affordable and reliable transition of the energy system to fit into a carbon neutral society. Thus, the concept itself is embedded in a theoretical understanding of the aim and concept which can be expressed in the following important hypotheses:

1. Given the complexity of the situation, one cannot find the best solutions for affordable and reliable transitions of the energy system into a carbon neutral society solely within each subsector of the energy system. One must approach the transition in a holistic and cross-sectoral smart energy system perspective in order to be able to identify the best solution for the overall energy system and for society as a whole.
2. Subsector studies (no matter if they concern the role of a specific technology or the role of a region or country) should aim at identifying the role to play in an overall transition of the whole system, rather than aim at decarbonizing the subsector on its own.

The analysis of smart energy systems calls for tools and models that can provide similar and parallel analyses of electricity, thermal, and gas grids as well as rest of the energy system. Moreover, it should be able to align to the context of a carbon-neutral society, as will be elaborated in Chapter 8. As described in Chapter 4, EnergyPLAN is one such tool, since it can do hour-by-hour analysis of all these grids including the storage and interaction between them.

The following sections describe a number of studies relevant to the understanding of smart energy systems and infrastructures, in which the EnergyPLAN tool has been used.

2 The role of district heating[b]

This section is based on Lund et al. (2010) article "The Role of District Heating in Future Renewable Energy Systems." Based on the case of Denmark, the paper analyzed the role of district heating infrastructure in future renewable energy systems. It defined a scenario framework in which the Danish system was converted to 100 percent renewable energy sources by the year 2060 including reductions in space heating demands by 75 percent. Through a detailed EnergyPLAN energy system analysis of the complete national energy system, the consequences in relation to fuel demand, CO_2 emissions, and cost were calculated for various heating options, including district heating as well as individual heat pumps and micro-CHPs.

The study included the entire heating sector with a focus on the 24 percent of the Danish building stock that had individual gas or oil boilers at the time, and which was located relatively close to existing district heating areas. These could be substituted by district heating or a more efficient individual heat source. As elaborated in the following, in this overall perspective, the best solution will be to combine a gradual expansion of district heating with individual heat pumps in the remaining houses. This conclusion is valid in the reference systems, which are mainly based on fossil fuels, as well as in a potential system based 100 percent on renewable energy.

In many countries around the world, the ability to heat and supply hot water to buildings is essential. It is being discussed intensively how to do so in the best way in future energy systems in which the combustion of fossil fuels should be reduced or completely avoided. In the discussion, one can identify at least two different views. One view states that future low-energy buildings could completely remove the need for heating or even, by the use of, e.g., solar thermal energy, be plus energy houses producing more heat than they demand. The other view states that the excess heat production from industries, waste incineration, and power stations may also be used together with geothermal energy, large-scale solar thermal energy, and large-scale heat pumps to utilize excess wind energy for house heating. Recently, it has sometimes also been expressed that natural gas in individual boilers may simply be replaced by hydrogen. In the first and last case, a district heating network may not be needed, while, in the second case, a district heating network becomes essential.

Regardless of the view adopted, the main point here is that it cannot be concluded from purely a house heating perspective whether one district heating strategy fits

---

[b] Excerpts reprinted from *Energy* 35, Henrik Lund, Bernd Möller, Brian Vad Mathiesen, and Anders Dyrelund, "The Role of District Heating in Future Renewable Energy Systems", pp. 1381–1390 (2010), with permission from Elsevier.

better than the other in terms of implementing future renewable energy systems. One has to include the rest of the energy system to evaluate how to use the available resources in the overall system in the best way, and how to combine energy savings and efficiency measures with renewable energy to eliminate fossil fuels at the lowest possible cost for society. Consequently, Lund et al.'s paper seeks to perform an advanced energy system analysis of the whole national energy system comparing different options of house heating in a traditional reference as well as in future energy systems to evaluate the impact of different heating options on the total fuel demand and CO_2 emissions.

Geographical Information System tools were used to create a heat atlas and identify potential scenarios and the cost of expanding district heating. The methodology and data are described further in Möller (2008) and Möller and Lund (2010). The year 2006 was chosen as a starting point for the study. The total house heating consumption was identified as 60.1 TWh/year, of which 27.9 TWh were supplied from district heating and 32.2 were supplied from individual boilers and heaters. Compared to the situation in 2006 (in the following referred to as the reference situation), the following scenarios of potential expansion of district heating were defined and identified:

- *Scenario 1*: All buildings within areas defined as existing or planned district heating areas are connected to the system, which increases the district heating demand from 27.9 to 31.6 TWh/year.
- *Scenario 2*: All areas supplied by natural gas for individual boilers in the direct vicinity of existing district heating areas are converted to district heating, which increases the share further from 31.6 to 37.6 TWh/year.
- *Scenario 3*: Further natural gas areas of a distance of up to 1 km from existing grids and up to 5 km from large central district heating plants are converted to district heating, which increases the share from 37.6 to 42.3 TWh/year.

The analysis of district heating versus various kinds of individual heating was carried out with regard to the energy system from 2006, as well as future energy systems, leading to a vision of an energy supply based on 100 percent renewable energy. The energy system analysis of the complete Danish energy system was carried out by means of the EnergyPLAN model. First, the energy system model was calibrated in order to adjust it to the output of Danish energy statistics from 2006 as well as a business-as-usual projection made by the Danish Energy Agency (January 17, 2008). Compared to the 2006 situation, the future energy systems include more wind power, heat savings, and better CHP and power plants, etc.

It must be emphasized that the reference scenario by no means represents a comprehensive identification of the optimal solution for a Danish 100 percent renewable energy system, nor a fully carbon-neutral society. These will be elaborated in Chapters 7 and 8. Here, the scenario solely serves as a proper framework for analyzing whether conclusions with regard to district heating in the reference system will also be valid in a probable future 100 percent renewable energy system. The focus is on the framework conditions related to the heat supply, and the scenario is not comprehensive regarding transportation and industrial sectors.

In the analysis, special attention was paid to the hourly modeling of district heating demands in relation to reductions in the demand for space heating. The starting point

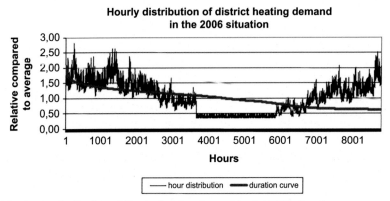

Fig. 6.3 Hourly distribution of district heating demand in the 2006 situation.

was the annual district heating demand in 2006 of 35.77 TWh, divided into a net heat demand of 28.35 TWh and grid losses of 7.42 TWh. This demand was subject to a typical hourly distribution, as shown in Fig. 6.3. In the scenarios of reduced space heating demands, the shape of the duration curve as well as the hourly distribution were adjusted, as shown in Fig. 6.4, representing the case of a 75 percent reduction in the space heating demand. In this case, the grid loss and the demand for hot water were not adjusted in the same way as the space heating demand.

The analysis defined and compared the 10 following heating technologies:

1. *Ref.*: Individual oil, natural gas, and biomass boilers.
2. *HP-gr*: Individual heat pumps using ground heat including electric heating for peak load assuming an average coefficient of performance (COP) of 3.2. (In the case of space heat reductions, the COP decreases due to an increasing share of hot water demands to 3.1 at 25 percent savings, 3.0 at 50 percent savings, and 2.8 at 75 percent savings, respectively.)
3. *HP-air*: Individual heat pumps using air including electric heating for peak load assuming an average COP of 2.6 (in the case of savings reduced to 2.5, 2.4, and 2.3, respectively).

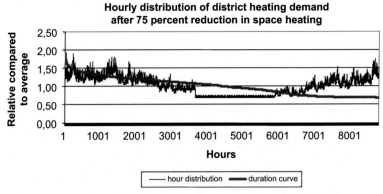

Fig. 6.4 Hourly distribution of district heating demand in a situation in which space heating demands have been reduced by 75 percent.

4. *EH*: Individual electric heating with a COP of 1.
5. *MiCHP*: Individual fuel cell natural gas micro-CHP units with an electric output of 30 percent and a heat production of 60 percent. The CHP unit supplies 60 percent of the peak demand. The rest is covered by a natural gas boiler.
6. *H2-CHP*: Individual micro-CHP units based on hydrogen assuming 45 percent electric output and 45 percent heating output. The CHP unit supplies 60 percent of the peak demand. The rest is covered by a boiler. Hydrogen is supplied via a gas pipeline system and produced on electrolysis assuming an efficiency of 80 percent. The system makes use of hydrogen storage equal to one week's average production.
7. *DH-Ex*: District heating without investment in new production units apart from increasing the capacity of peak-load boilers.
8. *DH-chp*: District heating in combination with an expansion of the CHP capacity at existing CHP plants.
9. *DH-HP*: District heating in combination with adding large-scale heat pumps to the CHP plants assuming a COP of 3.5.
10. *DH-EH*: District heating in combination with adding electric boilers to the CHP plants.

As explained in Chapter 5, the 2006 situation with increasing imbalances in the electricity supply caused by wind power and CHP called for solutions like heat pumps and electric boilers to increase the flexibility of the system. This is the reason for including the different district heating alternatives. These alternatives are only used for the 2006 system. In the future scenarios of 2020, 2040, and 2060, it is assumed that a good balance between CHP units, heat pumps, and peak-load boilers has been implemented and that, consequently, a potential increase in district heating has been followed by a marginal increase in all three types of units.

The cost estimate was made as a socioeconomic calculation excluding taxes and subsidies as explained in Chapter 3. It was based on a simple calculation of saved fuel and maintenance costs compared to additional investment cost by use of a real interest of 3 percent. The cost of individual solutions was based on an estimate of actual prices in Denmark as shown in Table 6.1. The prices applied to a typical average house with a heat demand of 15 MWh/year. The prices shown in the table relate to the 2006 level of heat demand and were reduced in scenarios with reduced heat demands. Electrolyzers for hydrogen production were assumed to be community installations equal to an investment cost of 2700 EUR per household. For heat pumps based on ground heat, the heat pumps had an expected lifetime of 15 years, while the ground heat source pipes had a lifetime of 40 years.

For electric heating and heat pumps, an increased cost of expanding the electric grid was included based on the following estimate:

Investments in low-voltage grids accounted for 0.013 EUR/kWh and the increase in peak-load production was included as an additional demand for transmission and production, corresponding to 1000 EUR/kW for a lifetime of 30 years.

The cost of increasing district heating based on the heat atlas model's calculation of the scenarios is shown in Table 6.2 together with the cost assumption of additional production units that have to be added if district heating demands are increased.

Fuel cost was analyzed on the basis of world market prices plus the cost of transporting the fuels to the relevant end users. Three world price levels were identified equivalent to oil prices of 55, 85, and 115 USD/barrel, respectively. With regard

Table 6.1 Cost of individual heat technologies for a typical house with a 15 MWh/year heat demand.

Heat prod. technology		Unit	Central heating	Storage/ electrolyzer	O&M (fixed) EUR/year	O&M (fixed) percent of investment
Oil boiler	EUR/unit	6000	5400	1300	320	2.5
	lifetime (year)	15	40	40		
Biomass boiler	EUR/unit	6700	5400	1300	380	2.8
	lifetime (year)	15	40	40		
Natural gas boiler	EUR/unit	4000	5400		200	2.1
	lifetime (year)	15	40			
Micro-FC CHP on natural gas	EUR/unit	6700	5400		330	2.8
	lifetime (year)	10	40			
Micro-FC CHP on hydrogen	EUR/unit	6000	5400	2700	270	2.4
	lifetime (year)	10	40	15		
District heating excellent pipes	EUR/unit	2000	5400		70	0.9
	lifetime (year)	20	40			
Electric heating including hot water	EUR/unit	1100	2700		30	0.9
	lifetime (year)	20	40			
Heat pump ground heat	EUR/unit	13,400	5400		110	0.6
	lifetime (year)	15/40	40			
Heat pump air	EUR/unit	6700	5400		110	0.6
	lifetime (year)	15	40			

For scenarios with reduced space heating demand the cost has been reduced.

Table 6.2 Cost of expanding district heating networks and of adding production units.

Unit	Investment MEUR	Lifetime year	O&M (fixed) percent of investment	O&M (variable) EUR/unit
Peak-load boilers	0.15 per MW_{th}	20	3.0	0.15 EUR/MWh_{th}
Small CHP plants	0.95 per MW_e	20	1.5	2.70 EUR/MWh_e
Large CHP plants	1.35 per MW_e	30	2.0	2.70 EUR/MWh_e
Heat pumps	2.70 per MW_e	20	0.2	0.27 EUR/MWh_e
Electric boilers	0.15 per MW_e	20	1.0	1.35 EUR/MWh_e
District heating Scenario 1	1070 in total	40	1.0	0
District heating Scenario 2	4430 in total	40	1.0	0
District heating Scenario 3	10,470 in total	40	1.0	0

to biomass, the prices were assumed to follow variations in coal prices. The analysis used the price level of 85 USD/barrel as a base level with the other two levels added as sensitivity factors. However, in the future 100 percent renewable energy scenario in which no fossil fuels are left, the analysis was based on the high price level assuming biomass prices equivalent to similar types of fossil fuels. Consequently, biomass for individual houses was assumed to have the price of wood chips, while biogas/syngas was assumed to have a price equivalent to that of light oil.

It should be noted that, in such projections, the coal price is expected to be substantially lower than the price of oil and natural gas. In the Danish system, a combination of large coal-based and small- and medium-sized natural gas-based CHP plants are operated at the Nord Pool electricity market, which means that coal will replace natural gas in certain situations.

The cost calculation did not include external costs related to, e.g., pollution and health, apart from a CO_2 emission trade cost of 23 EUR/ton. With regard to the exchange of electricity on the Nordic Nord Pool market, the analysis was, as a starting point, based on the expectations of the Danish energy authorities, which stated that the future average price level would be 47 EUR/MWh in combination with CO_2 trading prices of 23 EUR/ton. In the energy system analysis conducted in the EnergyPLAN model, this average price was distributed on an hourly basis using the hourly distribution of the year 2005 and following the same methodology as described in Chapter 5.

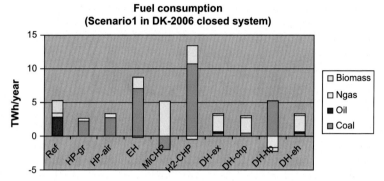

Fig. 6.5 Fuel demands of 10 options of supplying scenario 1 houses with heat.

An analysis was made of the reference system as well as the potential future energy systems of 2020 and 2060. First, a comparison of the consequences of applying the 10 different heating options to scenario 1 is shown. This scenario involved the houses within district heating areas that were not connected to the network when the study was conducted. The resulting fuel demand of each option is illustrated in Fig. 6.5.

As illustrated in Fig. 6.5, the reference (pillar 1, Ref), the houses of scenario 1 were supplied by heat from individual boilers based on oil, natural gas, or biomass. The resulting fuel demand was 5.25 TWh/year. If the supply of all houses was converted to heat pumps (pillars 2 and 3, HP), the resulting fuel demand of the system would be reduced to 2.55 and 2.23 TWh/year, respectively. This conversion would replace the fuel demand in individual boilers by a demand for electricity that would mostly be produced by coal-fired power plants because of the price relation between coal and natural gas. However, this electricity demand would also increase the possibility of utilizing existing CHP plants (coal as well as natural gas) in a better way. In principle, the same would be the case if electric heating supplied all buildings (pillar 4, EH). However, the fuel demand would increase to 8.44 TWh/year due to the inefficiency of electric heating compared to heat pumps. A very small amount of biomass would be saved because CHP plants can be operated for a longer period of time and can save fuel on peak-load boilers, some of which are fueled by biomass.

If the supply of all buildings was converted to micro-CHP units on natural gas (pillar 5, MiCHP), the demand for natural gas would increase and the demand for coal would decrease, since the electricity produced by the micro-CHP units would reduce the production at the coal-fired power stations. Altogether, the net fuel demand was reduced to only 2.95 TWh/year. If, instead, the micro-CHP units utilized hydrogen (pillar 6, H2-CHP), the resulting fuel demand would increase to as much as 12.87 TWh/year because of the demand for electricity for the electrolyzers. Furthermore, the CHP units produced electricity, which would save coal, but the demand for electricity in this alternative by far exceeded the production. It should be noted that both micro-CHP options assume the existence of a gas distribution network, which was not present in most areas of the scenario.

If all buildings were connected to the district heating network in which they were located (some to small CHP plants fueled by natural gas and others to large CHP plants fueled by coal), the general picture is that the fuel demand would decrease. This is a consequence of expanding the use of CHP. If no additional investments were made in production units, except increasing the peak-load boiler capacity, the fuel demand would be 3.20 TWh/year (pillar 7, DH-ex). It could be further reduced to 2.86 TWh/year if a total CHP capacity of 400 MWe was added (pillar 8, DH-chp). If heat pump capacity was added instead (pillar 9, DH-hp), the fuel consumption would be 2.93 TWh/year. However, in such a scenario, due to the price relation between coal and natural gas, the system would save natural gas at the small CHP units and increase the electricity production at the large coal-fired plants. In this case, the investment in electric boilers (pillar 10, DH-eh) would result in the same fuel demand as the investment in peak-load boilers. This was caused by the design of the reference system, which would result in almost no cheap excess electricity production.

In Fig. 6.6, the CO_2 emissions are shown for the same analysis. As can be seen, the overall picture is the same as for the fuel demand. The exception is micro-CHP units based on natural gas, which show a remarkable reduction in CO_2 emissions. This was caused by the combined effect of increasing CHP while, at the same time, replacing coal by natural gas.

In Fig. 6.7, the cost is shown for the same analysis. Again the overall picture is very much the same. The district heating solutions were among the cheapest options and the high cost of district heating networks was not dominating when compared to the total costs of all options.

The same calculations have been made for all three district heating scenarios, and the results show the same overall picture in all cases. However, the cost-effectiveness of district heating decreased along with increased costs in the district heating networks in scenarios 2 and 3 compared to scenario 1. Gradually, the heat pump option became competitive with the district heating solutions.

The next step was to make the same calculations for the 100 percent renewable energy system in the year 2060. The results are shown in Fig. 6.8. Here, the district heating option was calculated for only one solution, since it is expected that a suitable

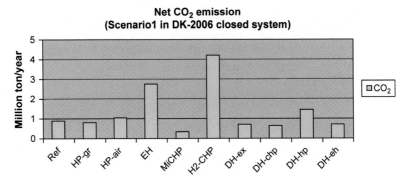

Fig. 6.6 CO_2 emission of the 10 heating options applied to scenario 1.

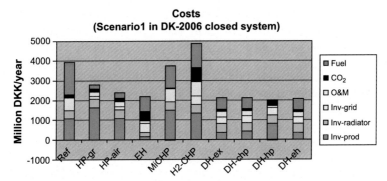

Fig. 6.7 Total annual cost of the 10 heating options applied to scenario 1.

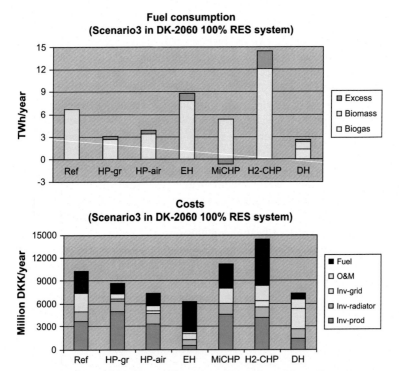

Fig. 6.8 Fuel demands and total cost of different heating options of scenario 3 seen in a future 100 percent RES system year 2060.

combination of heat pumps, CHP units, and peak-load boilers has already been established in the future. Consequently, the district heating option involves a coordinated investment in the expansion of all three types of production units.

In the 100 percent renewable energy system, the expansion of wind power leads to an excess electricity production of 3.2 TWh. The energy system analysis includes the factor that some of the heating options provides a better utilization of the excess production. Moreover, it should be highlighted that the space heating demand has decreased by as much as 75 percent.

Compared to the reference 2006 system, the overall picture for the year 2060 shows that electricity-consuming options (electric heating and heat pumps) improve, while electricity-producing options (CHP) deteriorate. In general, this change is caused by the excess electricity production from wind turbines. It should be emphasized that the ability of the hydrogen fuel cell option depends on the hydrogen storage, which has been defined here as an electrolyzer capacity of 3200 MWe in combination with a hydrogen storage of 200 GWh, corresponding to the production of approximately 14 days. Electric heating seems to have a low cost. However, the fuel demand is high and, consequently, such an option is extremely sensitive to shifting fuel prices. Moreover, this solution puts pressure on the need for biomass and other renewable energy sources.

Consequently, the best solutions again seem to be individual heat pumps and district heating, while individual CHP options do not seem to be desirable neither in terms of fuel efficiency nor from an economic point of view.

All in all, the analysis shows that a substantial reduction in fuel demands and CO_2 emissions as well as cost can be achieved by converting to district heating. This conclusion seems to be valid for the reference energy systems as well as for a future scenario aiming at a 100 percent renewable energy supply in 2060, even if the space heating demand is reduced to as much as 25 percent of the reference demand. However, other options than boilers exist, which have also been analyzed as in the following:

- *Micro-CHP based on fuel cells on hydrogen.* This solution does not seem to be able to reduce fuel demands, CO_2 emissions, or cost in the reference system or in a future 100 percent renewable energy system. The efficiency is simply too low and the cost too high. Moreover, better and more cost-effective solutions can be found to deal with the problem of excess electricity production from wind power and CHP.
- *Micro-CHP based on natural gas* seems to be an efficient way to reduce fuel demands and especially CO_2 emissions in the short term. CO_2 emissions are reduced both by expanding CHP and by converting from coal to natural gas in the overall system. The solution is, however, very expensive compared to district heating because of the substantial investments in micro-CHP units in various buildings. In the long-term perspective, in a 100 percent renewable energy system, the solution is not competitive regarding fuel, CO_2 emissions, or cost reduction compared to district heating and not even compared to individual boilers based on biomass.
- With the high oil and gas prices of 2008 and, at the same time, low coal and Nord Pool electricity prices, *electric heating* is a socioeconomically reasonable alternative, mainly because of the saved central heating system cost. In the short term, this is not valid for houses that already have central heating. In the long-term perspective, electric heating is bad for fuel demands and CO_2 emissions. Moreover, this alternative becomes very sensitive to potential increases in future fuel demands and prices.
- *Individual heat pumps* seem to be the best alternative to district heating. In the short term, heat pumps are at the same level as district heating in terms of fuel efficiency, CO_2 emissions, and cost. The cost is a little higher in areas close to the district heating system

but a little lower in houses farther away. In the long-term perspective, in a 100 percent renewable energy system, the fuel efficiency is high and, regarding cost, the solution is more or less equal to district heating. However, it is highly dependent on the distance to existing district heating grids.

For all the alternative options, it is relevant to add solar thermal energy. However, this option has not been included in the analysis presented here.

In an overall perspective, the conclusion seems to be that the best solution will be to combine a gradual expansion of district heating with individual heat pumps in the rest of the areas. The analysis indicates that the optimal solution will be to expand district heating from the 46 percent of the reference to a level between 63 and 70 percent.

It must, however, be emphasized that the analysis is based on the assumption that a gradual improvement of district heating technologies takes place in accordance with the concept of smart thermal grids. This involves, among other initiatives, a decrease in temperature in combination with a reduction in space heating demands including a reduction in the return temperature of the heat from consumers. Therefore, it is crucial to continue the present development in such a direction. Moreover, the expansion of district heating will help utilize the heat production from waste incineration and industrial excess heat production, which has been included in the analysis. Additionally, district heating helps the integration of geothermal heating, biogas production (supply of heat), and solid biomass such as straw.

In recent years, growing attention has been paid to the substantial future potential of excess heat from power-to-X and data centers. As further detailed in Sorknæs et al. (2022), the harvesting of excess heat from data centers is substantial, but the utilization is strongly influenced by the transition to a 4G district heating system.

3   Economic crisis and infrastructure investments[c]

This section is based on Lund and Hvelplund's (2012) article "The Economic Crisis and Sustainable Development: The Design of Job Creation Strategies by Use of Concrete Institutional Economics." The paper presents concrete institutional economics (see Chapter 3) as an economic paradigm to understand how the request for renewable energy and related infrastructures in times of economic crisis can be used to generate jobs as well as economic growth. In most countries, including European countries, the United States, and China, the implementation of renewable energy solutions involves the replacement of imported fossil fuels by substantial investments in energy conservation and renewable energy. In this situation, it becomes increasingly essential to develop economic thinking and economic models that can analyze the concrete institutions in which the market is embedded. As a follow-up on the previous section, this section presents such tools and methodologies and applies them to the case of the Danish heating sector in terms of a specific investment program for district heating.

---

[c] Excerpts reprinted from *Energy* 43, Henrik Lund and Frede Hvelplund, "The Economic Crisis and Sustainable Development: The Design of Job Creation Strategies by Use of Concrete Institutional Economics", pp. 191–200 (2012), with permission from Elsevier.

The case shows how investments in this kind of infrastructure can be made in a way that has a positive influence on job creation and economic development, as well as public expenditures.

As explained in Chapter 3, the use of choice awareness methodologies with regard to concrete institutional economics involves the following three-step procedure:

Step 1: Analyze the technical scenarios and find the best ones. Concrete institutional economics recognizes that an economic balance of products and production factors such as labor will not be established automatically. Unemployment can develop and persist for years and decades. Deficits on the state budget can increase and lead countries into a debt trap without any automatic processes re-establishing financial balances. Therefore, feasibility studies of energy scenarios should also take into account and evaluate the effects of different technical scenarios on employment and public finances. Furthermore, positive effects on these macroeconomic indicators should be a part of the project evaluation process when looking for the best alternative.

Step 2: Analyze the present institutional and political contexts and find the hindrances to the best technical scenarios. This could be tariff and tax conditions supporting increased energy consumption, ownership design that hinders the local acceptance of wind power projects, the lack of financial possibilities for people who want to improve the energy standard of their houses, tax deduction rules supporting accelerating traffic development, etc. Altogether, these different institutional conditions may identify projects that are not implemented under the present institutional conditions, despite being both economically and environmentally feasible from social and economic points of view.

Step 3: Design the required institutional scenarios to implement the best technical solutions as described in step 1.

To illustrate this theoretical point of departure and these methodologies, a case was used based on the previous section on district heating. As already explained, the case involved the heating of the 24 percent of the Danish building stock which was heated by individual boilers fueled by oil, natural gas, or biomass and which was located relatively close to existing district heating areas. Based on the findings in the previous section, the case concerned the replacement of these boilers by district heating in urban areas in combination with individual heat pumps in the remaining buildings during a period of 10 years. This replacement involved the following:

- Expansion of district heating from 46 to 65 percent of the Danish heat market, equal to 80 percent of the buildings in question
- Individual heat pumps in the remaining buildings, equal to 20 percent of the buildings in question
- Gradual improvement of the district heating technology and operation introduced, among others by lowering the temperature in the distribution system along with implementing energy conservation in the buildings
- Additional investments in the district heating production plants including the addition of heat pumps and solar thermal, geothermal, and biomass boilers to the existing CHP plants

The economic calculation was based on the same investment costs as in the previous section, with the addition of new data to include the calculation of consequences for the balance of payment and job creation. All investments were assumed to have an import share of 40 percent, operation and maintenance an import share of 20 percent, and fossil fuels an import share of 80 percent. The remaining costs were assumed to generate jobs in Denmark at an average of 2 person-years per 1 million DKK (equal to approximately 135,000 EUR). Salary constituted 80 percent of this and the rest was capital income or savings.

In the analyses of the influence on governmental expenditures, including saved unemployment benefits as well as increases in income taxes, a net plus of 300,000 DKK (40,000 EUR) was used per person-year, equal to the general expectations used by the Danish Ministry of Finance. This amount included any additional effects arising from VAT, etc.

Different versions of the investment plan in terms of different production units were analyzed by use of the EnergyPLAN model and are reported in Lund and Hvelplund (2012). Along the lines of the previous section, the result shows that Danish society in general would be able to decrease the cost of heating the buildings in question by investing in district heating and individual heat pumps. Fuel costs were replaced by investments, but the annual investment costs were lower than the costs of the fuel saved when paid during their technical lifetime using a real interest of 3 percent.

The total additional net investment amounted to approximately 9 billion EUR; 13 billion EUR must be invested, while 4 billion EUR of investment could be saved compared to the reference. Table 6.3 illustrates the annual investment and operation costs, given that the alternative was implemented over a period of 10 years from 2011 to 2020. As shown, this implementation plan required substantial net investments, which would gradually result in substantial fuel cost savings. The net investments would have a negative influence (positive net import) on the balance of payment in the beginning. However, due to the saved import of fossil fuels, this influence would be reduced and would end up being positive (negative net import). When measured over the total lifetime of the investment, the effect was both substantial and positive. The net investments made during the 10-year implementation period would also result in the creation of approximately 7000–8000 jobs each year during the entire 10-year period.

However, the implementation of the alternative would have an influence on the governmental expenditures in several ways. First of all, the alternative could not be implemented without the formulation of an active energy policy, since some of the investments were not feasible to the investors with the current taxes and subsidies. Second, since oil and natural gas consumption was taxed, the government would lack this income when these fuels were replaced. Finally, the creation of jobs would generate additional income taxes.

In Table 6.4, an estimate is made of the extent of the different consequences. This is an estimate, since VAT and multiplication effects have not been included. Moreover, due to the very complex taxation system in Denmark, not all effects have been calculated in detail. However, the table provides a good overview of the magnitude of the influence of the different measures.

In Table 6.4, the top lists the taxes as input; next, all the changes in relevant fossil fuel consumption are listed and divided into the relevant taxation categories. Based on these two types of input, a calculation was made: first, of the decreases in taxes on oil and natural gas for individual boilers compared to the reference, and, second, of the

Table 6.3 Costs and job creation.

MDKK	All years	2011	2012	2013	2014	2015	2016	2017	2018	2019	2020	2021
District heating ^{g}ncl	58,000	5800	5800	5800	5800	5800	5800	5800	5800	5800	5800	0
DH house installations	9240	924	924	924	924	924	924	924	924	924	924	0
Individual heat pumps	11,550	1155	1155	1155	1155	1155	1155	1155	1155	1155	1155	0
Large-scale heat pumps	6000	600	600	600	600	600	600	600	600	600	600	0
Peak-load boilers	4000	400	400	400	400	400	400	400	400	400	400	0
Biomass boilers	5000	500	500	500	500	500	500	500	500	500	500	0
Solar thermal	5600	560	560	560	560	560	560	560	560	560	560	0
Geothermal	1400	140	140	140	140	140	140	140	140	140	140	0
Total new investments	100,790	10,079	10,079	10,079	10,079	10,079	10,079	10,079	10,079	10,079	10,079	0
Saved oil boilers	11,400	1140	1140	1140	1140	1140	1140	1140	1140	1140	1140	0
Saved biomass boilers	8400	840	840	840	840	840	840	840	840	840	840	0
Saved Ngas boilers	10,460	1046	1046	1046	1046	1046	1046	1046	1046	1046	1046	0
Total saved investments	30,260	3026	3026	3026	3026	3026	3026	3026	3026	3026	3026	0
Net investments	70,530	7053	7053	7053	7053	7053	7053	7053	7053	7053	7053	0
Increased O&M	8438	70	211	352	492	633	774	914	1055	1195	1336	1406
Saved O&M	−9609	−80	−240	−400	−561	−721	−881	−1041	−1201	−1361	−1521	−1602
O&M net change	1171	−10	−29	−49	−68	−88	−107	−127	−146	−166	−185	−195
Fuel net change	−19,722	−164	−493	−822	−1150	−1479	−1808	−2137	−2465	−2794	−3123	−3287
Total costs	49,637	6879	6531	6182	5834	5486	5138	4790	4441	4093	3745	3482
Import costs	12,200	2688	2421	2154	1887	1620	1353	1087	820	553	286	−2669
Employment	74,874	8382	8220	8057	7894	7731	7569	7406	7243	7081	6918	−1627

Table 6.4 Net effects on the governmental expenditures.

Input data		DKK/M3	DKK/kWh	MDKK/GWh
Saved unemployment benefits:	0.12 MDKK/Man-year			
Increased income tax	0.18 MDKK/Man-year			
Individual natural gas		2.629	0.239	0.239
Industry natural gas		2.629	0.239	0.239
Heat from heat pump			0.208	0.208
Heat from boilers			0.208	0.208
Heat from CHP		2.620	0.238	0.238
Individual gas oil		2.469	0.247	0.247
Industry oil		2.469	0.274	0.247
Individual HP (elec.)			0.545	0.545

MDKK	All years	2011	2012	2013	2014	2015	2016	2017	2018	2019	2020	2021
Subsidy (solar and HP)		20	20	15	15	10	10	5	5	0	0	0
Employment (GWh/year)	74,874	8382	8220	8057	7894	7731	7569	7406	7243	7081	6918	−1627
Natural gas (individual)	−24,600	−205	−615	−1025	−1435	−1845	−2255	−2665	−3075	−3485	−3895	−4100
Natural gas (industry)	−15,240	−127	−381	−635	−889	−1143	−1397	−1651	−1905	−2159	−2413	−2540
Total natural gas CHP	−3060	−26	−77	−128	−179	−230	−281	−332	−383	−434	−485	−510
Tax free part (0.38/0.62)	−1875	−16	−47	−78	−109	−141	−172	−203	−234	−266	−297	−313
Taxed natural gas CHP	−1185	−10	−30	−49	−69	−89	−109	−128	−148	−168	−188	−197
Natural gas (boiler)	−9360	−78	−234	−390	−546	−702	−858	−1014	−1170	−1326	−1482	−1560
Gas oil (boiler)	−12,420	−104	−311	−518	−725	−932	−1139	−1346	−1553	−1760	−1967	−2070
Oil (industry)	−17,160	−143	−429	−715	−1001	−1287	−1573	−1859	−2145	−2431	−2717	−2860
Heat from heat pump	38,700	383	963	1613	2258	2903	3548	4193	4838	5483	6128	6450
Electricity to individual HP	4878	41	122	203	285	366	447	528	610	691	772	813
MDKK losses in taxes		−140	−421	−701	−981	−1262	−1542	−1823	−2103	−2384	−2664	−2804
Additional taxes		71	212	353	495	636	777	919	1060	1201	1343	1413
Net decrease in taxes		−70	−209	−348	−487	−626	−765	−904	−1043	−1182	−1321	−1391
Saved benefits		1006	986	967	947	928	908	889	869	850	830	−195

Table 6.4 Continued

MDKK	All years	2011	2012	2013	2014	2015	2016	2017	2018	2019	2020	2021
Increased income taxes		1509	1480	1450	1421	1392	1362	1333	1304	1275	1245	−293
Net influence before policy means		**2445**	**2257**	**2069**	**1881**	**1693**	**1506**	**1318**	**1130**	**942**	**754**	**−1879**
Subsidy solar and heat pumps		−232	−232	−174	−174	−116	−116	−58	−58	0	0	0
Compensation natural gas companies		−56	−56	−56	−56	−56	−56	−56	−56	−56	−56	−56
Tax release individual heat pumps		−14	−41	−68	−96	−123	−151	−178	−205	−233	−260	−274
Tax release large heat pumps		−23	−68	−114	−159	−204	−250	−295	−341	−386	−431	−454
Policy means total		**−325**	**−397**	**−412**	**−485**	**−499**	**−573**	**−587**	**−660**	**−675**	**−747**	**−784**
Net influence on government expenditures		2120	1860	1657	1396	1194	933	731	470	267	7	−2663

increases in other taxes, i.e., on electricity for heat pumps. As can be seen, the government would lose 140 MDKK in taxes on oil and natural gas in the first year, gradually rising to 2800 MDKK after 10 years, when the strategy would be fully implemented. Approximately 50 percent of this loss was compensated for by increases in the taxation of electricity for heat pumps.

However, in return for the net loss in fuel taxes, the government benefited from the job creation in two ways. First, governmental contributions to unemployment benefits were saved. Next, the income taxes were increased. In total, this effect raised the income by 2500 MDKK/year in the beginning, slightly decreasing to 2000 MDKK in 2020.

The net effect (if the plan could be implemented without any subsidies or similar) was a positive contribution to the governmental expenditures of approximately 2500 MDKK in 2011, decreasing to approximately 750 MDKK in 2020.

A survey of the barriers to making the investments feasible on the market resulted in the following public regulation measures to be taken:

- A subsidy for heat pumps and solar thermal power of 20 percent in the first 2 years should be introduced and gradually decreased to 15, 10, and 5 percent, respectively, over the period.
- The natural gas companies should be compensated for loans in the natural gas grid that had not yet been paid, so the remaining consumers would not have to pay for those leaving the system.
- Taxes on electricity for small- as well as large-scale heat pumps should be reduced.

All in all, these subsidies and tax reductions would increase the governmental expenditures by approximately 300 MDKK in the first year, increasing to approximately 800 MDKK/year after 10 years.

As shown in Table 6.4, when all these measures were introduced and calculated, the net consequences for the government ended at a profit of approximately 2 billion DKK in the first year, gradually decreasing to zero over a period of 10 years.

Consequently, the case illustrates how an economic crisis such as the one from 2008 and onwards enables the implementation of essential elements of future renewable energy solutions and infrastructures which generate jobs without having a negative influence on the governmental expenditures.

A core element of concrete institutional economics is the analysis of the specific institutional context of the market in question including the fact that such institutions differ from one country to another and change from one period to another. Consequently, even though the need for concrete institutional economics may apply to many different cases, the results of the Danish case under the given conditions cannot necessarily be applied to other countries and/or other situations. One needs to make a specific analysis of the country in question and may find it relevant to include issues such as the involvement of foreign company laborers, or foreigners in general, as well as a commitment to sustainable development and anti-corruption. However, in times of the economic crisis, a number of different countries are likely to come to similar conclusions regarding the possibilities of designing strategies that further the combination of economic growth and the implementation of sustainable development.

4   Zero energy buildings and smart grids[d]

This section is based on Lund et al.'s (2011b) article "Zero Energy Buildings and Mismatch Compensation Factors." The paper takes an overall energy system approach to analyzing the integration of zero energy and zero emission buildings (ZEBs) into the electricity grid. The integration issue arises from hourly differences in energy production and consumption at the building level, and these differences result in the need for an exchange of electricity via the public grid, even though the building has an annual netexchange of zero.

A ZEB combines highly energy-efficient building designs and technical systems and equipment to minimize the heating and electricity demand with on-site renewable energy generation, typically including a solar hot water production system and a rooftop photovoltaic (PV) system. However, heat pumps and small micro-CHP units, preferably based on biomass fuels, have sometimes been taken into consideration as well.

A ZEB can be off-grid or on-grid. For the grid-connected ZEBs, the combination of a reduced demand and an on-site production of heat and electricity to reach zero raises the issue of the hourly difference between demand and production and how to deal with this difference. How do you solve the problem that a building that combines conservation with production, such as PV, may have a zero net energy input on an annual basis, but at the same time exchanges huge amounts of electricity with the public grid? Should the building itself compensate for the need for exchange or should the problem be solved at the aggregated level?

Measures at the individual building level could be either flexible demand or energy storage. However, this section argues that, when seen in the view of optimizing the complete overall energy system, these differences should not be dealt with at the individual building level, but rather at an aggregated level. Compared to the aggregated level, a solution at the individual level is not economically feasible. Moreover, individual solutions involve a risk of making things worse. Of course the measure "flexible demand" should be carried out at the building level, but the aim should not be to level out the need for exchange at the individual building in question. Instead, the flexible demand should aim at contributing to the compensation of the aggregated exchange of many buildings.

To be able to quantify the need for exchange, the following four types of ZEBs were defined:

1. *PV ZEB*: Building with a relatively small electricity demand and a PV installation
2. *Wind ZEB*: Building with a relatively small electricity demand and a small on-site wind turbine
3. *PV-solarthermal-heatpump ZEB*: Building with a relatively small heat and electricity demand and a PV installation in combination with a solar thermal collector, a heat pump, and heat storage
4. *Wind-solarthermal-heatpump ZEB*: Building with a relatively small heat and electricity demand and a wind turbine in combination with a solar thermal collector, a heat pump, and heat storage

---

[d] Excerpts reprinted from *Energy and Buildings* 43, Henrik Lund, Anna Marszal and Per Heiselberg, "Zero Energy Buildings and Mismatch Compensation Factors", pp. 1646–1654 (2011), with permission from Elsevier.

An existing building aiming for zero emissions was used to quantify the relation between heat and electricity demand, namely, a single-family house constructed in the town of Lystrup, Denmark. The building was constructed in 2009 as a demonstration house of the project "Active houses" built by VHR Holding. The house is $190\,m^2$ with the following energy demand:

- Domestic hot water: $18.3\,kWh/m^2$ ($66\,MJ/m^2$) per year
- Space heating: $15\,kWh/m^2$ ($54\,MJ/m^2$) per year
- Electricity for operating the house: $6.7\,kWh/m$ per year
- Electricity for household: $13.2\,kWh/m^2$ per year
- PV electricity production: $29.1\,kWh/m^2$ per year
- Solar thermal: $11\,kWh/m^2$ ($40\,MJ/m^2$) per year
- Heat pump thermal output: $22.4\,kWh/m^2$ ($81\,MJ/m^2$) year

Based on the data for the ZEB in Lystrup, the following expected annual figures were used for the PV-solarthermal-heatpump ZEB:

- Heat demand: 6.3 MWh/year
- Electricity demand: 3.8 MWh/year
- Solar thermal: 2.1 MWh/year
- PV production: 5.5 MWh/year
- Heat pump: converting $1.7\,MWh_e$/year (5.5–3.8) into $4.2\,MWh_{th}$, i.e., having a COP of 2.5
- In the calculation, a heat storage capacity of 17 kWh, equaling one day's average heat demand, was added to level out variations in the hot water consumption and the solar thermal production.

The same figures were used for the wind-solarthermal-heatpump ZEB only replacing the PV production with a wind turbine. It could be a small on-site wind turbine or a share of a larger wind turbine depending on the definitions of ZEB. However, the following calculation represents both cases. A principle of these two ZEBs is shown in Figs. 6.9 and 6.10. The other two buildings, i.e., the PV ZEB and the Wind ZEB, were based on the same figures excluding the heat demand, i.e., an electricity demand of 3.8 MWh/year and a similar PV or wind production.

The connection of ZEBs to the public electricity grid and the need for exchange and compensation should not be dealt with at the individual level. It should be compensated for at the aggregated level for the following reasons:

First, the cost of investing and operating one large battery of 10,000 kWh is substantially lower than operating a thousand 10 kWh batteries, among others, because large batteries make it possible to utilize new battery technology, like vanadium redox flow batteries.

Second, other options exist at the system level that can provide the same regulation at even lower cost. As explained carefully in Chapter 5, such options include changing the regulation of existing small CHP plants, introducing large-scale heat pumps at existing CHP and district heating supplies, using electricity in the transportation sector, or even introducing electricity storage systems such as compressed air energy storage. Consequently, one will save money dealing with these problems at the aggregated level.

Third, the influence of the individual building is leveled out at the aggregated level. One may compare it to the design of power supply systems. Power plants are not designed to meet the needs of the number of consumers multiplied by the maximum consumption of each

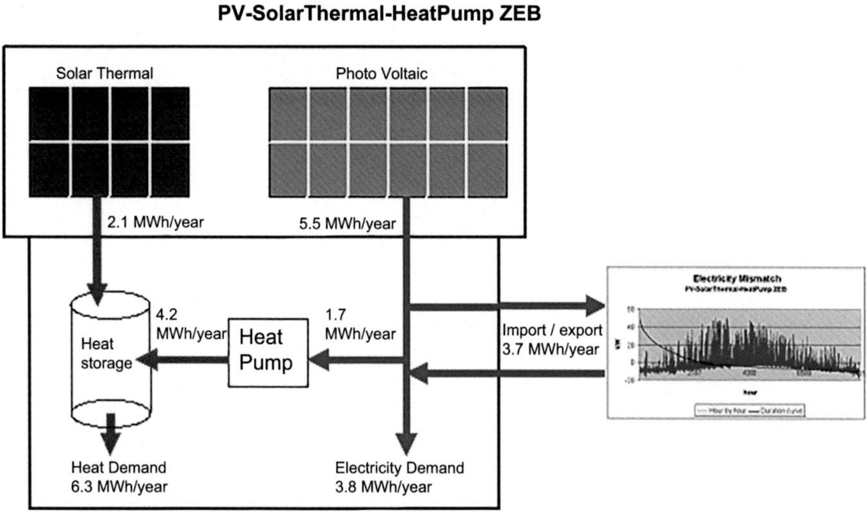

PV-SolarThermal-HeatPump ZEB

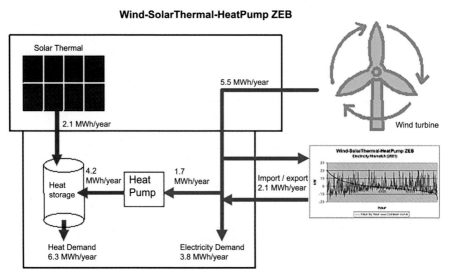

Fig. 6.9 Principle diagram of the PV-solarthermal-heatpump ZEB with an expected annual net heat and electricity demand of zero, but with a substantial exchange of electricity. The electricity exchange has been calculated as the average ZEB contribution at the aggregated level.

Wind-SolarThermal-HeatPump ZEB

Fig. 6.10 Principle diagram of the wind-solarthermal-heatpump ZEB with an expected annual net heat and electricity demand of zero, but with a substantial exchange of electricity. The electricity exchange has been calculated as the average ZEB contribution at the aggregated level.

consumer. In such case, one would make huge overinvestments in transmission lines and power stations. The sum of maximum consumption never happens for the simple reason that not all consumers peak in consumption at the same time. At the aggregated level, individual consumptions are leveled out. The same concerns differences created by changes in electricity demand and PV or wind power production at the individual building level. The exchange of one building is partly compensated for by exchanges of other buildings, and the attempt to compensate for each need for exchange individually will lead to situations in which one building is charging a battery at the same time as another building is discharging a battery, leading to unnecessary losses. Seen from the viewpoint of the electricity supply system, it is not the individual "one building exchange" that is interesting; it is the sum of the exchange from all buildings that counts. One would make significant overinvestments and inefficient operation of the system if one tried to compensate for each exchange at the individual level compared to the aggregated level.

Fourth, one risks making things worse. The reason is that exchange, from the individual building point of view, per definition is looked upon in a negative light. If there is a need for exchange, it is defined as a problem that has to be solved. However, from the viewpoint of the electricity supply system, exchange is not necessarily negative; it may also be positive.

This is illustrated by Fig. 6.11, which shows the variation of the electricity demand in western Denmark in a week in February 2001. The electricity consumption was high during the day and low during nights and weekends. This variation is typical for all countries even though the specific shape of the electricity demand curve differs from one country to another. From the electricity supply point of view, any exchange that decreases the demand during the night and increases it during the day is negative. This exchange will increase the demand for capacity and increase the production of expensive units during peak hours and only save less expensive units during base load hours. However, for the same reason, exchange resulting in the opposite, i.e., a decrease during peak load and an increase during base load, creates a positive exchange for the system. Consequently, such an exchange should not be compensated for, and if investments in flexible demand or storage systems are made at the building level to minimize this exchange, this will only make things worse.

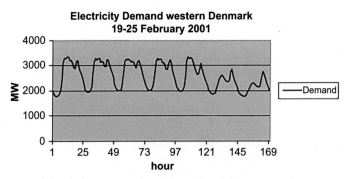

Fig. 6.11 Hourly fluctuations in the electricity demand of western Denmark in February 2001. A mismatch arising from a ZEB that consumes electricity during night hours and produces electricity during day hours is a positive mismatch that should not be compensated for.

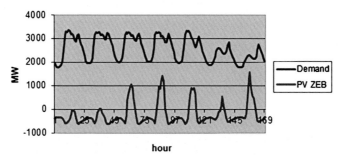

Fig. 6.12 The mismatch of approximately 1 million PV ZEBs compared to the hourly fluctuations in the electricity demand of western Denmark in February 2001. It is based on actual PV production during the same week in February, taking into account the leveling out between approximately 267 PV installations.

Figs. 6.12 and 6.13 compare the fluctuations of the same electricity demand with the exchange of PV ZEB and Wind ZEB, respectively, with an annual electricity demand of 3.8 MWh. The calculation of the need for exchange was based on the actual electricity production from PV collectors and wind turbines in western Denmark in the same week of February 2001, using the same data as explained in Chapter 5. The fluctuations are shown at the aggregated level and represent approximately 1 million ZEBs. As one can see, the difference between the buildings' electricity demand and PV production generally results in a positive exchange for the system, while the difference between demand and wind power gives both a positive and a negative exchange. It should be noted that 1 million ZEBs is an unrealistic number in Denmark

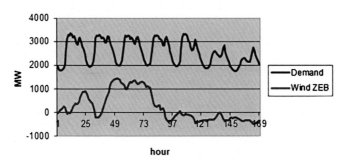

Fig. 6.13 The mismatch of approximately 1 million Wind ZEBs compared to the hourly fluctuations in the electricity demand of western Denmark in February 2001. It is based on actual wind power production during the same week in February, taking into account the leveling out between the numbers of wind turbines in western Denmark the same year.

in the foreseeable future, and here this number only serves to illustrate the principle in the diagrams. Moreover, the discussion here does not take into account possible limitations in local distribution grids leading to voltage drops.

The above case illustrates how important it is to include the design and operation of ZEBs into the coherency of the overall energy system if the aim is to identify the most efficient and least-cost integration of combinations of energy efficiency and renewable energy measures. An attempt to deal with this integration without taking the overall energy system into account is likely to lead to overinvestments and inefficient operation at the individual building level, as well as in the overall system as a whole.

5 Future power plants and smart energy systems[e]

This section is based on Lund et al.'s (2012) article "From Electricity Smart Grids to Smart Energy Systems—A Market Operation Based Approach and Understanding." By describing the case of Skagen CHP plant located in the northern part of Denmark and how this plant is designed and operated, the paper provides an example of what future power plants could look like according to the approach of smart electricity grids as part of a future smart energy system based on renewable energy. An interesting point about the case is that some of the major requirements for smart grids and smart energy systems have already been implemented and have been in operation for several years now. It is also interesting to note that these changes were implemented as a consequence of Skagen CHP plant making active use of the opening of the Nordic electricity market Nord Pool.

The step from the traditional fossil fuel energy system via large-scale integration of renewable electricity toward 100 percent renewable energy poses a challenge to the operation of the electricity grid, as well as to the CHP and power plants, which have to produce power when the wind is not blowing and the sun is not shining. As already described in Chapter 5, to meet this challenge, improvements will be necessary in grid stability, while also creating more flexible electricity production. This means that future power plants will have to look quite different from the way they do now. The main challenges are described as follows.

First, there is a much greater need for technical *flexibility*. Future power plants will have to be able to change production much faster than most nuclear and coal-fired steam turbines can. Moreover, they have to be able to stop production within 1 hour and run again soon after; again something that traditional steam turbines cannot do.

Second, power plants need to be able to survive *financially* with reduced annual production. Along with increasing the share of RES to 50 percent in 2020 as planned and implemented in Denmark, and further in a 100 percent renewable energy system, the annual production hours of the power plants will decrease accordingly, as will be elaborated in Chapters 7 and 8. In Lund and Mathiesen (2012), it was calculated that a power station that previously typically

[e] Excerpts reprinted from *Energy* 42, Henrik Lund, Anders N. Andersen, Poul Alberg Østergaard, Brian Vad Mathiesen, and David Connolly, "From Electricity Smart Grids to Smart Energy Systems—A Market Operation Based Approach and Understanding", pp. 96–102 (2012), with permission from Elsevier.

had approximately 4000 production hours/year on average, may only have 1200 hours/year or less in the future. This poses a financial challenge to the power stations. How can they survive and make a profit with so few hours of production?

Third, it will be a challenge to maintain *grid stability* and provide similar auxiliary services such as regulating power. Today, in most countries, the task of maintaining grid stability (frequency and voltage) is handled by the large steam turbines and/or hydro power generation. In the future, however, grid stabilization should be managed also when the power plants are not producing.

Fourth, preferably, power plants should be mostly *CHP plants*. To achieve the most efficient use of fuels in the system, power stations producing only electricity should be at a minimum. All these units should be able to supply heat to district heating and/or cooling as well as biomass conversion and industrial purposes. Consequently, power stations should be CHP plants while also meeting the challenges above.

The question is whether it is possible to design a future power plant that can meet all of these challenges and at the same time be feasible. The case of Skagen can be used to illustrate how this might be done. Skagen was already a CHP plant, and it was supplemented with a waste incineration boiler and an electric boiler as well as contribution from industrial excess heat. Consequently, the case of Skagen illustrates how a CHP plant can be supplemented with other heat-producing units. Moreover, the case illustrates the significance of including distributed CHP and renewable power production units in the task of grid stabilization, i.e., securing voltage and frequency stability for the electricity supply.

The case of Skagen presents a technical design for potential flexible energy systems that are able to balance production and demand, while also fulfilling voltage and frequency stability requirements of the grid. It also illustrates how this operation has already been implemented for years in some places in Denmark.

Skagen CHP plant had three gas engines, heat storage, a gas peak-load boiler, and an electric boiler, as listed in Fig. 6.14. Moreover, Skagen CHP plant received heat from an incineration plant and waste heat from industry, and was considering investing in a large-scale heat pump.

The organization of the Danish electricity markets, as part of the Nordic system, is shown in Fig. 6.15. As shown, the market is divided into a day-ahead spot market and a number of regulating power markets. The specific organization varies from one European system to another, but the principle shown in Fig. 6.15 is typical for most countries.

Access to the different markets was granted for small CHP plants, like the one in Skagen, along the time line listed below:

- Day-ahead spot market in January 2005
- Regulating power market in 2006
- Automatic primary reserve market in November 2009—Now called Frequency Containment Reserve (FCR).

Skagen CHP plant had been operating in the day-ahead spot market for several years and was one of the first small CHP plants to enter the regulating power market. Since November 2009, Skagen CHP plant had also been operating in the automatic primary reserve market.

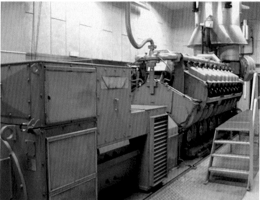

Unit	Size
CHP capacity	13 MW$_e$ and 16 MJ/s (three 4.3 MW$_e$ natural gas units)
Heat storage	250 MWh
Peak load boilers	37 MW
Electric boiler	10 MW
Compression heat pump	Under consideration

Fig. 6.14 Technical specifications and illustration of Skagen CHP plant.

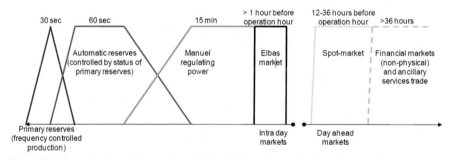

Fig. 6.15 The main electricity markets (typical for many market-based systems).

The simultaneous operation of the plant in all these markets was done in the following sequence. Bids were given a day ahead on the spot market. Bids for electricity production from the CHP units were given on the basis of alternative costs of supplying heat from the gas boiler or the electric boiler. In the calculation of the bids, the heat storage option was carefully taken into account. The calculation of the bids is described in Andersen and Lund (2007) and considerations to optimize the heat storage investment are described in Lund and Andersen (2005).

The CHP units could be operated in the regulating power market in the following two ways: If operation in the spot market was won, a downward regulation could be offered; if not, an upward regulation could be offered. The reverse situation applied to the electric boiler. Additionally, the CHP units could be operated in the automatic primary reserve market. This was done by offering the CHP plants to the spot market at full capacity minus 10 percent. If the bid was won, the same unit could be offered for a ±10 percent operation in the primary automatic reserve market. The same principle could be applied to the electric boiler. Fig. 6.16 illustrates the operation of the plant on a day in May 2010.

Fig. 6.16 shows that on Thursday, May 13, 2010, the three CHP units traded their full load in the spot market during the well-paid hours in the middle of the day and in the evening. On Friday, May 14, the three CHP units were traded in the spot market during the well-paid hours in the middle of the day; however, it was not at full load, and the remaining capacity was traded in the primary reserve market. On Sunday, May 16, the electric boiler ran at half load, allowing it to be both positive primary reserve (reducing consumption) and negative primary reserve (increasing consumption). On Tuesday, May 18, all three CHP units were activated in the regulating power market for an upward regulation. The following day, the 10 MW electric boiler was activated in the regulating power market for a downward regulation.

Another interesting example of Skagen CHP plant's regulating potential occurred on March 25, 2011, as displayed in Fig. 6.17. In the first four hours of the day, Skagen won the negative primary reserve with the 10 MW electrical boiler. Hence, it operated below full capacity. A little before 3 a.m., Skagen won a downward regulation in the regulating power market, so the electric boiler increased its output to approximately 4 MW. At the same time, Skagen still performed the frequency regulation, which it had won in the primary reserve market. After 4 a.m., Skagen had not won any additional primary reserve, so the electric boiler was offered at full capacity (i.e., 10 MW) for downward regulation in the regulating power market, winning it for a full hour. From 4 to 8 p.m., only part of the CHP units were sold in the spot market, which made it possible to offer both positive primary reserve and negative primary reserve during these 4 hours.

The online operation of Skagen CHP plant and the prices of the spot market and the regulating power market can be seen at https://www.energyweb.dk/skagen. The investment cost related to the involvement of Skagen CHP plant in the primary reserve market was surprisingly low. The cost of control equipment for the existing CHP units, which made it possible to offer ±1.4 MW, was only 27,000 EUR and the cost of the 10 MW electric boiler was 0.7 million EUR.

As illustrated above, Skagen CHP plant is an example of a power plant that, to a large extent, can meet the requirements of future power stations listed in the beginning of this section. Skagen is a CHP plant and it provides flexible production, which can change from one hour to another between producing and consuming electricity. At the same time, the plant plays an active role in the grid stabilization and regulating power tasks. Skagen has done this with relatively few utilization hours on the CHP unit, and this set-up allows substantial inputs of wind in the electricity supply. It also enables the use of waste incineration and industrial excess heat in the heat supply, while the plant can still survive financially. This ability to provide high power inputs for only a few hours may even be expanded substantially in the future, if Skagen saves the existing engines for the power reserve market when these have to be replaced by new ones. Considering the fact that Denmark has a huge number of small CHPs of similar

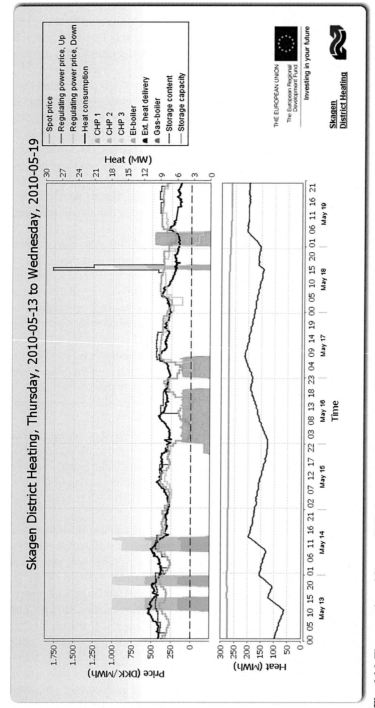

Fig. 6.16 The operation of Skagen CHP plant on the spot market, the regulating power market, and the primary reserve market on May 13–19, 2010; https://www.energyweb.dk/skagen.

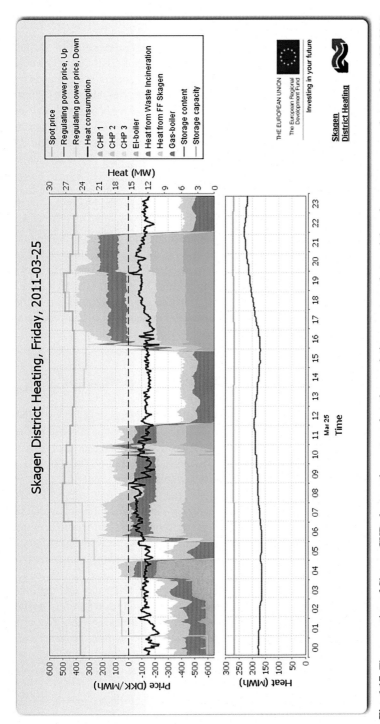

Fig. 6.17 The operation of Skagen CHP plant on the spot market, the regulating power market, and the primary reserve market on March 25, 2011; https://www.energyweb.dk/skagen.

capacity in its system, this solution can be copied and together these plants may form a coherent solution for the whole system in the future.

Regarding smart grids and smart energy systems, the flexible operation of Skagen CHP plant illustrates the significance of including distributed CHP and renewable power production units in the task of grid stabilization, i.e., ensuring voltage and frequency stability of the electricity supply. Today, in most countries, electricity is produced either by hydro power or by large steam turbines on the basis of fossil fuels or nuclear power. Variable renewable power production constitutes only a small part of the production. Until now, the tasks of balancing supply and demand and securing frequency and voltage on the grid have been managed only by such large production units.

However, the opening of the spot market, and later both the regulating power market and the primary reserve market, has made it possible for small distributed CHP plants to enter these markets. The case of Skagen CHP plant equipped with CHP units, heat storage, and electric boilers illustrates how small plants can provide valuable grid stabilization at very low additional investment and operating costs. Moreover, this case illustrates how the perception of the electricity sector as part of a complete renewable energy system paves the way for better and more cost-effective solutions to smart grid applications, compared to looking at the electricity sector as a separate part of the energy system. Consequently, a smart energy system approach may prove beneficial compared to a sole smart electricity grid approach.

6   Renewable energy transportation fuel pathways[f]

This section is based on a chapter in Mathiesen et al.'s (2014) report "CEESA 100% Renewable Energy Transport Scenarios Towards 2050." The report is a result of the Coherent Energy and Environmental System Analysis (CEESA) project partly financed by the Danish Council for Strategic Research. Furthermore, the report is a background report to the CEESA 100 percent renewable energy scenario, which is presented in Chapter 7. As explained in Chapter 7, the basic finding in the CEESA project—as well as similar studies shown in Chapter 7—is that the biomass available for energy use is limited, while biomass-based gas or liquid fuels are required in the transportation sector to supplement the direct use of electricity. This forms a challenge to the system design in terms of identifying suitable renewable energy transportation fuel pathways.

Considering the renewable resources available to produce electricity and the limitations associated with biomass, maximizing the use of electricity and minimizing the use of bioenergy in the transportation sector are key considerations. Overall, five distinct pathways were analyzed in detail (see Table 6.5): electrification, fermentation, bioenergy hydrogenation (includes biomass and biogas), CO_2 hydrogenation, and co-electrolysis. Later, the three pathways were also known as electrofuel, e-fuel or

[f] Excerpts reprinted from Brian Vad Mathiesen, David Connolly et al. (2014) "CEESA 100% Renewable Energy Transport Scenarios Towards 2050".

Table 6.5 Transportation fuel pathways considered in CEESA and their principal objective.

Pathway considered	Principal objective
Direct electrification	Use electricity as the primary transportation fuel
Fermentation	Convert straw to a fuel suitable for transportation (i.e., ethanol) using a fermenter
Bioenergy hydrogenation	Gasify a biomass resource OR use anaerobic digester to produce biogas; afterward boost its energy potential as a transportation fuel using hydrogen from steam electrolysis
CO_2 hydrogenation (CO_2Hydro)	Create a fuel without any direct biomass consumption using hydrogen from steam electrolysis and sequestered carbon dioxide
Co-electrolysis	Create a fuel without any direct biomass consumption by coelectrolyzing steam and sequestered carbon dioxide

Power-to-X (PtX) pathways. All these pathways are described in detail in the report along with an overall comparison. A separate energy flow diagram is available for each pathway outlining the electricity and biomass required to produce 100 PJ of the primary fuel. However, for practical reasons, only the principal pathway diagrams have been included in the following.

At the current stage, it is uncertain to which degree the future transportation system ends up using liquid fuel or gas (or a combination of both) to supplement the direct use of electricity. For that reason, the following pathways were made for both gas and liquid fuel. However, in order to create the pathways, the final fuel had to be defined to identify the conversion losses. For the gas alternative, methane was chosen, which is very close to natural gas. Natural gas-based vehicles are already well established, with over 10 million vehicles worldwide.

For various reasons, methanol is assumed to be the preferred liquid fuel in a 100 percent renewable energy system. Methanol is the simplest alcohol with the lowest carbon content and the highest hydrogen content of any liquid fuel. Furthermore, methanol can be used in internal combustion engines as a replacement for petrol with relatively few modifications. This has already been proven in the United States when ~20,000 methanol cars and 100 refueling stations were in use in the mid-1990s (Bromberg and Cheng, 2010). At the time of the CEESA study, it was also proved in China, where over 200,000 methanol vehicles were introduced over 5 years (Methanol Institute China, 2011). It is worth noting, however, that dimethyl ether (DME) could also be used, since it is the first derivative of methanol and it is very suitable as an alternative to conventional diesel (Pontzen et al., 2011). The efficiency lost when choosing DME compared to methanol can be gained due to the higher efficiencies of diesel engines compared to petrol engines. Therefore, the transportation demands displayed in the flow diagrams are similar for both methanol and DME from a well-to-wheel perspective.

For all fuels, a passenger transportation demand (pkm) and a freight transportation demand (tkm) are displayed, since they can be used for either one or the other demand in all cases except one (i.e., battery electrification). The specific energy consumption

Table 6.6 Specific levels of energy consumption used to estimate the transportation demand that can be met by the transportation fuels produced.

Fuel	Passenger transport		Freight transport	
	Load factor (p/vehicle)	Specific energy consumption (MJ/pkm)	Load factor (t/vehicle)	Specific energy consumption (MJ/tkm)
Electric rail	84.00	0.34	278	0.31
Electric car	1.50	0.32	n/a	n/a
Methanol/DME	1.50	1.15	12	1.90
Methane	1.50	1.57	12	2.65
Ethanol	1.50	1.50	12	3.30

Based on data from the 2010 reference and vehicle efficiency estimates by the Danish Energy Agency (2008).

is shown in Table 6.6 based on vehicle efficiencies from the Danish Energy Agency (2008). By assessing the energy losses from production to consumption, it was possible to compare each of the pathways in terms of the resources that they would require and the transportation demand that they would meet. The following sections describe some of the pathways individually and these are then compared to one another.

Direct electrification

Electricity can be used as a direct transportation fuel in two ways: by delivering it to the end user or by using batteries as a storage medium. To date, rail and buses (i.e., trolleybuses) are the only modes of transportation where electricity is delivered directly to the end user. The key limitation is the infrastructure required since a cable must be available to the end user at all times. This requires high initial investment costs and also restricts the feasible routes. However, once the infrastructure is in place, due to the high efficiency of rail, a relatively high transportation demand of 300 Gpkm or 325 Gtkm can be met when 100 PJ of electricity is available (see Fig. 6.18).

To increase the route flexibility of electrification, batteries can be used. Several electrical and hybrid electrical vehicle technologies are already commercially available today (Hansen et al., 2011). Thus, from a technical point of view, it seems to be realistic to implement these technologies in the near future. As outlined in Fig. 6.19, this is also a very efficient pathway. However, batteries come with a number of key limitations, particularly their energy density.

In comparison to liquid fuels, batteries can store very little energy relative to their weight. Direct electrification is thus not suitable for all modes of transportation such as trucks, aviation, and marine transportation. In a 100 percent renewable energy system, some form of fuel with high energy density (such as methanol/DME or methane) will also be necessary to supplement direct electrification.

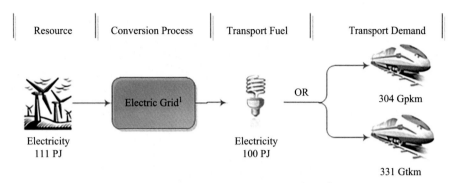

Fig. 6.18 Direct use of electricity by the end user for transportation. [1]Assuming 10 percent loss for the electric grid.

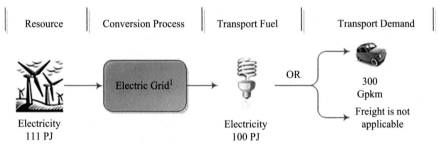

Fig. 6.19 Direct use of electricity for transportation using battery storage. [1]Assuming 10 percent loss for the electric grid.

Fermentation

The principal objective in the fermentation pathway is to convert straw to ethanol by use of a fermenter. Even though this process itself has a limited efficiency of approximately 25 percent, as only the cellulose is fermented, a number of byproducts produced from hemicellulose and lignin can subsequently be used to create other fuels. Since a variety of options are available, two distinct pathways are presented here.

The first fermentation pathway is the "fuel optimized" option. It is designed to produce the maximum amount of useful fuel with minimum input. For example, in this pathway, the lignin and residual sugars from the fermenter are hydrogenated to create an "oil slurry," which is well suited as a fuel for marine diesels. In addition, the C5 sugars can be converted into conventional diesel and be used for trucks. From the hydrogenation process, there are also byproducts of coke and inorganic materials, which can be utilized in a number of ways, i.e., gasified and hydrogenated, to produce more fuel such as methanol/DME or simply burned in a power plant to produce electricity. At present, it is unclear which option would be most suitable in a 100 percent renewable energy system. However, in Chapter 8, a proposal on how best to design an overall solution for biomass conversion in a fully decarbonized Danish society is shown.

The second fermentation pathway is the "energy optimized" option. It is designed to maximize the energy available in the fuels created. In this process, the CO_2 from the

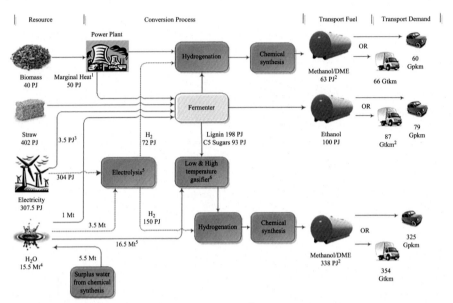

Fig. 6.20 Energy optimized fermentation pathway. [1]Assuming a marginal efficiency of
125 percent and a steam share of 12.5 percent relative to the straw input. [2]A loss of 5 percent was
applied to the fuel produced to account for losses in the chemical synthesis and fuel storage.
[3]Assuming an electricity demand of 0.8 percent relative to the straw input. [4]This is the net
demand for water, i.e., it is reduced by the water recycled from hydrogenation. [5]Assuming an
electrolyzer efficiency of 73 percent for the steam electrolysis. [6]Assuming the same conversion
process as for cellulous gasification and hydrogenation to methanol, but the round-trip losses
have been doubled since there are two gasifiers here (low and high temperature) and there is
uncertainty in relation to the gasification of lignin and C5 sugars. [7]Assumed that ethanol trucks
require approximately 25 percent more fuel than diesel equivalents, based on the difference
between ethanol and diesel cars.

fermenter is hydrogenated in the same way as in the fuel optimized process. However,
the lignin and residual sugars are gasified instead of hydrogenated. Due to the high salt
content present in all agricultural residues, it is assumed that this will require both low-
and high-temperature gasifiers. Once again, the losses associated with wood gasification are assumed for both low- and high-temperature gasifiers, which are optimistic
assumptions. After gasification, the gas is hydrogenated to produce syngas, which
can be converted to methanol/DME using chemical synthesis. The final energy flows
of the energy optimized fermentation process are displayed in Fig. 6.20.

Bioenergy hydrogenation

The principal objective of the bioenergy hydrogenation pathway is to create a transportation fuel from bioenergy, which is boosted by hydrogen from steam electrolysis.
In this way, the energy potential of the bioenergy resource is maximized. Here, three
different bioenergy pathways have been considered:

1. Biomass hydrogenation to methanol
2. Biomass hydrogenation to methane
3. Biogas hydrogenation to methane

A variety of biomass feedstocks can be used in this process; wood gasification is already being commercialized on a large scale, while the gasification of biomass from energy crops and straw is currently in the demonstration phase. Once biomass has been gasified, it is hydrogenated using hydrogen from steam electrolysis. Hydrogenating the biomass increases the energy content and the energy density of the original biomass, thus reducing the share of biomass needed. The resulting syngas is transformed into a transportation fuel using chemical synthesis, which is already a well-established technique used by the fossil fuel industry for converting coal and natural gas into liquid fuels.

In this study, the energy and mass balance assumed for biomass hydrogenation was based on the hydrogenation of cellulous to both methanol and methane. The resulting energy flow diagram for the case of methane is outlined in Fig. 6.21. The diagram for methanol is quite similar. In practice, additional conversion procedures could also be necessary for biomass gasification, since a wide variety of different technologies are to be utilized. For example, oxygen could be used to gasify the biomass and it is also assumed here that all of the carbon is utilized in the reaction. If this is not possible in practice, further losses may occur. However, the overall demand for biomass and hydrogen per unit of methanol produced is indicative of the future demand if this pathway is chosen.

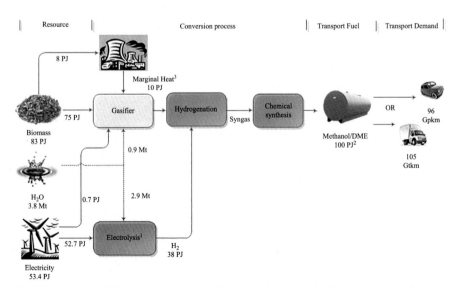

Fig. 6.21 Steam gasification of biomass, which is subsequently hydrogenated to methanol. [1]Assuming an electrolyzer efficiency of 73 percent for the steam electrolysis. [2]A loss of 5 percent was applied to the fuel produced to account for losses in the chemical synthesis and fuel storage. [3]Assuming a marginal efficiency of 125 percent and a steam share of 13 percent relative to the biomass input.

Biogas hydrogenation or biogas methanation is also included here based on two reactions: first, the gasification of glucose, which occurs in an anaerobic digester, followed by the hydrogenation of the resulting CO_2. When glucose is gasified, it results in a gas that contains approximately 50 percent methane and 50 percent CO_2 by volume. In practice, the mix is usually 55–70 percent methane, 30–45 percent CO_2, and 1–2 percent other elements, since the feedstock is never pure glucose. Hence, the estimates here assume a slightly higher CO_2 output than typically obtained from anaerobic digesters. Since methane is typically the major component in biogas from anaerobic digesters, the CO_2 contained in the biogas is hydrogenated to methane. The resulting flow diagram for biogas hydrogenation is presented in Fig. 6.22.

It is possible to convert the resulting methane to methanol by reforming it to a synthetic gas and then synthesizing the gas to methanol using high pressure. The reforming step is a very energy intensive process, since it is a strongly endothermic reaction. This process requires a large external energy supply, while the second phase of conversion from syngas to methanol is negligible, since it only requires a suitable catalyst. However, a total of approximately 20–30 percent of the fuel is lost during the transition from methane to methanol.

The CO_2 hydrogenation (CO_2Hydro) pathways combine carbon dioxide and hydrogen gases, followed by a chemical synthesis to produce a fuel for transportation. The principal objective of these pathways is to create a fuel that does not require any direct biomass input by using steam electrolysis and sequestered carbon dioxide. Separate pathways are included for methanol/DME and methane based on these energy and mass balances. The hydrogen can be produced by steam electrolysis, which requires electricity and water. To collect the carbon dioxide, carbon capture and utilization (CCU) from biomass power plants may be an option. However, one could also consider carbon trees (Lackner, 2009). Therefore, a total of four pathways can be described: two using CCU and two using carbon trees either producing methane or

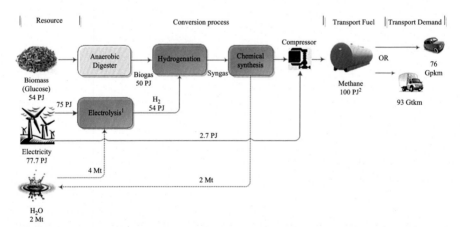

Fig. 6.22 Production of biogas from biomass, which is subsequently hydrogenated to produce methane. [1]Assuming an electrolyzer efficiency of 73 percent for the steam electrolysis. [2]A loss of 5 percent was applied to the fuel produced to account for losses in the chemical synthesis and fuel storage.

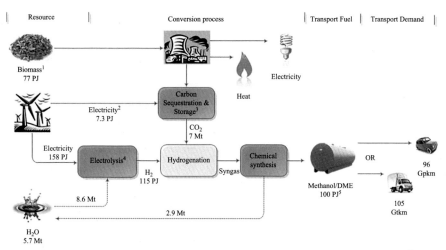

| Resource | | | Conversion process | | Transport Fuel | Transport Demand |

Fig. 6.23 Hydrogenation of carbon dioxide sequestered using CCU to methanol/DME. [1]Based on dry willow biomass. [2]Based on an additional electricity demand of 0.29 MWh/tCO$_2$ for capturing carbon dioxide from coal-fired power plants. [3]CCU is used in CEESA since it is currently a cheaper alternative to carbon trees. If carbon trees were used here, they would require approximately 5 percent more electricity. [4]Assuming an electrolyzer efficiency of 73 percent for the steam electrolysis. [5]A loss of 5 percent was applied to the fuel produced to account for losses in the chemical synthesis and fuel storage.

methanol/DME. The alternative of CO$_2$ hydrogenation to methanol/DME using CCU is illustrated in Fig. 6.23.

According to research (Danish Energy Agency, 2008; Lackner, 2009), there is only a 5 percent difference in the electricity demand required to sequester the carbon dioxide in these two pathways. The key difference is that carbon trees do not require any combustible fuel in the energy system. However, if biomass is already used in the power plants, these two pathways are almost identical in terms of energy consumption. The only key difference between the two carbon sequestration options is the cost: The estimated cost for CCU is approximately 30 EUR/tCO$_2$ (Danish Energy Agency, 2008), while for carbon trees, it is approximately 200 USD/tCO$_2$ (Lackner, 2009).

Co-electrolysis

The co-electrolysis pathways are quite similar to the CO$_2$Hydro pathways. Their principal objective is to create a fuel that does not require any direct biomass input. However, instead of using carbon dioxide, the co-electrolysis pathway combines hydrogen with carbon monoxide gas followed by a chemical synthesis to produce either methanol/DME or methane. To do so, steam and carbon dioxide are broken down at the same time in one electrolyzer unit, hence the name co-electrolysis. Like in the CO$_2$Hydro pathway, the carbon dioxide can be obtained using either CCU or carbon trees, which again leads to the definition of four pathways. The alternative of co-electrolysis to methanol/DME using CCU is illustrated in Fig. 6.24.

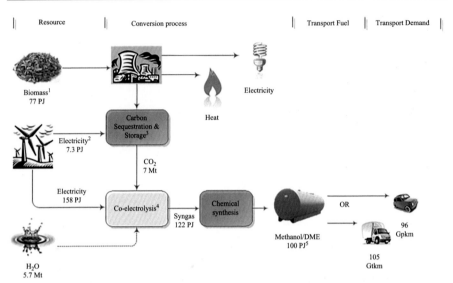

Fig. 6.24 Co-electrolysis of steam and carbon dioxide which is obtained using CCR to methanol/DME. [1]Based on dry willow biomass. [2]Based on an additional electricity demand of 0.29 MWh/tCO$_2$ for capturing carbon dioxide from coal-fired power plants. [3]CCR is used in CEESA since it is currently a cheaper alternative to carbon trees. If carbon trees were used here, they would require approximately 5 percent more electricity. [4]Assuming a co-electrolyzer efficiency of 78 percent: 73 percent for steam and 86 percent for carbon dioxide. [5]A loss of 5 percent was applied to the fuel produced to account for losses in the chemical synthesis and fuel storage.

Once again, if biomass is being utilized in the electricity and heat sectors, the key difference between these two forms of carbon sequestration is the cost. In comparison to the CO$_2$Hydro pathway, co-electrolysis requires a lower water input, but there is no excess water from the reaction. Hence, the net water demand is the same for both pathways.

Comparison

Using the energy flow diagrams presented in Figs. 6.20–6.24, it is possible to compare each of the pathways by identifying the electricity and bioenergy required to produce 100 Gpkm for passenger transportation and 100 Gtkm for freight transportation.

This comparison confirms that direct electrification is the most sustainable form of transportation in terms of the resources consumed. It requires the lowest amount of electricity compared to all of the pathways and it does not require any direct bioenergy consumption. Battery electrification is also a very efficient pathway, but as mentioned previously, in a 100 percent renewable energy system, this will need to be supplemented by some form of high energy density fuel.

The energy optimized fermentation process illustrated in Fig. 6.20 can be compared directly to the biomass hydrogenation illustrated in Fig. 6.21, since both use a combination of electricity and biomass. A quantitative comparison clearly shows that biomass hydrogenation requires slightly less electricity and biomass. However, this analysis of differences between the two is not sufficient to make a conclusive decision on which future pathway to utilize, which is similar to the conclusions drawn in previous research in Sweden (Grahn, 2004). From a qualitative perspective, however, the difference between these two pathways is much more significant. Both the fuel optimized and energy optimized fermentation processes are very complex and include numerous conversions that are very uncertain, especially the gasification of coke and lignin. Since the conversion loss figures for the gasification of coke and lignin are currently not available, the conversion losses for cellulous gasification were assumed here, which is an optimistic assumption. In addition, the fermentation processes include an array of interactions between different subpathways for its byproducts (i.e., hydrogenation and gasification). In comparison, the biomass hydrogenation process consists of only one principal interaction (i.e., gasification of biomass). Further, the biomass hydrogenation process finishes with a chemical synthesis, meaning that the choice of fuel produced is very flexible, whereas the fermentation pathway will always be partially restricted to ethanol and, more specifically for the fuel optimized process, to marine diesels. Based on this, the biomass hydrogenation process at present:

- Is more efficient
- Requires less bioenergy and therefore land
- Provides more flexibility to the energy system
- Is subject to less uncertainty

Therefore, in the CEESA 100 percent renewable energy scenario to be presented in Chapter 7, biomass hydrogenation was used to simulate the direct use of biomass as a liquid fuel and so the output methanol/DME was referred to as *bio-methanol DME*. It is essential to note that the principle of boosting the biomass resource with electricity is the same in both the fermentation and the biomass hydrogenation pathways. Hence, the modeling carried out in CEESA is indicative of the energy flows represented in fermentation as well. However, based on existing knowledge, biomass hydrogenation seems more likely to achieve its technological development targets.

The remaining pathways, CO_2Hydro and co-electrolysis, do not require any direct biomass input, and they both require the same amount of electricity. However, the CO_2Hydro pathway uses steam electrolysis, which is already a well-established technology, whereas co-electrolyzers are still under development. Therefore, CO_2Hydro was used in CEESA when simulating liquid fuel, which does not require any direct biomass input and so the output methanol/DME is referred to as *e-methanol/DME*. Both the CO_2Hydro and co-electrolysis pathways represent the same principle, which is the use of electricity and captured CO_2 to create liquid fuel. Therefore, although the CO_2Hydro pathway is used in CEESA, the results are indicative of those that would also be achieved with co-electrolysis.

It is critical to recognize that biomass hydrogenation and CO_2Hydro have been chosen to represent two uniquely different methods in the future transportation sector: one which boosts a biomass resource and one which uses captured CO_2. This does not mean that the biomass hydrogenation pathway will be used instead of bioethanol or that CO_2Hydro will be used instead of co-electrolysis. In the future, the ultimate decision will depend on the technological development and demonstration of these facilities on a large scale. It is clear that these two principles will need to be applied in some way to achieve a 100 percent renewable energy system, depending on the residual bioenergy resource available.

Looked upon from a smart grid and smart energy systems point of view, it is evident that all fuel pathways described in this section to supplement the direct use of electricity involve hydrogenation and/or biomass conversion technologies that produce gas or liquid fuel. The different pathways will typically benefit from integration with the rest of the systems in one or more of the following aspects:

- The conversion requires heat and/or produces heat and, consequently, will benefit from integration with CHP and/or district heating.
- The conversion involves electricity inputs and, consequently, benefits can be achieved by replacing electricity storage with gas or liquid fuel storage, which is typically more efficient and cheaper. Moreover, regulation and even overinvestment in capacities of the electrolyzers may be used to integrate more variable renewable electricity supplies.
- The gas produced needs to be stored and/or distributed and, consequently, one will benefit from the use of a smart gas grid. Since the production units may benefit from being distributed (e.g., to avoid the transportation of manure and/or to better integrate them with efficient heating solutions), a smart grid that can handle bidirectional flows will be relevant.
- Since the fuel pathways have the potential to participate in several smart grids and their related sectors, the identification of optimal solutions implies a smart energy systems approach.

7 Reflections

Reflections and conclusions on the analyses of smart energy systems and infrastructures in this chapter are made regarding principles and methodologies, as well as the implementation of renewable energy systems in Denmark and other countries.

Theory, tool and methodologies

As mentioned in the beginning of this chapter, the transformation into renewable energy systems poses a challenge of substantial changes to the infrastructures to carry the energy, i.e., the electricity grid, the gas and hydrogen grids, the district heating and cooling grids, as well as potentially CO_2 grids. All these grids face the common challenge of the integration of an efficient use of potential renewable energy sources as well as the operation of a grid structure allowing for distributed activities involving interaction with consumers and bidirectional flows. To meet this challenge, all grids will benefit from the use of modern information and communication technologies as an integrated part of the grids at all levels.

On such a basis, this chapter has defined the future concepts of a *smart electricity grid, smart thermal grids*, and *smart gas grids*. The three concepts have a lot of similarities; however, they also differ in the major challenges connected to each of them. Smart thermal grids face their major challenges in the temperature level and the interaction with low-energy buildings. Smart electricity grids face their major challenges in the reliability and in the integration of variable renewable electricity production.

Smart gas grids including hydrogen grids face their major challenges in mixing gases with different heating values and in the efficient use of limited biomass resources.

Each of the three types of smart grids provides important contributions to future renewable energy systems. However, each individual smart grid should not be seen as separate from the others or separate from the other parts of the overall energy system. First, it does not make much sense to convert one sector to renewable energy if this is not coordinated with a similar conversion of the other parts of the energy system. Second, this coordination makes it possible to identify additional and better solutions to the implementation of smart grid solutions within the individual sector, compared to the solutions identified with a sole focus on the sector in question. Consequently, this chapter has promoted the concept of smart energy systems defined as an approach in which smart electricity grids, smart thermal grids as well as smart gas grids are combined and coordinated to identify synergies among them and to achieve an optimal solution for each individual sector as well as for the overall energy system.

Moreover, two hypotheses have been formulated. First, that one must take a holistic and cross-sectoral smart energy systems approach in order to be able to identify the best solutions of affordable and reliable transitions of the energy system into a carbon neutral society. Next, that subsector studies (no matter if they consider the role of a specific technology or the role of a region or country) should aim at identifying the role to play in the context of the overall system transition rather than aim at decarbonizing the subsector on its own.

The analysis of smart energy systems calls for tools and models that can provide similar and parallel analyses of grids of electricity, heating, cooling, and gas. As described in Chapter 4, EnergyPLAN is one such model, since the tool can do hour-by-hour analysis of all four grids including storage and the interactions among them.

8 Conclusions and recommendations

As concluded in Chapter 5, the large-scale integration of renewable energy should be considered a way to approach renewable energy systems. The same conclusion can be made regarding smart energy systems and infrastructures. The long-term relevant systems are those in which renewable energy sources are combined with energy conservation and system efficiency improvements, and in this regard, the implementation of future energy infrastructures becomes crucial.

Seen from this perspective, Chapter 5 lists seven recommendations on how to deal with the integration of large-scale renewable energy. A main point is that the most efficient and least-cost solutions are not to be found within one sector but when sectors of electricity, heat, and transportation are combined. Moreover, the importance of including distributed flexible CHP plants and electricity-to-transportation systems in the task of grid stabilization is emphasized. This chapter adds to the points of Chapter 5 with the following recommendations:

1. District heating faces the challenge of supplying future low-energy buildings and at the same time utilizing low-temperature sources from, e.g., CHP and industrial processes. This challenge can be met by the development of low-temperature district heating grids, here defined as 4GDHs.

2. If the challenges of future district heating grids are met, the best heating solution, looked upon from an overall system point of view, seems to be a combination of district heating in urban areas with individual heat pumps in the rest of the system. In the current system, this solution shows substantial reductions in fuel demands and CO_2 emissions, while in a future 100 percent renewable energy system, it reduces the need for renewable energy sources including the need for biomass. This seems to be valid even if the space heating demand is reduced by as much as 75 percent.

3. Grid-connected low-energy houses such as zero energy buildings, which include on-site production in terms of PV, wind power, and solar thermal, should not try to balance supply and demand at the individual building level. Such problems are better dealt with at an aggregated level. At the individual level, one is likely to implement an expensive solution with greater losses and at the same time increase the risk of worsening instead of helping the integration of variable renewable energy into the electricity supply.

4. The nature of renewable energy systems calls for future power stations to be flexible CHP plants that can supply grid stabilization when they produce as well as consume electricity, at the same time as they can survive financially with small utilization hours on the CHP unit. Such *smart energy systems* and *smart electricity grid* requirements cannot be met by existing nuclear and coal-fired steam turbine technologies.

5. The opening of electricity markets for supplying regulating power and automatic primary reserve has created examples of the implementation of relevant future power plants. These examples show how small CHP plants can provide flexible production, which can change from one hour to another between producing and consuming electricity, while taking an active role in the grid stabilization and regulating power tasks. These CHP plants produce with relatively few utilization hours on the CHP unit and allow substantial inputs of wind in the electricity supply, along with inputs of waste incineration and industrial excess heat in the heat supply, while still surviving financially. However, it should be noted that some sort of capacity payment may have to support the CHP plants.

6. The ability to provide high power inputs for only a few hours may even be expanded substantially in the future if existing small CHP plants save their existing engines for the power reserve market when they have to be replaced by new ones. Considering the fact that Denmark has a large number of small CHPs in its system, together these may form a coherent solution for the whole system in the future.

7. Limitations in the biomass resources available in combination with the need for gas or liquid fuel in the transportation sector to supplement the direct use of electricity call for pathways in which the conversion of biomass is boosted with hydrogen from electrolysis. Looked upon from an energy system integration point of view, such pathways call for a new smart gas grids in which different gas inputs from local production are stored and distributed in a bidirectional flow handling gas inputs of potentially different heating values as well as the potential use of a hydrogen grid.

8. The need for power-to-X for transportation makes it possible to replace the potential long-term need for electricity storage with gas (or liquid) storage, which is both cheaper and more efficient. Consequently, the integration of renewable energy into the electricity supply should not only include the measure of direct use of electricity for transportation, but also the need for power-to-X.

9. In times of unemployment and economic crisis, the investment in the transformation toward renewable energy systems represents an option in which jobs can be created and economic growth can be increased, while public expenditures are not negatively affected but can even be improved. This is valid especially for projects with high investments and long lifetimes such as infrastructure projects. To be able to identify and implement these options, one will benefit from taking a concrete institutional economics approach.

100 percent renewable energy systems

7

The implementation of 100 percent renewable energy systems adds to the challenge of integrating renewable energy sources (RES) into existing energy systems on a large scale as well as to the implementation of smart energy infrastructures. Not only must the variable renewable energy production be coordinated with the rest of the energy system, but the size of the energy demand and the design of the system with regard to energy efficiency must also be adjusted to the realistic amount of potential and affordable renewable sources. Furthermore, this adjustment must address the differences in the characteristics of different sources, such as biomass fuels and electricity production from wind power.

The design of suitable energy systems must consider both conversion and storage technologies. Renewable energy will have to be compared not to fossil fuels, but to other sorts of renewable energy system technologies, including conservation, efficiency improvements, and storage and conversion technologies—for example, wind turbines versus the need for biomass resources. The selection of technologies is complex, not only considering the differences in hourly distributions of the technologies, but also in terms of the identification of a suitable combination of changes in conversion and storage technologies.

The design of renewable energy systems involves three major technological changes: energy savings on the demand side, efficiency improvements in energy production, and the replacement of fossil fuels by various sources of renewable energy. Consequently, the analysis of these systems must include strategies for integrating renewable sources into complex energy systems, which are influenced by energy savings and efficiency measures. The design of 100 percent renewable energy systems can be addressed at the project level as well as the national level, and, at some point, the global level. At the national level, three studies of Danish cases are presented. As already mentioned in Chapter 5, Denmark is a front-runner in that respect and therefore represents a suitable case for the analysis of large-scale integration as well as the development of 100 percent renewable energy systems. Subsequently, these examples are taken to the near-global level by discussing the options of applying the same approach and methodologies to China.

In Denmark, savings and efficiency improvements have been important parts of the energy policy since the first oil crisis in 1973. Hence, by means of energy conservation and the expansion of combined heat and power (CHP) and district heating, Denmark has been able to maintain the same level of primary energy supply for a period of 50 years, in spite of the fact that in the same period, transportation and electricity consumption as well as the heated space area have increased substantially.

Thus, Denmark provides an example of how renewable energy development strategies constituted by a combination of savings, efficiency improvements, and RES can

Renewable Energy Systems. https://doi.org/10.1016/B978-0-443-14137-9.00007-3

be implemented. As described in Chapters 5 and 6, Denmark is now facing two problems: how to integrate the high share of variable electricity from RES and how to include the transportation sector in future strategies when limitations on the biomass resource are taken into consideration. Taking this development of strategies a step further, the implementation of renewable energy systems is not only a matter of implementing savings, efficiency improvements, and RES; it also becomes a matter of introducing and adding flexible energy conversion and storage technologies and designing integrated energy system solutions. Therefore, the analysis of these strategies calls for an integrated *smart energy systems* approach, if one wishes to identify the best solutions.

According to estimations by the Danish Energy Agency from 1996, the realistic biomass potential for energy purposes in Denmark corresponded to 20–25 percent of the total primary energy supply. Meanwhile, Denmark has great potential for other sorts of renewable energy, especially wind power. In many ways, Denmark provides a typical example of the situation in many countries: the transportation sector is mostly fueled by oil, and although the biomass potential is not big enough to replace fossil fuels, the potential of variable renewable sources is substantial.

Based on the cases of Denmark and China, this chapter presents a series of studies that analyzes the problems and perspectives of converting the traditional fossil fuel-based energy system into a 100 percent renewable energy system. Three Danish case studies are presented. The first study is a one-person university study that applies the information presented in Chapter 5 to the analysis of a coherent renewable energy system. The second study is based on the technical inputs of members of the Danish Society of Engineers (IDA). The input to the study is the result of the organization's Energy Year 2006, during which 1600 participants at more than 40 seminars discussed and designed a model for the future energy system of Denmark. The third study is the result of the collaboration of researchers from five Danish universities. The study was partly financed by the Danish Council for Strategic Research and involved a coherent energy and environmental systems analysis (CEESA) of the transformation into 100 percent renewable energy systems. The study might be seen as a follow-up to the first IDA plan, in which an important further step was taken regarding the smart energy systems analysis of the integration of the transportation fuel pathways mentioned in Chapter 6. Among others, hour-by-hour analyses of electricity and district heating were supplemented with similar hour-by-hour calculations for gas.

All three Danish case studies analyzed the design of coherent and complex renewable energy systems, including the suitable integration of energy conversion and storage technologies. Furthermore, all studies were based on detailed hour-by-hour simulations carried out with the EnergyPLAN model.

The Danish cases at the national level are then discussed in relation to applying the same analyses to China. The energy demand in China has risen rapidly and reached an unprecedented level due to the massive economic growth and modern development. Since 1978, China's GDP has been increasing at a rate of 10 percent annually and the average energy consumption has risen by 5.2 percent. After 2001, the primary energy consumption soared through an average annual increase of 9 percent during 2001–2011, and the GDP increased by 11 percent in the same period. There is little doubt that the energy demand in China will continue to grow, driven by the country's highly energy-intensive economy and strong GDP growth.

1 The first approach to coherent renewable energy systems[a]

This section is based on Lund's (2007a) article "Renewable Energy Strategies for Sustainable Development." On the basis of the many studies described in Chapter 5 and the case of Denmark, this article discussed the problems and perspectives of converting the traditional fossil fuel-based energy system into a 100 percent renewable energy system. The article concluded that this conversion is possible. The necessary renewable energy sources are present, and if further technological improvements of the energy system are made, a 100 percent renewable energy system can be created. Most important are the technological conversion of the transportation sector and the introduction of flexible energy system technologies.

The article referred to the RES potential in Denmark as estimated by the Danish Energy Agency in 1996 as part of the Danish government's energy plan, Energy 21 (Danish Ministry of Environment and Energy, 1996). The estimate, which is shown in Table 7.1, dates back several years, and already when the study was made in 2007, it

Table 7.1 Potential RES in Denmark.

Renewable energy source	Potential
Wind power (onshore)	5–24 TWh/year
Wind power (offshore)	15–100 TWh/year
PV (10–25 percent of houses, 100–200 kWh/m^2)	3–16 TWh/year
Wave power	17 TWh/year
Hydropower	∼0 TWh/year
Total electricity	**40–160 TWh/year**
Solar thermal (individual houses)	6–10 PJ/year
Solar thermal (district heating)	10–80 PJ/year
Geothermal	>100 PJ/year
Total heat	**100–200 PJ/year**
Straw	39 PJ/year
Wood	23 PJ/year
Waste (combustible)	24 PJ/year
Biogas	31 PJ/year
Energy crops	65 PJ/year
Total biomass fuel	**182 PJ/year**

Data from Danish Energy Agency, 1996. Danmarks vedvarende energiressourcer. Danish Energy Agency, Copenhagen. Table is original content by the author.

---

[a] Excerpts reprinted from *Energy*, 32/6, Henrik Lund, "Renewable Energy Strategies for Sustainable Development", pp. 912–919 (2007), with permission from Elsevier.

seemed that some of the potential was underestimated. This was particularly true regarding the offshore wind potential, which is very dependent on technological development. The potential was considered higher already in 2007, and was expected to increase in the future along with the growth in the size of the wind turbines. Furthermore, it should be noted that the theoretical biomass potential in the survey from 1996 was estimated to be as high as 530 PJ/year, assuming that all farming areas were converted into energy crops, and 310 PJ/year, in the case that Denmark was self-supplied by food and the remaining areas were converted into energy crops. Again, this estimate was several years old, and since then, biomass resources in particular have been discussed, indicating that the potential may be even higher if a selection of crops is made with the concerted purpose of both producing food and energy. However, this increase in potential should be coordinated with the future needs for food and material as well as needs for negative carbon emissions in order to compensate for other greenhouse gasses. The latter will elaborated in Chapter 8. Nevertheless, the total potential of 180 PJ/year, including only a minor share of energy crops, is to be considered a business-as-usual scenario in terms of food production. All in all, the RES potential is sufficient, and only a small share is used today. In Fig. 7.1, minimum and maximum levels of potential are compared to the traditional primary energy supply in Denmark, represented by the year 2003.

The Danish energy supply is traditionally based on fossil fuels. Denmark has a very limited hydropower potential, and during the 1960s and 1970s, the electricity supply was dominated by large oil- and coal-fired steam turbines located near the big cities. However, after the first oil crisis in 1973, Denmark became a leading country in terms of implementing CHP, energy conservation, and renewable energy. Consequently, when the study was made, the Danish energy system had changed from a situation in 1972 in which 92 percent of a total supply of 833 PJ was based on oil into a situation in 2005 in which only 41 percent of 850 PJ was oil-based. In the same period, transportation and electricity consumption, as well as the heated space area, had increased substantially. Later, the share of electricity produced from CHP became as high as

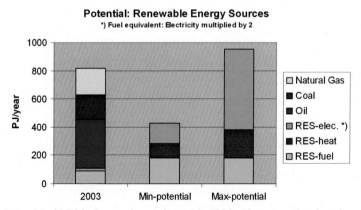

Fig. 7.1 Potential of RES in Denmark as estimated in 1996 and compared to the primary energy supply in 2003.

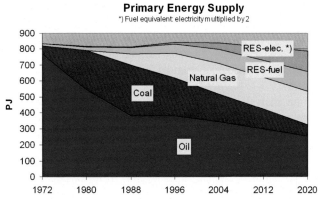

Fig. 7.2 Primary energy consumption in Denmark, including future expectations.

50 percent. Fig. 7.2 illustrates the development from 1972 until 2005, as well as future expectations resulting from the reference scenario described in Chapter 5: the 2020 projection made by the Danish Energy Agency in 2001. Fig. 7.3 shows the energy flow of the system in the reference situation.

When analyzing the possibilities of continuing the development and replacing more fossil fuels by RES, two problems arose. One was the transportation sector, which was almost totally fueled by oil. Consumption increased from 140 PJ in

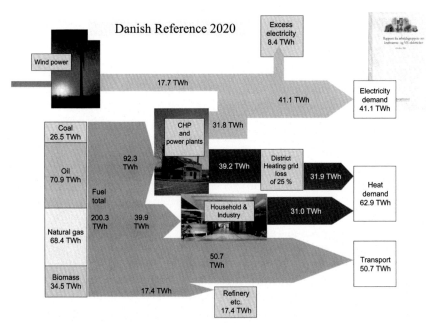

Fig. 7.3 Principal diagram of the energy flow in the Danish reference energy system in the year 2020.

1972 to 180 PJ or more in 2020. Thus, the transportation sector accounted for most of the expected oil consumption. Another problem was the integration of electricity produced from CHP and wind power. Previously, CHP stations were not operated to balance fluctuations in wind power. As a consequence, Denmark had problems of excess electricity production in periods of strong winds, if these coincided with the operation of CHP stations.

The aim of the analysis was to evaluate whether a 100 percent renewable energy system would be a possibility for Denmark and to identify key technological changes and suitable implementation strategies. As a starting point for the analysis, it was assumed that the design of renewable energy systems would involve three major technological changes: energy savings on the demand side, efficiency improvements in the energy production, and the replacement of fossil fuels by various RES. Consequently, the following technological changes of the reference (Ref 2020) were identified for the analysis of the first step (Step 1) of converting the Danish energy system into a renewable energy system:

- *Savings*: A 10 percent decrease in electricity demand, district heating, and heating for households and industry.
- *Efficiency*: A combination of better efficiencies and more CHP. Better efficiencies were defined as 50 percent electricity output and 40 percent heat output of CHP stations. This could be achieved either by partly implementing fuel cell technology or by improving existing steam turbine/engine technologies. More CHP was defined as the conversion of 50 percent of fuels for individual houses and industry into CHP, partly through district heating.
- *RES*: An increase in biomass fuels from 34 to 50 TWh/year (125–180 PJ/year) and the addition of 2.1 TWh of solar thermal to district heating and 5000 MW of PV to electricity production.
- *All*: A combination of the three preceding measures.

It should be noted that these technological changes were moderate compared to the maximum potential. Thus, it was both possible and realistic to save more than 10 percent as well as to replace more than 50 percent by CHP, and so forth.

As Step 1, the consequences of each of the three technological changes were analyzed as well as the combination of the three. The results are shown in Fig. 7.4 in terms of primary energy consumption. Fig. 7.4 shows that an increase rather than a decrease in fuel consumption is the main tendency. This is because the technological changes of Step 1 lead to a substantial increase in excess electricity production. More CHP, better efficiencies, less demand (savings), and more variable resources all create a higher excess production, unless measures are taken to prevent these problems. The resulting excess production is given in Table 7.2.

As a general tendency, the intention to decrease fossil fuels leads to increased excess electricity production. One way to avoid excess electricity production is to use it for domestic purposes. In Step 2, such an analysis was carried out in order of priority. In the case of excess production: (1) CHP units were replaced by boilers, (2) boilers were replaced by electric heating, and (3) the production from wind turbines and/or PV was simply reduced. This is a rather simple and inexpensive way of avoiding excess production. The results are shown in Fig. 7.5.

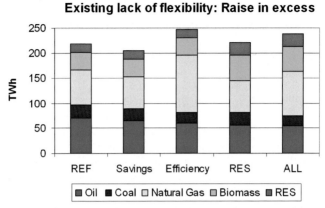

Fig. 7.4 Step 1: Primary energy supply, year 2020, in the reference Ref 2020 compared to the three technological changes of Step 1, including excess electricity production.

Table 7.2 Resulting primary energy supply and excess electricity production of the three Step 1 technological changes compared to the reference Ref 2020.

TWh/year	Ref	Savings	Efficiency	RES	All
Total fuel consumption	218	205	248	220	238
Excess electricity production	8.4	9.6	45.5	11.7	48.2

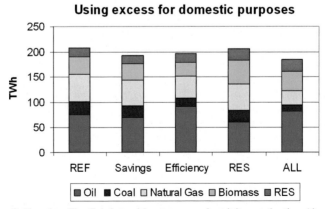

Fig. 7.5 Step 2: Equal to Fig. 7.4, but without excess electricity production (the excess production is used to replace fuels by simple and inexpensive measures).

All technological improvements resulted in a decrease in fuel consumption. However, the decrease was small, since most of the benefits derived from technological improvements were lost in the excess production. Another problem was the high share of oil used for transportation. This shows that the problem of integration becomes important when implementing savings, efficiency measures, and RES, as does the

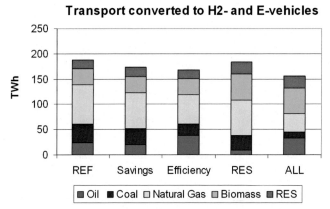

Fig. 7.6 Step 3: Primary energy consumption (same as Fig. 7.5) when oil for transportation is replaced by electricity for electric (E) vehicles and hydrogen (H₂) vehicles.

issue of including transportation. Consequently, in Step 3, the following changes were analyzed:

- Transportation: Oil for transportation was replaced by electricity, according to a scenario described by Risø National Laboratory and discussed in Chapter 5 (Nielsen and Jørgensen, 2000). Vehicles weighing less than 2 tons were replaced by battery vehicles and hydrogen fuel cell vehicles. In this scenario, 20.8 TWh of oil was replaced by 7.3 TWh of electricity. Here, the same ratio was used for converting the total oil consumption of 50.7 TWh in the reference system Ref 2020 into an electricity consumption of 17.8 TWh. The electricity demand was made flexible within the time period of one week and had a maximum capacity of 3500 MW. The results are shown in Fig. 7.6. In this case, both the reference Ref 2020 and all three Step 1 alternatives resulted in a decrease in fuel consumption.

Step 4 added more flexibility in terms of heat pumps and CHP regulation together with electrolyzers:

- *Flexible CHP and heat pumps*: Small CHP stations were included in the regulation together with heat pumps added to the system. A 1500 MW$_e$ heat pump capacity with a coefficient of performance of 3.5 was analyzed.
- *Electrolyzers and wind regulation*: Electrolyzers were added to the system and, at the same time, wind turbines were included in the voltage and frequency regulation of the electricity supply.

In Step 4, together with these measures of flexibility, wind power was added to the system until the resulting fuel consumption was equal to the available biomass resources of 180 PJ (50 TWh/year). The results are given in Fig. 7.7. In this case, the main question is how much wind power is needed to fulfill the objective. The resulting wind power capacity is given in Table 7.3.

As seen, the Danish energy system can be converted into a 100 percent renewable energy system when combining 180 TJ/year of biomass with 5000 MW of PV and between 15 and 27 GW of wind power. In the reference, 27 GW of wind power is required, while in the combination with savings and efficiency improvements, the

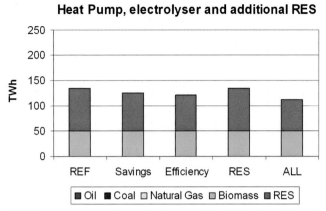

Fig. 7.7 Step 4: Primary energy consumption (same as Fig. 7.6) when adding flexible energy systems and converting to 100 percent RES.

Table 7.3 Resulting fuel consumption and required wind power capacity.

Capacity	Ref	Savings	Efficiency	All
Total fuel consumption (TWh/year)	134	125	121	112
Wind power (GW)	27.1	22.1	18.6	15.6
Annual wind investment (MW/year) Lifetime = 30 years	900	740	620	520

necessary capacity is reduced to 15 GW. With an expected average lifetime of 30 years, the total capacity of 15 GW of new offshore wind power can be reached by installing 500 MW/year. Subsequently, the 15 GW can be maintained by a continuous replacement of 500 MW each year. Since 3 GW have already been installed, the total capacity can be reached within approximately 25 years.

In Fig. 7.8 and the upper part of Fig. 7.9, the primary energy supply and the energy flow of such a system are illustrated. The two figures are comparable to Figs. 7.6 and 7.7. Altogether, the study indicates that a 100 percent renewable energy system based on domestic resources is physically and technically possible in Denmark. However, it should be emphasized that the proposal presented here was based on a conversion of the entire transportation sector into a combination of electrical and hydrogen fuel cell vehicles. Such a conversion may prove unrealistic for the entire sector. The obvious technological alternative is to convert into biofuels, as discussed in Chapter 5. In Fig. 7.9 (bottom diagram), the consequences of this conversion are shown, based on the solution already described in Chapter 5.

As the two diagrams in Fig. 7.9 show, the choice between electricity/hydrogen-based or biofuel-based transportation technologies has a major impact on the size of the resulting primary energy supply of the system. In particular, the amount of biomass required is affected. The results emphasize the importance of further developing electric

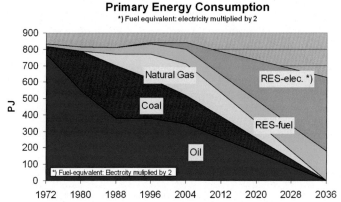

Fig. 7.8 Primary fuel consumption if the Danish energy system is converted into 100 percent RES.

vehicle technologies and indicate that biofuel transportation technologies should be reserved for the areas of transportation in which the electricity/hydrogen solution proves insufficient. This issue is included in the investigations of the next two examples.

2   The Danish Society of Engineers' energy plan[b]

This section is based on Lund and Mathiesen's (2009) article "Energy System Analysis of 100 Percent Renewable Energy Systems," which presents the methodology and results of an overall energy system analysis of a 100 percent renewable energy system. The input to the system is the result of the Danish Society of Engineers' project Energy Year 2006, in which a model for the future energy system of Denmark was discussed and designed.

As in the previous chapter, the energy system analysis was performed using the EnergyPLAN model, including hour-by-hour simulations, leading to the design of a flexible renewable energy system with the ability to balance the electricity supply and demand. The results were detailed system designs and energy balances for two energy target years: 2030, with 45 percent renewable energy, demonstrating the first important steps on the way toward a 100 percent renewable energy system (IDA, 2030), and 2050, with 100 percent renewable energy from biomass and combinations of wind, wave, and solar power (IDA, 2050).

The analysis concludes that a 100 percent renewable energy supply based on domestic resources is physically possible and that the first step toward 2030 is economically feasible to Danish society. However, Denmark will have to consider to which degree the country should rely mostly on biomass resources, which will involve the reorganization of the traditional use of farming areas, or rely mostly on wind

[b] Excerpts reprinted from *Energy*, 34/5, Henrik Lund and Brian Vad Mathiesen, "Energy System Analysis of 100 Percent Renewable Energy Systems", pp. 524–531 (2009), with permission from Elsevier.

Nuclear Fission occurs in the nuclei of elements such as uranium and plutonium. Fission is a process in which the nucleus of the element breaks down completely to form lower atomic weight elements. This is a chain reaction initiated by bombardment with neutrons to bring about the fission of the nucleus with the concurrent emission of additional neutrons. The neutrons formed can then initiate a further reaction, thus resulting in a nuclear chain reaction. The reaction therefore has to be very carefully controlled, for example by inserting graphitic carbon moderator rods, in order to capture the excess neutrons and to prevent explosion (such as that in an atomic bomb). The reaction of ^{235}U (one of the naturally occurring isotopes of uranium, found at an abundance of 0.72%) gives a series of reactions, the first of which is as follows:

$$^{235}U + n \rightarrow {}^{140}Ba + {}^{96}Kr + 3n + \sim 200\,mev$$

The 200 mev liberated per fission event is equivalent to $19.2 \times 1012\,J$ per mol of the U compound decomposed, a figure far in excess of the energies associated with normal chemical reactions. The fission process is depicted schematically in Fig. 3.3.

Plutonium 239 (^{239}Pu) is also used as a nuclear fuel. This element occurs in nature but it can also be produced from ^{238}U by bombardment with neutrons. Yet another fissionable isotope, ^{238}Pu, can be produced by bombarding ^{238}U with deuterons (deuterium atoms).

Radioactive decay occurs naturally: all radioactive elements gradually break down to give elements of lower atomic weight; this decay creates lower levels of energy and so the processes based on decay are only used in very specialised situations. Both fission and decay generally give rise to radioactive products and the storage and disposal of the radioactive waste is a very significant barrier that has led to the banning of the introduction of nuclear reactors in a number of countries, such as Ireland, or their phase-out, as in Germany.

Nuclear fusion involves the reaction of the nuclei of two lighter atoms to give a heavier element, the process creating significant amounts of energy (see Fig. 3.4). Fusion is an energy-creating technology which has not yet been commercialised although much research continues to be performed on the topic. The reaction is generally carried out with one of the heavier isotopes of hydrogen (deuterium or tritium) and requires very high temperatures such as those found in plasmas, tens of millions of degrees.[b]

---

[b] A hydrogen bomb using fusion achieves the required temperatures by the incorporation of a fission component.

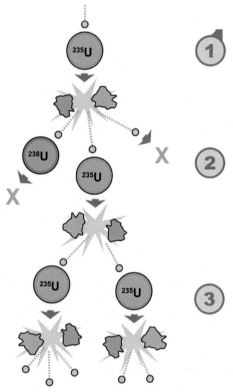

Fig. 3.3 Schematic representation of the fission of 235 U brought about by bombardment with neutrons, showing the chain reaction that occurs as a result of the creation of more and more neutrons. At each collision of a neutron with a uranium species, more than one additional neutron is emitted and so the reaction speeds up, ultimately giving an explosion, unless some of the neutrons are absorbed by other moderating species. *(From https://en.wikipedia.org/wiki/Nuclear_fusion)*

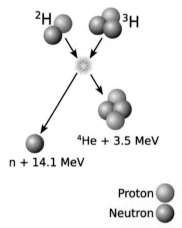

Fig. 3.4 The fusion reaction of deuterium (2H) or tritium (3H) to give helium (4He) plus large amounts of energy. *(From https://en.wikipedia.org/wiki/Nuclear_fusion)*

Fig. 3.5 The nuclear reactor at Calder Hall in Cumberland (UK) has been in operation since 1956. *(From https://en.wikipedia.org/wiki/Sellafield)*

The advantage of using such a process would be that there would be no radioactive waste. However, despite much research on the topic, no reactor has yet been developed to allow commercial operation since the energy input required to reach reaction temperature and pressure using currently available equipment is of the same order as the amount of energy produced. A further problem is that the neutrons emitted in the process degrade the reaction vessel with time and so the development of nuclear fusion will also require major innovations regarding construction materials.[c]

The Use of Nuclear Power. The first commercial reactor employing nuclear fission was that at Calder Hall in the United Kingdom (Fig. 3.5), first connected to the grid in 1956. Shortly after, several plants were commissioned in the United States and a significant number of plants were thereafter built in many other countries. France is a particularly noteworthy user of nuclear electricity since much of its demand is now satisfied by nuclear generation; as a result, France is also a very large exporter of electricity. The steady growth of France's nuclear electrical production is shown in

[c] The processes of fission, decay and fusion are well described on the web; see for example https://en. wikipedia.org/wiki/Nuclear_power.

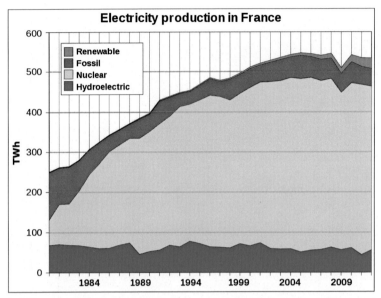

Fig. 3.6 The significant growth of the proportion of French nuclear electricity generation since 1980. *(From https://en.wikipedia.org/wiki/Nuclear_power_in_France)*

Fig. 3.6; French production is currently about 400 terawatt hours (TWh). The total global generation of electricity by nuclear reactors was 2563 TWh in 2018, this being about 10% of worldwide electricity generation. The USA is now the main producer of electricity using nuclear power, having a 30% share of global production. In December 2019, there were 4433 civilian nuclear reactors worldwide, these having a combined electrical capacity of 395 gigawatts (GW). An expansion of almost 50% in this power output is either in construction or planned, this growth occurring predominantly in Asia. It is clear that the production of this amount of electricity has saved the emission of significant amounts of CO_2 in the period since nuclear production was first commercialised.[d] However, as indicated above, there are many countries where there is considerable opposition to the use of nuclear generation as there are major concerns regarding the safety of nuclear reactors and about the safe disposal of nuclear waste. In the United States and in the United Kingdom, an expansion of nuclear power is planned as part of the future energy supply, together with energy from renewable resources.

---

[d] It has been claimed that the emission of CO_2 has been reduced by 64 billion tons since the 1970s by the introduction of nuclear power.

BOX 3.1 Cold fusion

Great scientific excitement was generated in 1989 when Pons and Fleischmann reported that they had produced excess heat during an experiment involving the electrolysis of heavy water, D_2O, using Pd electrodes. They reported that they had observed small quantities of neutrons as well as tritium, the expected products of a nuclear fusion reaction. Attempts by many scientists to replicate this work led to many papers on the topic, some supporting the original claims and others not. Finally, a recognition emerged that the reported results were faulty and it became clear that Pons and Fleishmann had not observed the products reported. 'Cold Fusion' therefore died a natural death.

Any energy-producing technology depends on there being a significant supply of the fuel used. Uranium is a reasonably common element, having an abundance similar to those of tin or germanium. It is currently economically extracted from deposits that contain relatively high concentrations. However, it is also present at low concentrations in many rocks and also in seawater; extraction from the seawater is possible but the cost would be significantly higher than that from currently used sources. Reserves of the most economically extracted uranium-containing materials are currently sufficient for about a century at the current rate of use. While accepting that the costs of extraction could increase if the easily recovered reserves are depleted, it could be argued that nuclear energy is a renewable resource as the total amount of uranium used is very small compared with the total level of reserves. However, the currently unsolved problems associated with the safe disposal of nuclear waste make this a very questionable approach.

Another potential source of renewable energy, from Cold Fusion, now totally discounted, is described in Box 3.1.

Geothermal energy

Geothermal energy is energy that is derived from within the earths crust. Most of this arises from radioactive decay occurring deep within the earth's core but part is residual energy remaining there from the original formation of the planet. The temperature difference at the boundary between the molten core and the earth's crust can reach values of over 4000°C. The movement of components of the mantle towards the surface (e.g. in volcanoes) brings some of this energy towards the surface. The water from hot springs

Fig. 3.7 Kafla geothermal power station, Iceland. *(From https://en.wikipedia.org/wiki/Geothermal_power)*

heated by such volcanic activity has been used for centuries in thermal baths and some of the sources of hot water have been used in thermal power stations. Fig. 3.7 shows one such station in Iceland, a country in which there are many such sources of geothermal energy situated near tectonic plate boundaries by which the energy can more easily reach the surface. Whether the heat supplied by a particular source is sufficient to produce electricity or is only usable for local heating purposes depends on the temperature of the water. Table 3.1 shows the geothermal electric capacities of the countries producing the largest amounts of electricity using geothermal sources.

It can be seen from these data that although Iceland produces only 5% of the world's supply of geothermally-generated electricity, this represents 30% of the country's own electricity use. The third country on the list, Indonesia, has the largest estimated reserves in the world (28,994 MW) and it is predicted that this country will soon overtake the USA in thermal electricity production. The main operation in the USA is at the Geysers in Northern California.

The efficiency of geothermal power plants such as that shown schematically in Fig. 3.8 is low, ranging from 10% up to 23%. Whether or not the heat from such a source can be used to generate electricity depends greatly on the temperature of the source. In many locations where the source

Table 3.1 Installed geothermal electricity capacities of the major producing countries.

Country	Capacity in 2010/MW	% of national electricity production	% of global geothermal electricity production
United States	3086	0.3	29
Philippines	1904	27	18
Indonesia	1197	3.7	11
Mexico	958	3.0	9
Italy	843	1.5	8
New Zealand	628	10.0	6
Iceland	575	30.0	5
Japan	536	0.1	5

Source: Condensed from a larger listing in Wikipedia.

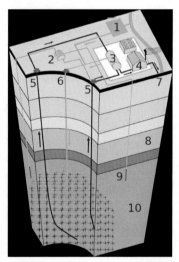

Fig. 3.8 Schematic representation of a geothermal electricity generation plant. 1, Reservoir; 2, Pump house; 3, heat exchanger; 4, Turbine hall; 5, Production well; 6, Injection well; 7, Hot water to district heating; 8, Porous sediments; 9, Observation well; 10, Crystalline bed rock. *(From https://en.wikipedia.org/wiki/Geothermal_power)*

temperature is too low, geothermal energy is used only for local heating purposes such as district heating and in greenhouses. In order to use hot aquifer sources (which can be at depths of up to 10 km) for electricity generation, the energy is extracted by a geothermal heat pump similar in operation to that shown in Fig. 3.9 of Box 3.2 so that temperatures high enough to operate the associated steam turbines are achieved. This is only economically

BOX 3.2 The operation of a heat exchanger for domestic heating purposes

Fig. 3.9 shows the operation of a domestic heat exchanger in which cold air is being delivered into a building for cooling purposes. The energy removed by the fan/evaporator system is taken up by the condenser which is then cooled either by the external air or by extraction to the surrounding ground. (This process is similar to that found in a domestic refrigerator but the latter operates without a fan.) By changing the direction of the operation of the fan, heat can be extracted from the exterior surroundings of the building (air or ground) and supplied via the condenser for heating purposes (as in so-called 'air-to-water' or 'ground-to-water' systems). Such central-heating systems are now in relatively common use; similar installations can also be used solely for ventilation purposes, with the transfer of the heat from the expelled air and its surroundings to the incoming air. Although such heat exchangers are generally most simply installed during the construction phase of a building, retrospective fitting is also now possible as a result of the development of suitable modules. Systems of this type require the building to be relatively 'leak free'; if not, more energy is required to circulate the air than is extracted from the surrounding air or ground and there is no reduction in greenhouse gas emission unless totally renewable electricity is used. Similar heat exchange units are also used in district heating systems recently introduced in a number of countries, particularly in Sweden. As with other forms of geothermal energy, the system is only 100% sustainable if renewable electricity is used for operating the system; if electricity derived from fossil fuels is used, the all-over generation of greenhouse gas emissions is generally reduced by only 60%–70%.

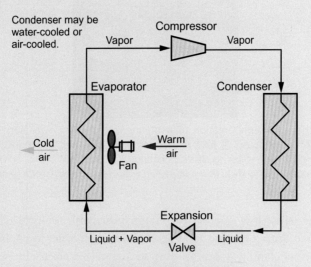

TYPICAL SINGLE-STAGE
VAPOR COMPRESSION REFRIGERATION

Fig. 3.9 The operation of a heat exchanger. *(From https://en.wikipedia.org/wiki/Heat_pump_and_refrigeration_cycle)*

feasible if the amount of electrical energy generated is significantly higher than that needed to operate the pumping and related systems. Not only does a plant such as that shown in Fig. 3.8 generate electricity but it also provides hot water for local district heating purposes.

Tidal energy

A number of different renewable technologies dependent on water rely on energy originally imparted by solar radiation, as discussed further below. However, one exception is tidal energy, this arising predominantly from gravitational forces generated by the moon as it orbits around the earth; gravitational forces from the sun also contribute to the magnitude of the tides but to a lesser extent. (The actual magnitude of each tide depends on the relative positions of the earth, the moon and the sun.) The presence of tides creates very large movements of water and these movements have associated with them very significant amounts of kinetic energy.[e] The dissipation of this energy does not occur evenly throughout the world's oceans and some areas have much higher tidal energies than others. Fig. 3.10 shows the global distribution of tidal energy; although a significant proportion of the total

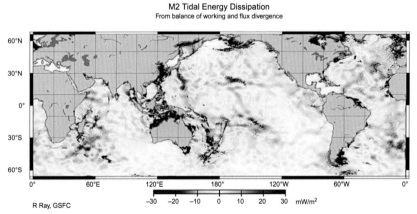

Fig. 3.10 The distribution of global tidal energy dissipation, showing those areas where tidal effects are greatest. *(From https://earthobservatory.nasa.gov/images/654/ dissipation-of-tidal-energy.)*

---

[e] A tidal flow rate of 10 mph gives an energy output equal to or greater than that for a wind speed of 90 mph.

Fig. 3.11 A tidal mill on the west coast of Ireland. Water entered and left the lagoon visible in the background through the mill-race on the right-hand side of the main structure; the mill wheel was mounted in this mill-race which is connected to the sea through a man-made channel. *(Photo: J.R.H. Ross.)*

tidal energy available is dissipated in the middle of the oceans, there are coastal regions where the effects are very high and others where the effects are much lower.

The tidal movement has for many centuries been used as a source of mechanical energy. An example of a tidal mill on the west coast of Ireland that began operation in about 1804 is shown in Fig. 3.11. This mill was built at the entrance to a large tidal lagoon which has an area of about 0.5 km^2. Water was trapped twice a day at high tide (tidal differences up to about 4 m) and allowed to flow out through the mill-race (visible in the photograph to the right of the main structure) at each low tide. The mill was used very profitably for grinding wheat for sale in England. (At that time, England had been cut off from European markets by the Napoleonic wars.) The mill ceased production when power from other sources became more generally available at the beginning of the 20th century. This technology is very similar to that now used in hydroelectric power generation, a topic that will receive further attention below.

Fig. 3.12 The SeaGen tidal stream generator situated in Strangford Lough, Northern Ireland that operated from 2006 to 2019. *(From https://en.wikipedia.org/wiki/SeaGen)*

Tidal energy is currently used for the generation of electricity in a limited number of locations, many of them in regions with high tidal energies (Fig. 3.10). In most cases, this is achieved not by the creation of tidal dams but by the use of submerged turbines. The earliest example of the use of a large commercial turbine assembly is shown in Fig. 3.12. This system, recently decommissioned following successful operation for 11 years, was located at the narrow entrance of a large tidal area, Strangford Lough, and the power produced was used in the geographical region close to the generation system: in the neighbouring town of Portaferry and its immediate surrounds. (This turbine was switched off if there was a large movement of fish into or out of the Lough.[f]) Power generation in such a system is at a maximum for several hours on either side of high tide so operation lasts only about 10 h per day. Hence, either storage is required or the power supply has to be integrated into the regional electricity supply system, using power from another source in periods when generation does not occur. We will return to the important topic of storage of electrical energy in Chapter 7.

The amount of energy associated with the tidal flow can be very high, depending on the location. Tidal flows along the Atlantic seaboard of

---

[f] Environmental arguments generally preclude the construction of tidal barriers of the type formerly used for the mill shown in Fig. 3.11. The is a significant danger of silting up and the effects on wild life can also be very damaging.

Europe are among the highest shown in Fig. 3.10, with the tidal flows around the east coast of Ireland being particularly strong. A report from Sustainable Energy Ireland[g] indicates that the use of the practically accessible tidal resources around the Irish coast could provide over 5% of the electric requirements for the whole island. The theoretical limit is much higher (up to 5 times the total electricity consumption) but there is still a need for considerable developments in the technology required, much of which is in its infancy. Environmental constraints may also have an important role in determining whether or not the necessary developments would ever be carried out. Similar arguments apply to potential developments in the other regions with high tidal energies shown in Fig. 3.10.

A closely related area of interest is the use of the energy associated with marine currents, these being a consequence of the movement of the tides. The energies associated with marine currents can be very high and they can be utilised using turbine devices similar to those described above. It has been estimated that the total worldwide power in ocean currents is about 5000 GW, these corresponding to power densities of up to $15\,kW/m^2$. An example of an ocean current that carries very high energies and that might be used for power generation is the Gulf Stream which flows past Florida and up the eastern coast of the United States. A serious obstacle to the development of such systems to harvest ocean currents is the distance of the currents to be utilised from land and the consequent need for long power cables. There are also potentially significant environmental problems. There do not appear to be any commercial examples of turbines using marine currents even though there has been a significant amount of research on the subject.

Wave power

Wave power, a closely related form of renewable energy, results entirely from solar radiation. Waves are created by the passage of winds, generated by solar radiation, passing over the surface of the ocean. The energy contained in the waves depends on many factors such as the wind speed, the depth of the water and the topography of the sea-bed. A moderate ocean swell in deep water some distance from a coastline with a wave height of 3 m and an average period of 8 s has energy approximately equivalent to 36 kW per metre of wave crest. In a major storm, with wave crests of about 15 m, the energy of a wave crest can rise to about 1.7 MW per metre. As shown in Fig. 3.13, some coastal regions of the world have higher wave

[g] https://www.seai/publications/Tidal_Current_Energy_in_Ireland_Report.pdf.

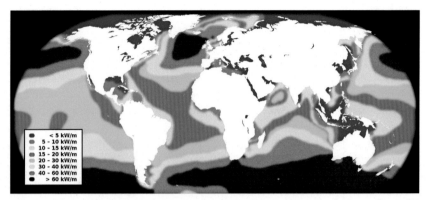

Fig. 3.13 Global distribution of wave energies. The figure shows wave energy flux in kW per meter wavefront. *(From https://en.wikipedia.org/wiki/Wave_power)*

energies than others; the most viable areas for wave power generation are on the western seaboard of Europe, the northern part of the UK, parts of the Pacific coastlines of both North and South America, Southern Africa, the southern coasts of Australia and New Zealand. Further information on wave energy and related topics can be found on the web, for example at https:// en.wikipedia.org/wiki/Wave_power.

A series of different types of devices have been developed to utilise wave energy, for example using turbines of the type discussed above for tidal power or some type of floating device tethered to the ocean bed. As with tidal energy, these technologies are generally at a relatively early stage of development. An important aspect of wave energy is that it is not as consistent and predictable as is tidal energy. The use of wave energy also presents a range of environmental constraints similar to those associated with tidal energy.

Hydroelectric power

The three most important methods of generation of renewable energy from a practical point of view now in use are hydroelectric systems, wind turbines and solar generators and these will be considered in the following three sections. Hydropower has been used for several centuries. It was an important basis for the development of the industrial revolution when it was used to provide the mechanical energy required to operate textile mills (see Chapter 2). In the twentieth century, hydroelectric power generation became a very important source of renewable electricity in countries where there are large rivers in which such generating stations can be operated. One of the world's first large-scale generating stations was on the River Shannon in Ireland and this started generation in 1929 (see Box 3.3) but

BOX 3.3 The Shannon hydroelectric scheme

When Ireland became independent of the UK in the early 1920s, it was recognised that the country had very limited natural resources such as coal. Natural gas reserves were only found in 1973 (coming on stream in 1978) and so the new nation had been largely dependent on the use of large reserves of peat for heating purposes. (Peat was also used later in a limited way for electricity generation.) It was recognised that the River Shannon, the longest river in the British Isles, was a potential energy resource and this led to the creation of the Shannon hydroelectric scheme. The Ardnacrusha power station that was then built (Fig. 3.14) was opened in 1929. The plant was built by the German company Siemens-Schuckert and generated 85 MW of electricity, sufficient (after the construction of the appropriate grid system) for the requirements of the whole country. At the time, it was the largest hydroelectric power plant in the whole world. The new power plant also required the construction of a series of dams and bridges as well as the diversion of the majority of the water from the upper reaches of the Shannon and its associated lough system through a new canal leading to the plant, the water having an average head of 28.5 m. It is interesting to note that the cost (£5.2 million) was at that time an enormous sum, thus representing more than 20% of the budget of the newly created state.

Fig. 3.14 The Ardnacrusha power plant of the Shannon Hydroelectric scheme. *(Photograph reproduced with the kind permission of ESB Archives.)*

Continued

BOX 3.3 The Shannon hydroelectric scheme—cont'd

The diversion of the main flow of the Shannon that was carried out had enormous environmental consequences which at the time of construction were largely disregarded. One of the immediate consequences was that the reaches of the Shannon above Ardnacrusha, which had been one of the most important angling locations in the West of Ireland, were no longer accessible to breeding salmon. Navigation was also affected and there were problems of flooding up-river. Some of these difficulties have since been somewhat alleviated, for example by the construction of a fish pass for the migrating salmon, but flooding problems still persist and navigation is very difficult. The Shannon Hydroelectric scheme still supplies renewable electricity but its output now represents only about 2% of the country's requirements.

Table 3.2 Hydroelectric generation in the top producing countries in 2014.

Country	Annual hydroelectric production/TWh	% of total electricity production
China	1064	18.7
Canada	383	58.3
Brazil	373	63.2
United States	282	6.5
Russia	177	16.7
India	132	10.2
Norway	129	96.0

many larger-scale stations have since been built. In 2015, hydropower generated some 16.6% of the world's electricity requirements, this corresponding to more than 2/3 of all renewable electricity. Table 3.2 gives data for 2014 for the world's top hydroelectric generating countries, showing the % hydroelectric generation compared with total electricity production for each of these countries. China is top of the list and it would appear that its share of production is still increasing rapidly as a result of continued new construction. Canada, Brazil and, in particular, Norway all produce a significant proportion of their national demands of electric power from hydroelectric generation. Although the United States is a large producer, the proportion from this source is relatively low. Hydroelectricity is a very reliable source of energy as it is dependent in most cases on the creation of a reservoir of water above the power station and this ensures continuous generation under all but the most extreme drought conditions.

Fig. 3.15 The 124 m high Krasnoyarsk Dam (Siberia, Russia). *(From https://en.wikipedia. org/wiki/Krasnoyarsk_Dam)*

There have been significant environmental aspects of the creation of large dams for hydroelectric generation. As discussed in Box 3.3, damming the River Shannon had some significant effects, the most lasting of which has been continued flooding of the upper reaches of the river almost every winter. Another example of environmental problems is the microclimatic changes that occur in the Siberian city of Krasnoyarsk due to the construction, completed in 1972, of a large 124 m high concrete dam on the River Yenisey at Divnogorsk, 30 km upstream of Krasnoyarsk; see Fig. 3.15. This dam was created largely to provide energy for the large aluminium plant in Krasnoyarsk (at that time a closed city inaccessible to Westerners). When the electricity generation plant started operation, it supplied an enormous 6000 MW of power, a level only since exceeded by the Grand Coulee Dam in Colorado, USA, which reached 6181 MW in 1983. The completion of the Krasnoyarsk Dam resulted in the creation of a large lake, now known as the Krasnoyarsk Sea, this having an area of approximately 2000 km². (A fascinating aspect of this dam is that shipping using the River Yenisey can be transported past the dam using a very large electric rack railway specially constructed for the purpose.) The volume of water in the Krasnorarsk Sea is more than 70 km³ and it passes the dam continuously during all seasons as a liquid despite the fact that the river would have previously frozen over during the winter months. This enables the year-round passage of shipping in the lower reaches of the river, something that was not possible before its construction. However, the outflow of cold water at all seasons of the year also affects the climate in the vicinity by causing cylindrical rotation of the weather in the river valley, the direction of rotation depending on whether the land is hotter or colder than the river. Krasnoyarsk therefore suffers from severe environmental problems since the emissions from its industrial activities are trapped in its vicinity by these unusual air vortices rather than being released into the surroundings.

Wind power

The generation of electricity using wind power is now familiar to everyone and land-based wind turbines such as that shown in Fig. 3.16 are common. Wind generation has increased very significantly over the last two decades (see Fig. 3.17) and additional capacity is added yearly, the main limitation to such expansion being the capacity of the electrical network (grid) of the country in question. Table 3.3 shows the installed wind capacity of the main

Fig. 3.16 A wind turbine in Co. Offaly, Ireland; its expected output approaches 10 GWh per year. *(Photo: J.R.H. Ross.)*

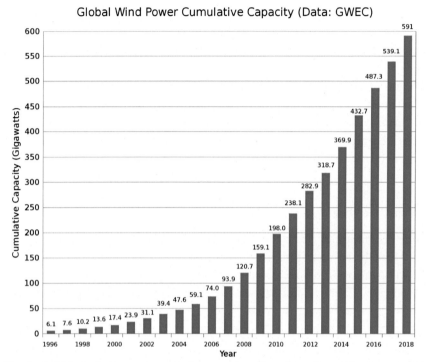

Fig. 3.17 Global wind power cumulative capacity. *(From https://en.wikipedia.org/wiki/Wind_power_by_country)*

Table 3.3 Installed wind power capacities in the major producing countries.

Country	2006	2019	% of national electricity production (2019)
China	2.6	236.4	5.4
United States	11.6	105.5	6.9
Germany	20.6	61.4	20.6
India	6.3	37.5	4.1
Spain	11.6	25.8	20.4
United Kingdom	2.0	23.5	19.8
France	1.6	16.6	6.2
Brazil	0.24	15.4	8.9
Canada	1.5	13.4	5.2
Italy	2.1	10.5	7.1
Denmark	3.1	6.1	53.0
Netherlands	1.6	4.6	9.5
Ireland	0.75	4.2	31.6
Belgium	0.19	3.9	10.2

Major producers of wind energy arranged in order of total capacity installed in 2019.
Data from Wikipedia and ourworldindata.org.

current users of wind power in 2019 as compared with the capacities of these countries in 2006. China is well ahead of other countries, having increased its capacity from only 2.6 GW in 2006 to almost 100 times as much, 236 GW, in 2019. Another country where there has been a significant increase in the use of wind energy is Brazil for which the comparable figures are 0.24 and 15.4 GW. An example of a country which now produces more than half of its electricity from wind is Denmark while another country with a high proportion of wind energy is Ireland for which the proportion is 31%.

There are many advantages of wind power over other technologies but there are also several disadvantages. The main advantage is that the installation and subsequent maintenance costs for wind turbines are relatively low and that the electricity can often be generated close to where it is needed, thereby reducing transmission losses. The main disadvantage is that wind power is very variable and there is therefore a need to have alternative standby generation capacity for periods when demand is high but the wind velocity is too low to provide significant electricity. Another problem is that turbines are relatively noisy and so they cannot be sited too close to populated areas. Wind turbines can also present visual problems, although opinions are divided on this topic, wind farms being liked by some and abhorred by others. One solution to the problem of visibility is to install them in the sea, some distance off-shore, where the winds can also be more powerful; however, the installation and maintenance costs then become more significant. Environmental problems also exist; for example, the installation of a wind farm at Derrybrien in the West of Ireland in 2003 caused dangerous land-slides with the result that $450,000 \, m^3$ of peat descended on the valley below, causing much environmental damage.

Solar power

As shown schematically in Fig. 3.18, when light hits a solid, electrons are emitted from the surface if the wavelength of the incident radiation is sufficient to detach the electrons from the material that is radiated. (In Fig. 3.18, the material radiated is shown as a metal but it could equally have been a semiconductor.) Photoemission has many applications but the use of relevance to the present discussion is in the operation of photovoltaic (PV) devices used to generate electricity when exposed to daylight. In a photovoltaic device, the electron, rather than being ejected into the surrounding space, is excited to a higher energy level and can be collected by a suitable collector. The first example of a solar cell operating on this principle was

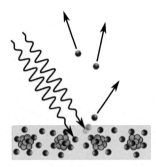

Fig. 3.18 Schematic depiction of photoemission. The emission of electrons from a metal plate caused by light quanta. *(From https://en.wikipedia.org/wiki/Photoelectric_effect)*

demonstrated by the American inventor, Charles Fritts, in 1883; his system, which exhibited a very poor efficiency, consisted of a layer of selenium covered by a thin conducting layer of gold and the first rooftop solar array using this combination was installed in New York as early as 1884.[h]

Solar cells are now generally made using crystalline silicon and in modern units these are usually in the form of thin films. The construction of a single cell is similar to that of a solid-state diode and consists of a thin P-type Si layer on top of an N-type Si layer. These are sandwiched between metal plates, the top layer having an open structure to allow light through. This structure is then coated by a thin glass window. A series of cells are then mounted together to then form a module (see Fig. 3.19), these being arranged either in series or parallel, depending on the desired output voltage. A number of these modules are then combined to form a solar panel and then a complete PV system. (As the panels will also gradually heat up, the removal of heat from the system also has to be taken into account.) As a PV system produces direct current, the output is fed to an inverter to convert it to AC for use in normal electrical systems or to feed it to the electrical grid. For domestic use of the electricity, it may be necessary to store the electricity in batteries for later use since power is produced only during daylight hours (see Fig. 3.20). The great advantage of solar power compared with wind energy is that it is much more consistent and predictable, not being dependent on direct

---

[h] It is interesting to note that such early cells, although consisting of expensive materials and having low efficiencies, found use as exposure-timing devices in more complex photographic cameras up until the 1960s.

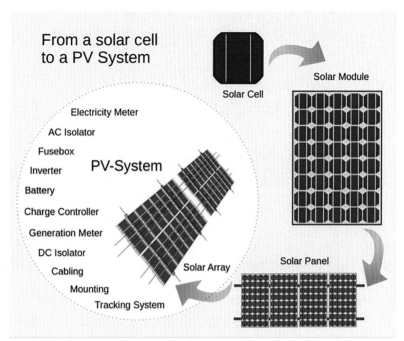

Fig. 3.19 From a solar cell to a PV system. *(From https://en.wikipedia.org/wiki/Solar_cell)*

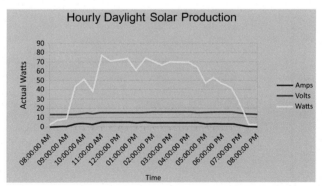

Fig. 3.20 Typical power production from a nominal 100 W solar module during daylight hours in August. *(From https://en.wikipedia.org/wiki/Solar_panel)*

sunlight but operating also on cloudy days, albeit with reduced efficiency on such days.

As a single PV cell can only be excited by only a limited range of wavelength of the incident light, a module may contain a variety of cells each accepting light of different wavelengths. The efficiency of an array can also be improved by focussing the light source in a so-called 'concentrator pho-tovoltaic cell' (CPV). Other semiconductor materials are also used in modern devices, the most common of these being cadmium telluride and gallium arse-nide. Many different formats have also been examined, ranging from solid materials to flexible films with suitable coatings. The efficiency of solar cells (defined as the percentage of incident radiation converted to electricity) is generally about 20% but this value is steadily being improved and higher values can be obtained for a given cell composition using a concentrator cell design. Fig. 3.21 shows that there has been a gradual improvement of module efficiencies of the various different types as a function of time. The cost of such modules are also gradually decreasing, as shown very schematically in Fig. 3.21.

A typical photovoltaic power plant is shown in Fig. 3.22. Such large-scale photovoltaic power plants (or colloquially 'solar farms') are becoming more and more common and one reads daily of new licences being granted for their construction. A number of governments worldwide have policies favouring such installations. However, as with wind energy, the rate at which installation can be achieved depends very much on the local capacities of the grid system in that country. There is no doubt that solar power will play an increasing role in decreasing universal greenhouse gas emissions.[i]

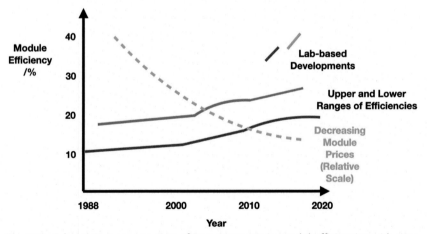

Fig. 3.21 Schematic representation of improvements in module fficiencies with time, showing upper and lower limits only. For fuller details, see https://en.wikipedia.org/wiki/Solar_cell_efficiency *(Redrawn from https://en.wikipedia.org/wiki/Solar_cell.)*

---

[i] The interested reader can obtain a free daily update of developments in the worldwide use of PV technology by subscribing to an electronic copy of PV Magazine (https://www.pv-magazine.com).

Fig. 3.22 Mount Komekura Photovoltaic power plant, January 2012. *(From https://en. wikipedia.org/wiki/Komekurayama_Solar_Power_Plant.)*

Solar power is also used for heating purposes using either reflective solar concentrators or via water heating. We will return in Chapter 4 to a discussion of the use of solar furnaces for hydrogen production and so will here only briefly discuss solar water heating systems. Fig. 3.23 shows a typical roof-mounted solar heating system used for domestic water heating. This consists of a series of tubes containing a high-temperature boiling fluid (in this case, a glycol/water solution) arranged to give optimum exposure to sunlight. The tubes absorb solar radiation, heating them to relatively high temperatures. The heated fluid in the tubes is then circulated through a specially constructed domestic hot water tank where heat is transferred to the water. Such a system works at all times of the year but has a much higher effect in the summer months. However, unlike solar PV panels, solar heating requires direct sunlight. On the day when the photograph was taken (just after the Autumn equinox), the roof fluid temperature rose to 55°C and the temperature at the base of the domestic tank was 49°C; in the summer months, these values can be much higher and the system in question is regulated to stop circulation if the hot water tank temperature exceeds 80°C. Electrical energy is required for the circulation of the fluid but the

Fig. 3.23 A solar water-heating installation. *(Photo by J.R.H. Ross.)*

circulation pump only operates if the temperature on the roof exceeds that in the storage tank; hence, assuming that the water would otherwise have been heated by electricity or by the combustion of fossil fuel, the savings in the emission of CO_2 associated with solar water heating can be quite significant. It is claimed that such systems can give a saving of up to 70% of the energy needed to provide a household's hot water requirements.

A recent development is the creation of dual-purpose modules which not only contain PV cells but also water-heating elements. As an example of such systems, the Italian company Greenetica Distribution has announced plans to build and sell a parabolic trough concentrating photovoltaic-thermal (CPVT) system. In this system, four parallel parabolic mirrors concentrate the solar radiation on two linear photovoltaic-thermal modules, the total width being 1.2 m. The photocells are based on InGaP, GaAs and Ge and work best at temperatures about 80°C. The parabolic mirrors give a concentration ratio of nearly 130 and approximately 91% of the incident sunlight is converted to heat or electricity, with an optimum output of 1 kW of electricity and 2.5 kW of thermal energy.[j]

[j] E. Bellini, PV Magazine, 6th October 2020.

Concluding remarks

It is clear that there have recently been some very significant advances in the use of renewable resources for the generation of heat and electricity and that many countries now produce significant amounts of renewable energy. However, as can be seen from Fig. 3.24, which shows the current level of renewable energy production in the European states, some countries are much more advanced than others. While some EU member states, notably Sweden, Finland and Denmark, exceeded their targets (these being shown as black lines for each country), a number of states, particularly those on the right-hand side, have failed to do so and are well below the EU average of just below 20% renewable energy production.

It is to be expected that there will be a very significant increase in the use of some of the methods that have been discussed in this chapter over the next few decades, this as a result of a strong worldwide pressure to overcome the problem of global warming and to approach net-zero greenhouse gas emissions. The following chapter, which describes the production and uses of hydrogen, includes a discussion of some of the ways in which renewable energy can be used to improve the emissions from some important chemical processes involving the use of hydrogen.

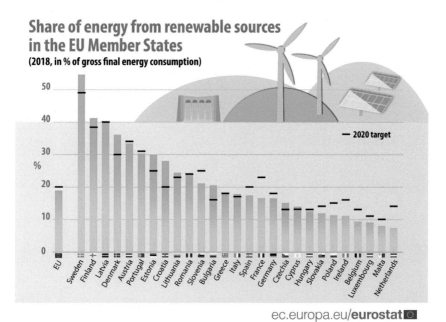

ec.europa.eu/**eurostat**

Fig. 3.24 The use of renewable energy in the EU. The black bars for each country denote their stated targets for 2020. *(From https://ec.europa.eu/eurostat/cache/infographs/energy/bloc-4c.html)*

CHAPTER 4

The production and uses of hydrogen

Introduction

Much has been written in both the technical literature and the popular press about a 'hydrogen economy' and the use of hydrogen as an energy carrier. We have all heard, for example, of the introduction of fleets of hydrogen-fuelled buses. Hydrogen as a fuel has the great advantage that only water is formed when it is combusted. However, as with many of the possible scenarios that are being considered as ways to reduce greenhouse emissions, there are a number of problems associated with the widespread introduction of hydrogen as a fuel. These include not only the fact that the most economical method of hydrogen production (steam reforming of methane) involves the production of large quantities of CO_2 but also that hydrogen transportation and storage both present problems. This chapter first discusses the production of 'grey' hydrogen by steam reforming, the currently most economical route, and also the use of CO_2 capture and storage (CCS) to produce 'blue hydrogen'. Also included is a discussion of the related processes used for the production of ammonia, methanol and hydrocarbon fuels, all of these currently being based on natural gas reforming. The current status of the production of hydrogen by electrolysis will then be discussed: if renewable electricity is used to provide so-called 'green hydrogen' using an electrolysis system, the derived products can also become 'green'. Hydrogen also has an important potential use as a fuel, in either direct combustion applications or in fuel cells, particularly in transport applications. The chapter, therefore, concludes with sections on transporting and storing hydrogen and on its use as a fuel in transportation applications.

The production of hydrogen from natural gas by steam reforming

In addition to its use as a fuel as discussed in Chapter 2, methane is currently the predominant source of hydrogen for industrial use. The majority of the hydrogen in general use is produced by the well-established steam reforming route although some is now produced by 'autoreforming', a topic to be discussed further below. The annual global production of hydrogen is about 70 million tonnes, of which approximately three quarters is formed from natural gas and this route uses approximately 6% of global natural gas production. (See Box 4.1 for a classification of the routes by which hydrogen is produced; some hydrogen is still made by coal gasification, as described in Chapter 2.) An alternative route for hydrogen production is the electrolysis of water (also discussed further below and in more detail in Chapters 7 and 8) but the cost of the hydrogen so produced is currently too high to make this method of production competitive except when it is used in very specialised situations. The data in Fig. 4.1 show that there is a continuously increasing use of pure hydrogen. Approximately 50% is used for refinery purposes (light blue bars), and much of the remainder (dark blue) is used for ammonia production; the small amount remaining ('other', ca. 5%) finds a variety of uses, including as a fuel for space-craft. The data in Fig. 4.1 do not include the hydrogen that is produced as a component of *syn*-gas (a mixture of H_2 and CO). In 2018, this additional quantity amounted to about 42 million tonnes of which about 12 million tonnes were used for methanol production, about 4 million tonnes were used for direct reduction of iron in steel production and the remaining 28 million tonnes was used for fuel

BOX 4.1 Colour-coded nomenclatures for hydrogen production

In some of the literature on hydrogen production, hydrogen is listed under a set of colour-coded headings:

Black hydrogen—produced by gasification of coal
Brown hydrogen—produced by gasification of lignite
Grey hydrogen—produced by steam reforming of natural gas without CCS
Blue hydrogen—produced by steam reforming of natural gas with CCS
Green hydrogen—produced by electrolysis using renewable electricity

The most commonly encountered of these are *grey*, *blue* and *green* hydrogen since coal and lignite gasification are gradually being phased out. Nevertheless, *black* hydrogen will still be produced for many years to come in countries such as China and India that have low reserves of natural gas.

The term CCS ('carbon capture and storage') is sometimes replaced by the term CCUS ('carbon capture, use or storage') in situations in which any CO_2 formed is used in another related process rather than being stored.

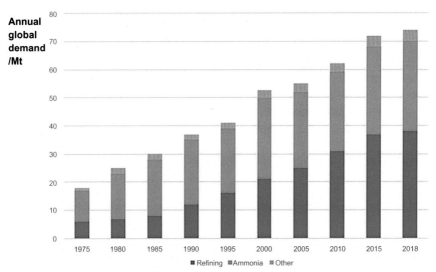

Fig. 4.1 Global demand for pure hydrogen since 1975. *(IEA, Global demand for pure hydrogen, 1975–2018, IEA, Paris https://www.iea.org/data-and-statistics/charts/global-demand-for-pure-hydrogen-1975-2018.)*

and feedstock purposes (e.g. for Fischer Tropsch synthesis).[a] Some of the methods currently available for hydrogen production are now described and subsequent sections describe some other methods of production as well as some of the current uses of hydrogen, these including the synthesis of ammonia, methanol and synthetic fuels (See Box 4.2).

The arrangement of a typical plant of the type that is used for pure hydrogen production is shown schematically in Fig. 4.2.

The steam reforming reaction may be depicted as follows:

$$CH_4 + H_2O \rightleftharpoons CO + 3\,H_2 \quad \Delta H^\circ = +206.2\,kJ$$

This endothermic reversible reaction is shown as being reversible as the conversion only approaches 100% at the very high temperatures required for the process, 850–900°C.[b] The reaction is in most applications catalysed by a Ni-containing material and the reaction is operated in the presence of excess steam to prevent the deposition of carbon by either the methane decomposition reaction:

---

[a] Data from 'The Future of Hydrogen: Seizing Today's Opportunities', IEA Report prepared for the G20 meeting in Japan in 2019 (https://www.iea.org/topics/hydrogen/).

[b] Because the hydrogen produced in the reaction is generally required to be at high pressures, the steam reforming reaction is also carried out at high pressures, typically above 25 atm. The consequence is that the temperature needed for the conversion desired has to be increased further.

BOX 4.2 Carbon capture and storage (CCS)

As discussed in the main text, the CO_2 emissions from many different industrial processes can be very high. In many cases, such as in ammonia synthesis, CO_2 is removed from the product gases resulting from steam reforming of natural gas by well-established amine scrubbing processes; some of the CO_2 is then used in a variety of downstream processes such as in methanol synthesis, even though large proportions are still emitted to the atmosphere. In other processes such as electricity generation, the CO_2 is currently in almost all cases emitted to the atmosphere. In order to improve these situations, methods of storing the CO_2 are being developed, these mostly involving storage in deep subsurface geological formations such as spent oil or gas fields or disused salt or coal mines. Most of this gas that is trapped will be stored indefinitely but some may also be re-used, for example for enhanced recovery of oil, coal-bed methane and gas. The requirements for successful storage include that they should have a rock structure to prevent the escape of CO_2, preferably at a depth of at least 800 m, and that the chambers should be large enough to store the required amount of CO_2 for the lifetime of the site. CCS using such storage facilities present many problems, including the fact that the CO_2-rich gases to be stored must be transported to the storage site. This will normally require the use of suitable pipelines; although many such pipelines already exist, it has been estimated that an additional 70,000 to 120,000 km of the pipeline would be required globally by 2030 if CCS was to be widely introduced. Any storage facility would also need to be monitored continuously for leakage and for any environmental effects throughout its lifetime. Perhaps one of the most important factors is that there is worldwide only a relatively small number of suitable storage sites available and so there is a finite limit to the amount of storage that can be achieved. One of the consequences of this is that there will be a limit to the amount of 'blue hydrogen' that can be produced.

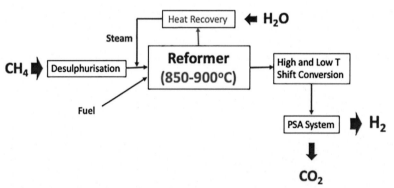

Fig. 4.2 Schematic representation of a plant for the steam reforming of methane to produce pure hydrogen.

$$CH_4 \rightleftharpoons C + 2H_2 \quad \Delta H° = 74.9\,kJ$$

(a reaction that predominates at the high temperatures of the steam reforming process) or the Boudouard reaction:

$$2CO \rightleftharpoons C + CO_2 \quad \Delta H° = -172.4\,kJ$$

(a reaction that predominates at lower temperatures such as those at the exit of a steam reformer system). At the high temperatures required for steam reforming, CO is the main additional product formed but the reactor effluent can also contain small quantities of CO_2, this being formed in the water-gas shift equilibrium reaction:

$$CO + H_2O \rightleftharpoons CO_2 + H_2 \quad \Delta H° = -41.2\,kJ$$

When hydrogen is the main desired product, for example for use in ammonia synthesis, the product gas from the steam reformer is then passed through two water-gas shift reactors operating at lower temperatures; these convert the majority of the CO into CO_2 which is then removed in a pressure-swing absorption (PSA) system. (Any remaining traces of CO, a poison for ammonia synthesis catalysts, are then removed by methanation, the reverse of the steam reforming reaction.) Box 4.3 gives some more detail on the catalytic processes occurring in the steam reforming reaction.

Fig. 4.4 shows the exterior of a typical steam reforming plant. The scale of this plant can be seen from the size of the reactor housing. This housing contains approximately 300 parallel reactor tubes, each typically 10 m in length, that are heated by the combustion of natural gas in burners spaced along the tubes. As discussed in Box 4.3, the reaction rate is determined by the rate at which the reactants reach the active catalyst surface. Hence, the process can easily be scaled down (although further scaling up is not economically justifiable due to reactor costs) so that it is possible to produce hydrogen by steam reforming in relatively small plants compared with ones such as that shown in Fig. 4.4. Hence, for example, it is possible to envisage a situation in which hydrogen to be used for transportation purposes could be generated at the fuelling station rather than having to be transported to that station using sophisticated high-pressure transporters. Fig. 4.5 shows a small-scale steam reforming plant designed for this purpose. This plant contains all of the elements of the full-scale plant shown in Fig. 4.4 (hydrodesulphurisation, water-gas shift, pressure swing adsorption) but the reactor tubes are now only approximately 2 m in length. HyGear, the company making this unit, says that their plants of various different sizes produce hydrogen up to 99.9999% purity at a flow rate of up to 500 Nm^3/h and at pressures

BOX 4.3 Catalysis and the steam reforming of methane

What is a catalyst? A catalyst is a material which, when added to a chemical reaction mixture, increases the rate of the chemical reaction but is itself not used up in the reaction.[c] In the case of the steam reforming of methane, the reaction would require some extremely high temperatures in order to proceed at an acceptable rate if no catalyst was present. Using a catalyst, the steam reforming reaction occurs at much lower temperatures, at ca. 450°C and above, by adsorbing the reactants (and products) on the surface of active material, in this case generally metallic nickel, in such a way that the all-over activation energy for the process is significantly reduced. The attainment of good conversions in the steam reforming reaction is however further complicated by the fact that the reaction is very endothermic, this having the consequence that high temperatures are necessary in order to obtain the desired conversions, particularly as the pressures are also relatively high (see main text).

The steps occurring during the catalytic conversion of gaseous methane and water (the feed) to CO and hydrogen (the products) are shown very schematically in Fig. 4.3. The feed enters the reactor, in this case a long tubular packed-bed (or more frequently, a large assembly of such tubes in parallel, as discussed in the main text) containing a suitable catalyst. The catalyst generally comprises of nickel on a temperature-resistant support material such as α-alumina or magnesium aluminate spinel in the form of Raschig rings (or even more complicated shapes). Such shapes are used to optimise gas flow through the reactor and to enable the reaction mixture to reach all parts of the external surfaces of the support materials. These generally have relatively large pores and the feed molecules diffuse to the nickel surfaces within the support through these pores. Adsorption and dissociation of the methane and water then occur and reaction occurs between these species on the Ni surface. Finally, the products (carbon monoxide and hydrogen) desorb from the nickel and diffuse out of the pores and hence towards the exit of the reactor. The steam reforming reaction is one of the very few catalytic reactions in which the rates of the individual chemical surface processes occurring do not determine the rate of the reaction; instead, the conversion achieved in the tubular reactor depends solely on the rates of external and internal diffusion, the majority of the reaction occurring near the entrances to the pores where equilibrium conversions are rapidly achieved. Everything that occurs in the reactor is therefore determined by the thermodynamics of the various reactions taking place and the kinetics of the reactions involved have little or no influence. It is for this reason that the reaction must be operated using excess steam in order to avoid conditions that would thermodynamically permit carbon deposition to occur in any position in the reactor.[d]

Many of the transition metals of Group 8 will bring about the steam reforming reaction. However, of these, Ni is the metal most commonly used, mainly because it

Continued

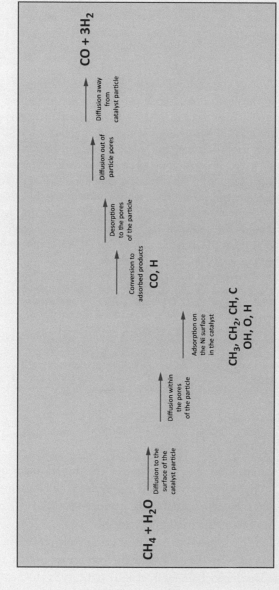

Fig. 4.3 Schematic representation of the stepwise conversion of natural gas to hydrogen in the steam reforming reaction. Note that the reaction will occur in the reverse direction (methanation of CO) if the temperature of the reaction is much lower. The possible existence of the water-gas shift reaction (giving CO_2) is not included in this scheme as the equilibrium proportion of CO_2 under these reaction conditions is negligible. Note that species shown below the level of the feed CH_4 and H_2O are formed exothermically (release energy), while those above this level are formed endothermically (require energy).

Continued

BOX 4.3 Catalysis and the steam reforming of methane—cont'd

is much cheaper and more abundant than the other Group 8 metals. It is also important in the choice of active component that the active metal is not oxidised by the water present under reaction conditions according to the reversible reaction:

$$M + H_2O \rightleftharpoons MO + H_2$$

For a nickel, the metal remains in its reduced form as long as the ratio P_{H2}/P_{H2O} is greater than about 3×10^{-3}. Cobalt will also remain in its reduced state as long as this ratio is greater than 2×10^{-2}. However, iron requires significantly higher hydrogen partial pressures to ensure that the metal remains reduced; if the oxidised state considered is FeO, the ratio has to be greater than 5.0. For the noble metals, the potential oxidation of the metal is not an important limitation as they all remain in the reduced form under these conditions. It has been found that they all carry out the reforming reaction very efficiently. However, the costs of the noble metals are too high for them to be used in practice. Rostrup Nielsen has reported that the order of activities of the noble metals for the steam reforming of ethane is: Rh, Ru > Ni, Pd, Pt > Re.[e]

[c] A full discussion of heterogeneous catalysis and of the preparation, characterisation and application of catalysts is beyond the scope of this book. A reader who does not have a basic knowledge of these topics is advised to refer to a textbook such as 'Contemporary Catalysis - Fundamentals and Current Applications' Julian R.H. Ross, Elsevier, 2019; ISBN: 978-0-444-634740-0.

[d] The present author has reported work on the steam reforming of methane carried out at low pressures (ca. 0.03 atm.) in the temperature range 500–680°C when CO_2 was also a possible product. The work showed that CO was the primary product of the reaction and that the adsorption of methane on the Ni surface was the rate determining step under these conditions. (see J.R.H. Ross and M.C.F. Steel, 'Mechanism of the Steam Reforming of Methane over a Coprecipitated Nickel-Alumina Catalysts', J. Chem. Soc., Faraday Trans. I, 69 (1973) 10–20.)

[e] J.R. Rostrup Nielsen, J. Catal., 31 (1976) 110.

up to 300 bar (g). The system illustrated is a HyGEN 150 model that is built in a 40 ft. container. The hydrogen production rate is described as being in the range 123 to 141 Nm3/h, depending on the purity required (99.5 to 99.999999%), at a pressure from 1.5 bar(g) to 7 bar(g).

It is possible to produce hydrogen on an even smaller scale using microreactor technology. As discussed in Box 4.4, such microreactors contain thin layers of catalyst confined in microchannels in close proximity to the heating elements used to provide the necessary energy for the reaction. Such reactors give much higher catalyst effectiveness than the more conventional system described above by minimising diffusion limitations. Many different designs of such microreactors have been described in the literature but there is as yet no indication of their commercial application.

Fig. 4.4 A typical large-scale plant for the steam reforming of natural gas to produce pure hydrogen This is a plant for the production of ammonia at a rate of 1500 MTPD; the secondary reformer is also visible. Photograph reproduced with kind permission of Haldor Topsoe. (See footnote 't'). *(Photograph reproduced with permission from Haldor Topsoe's White Paper on the New SynCOR Ammonia process; www.topsoe.com.)*

Fig. 4.5 A small-scale steam reforming unit for the local production of hydrogen at hydrogen-refueling stations. *(Photograph kindly supplied by HyGear Ltd.)*

BOX 4.4 Microreactors and their use for hydrogen production

There has been a great deal of work on the use of microreactor systems for reactions such as the steam reforming of methane. An important aspect of a microreactor is that heat exchange to and from the active catalyst is easily achieved as the catalyst is generally situated very close to the wall of the channel in which it is situated and heat is easily transmitted in either direction at that wall, reducing heat transfer limitations. The topic of microreactors has been well reviewed by a number of groups active in the field. A particularly useful review is one by Kolb and Hessel in which a description is given of some of the various reactor structures that had been reported in the literature up until 2004.[f] Fig. 4.6 is a diagram taken from the review that shows the structure of a typical microreactor used as part of a complete system for the steam reforming of methanol, showing typical dimensions of such a reactor. The length of each reaction channel is only several cm. In this example, a hydrogen-rich gas containing traces of CO produced by the steam reforming of methanol enters on the right-hand side and passes through the appropriate heat exchangers as well as the central section containing a selective oxidation catalyst before passing to a fuel cell. The important aspect of this type of reactor is that heat and mass transfer limitations are minimised, vastly improving the efficiency of the process involved, and that the quantity of catalyst required is also therefore significantly reduced.

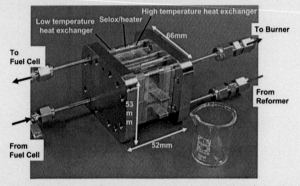

Fig. 4.6 Typical dimensions of a microreactor An example of the structure of a microreactor designed for the selective oxidation of traces of CO from a hydrogen stream emerging from an ethanol steam reforming reactor prior to its admission to a fuel-cell system. *(From 'Micro-structured Reactors for Gas Phase Reactions'. (2004). Chem. Eng. J., 98, 1–38. Reproduced with the kind permission of Elsevier.)*

[f] G. Kolb and V. Hessel, 'Micro-structured Reactors for Gas Phase Reactions', Chem. Eng. J. 98 (2004) 1–38.

The energy required to control the temperature of the catalyst assembly in a microreactor used for steam reforming of methane can be applied by electrical heaters. However, an alternative is to insert between each steam reforming element a second catalyst layer in which methane combustion takes place. Fig. 4.7 shows such a possible structure as described in a paper by Junjie Chen and colleagues that concerns the modelling of a thermally integrated microreactor system in which a very active rhodium catalyst, used for steam reforming, is contained in one set of channels and an equally active Pt catalyst, used for methane combustion, is contained in the others, there being good thermal conductivity between the two sets of channels.[g] (The authors stress the importance of the presence of highly active catalysts; in such a micro-reactor construction, the surface reaction becomes the important step and diffusion limitations are negligible.) Fig. 4.8 shows the results of modelling the temperature profile through a single reforming channel and the corresponding methane mole fractions, the methane conversions and the fluid velocities through the same channel. The conversions of methane are greatest near the walls of the channel and the highest conversion is attained at the end of the channel (The rate of flow is also shown, this being greatest at the end of the channel as a result of the increased number of molecules in the product gases compared with the reactants.)

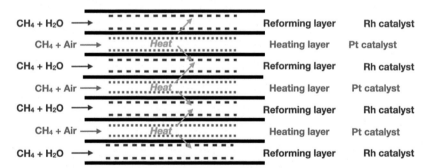

Fig. 4.7 A microreactor system containing parallel steam reforming and methane oxidation channels used for modelling the reaction; there is good heat transfer between the channels. *(Derived from 'Compact Steam-Methane Reforming for the Production of Hydrogen in Continuous Flow Microreactor Systems.' (2019). A.C.S Omega, 4, 15600–15614.)*

[g] J.J. Chen, W.Y. Song and D.G. Xu, 'Compact Steam-Methane Reforming for the Production of Hydrogen in Continuous Flow Microreactor Systems', ACS Omega, 4 (2019) 15600–15,614.

Reforming channel

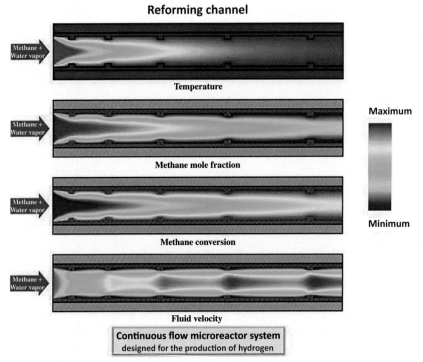

Maximum

Minimum

Continuous flow microreactor system
designed for the production of hydrogen

Fig. 4.8 The results of modelling the temperature profile through a single reforming channel and the corresponding methane mole fractions, the methane conversions and the fluid velocities through the same channel. *(From 'Compact Steam-Methane Reforming for the Production of Hydrogen in Continuous Flow Microreactor Systems.' (2019). A.C.S Omega, 4, 15600–15614, with kind permission of American Chemical Society.)*

A significant addition to the cost of operating steam reforming plants of the type described above is related to the cost of the removal of CO_2 from the effluent of the reactor. One method of avoiding this problem is the use of a reactor system incorporating a membrane separation system of the type shown schematically in Fig. 4.9. It is well established that membranes of palladium or its alloys (e.g. Pd/Ag) can be used to separate out the hydrogen formed. The hydrogen is dissociatively *adsorbed* on the Pd surface before being *absorbed* in the bulk of the Pd; the absorbed atomic hydrogen then diffuses through the bulk of the palladium to the other side of the membrane where it is desorbed as molecular hydrogen. The driving force for this diffusion is the hydrogen concentration gradient through the membrane.

There has been significant effort in designing and testing systems that allow membrane separation to occur within the reformer system. Fig. 4.10, taken from a review by Nikolaidis and Poullikkas, shows one such

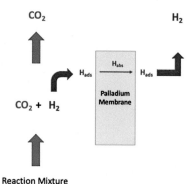

Reaction Mixture

Fig. 4.9 Schematic representation of a process for hydrogen separation from a steam reforming product stream using a membrane involving a metal such as Pd or a Pd/Ag alloy.

system.[h] In this design, the catalyst particles are placed in the spaces between the membrane tubes. As hydrogen is withdrawn from the product mixture, the equilibrium of the steam reforming reaction is driven to the right, allowing high equilibrium conversions to be achieved at lower temperatures than in conventional systems.

This approach is illustrated schematically in Fig. 4.11. The effectiveness of the combination of catalyst and membrane will depend on both the catalytic activity attained as well as the permeability of the membrane under the reaction conditions. Such a system will thus require careful control of the reaction conditions used, particularly the temperature of both the catalyst and membrane.

The use of other metals for membrane systems is discussed briefly in Box 4.5.

The production of hydrogen from natural gas by other methods

Syngas can be produced from methane by two other reactions, partial oxidation, often now referred to as 'autoreforming':

$$CH_4 + 0.5\,O_2 \rightarrow CO + 2\,H_2$$

and CO_2 reforming, better known as 'dry reforming':

$$CH_4 + CO_2 \rightleftharpoons 2\,CO + 2\,H_2$$

These two processes will now be discussed briefly in turn.

[h] P. Nikolaidis and A. Poullikkas, 'A Comparative Overview of Hydrogen Production Processes', Renewable and Sustainable Energy reviews, 67 (2017) 597–611.

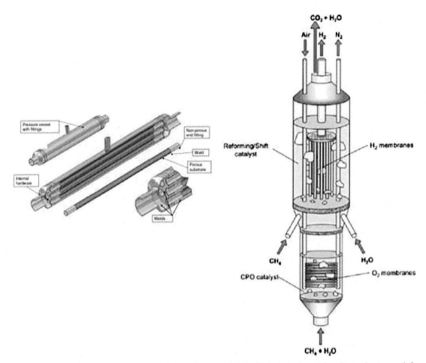

Fig. 4.10 A membrane reactor containing multiple Pd membrane tubes designed for the steam reforming of methane. *(From 'A Comparative Overview of Hydrogen Production Processes.' (2017). Renewable and Sustainable Energy Reviews, 67, 597–611. Reproduced with kind permission of Springer.)*

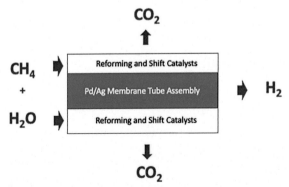

Fig. 4.11 Schematic representation of the combination of a methane steam reforming catalyst and a Pd-based membrane for simultaneous selective hydrogen removal.

BOX 4.5 The possible use of other metals such as tantalum as a membrane material

The composition of the membranes used for hydrogen separation has also received significant attention. For example, Rothenberger et al. have reported on the hydrogen permeability of tantalum-based membrane materials; they found that the rate of permeation was low, probably as a result of surface contamination.[i] In this context, it is probably worth noting some of the work from the PhD thesis work of the present author.[j] As part of a study of the adsorption of hydrogen sulphide and hydrocarbons such as neopentane on evaporated metal films, he showed that the uptake of hydrogen in clean films of tantalum evaporated under high vacuum conditions was facile and that the quantities absorbed increased with temperature in the range studied, 70 to 132°C; the uptakes corresponded to up to 0.14H per Ta atom in both the evaporated film and the filament from which the film had been evaporated, this being maintained at the same temperature. (As part of this work, it was also shown that uptakes of hydrogen in Pd of up to 0.6 H/Pd could be obtained in prolonged experiments but the total uptake possible for Ta was not measured.) The work showed the importance of having a clean metal surface to dissociate the hydrogen[k]: the effect on hydrogen absorption of adsorbing sulphur from the dissociation of H_2S was examined and it was found that a monolayer of S species completely stopped hydrogen absorption with both metals. However, desorption of hydrogen from one or other bulk hydride through the S-layer was still possible, this indicating that hydrogen atoms could still transfer through the contaminating S layer. Further, it was shown that it was possible to have continued absorption of hydrogen from the gas phase if the hydrogen gas was dissociated into atoms at the Ta filament from which the film had been evaporated when this filament was heated in the presence of hydrogen.

[i] K.S. Rothenberger et al., 'Evaluation of Tantalum-based Materials for Hydrogen Separation at Elevated temperatures and Pressures', J. Membrane Sci. 218 (2003) 19–37.

[j] J.R.H. Ross, PhD Thesis, Queen's University of Belfast, 1966; see also M.W. Roberts and J.R.H. Ross, 'The Interaction of Hydrogen Sulfide and Hydrogen with Palladium and Tantalum Films', in Reactivity of Solids, Edited by J.W. Mitchell et al., John Wiley and Sons, N.Y., 1969.

[k] Until the 1960s, work on the Pd/H and Ta/H systems had used electrochemical methods to incorporate the hydrogen, under which conditions possible surface contamination was less critical.

Autoreforming

The partial oxidation reaction is shown above as being irreversible as 100% conversion is thermodynamically possible at the higher reaction temperatures used (typically up to 1000°C) and the product has a syngas ratio of 1:2. No carbon deposition is allowed thermodynamically under normal reaction temperatures.[1] If pure hydrogen is required as the product gas, steam is added to the reaction mixture and this is available downstream to strip out the CO formed by carrying out the water-gas shift conversion as is shown in Fig. 4.12. In this case, a total of three hydrogen molecules are formed from the reaction of one molecule of methane. Many major plant construction companies are now installing autoreforming units in place of conventional steam reforming plants for hydrogen production. For example, Haldor Topsøe has published very informative 'White Papers' on the use of autoreforming in ammonia and methanol manufacture.[m] Both of these processes will be discussed further below in relation to some of the uses of hydrogen. Suffice it to say here that the air separation unit supplies a ready supply of pure nitrogen for the ammonia synthesis process while the CO_2 formed in the water-gas shift reactor is an excellent feed material for the synthesis of methanol. (Omitting a water-gas shift system would mean that the effluent gas could be used directly for the more conventional methanol synthesis route using CO as a reactant). Another important point to be made here is that the total size of an autoreforming plant is small compared with a conventional

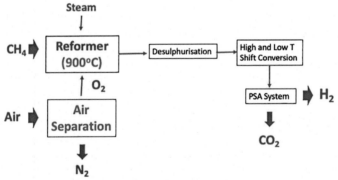

Fig. 4.12 Schematic representation of an autothermal reactor for the partial oxidation of methane to produce pure hydrogen.

---

[1] In practice, the methane entering the reactor is first oxidised exothermically to give water and CO_2; this is followed by the endothermic reactions, steam reforming and dry reforming, in the rest of the reactor.'

[m] 'New Syncor ammonia process'; 'Methanol for a more sustainable future - Electrified chemicals'; 'Lower your carbon footprint while boosting your CO output using Topsoe's unique ReShift technology'. Haldor Topsoe white papers, www.topsoe.com.

steam reforming plant. This has the important result that investment costs are lower. Additionally, the total capacity of the plants can also be much higher. We will return to this topic below when discussing the uses of hydrogen in ammonia and methanol synthesis.

Dry reforming of methane

As shown in the equation above, the dry reforming reaction produces a mixture of CO and H_2 in the ratio 1:1. Dry reforming is a topic that has received much research attention for a number of years but the reaction is not used in practice, partly because of problems with carbon deposition on the catalysts used and partly because there are few uses for the product gases in the ratio 1:1. The lack of such uses of the products in chemical processes precludes the potential use of the reaction for CO_2 mitigation, a factor generally overlooked by authors of research papers on the subject. However, a combination of steam reforming and dry reforming (i.e. co-feeding H_2O and CO_2) can be used as the product ratio is then much more usable. However, the problem of carbon deposition on the catalysts still persists.[n]

Methane pyrolysis

The technology for the production of hydrogen and syngas using the pyrolysis of coal has been known for many years. More recently, the method has been applied to the pyrolysis of hydrocarbon materials in which the hydrocarbon is the sole source of the hydrogen produced:

$$C_nH_m \rightarrow nC + m/2\ H_2$$

For low molecular weight hydrocarbon feeds (boiling point up to 200°C), hydrogen is produced directly. For higher molecular weights, the hydrocarbon is gasified in two steps; in the first, hydrogasification of the hydrocarbon feed to form methane is carried out:

$$C_nH_m + H_2 \rightarrow nCH_4$$

---

[n] The present author has reviewed some of the recent literature on dry reforming catalysts in an article entitled: 'Syngas production using carbon dioxide reforming: Fundamentals and perspectives'. (In 'Transformation and Utilization of Carbon Dioxide', Edited by B.M. Bhanage and M. Arai, pages 131–161, Springer, (2014).

and this is then followed by pyrolysis of the methane produced:

$$CH_4 \rightarrow C + 2H_2$$

It should be recognised that the hydrogen so produced can be considered as 'blue' (Box 4.1) since the carbon formed can easily be stored or alternatively can be used for soil remediation.

Electrolysis of water

The electrolysis of water to produce hydrogen is a well-established method of producing pure hydrogen:

$$2H_2O \rightarrow 2H_2 + O_2$$

As is discussed in greater detail in Chapter 7 the value of $\Delta G°$ for this reaction is positive and the decomposition of water would require very high temperatures for it to occur spontaneously; the reaction can only be brought about at and around room temperature by applying an electrical potential sufficient to exceed this positive value of $\Delta G°$.° Fig. 4.13 gives a schematic representation of an electrolysis system used for the reaction. In order to safely segregate the hydrogen and oxygen formed in the electrolysis reaction, it is necessary to introduce some sort of barrier (a membrane) between the anode and cathode of the electrochemical reactor. Common systems used are proton exchange membrane (PEM) cells and solid oxide electrolysis (SOEC) cells. PEM cells operate by transferring hydrogen ions (formed at the anode) from the anode compartment by way of an ion-conducting polymer membrane to the cathode compartment where hydrogen gas is then liberated at the cathode. With the SOEC cells, hydroxyl species formed at the cathode are transformed to oxygen ions at the solid membrane surface and are then transferred through the membrane, finally reacting at the anode to form molecular oxygen. These systems therefore differ in the species that are transferred from one side of the cell to the other, hydrogen or oxygen. Further, as SOEC cells operate most effectively at elevated temperatures, part of the energy

° An electrolysis reaction occurs at equilibrium if Go = − nFE, where n is the number of electrons involved in the process, F is the Faraday equivalent (the charge carried by one mole of electrons, 96,494 coulombs per mole) and E is the potential applied to the cell. In practice, a higher potential (an 'overpotential') has to be applied to cause the electrolysis reaction to proceed at a significant rate. (See Chapter 6 for a fuller discussion.)

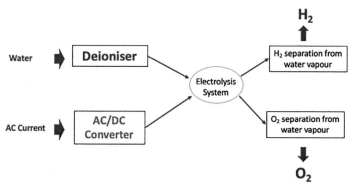

Fig. 4.13 Schematic representation of a system used for the electrolysis of water to produce both pure H_2 and pure O_2.

needed to drive the reaction must be supplied thermally rather than electrically.[P] In practice, an installation used to produce hydrogen requires water purification and an AC/DC converter to provide the necessary direct current voltage, as well as systems to dry the product gases (Fig. 4.13). The purities of both the hydrogen and oxygen produced are very high.

Electrolysis as a method for the production of hydrogen is currently seen only as a method to be used in very specialised applications since the cost of operation is high as a result of the high cost of electricity. The consequence is that only approximately 5% of current global hydrogen production is by electrolysis. However, the outlook is that we shall soon see a change in this scenario with the movement towards renewable electrical energy generation. If the costs of renewable electrical power can be reduced still further and the demands for local supplies of hydrogen increase, we are likely to see a steady move towards the production of 'green' hydrogen by electrolysis. This topic is handled further in Chapter 7.

[P] A detailed description of electrochemical processes is beyond the scope of this book. For further information, an interested reader should consult a modern textbook on Physical Chemistry or Electrochemistry.

The generation of hydrogen from biomass by various processes

As will be discussed in much greater detail in Chapter 5, the term *biomass* applies to a wide variety of materials, ranging from commonly available commodities such as grass and corn stover to timber and agricultural waste. Many of these materials contain bio-polymers that consist of cellulosic and hemicellulosic species as well as lignin and other molecules. All of these contain some level of hydrogen and a variety of methods, to be discussed later, have been used either to extract this hydrogen or to convert the biomass components into useful chemicals. As long as the biomass used is fully 'renewable', the products are generally considered as 'green', not contributing to greenhouse gas emissions when finally reaching the stage of disposal.[q] Other processes for biomass utilisation are also described in Chapter 5.

Comparison of hydrogen production costs for different processes

Nicolaides and Poullikka have assembled a series of useful estimates of the costs of hydrogen production by various different methods including those described above. Table 4.1 summarises some of their most relevant data. The interested reader should consult the original reference for details of the sources from which these data have been extracted. It can be seen that auto-reforming consistently gives among the lowest costs. As these data were assembled during the period 1992 to 2007, it is probable that the costs of production of hydrogen using autoreforming and electricity from solar PV will now both be significantly lower due to improvements in autoreforming plant design and improved PV efficiencies. For an explanation of the term Carbon Capture and Storage (CCS), see Box 4.2 above.

Methanol production

We now discuss a number of important industrial processes that use the hydrogen and syngas produced by the various routes. Methanol synthesis makes use of a syngas mixture according to the following reaction:

$$CO + 2H_2 \rightarrow CH_3OH$$

The proportions of CO and hydrogen in the syngas are adjusted to the correct value by making use of the water-gas shift reaction, as described above. The catalyst used for the synthesis process is generally a Cu-

---

[q] Forestry operations are considered to be renewable if equivalent new plantation occurs following harvesting. However, destruction of forestry in the process of land clearance is to be abhorred.

Table 4.1 Hydrogen production costs using different methods.

Process	Energy source	Feedstock	Capital cost/ m$	H_2 cost/\$ kg^{-1}
Steam reforming with CCS	Fossil fuels	Natural gas	226.4	2.27
Steam reforming without CCS	Fossil fuels	Natural gas	180.7	2.08
Coal gasification with CCS	Fossil fuels	Coal	545.6	1.63
Autoreforming with CCS	Fossil fuels	Natural gas	183.8	1.48
Methane pyrolysis	Internally generated steam	Natural gas	–[a]	1.59–1.70
Biomass pyrolysis	Internally generated steam	Woody biomass	53.4[b]	1.25–2.20
Biomass gasification	Internally generated steam	Woody biomass	149.3[c]	1.77–2.05
Solar PV and electrolysis	Solar	Water	12–54.5	5.89–6.03
Wind and electrolysis	Wind	Water	504.8–499.6[d]	5.89–6.03
Nuclear power and electrolysis	Nuclear	Water	–[a]	4.15–7.0

[a]No cost given.
[b]For a plant capacity of 72.9 tons per day.
[c]For a plant capacity of 139.7 tons per day.
[d]The higher cost presupposes the generation of electricity along with hydrogen while the lower cost is for hydrogen alone.
Data extracted from the reference of footnote 'g'.

ZnO-Al_2O_3 material similar to that used for the water-gas shift reaction. There are many large-scale methanol plants worldwide, a particularly high concentration of these being in Asia; the total global production of this important chemical was about 140 million tonnes in 2018 and it is predicted that that figure will double by 2030. There has recently also been an interest in producing methanol from CO_2; see Box 4.6.

Production of fuels using the Fischer Tropsch process

Another important use of syngas is in the Fischer-Tropsch (F.T.) process for the production of synthetic fuels. The reaction can be represented very roughly by the chemical equation:

$$nCO + (2n + 1)H_2 \rightarrow CnH_{2n + 2} + nH_2O$$

BOX 4.6 Green methanol

There is currently significant interest in producing methanol by the hydrogenation of CO_2 rather than of CO. The process proceeds in two steps, the first of these being the reverse of the normal water-gas shift reaction discussed above:

$$CO_2 + H_2 \rightleftharpoons CO + H_2O$$

This is followed by the reaction of the CO with hydrogen to give methanol as described in the main text:

$$CO + 2H_2 \rightarrow CH_3OH$$

The all-over reaction is therefore:

$$CO_2 + 3H_2 \rightarrow CH_3OH + H_2O$$

The production of methanol from CO_2 therefore requires the use of more hydrogen than does its production from CO. The process described here could be used as a method of utilising CO_2 but it would only have any significant value if the hydrogen could be produced using renewable energy, for example by an electrolysis route using electricity produced using a renewable route such as one of those discussed in Chapter 2.[r] The methanol so produced has been termed 'green methanol'. See also Chapter 7 where an alternative method of using CO_2 for this reaction is discussed.

[r] Haldor Topsøe have recently described their work on the production of 'e-Methanol', otherwise called 'electrified methanol', in a brochure entitled: 'Methanol for a More Sustainable Future' (topsoe.com/emethanol). Their process uses a specially designed Cu-containing methanol catalyst 'MK-317 SUSTAIN' which they claim gives high selectivity for methanol synthesis from CO_2 and hydrogen, the latter being produced by electrolysis. Their process uses ca. 500 kWh of electricity to produce one ton of high-purity methanol. Topsøe's work on the production of hydrogen using electrolysis is discussed in Chapter 7.

The range of products obtained is quite large, ranging from methane to waxes, and the resultant mixture may also contain some oxygenated species. The actual product composition obtained in a particular plant is very dependent on the reaction conditions and the catalyst used.[s]

The F.T. process was initially developed in Germany during the 1930s for the production of fuels. The original process used coal as raw material for syngas production using Lurgi gasifiers. The process was then further developed in South Africa by Sasol during the years of Apartheid, this having been essential since South Africa was at that time solely dependent on domestic resources. Following the discovery of offshore gas fields in Mossel Bay

[s] For further information on this topic, the reader should refer to Chapter 12 of the book listed in footnote 'b'.

(400 km. east of Capetown) in 1969, the South African Petroleum, Oil and Gas Corporation (PetroSA) started to use syngas produced from natural gas as a feed for the FT process. This installation currently produces 15% of South Africa's transport fuel requirements.

The Fischer Tropsch Process is now also used a number of major 'Gas to Liquids' (GTL) plants in which the syngas required is produced from natural gas by partial oxidation:

$$CH_4 + 1/2\,O_2 \rightarrow CO + 2\,H_2$$

This reaction can be carried out using either air, when the product gas will contain nitrogen, or pure oxygen (Auroreforming), when a preliminary air separation step is required, as described above. Shell uses partial oxidation in several major commercial GTL plants; the first of this started operation in Bintulu, Malaysia in 1993 and another, which Shell claims is the world's largest, started operation in Qatar in 2011 ('Pearl GTL').

Production of ammonia

Traditionally, ammonia is manufactured using pure hydrogen produced by steam reforming in a reactor system such as that of Fig. 4.2 and a plant such as that shown in Fig. 4.4. The hydrogen is then fed together with the nitrogen to an ammonia synthesis reactor. The synthesis of ammonia:

$$N_2 + 3H_2 \rightleftharpoons 2NH_3$$

is an exothermic reaction and high yields are thermodynamically favoured by operation at low temperature and low pressures. However, since the ammonia product is required at higher pressures (ca. 30 atm.), the synthesis is carried out at this pressure in a sequence of reactors operating at a series of decreasing temperatures, the final conversion being determined by the temperature of the final reactor.[t]

Haldor Topsøe has recently introduced a new technology for ammonia production, the 'Syncor Ammonia Process', this being based on the production of hydrogen from natural gas using their 'Syncor' reactor (see

---

[t] For a full description of the operation of some typical ammonia synthesis reactors, the reader should consult the book listed under footnote 'c'.

Fig. 4.14).[u] This reactor is a variant of the more usual autothermal reforming system (Fig. 4.12). Following a pre-reforming step in which any higher hydrocarbon present in the feed gas is converted by steam reforming to syngas at lower reaction temperatures, a mixture of natural gas and oxygen (the latter produced using an air separator), together with the products of the pre-reformer and some steam, is ignited at the top of the reactor. The flame of combustion gases (now a mixture of unburnt methane, CO, CO_2 and H_2O) heats the catalyst bed at the base of the reactor up to reaction temperature. The mixture is then converted completely to syngas without the need for any additional heat input.

Following two-shift reactors, both now operated at high temperature, and CO_2 removal, nitrogen from the air separator is added to the hydrogen stream and the mixture is admitted to a three-stage ammonia synthesis reactor. Fig. 4.15 shows a schematic representation of the complete plant. A full description of all the components of the plant is given in reference of footnote 't'.

Because there is no need for external heating, the plant with the Syncor reactor (Fig. 4.16) is much smaller than the equivalent conventional plant using tubular steam reforming reactors (Fig. 4.4). The consequence is that the maximum capacity for a Syncor plant can be up to 6000 metric tons

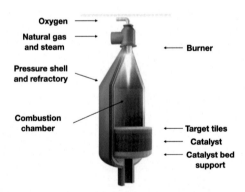

Fig. 4.14 Cutaway diagram of the Haldor Topsoe Syncor reactor. *(Diagram reproduced with permission from Haldor Topsoe's White Paper on the New SynCOR Ammonia process; www.topsoe.com. Reproduced with the kind permission of Haldor Topsoe.)*

---

[u] P.J. Dahl, C. Speth, A.E. Kroll Jensen, M. Symreng, M.K. Hoffmann, P.A. Han, S.E. Nielsen, 'New Syncor Ammonia Process', Haldor Topsoe white paper, cvr 41853816/CCM/0242.2017 (www.topsoe.com).

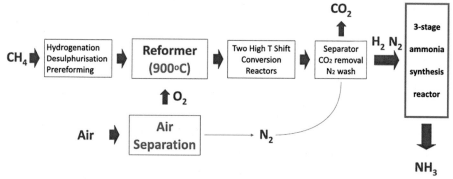

Fig. 4.15 Simplified schematic diagram of a complete ammonia synthesis plant incorporating a Topsoe Syncor reactor. *(Adapted from the white paper of footnote 'n'.)*

of ammonia per day compared with a more conventional plant for which the maximum capacity is only about 1500 metric tons per day. Another very important aspect of Syncor technology is that it operates at steam to carbon ratio of about 0.6. This is much lower than that in a conventional plant for

Fig. 4.16 An ammonia plant with a Syncor Autoformer. *(Reproduced from Haldor Topsoe white paper on Syncor Ammonia Process. (n.d.). www.topsoe.com, with kind permission. Figure kindly supplied by Haldor Topsoe.)*

which the steam/carbon ratio is about 3.0 and there is therefore a much-reduced requirement for steam handling equipment. Most importantly, as there is no requirement for reactor heating, the all-over production of CO_2 is much reduced when compared with the conventional plants involving steam reforming. Further, because of there being no need for steam recycling, the all-over site area ('footprint') required for the complete installation is also much lower. (Compare Fig. 4.4 and Fig. 4.16). Further developments related to this approach but using hydrogen produced by electrolysis are discussed in Chapter 7.

Conclusions

This chapter describes the methods used for the production of hydrogen, methanol, and ammonia. Depending on whether or not carbon dioxide capture and storage (CCS) is applied, these molecules can be termed as "Blue" or "Grey" (or even "Black" if the hydrogen is produced from coal). Use of the "Blue" products could make a very significant contribution to a reduction in total global green-house gas emissions. However, the success of such an approach will depend strongly on the success achieved in developing reliable CCS facilities. A more sustainable approach will be to make use of hydrogen and other products formed making use of renewable resources. In this context, Chapter 5 examines the use of biomass as a source of hydrogen and other chemicals while Chapter 7 includes a discussion of the developing opportunities related to the use of green hydrogen formed by electrolytic methods for transportation and chemical processing.

CHAPTER 5

Biomass as a source of energy and chemicals

Introduction

This chapter is devoted to the increasingly important topic of biomass utilisation. Biomass is formed by processes in which growth occurs by photochemical reactions of carbon dioxide and water, these giving products in which energy is stored. All the reactions of biomass which will now be discussed involve the release of some or all of this stored energy; the most stable products from biomass conversion are carbon dioxide and water formed in combustion reactions but many other less stable products are also possible. Although the various fossil fuels such as coal and oil discussed in earlier chapters were originally derived from biomass materials, the term biomass as used in this chapter refers to material originating recently, or relatively recently, from living vegetation, sources ranging from trees and grasses to seaweed and algae. Since both animals and humans depend either directly or indirectly on one or other type of vegetable matter as a source of food, the term also covers various forms of the resultant agricultural and human waste ('organic residues'). Grasses, starch crops and sugar crops can all be converted to sugars and hence to other valuable products by straightforward hydrolysis. Oil crops can be converted to biodiesel while lignocellulosic materials, lignocellulosic residues and algae can be converted by more exacting hydrolysis processes and also by fermentation or pyrolysis.[a] Fig. 5.1 shows some of these processes in a schematic form. The main chemical elements contained in all these biomass materials are carbon, hydrogen and oxygen, these resulting from CO_2 and H_2O. However, biomass sometimes also contains significant proportions of sulphur and nitrogen as well as inorganic elements that are incorporated during the growing process. As will be discussed further in later sections of this chapter, the

---

[a] F. Cherubini, G. Jungmeier, M. Wellisch, T. Wilke, I. Skiadas, R. van Ree and E. de Jong, 'Towards a common classification approach for biorefinery systems', Biofuels, Bioproducts and Biorefining, 3 (2009) 534–546.

Sustainable Energy
https://doi.org/10.1016/B978-0-12-823375-7.00004-4

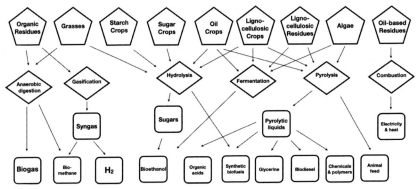

Fig. 5.1 Types of biomass and some of their products. *(Modified from reference of footnote 'a'.)*

predominant structural organic components of many biomass types are cellulose (a polymer composed of glucose units), hemicelluloses (a range of polymers composed of pentoses, especially xylose, and of other hexoses, glucose and mannose) and lignin (a heterogeneous phenolic polymer); in most cases, these are combined with lipids, proteins, starches and hydrocarbons. This chapter first discusses two long-standing uses of wood: the *combustion* of timber as a source of energy; and the production of paper from wood by pulping methods. It then covers routes by which various forms of biomass, including timber, can be converted into some important chemical products.

Wood as a source of energy and paper[b]

Biomass has been a source of energy for many centuries and the combustion of wood was the only source of energy for both heating and cooking that was used in many ancient civilisations before the use of coal became widespread. Biomass combustion still predominates for these purposes in many of the less advanced countries around the world. The combustion of wood gives rise predominantly to water and CO_2. However, there can also be significant emissions of SO_2 and NO_x and these emissions can be accompanied by significant proportions of non-combusted organic components as well as of particulates. These emissions are now known to cause severe health problems, something that is particularly problematic in developing countries where biomass combustion is used for domestic purposes, often

---

[b] Timber is also an important building material but that use is outside the scope of this book.

in situations where the ventilation is inadequate. Such emissions are particularly problematic in situations in which other materials such as cow dung are used as fuel. There has been a resurgence in the use of wood as a fuel, the argument being that it is a sustainable resource. The growth of biomass requires a combination of CO_2, water and sunlight and so plant growth is seen as sequestering CO_2. Hence, when the biomass is burnt, the resulting CO_2 is reckoned to be replacing that which had previously been sequestered. It is therefore commonly argued that a carbon balance is maintained as long as any material harvested for combustion is immediately replaced by establishing the growth of new plants. This argument fails to consider the whole picture and does not take into account the fact that the new growth does not immediately absorb as much CO_2 as was liberated during the combustion process. This is because the carbon accumulated in a plant or tree is stored not only in the growing vegetation but enters the surrounding soil via the roots where it accumulates in a time-dependent process as humic materials. Older growths accumulate these carbon-storing humic materials much more effectively than does new growth and so the carbon sequestering properties of new replacement plants is significantly lower than that of established plants. It will thus require many years of growth before any replacement forestation provides the same level of carbon sequestration as did the original trees. Another important factor in such a replacement programme, often ignored, is that the harvesting and planting processes invariably require energy inputs (farm transport, production and use of fertilisers, etc.). Hence, the all-over carbon balance associated with biomass combustion may be significantly negative, i.e. severely detrimental, rather than being carbon neutral. It is clear that practices such as forest clearance without replacement of the original vegetation are highly undesirable: even if the trees harvested are replaced by the growth of annual crops, such clearance decreases the allover sequestering ability of that region and contributes to significant increases in the greenhouse gas content of the atmosphere. The combustion of wood for heating purposes is therefore only justified if the wood is harvested as part of a fully sustainable forestry programme and if the predominant materials used are the result of forest thinning and pruning activities. Further, the timber that is used must be well dried before combustion or the temperature will be inadequate to provide complete combustion and hence the emissions will contain non-combusted organic molecules and particulates. The latter contaminants can be removed by fitting a catalytic converter to the stove used (see Box 5.1) but unfortunately, this technology is little used. One of the associated problems is that the catalysts of such devices need to be

BOX 5.1 Catalytic Wood-burning Stoves

Modern wood-burning stoves are constructed in such a way as to improve combustion efficiency and reduce undesirable emissions. They are generally designed for the combustion of uniform pellets with very low moisture contents that can be made from many different types of well-aged wood. The pellets can also be made from other forms of biomass such as coconut shells. A modern stove allows preheating of the air feed to the combustion chamber and good thermal emission from the stove to its surroundings. A catalytic converter can be included in the exit flue of the burner, this generally containing a Pd-containing element that enables further oxidation of unburnt components and of any CO formed in the burning process. Fig. 5.2 shows schematically the construction of such a stove. Cold air (feed, shown in blue) enters the base of the stove and is preheated as it passes the front of the stove on its way to the combustion region. The product gases leave the top of the stack of burning pellets and pass through the catalytic converter where combustion is completed utilising some of the excess oxygen of the feed. The stove in the diagram includes a baffle to the righthand side of the catalytic element that can be opened to allow the flue gas to bypass the catalyst during the start-up of the combustion process.

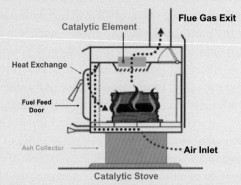

Fig. 5.2 A wood-burning stove with a catalytic converter. *(Modified from www. epa.gov/burnwise/choosing-right-wood-burning-stove#catalytic; this website gives useful information on the advantages of installing a wood stove containing a catalytic element.)*

protected from exposure to the high water vapour contents of the gases emitted in the initial combustion processes by opening the bypass baffle. The catalyst also needs to be well maintained and replaced as needed and so the effective operation of a stove of the type shown in Box 5.1 requires some care.

The use of wood in paper production

The production of paper from wood has been practised for many centuries and the chemistry of the breakdown of the lingo-cellulosic structure of the wood to give the cellulose fibres ('paper pulp') for use in the paper-making process is well understood. A total of 187 million tonnes of paper pulp was produced in 2018 (see Table 5.1).[c] The paper pulp industry is a significant energy user, consuming of the order of 7×10^{18} J (7 EJ) per year (about 6% of the world's energy consumption), and it is thus the fourth largest industrial energy user worldwide.

The extraction of cellulose fibres from wood is carried out using 'pulp mills' such as those used in the Kraft process (Fig. 5.3). The raw material used in such a mill is dry wood chips, often accompanied by recycled paper (see the next section). In the initial step of the pulping process, the feed material is treated in the digester with an aqueous solution containing sodium hydroxide (NaOH) and sodium sulphide (Na_2S). This treatment breaks down the lignocellulosic structure of the wood chips by depolymerising the lignin species that bond together the linear fibrous cellulosic and hemicellulosic polymers of the structure, leaving them relatively intact. (See Figs. 5.4–5.6 of Box 5.2.) By-products of the pulping process include turpentine (5–10 kg per ton of pulp) and 'tall oil' soap (30–50 kg per ton of the pulp). For more information on the various pulping processes used, the reader should consult one of the references in footnote.[d]

Table 5.1 Worldwide production and consumption of paper pulp in 2018.

Region	% Production	% Consumption
Europe	25.1	27.1
North America	33.9	28.9
Asia	22.2	36.3
Latin America	16.3	5.4
Rest of World	2.5	2.3
	Total World Production 187.2 million tonnes	Total World Consumption 186.6 million tonnes

Based on Cepi statistics. (See footnote 'c').

[c] 'Key Statistics 2019' published by the European Pulp and Paper Industry (CEPI); see https://www.cepi.org/wp-content/uploads/2020/07/Final-Key-Statistics-2019.pdf. This report gives a detailed breakdown of the European paper industry.

[d] For a full description of the Kraft process, the reader should consult sources such as https://encyclopedia.pub/976; https://en.wikipedia.org/wiki/Kraft_process.

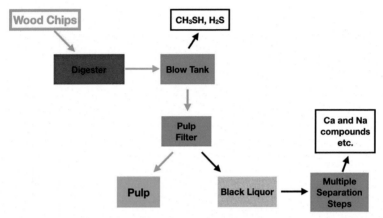

Fig. 5.3 The layout of a Kraft wood-pulping plant. The main depolymerisation reaction takes place in the digester and the pulp is then cleaned further and collected. The gaseous products are separated out and various other compounds are prepared from the black liquor, these including CaO and sodium salts. *(A schematic depiction of the plant is to be found at https://en.wikipedia.org/wiki/File:Pulp_mill_2.jpg.)*

BOX 5.2 The structure of lignocellulosic materials found in wood and other forms of biomass.

The main components of lingo-cellulosic materials are cellulose, hemicellulose and lignin. Cellulose is made up of a series of linear chains of D-glucose monomer ($C_6H_{12}O_6$) that are linked by *beta*-1,4-glycosidic bonds between successive glucose rings as shown in Fig. 5.4.

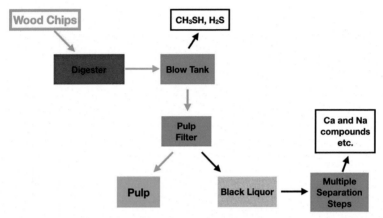

Fig. 5.4 The structure of cellulose. *(From Aqueous-phase hydrolysis of cellulose and hemicelluloses over molecular acidic catalysts: Insights into the kinetics and reaction mechanism. (2016). Applied Catalysis B: Environmental, 184, 285–298. Reproduced with the kind permission of Elsevier.)*

The dotted lines in this figure represent hydrogen bonds, the presence of which contributes to the rigidity of the linear structure; in particular, the hydrogen bond between successive glucose rings ensures the linear nature of the polymer. The degree of polymerisation depends on the source of the lignocellulose: plant cell

Continued

BOX 5.2 The structure of lignocellulosic materials found in wood and other forms of biomass—cont'd

walls generally contain cellulose chains with 5000 to 7500 glucose units while wood and cotton materials contain 10,000 to 15,000 units. Approximately 1.5×1012 tons of cellulose are produced annually.[e] As will be discussed in a subsequent section, the cellulose structure can be hydrolysed to form the individual sugars. Because of the complex structure of the lignocellulosic materials in which it is present, this hydrolysis occurs much less easily than if the chain existed alone.

Hemicellulose is also a polysaccharide but instead being composed of only the six-carbon D-glucose units as in cellulose, it is made up of a variety of different sugars including the five-carbon sugars xylose and arabinose, the six-carbon sugars glucose, mannose and galactose and the six-carbon deoxy-sugar rhamnose. The linkages between the sugar entities have some similarities to those in cellulose. Particularly important is that the structure is generally non-linear, having a number of side-groups. The composition of hemicellulose depends strongly on the plant type in which it is found; most contain predominantly xylose but soft-woods can contain predominantly mannose. Fig. 5.5 shows the structure of rabinoxylan, one of the most predominant forms of hemicellulose polymer. In this, the polymer chain is made up of the five-membered sugar, xylose, with side chains of the five-membered sugar arabinose. (The structures of some of the other varied forms are shown in the review cited for Figs. 5.4–5.6.[e]) Hemicelluloses consist of chains of between 500 and 3000 sugar units, much shorter than those in cellulose; the length of the polymer chain depends also on the degree of growth of the plant. The hemicellulose is connected to the cellulose chains by hydrogen bonds and Van der Waals forces rather than by means of chemical bonds. Hemicelluloses can also be hydrolysed to form the component sugars.

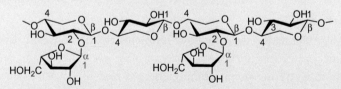

Fig. 5.5 The structure of rabinoxylan, one of the components of hemicellulose, showing the existence of side-chains. *(From Aqueous-phase hydrolysis of cellulose and hemicelluloses over molecular acidic catalysts: Insights into the kinetics and reaction mechanism. (c. 2016)., 184, 285–298. (2016). Applied Catalysis B: Environmental, 184, 285–298. Reproduced with kind permission of Elsevier.)*

Lignin, a complex organic polymer, is the third component of lignocellulose. Its composition varies from plant to plant but common to many of the structures is a phenylpropane entity. The lignin molecules provide the linkages between the cellulose and hemicellulose chains of the lignocellulose structures

Continued

BOX 5.2 The structure of lignocellulosic materials found in wood and other forms of biomass—cont'd

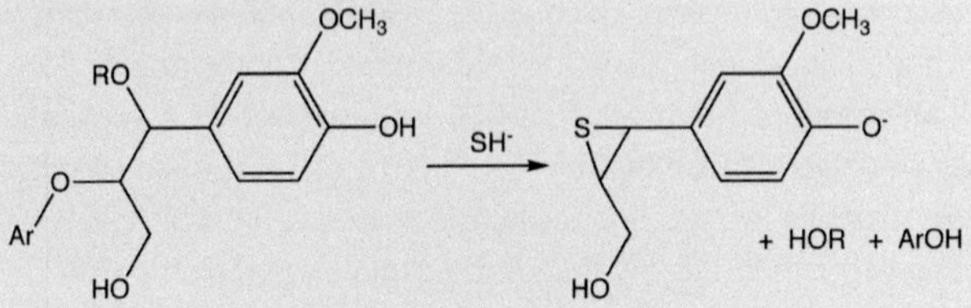

Depolymerisation of lignin by SH⁻ species

Ar = aryl group; R = alkyl group

Fig. 5.6 The net reaction for the depolymerisation of lignin using bisulphide (SH—) anions. (Ar=aryl; R=alkyl group). *(From Aqueous-phase hydrolysis of cellulose and hemicelluloses over molecular acidic catalysts: Insights into the kinetics and reaction mechanism. (2016). Applied Catalysis B: Environmental, 184, 285–298. Reproduced with the kind permission of Elsevier.)*

and are the main contributor to the strength of these structures, making them relatively stable to break down by hydrolysis under normal conditions. In the paper-making process discussed above, the preliminary hydrolysis is carried out using HS- ions, as shown in Fig. 5.6. The remaining traces of lignin are responsible for the ageing of the low-quality paper used for newspapers; the lignin must be removed completely from the pulp to be used for the production of higher-grade papers.

[e] L. Negahdar, I. Delidovich and R. Palkovits, Aqueous-phase hydrolysis of cellulose and hemicelluloses over molecular acidic catalysts: Insights into the kinetics and reaction mechanism Applied Catalysis B: Environmental 184 (2016) 285–298.

Paper recycling

Paper waste is a major constituent of municipal solid waste and although a significant proportion of this waste, about 50%, is recovered and recycled, paper waste is still the major component; see Box 5.3. It has been argued that the recycling of 1 ton of newsprint saves 1 ton of wood; the figures estimated for the reduction of energy required range from 40% to 64%. Recycling enables the cellulosic fibres recovered from the recycled paper to be extracted and reused in making new paper. However, the extraction process that is carried out in what is often a specialised pulping plant requires that any additives incorporated in the original paper-making process are extracted and that any ink residues are also removed. The recycling process often means that the cellulosic fibres are also partially disrupted by the extraction process and so the recycled fibres are often used in the preparation of lower quality products such as tissues rather than of paper

BOX 5.3 Paper recycling in the Netherlands

The recycling of paper and board products is a well-established procedure. In Europe (i.e. the EU plus Norway and Switzerland), 72% of such waste was recycled in 2019, this figure reflecting a significant increase since 1991 when only 40.3% was recycled.[3] In the Netherlands, 86% of the paper and board produced is made from recycled materials. A recent Dutch report[9] provides some very readable information on the processes involved in recycling and also discusses the structure of the Dutch industry and the types of products being produced. The majority of material recycled (85%) is termed 'coloured paper' and this is mostly used for the production of the corrugated and solid board for packaging purposes. Between 10% and 13% of the recycled material is white paper and that is re-used in the production of higher quality papers and board. The remaining 2% consists of items such as coated beverage containers that require specialised re-treatment. According to the Dutch report, approximately 15% of the paper for recycling consists of non-fibrous materials; of this, an average of 20.1% is clay, 57.6% is $CaCO_3$ and 11.2% is starch; the remaining 11.1% consists of other components such as inks, the actual proportions of these components depending on the material being processed.[7] The Dutch report emphasises the importance of careful formulation of all the paper and board that will be used commercially so that they are easily recycled. The printing processes used on the final products must also be carefully controlled, bearing fully in mind the ease with which the inks can later be removed during re-pulping; it is preferable that all the inks are water-soluble. Further, the proportion of waste that remains after re-pulping procedures have been carried out must also be recyclable or combustible and for this reason, the use of components such as PVC in paper coatings is undesirable.

[g] https://circpack.eu/fileadmin/user_upload/Recycling_of_Paper_and_Board_in_The_Netherlands_in_2019_-_final.pdf.

of the quality required for printing. Further, high–quality paper often contains additives such as china clay to provide a suitable surface texture and the clay materials separated out in such repulping activities can also be reused in other processes. As an example of such a recycling activity, the clay minerals extracted have been used for the manufacture of a cement substitute. In another example, the short cellulose fibres obtained from recycled toilet paper have been used as a component in the surface coating of bicycle lanes in the Netherlands.[f]

[f] https://www.mentalfloss.com/article/504999/netherlands-paving-its-roads-recycled-toilet-paper.

Non-traditional uses of biomass: First and second generation bio-refinery processes

We now consider some of the various routes outlined in Fig. 5.1 that are either currently available or are being developed for the conversion of biomass into useful products in so-called 'First and Second Generation Bio-Refineries'. Much of the biomass produced in the world is used either directly or indirectly as food for humans and animals.[h] The production of chemicals and biofuels (predominantly biogas, ethanol and biodiesel) from traditional food crops in so-called 'First Generation Biofuel Refineries' will first be discussed. The crops required for all these products are grown on arable land that might otherwise be used for food production, either indirectly for the production of animal feed or directly for human consumption. Hence, there is considerable concern regarding the competition between the production of products in first-generation bio-refineries and food production.

Organic residues and grasses

Fig. 5.7, an expansion of the left-hand section of Fig. 5.1, shows the predominant routes available for the conversion of organic residues and grasses. Organic residues are materials such as agricultural, human and food waste but they also include items such as the cellulosic fibres from paper recycling that cannot be re-used in other processes such as paper-making. In many countries, all these various residues are often either sent to landfill or are burnt for energy production in municipal incinerators. Alternative and more environmentally responsible processes are either *anaerobic digestion*[i] to form biogas or *gasification* to produce syngas and hence bio-methane or hydrogen. The biogas formed in the anaerobic digestion process contains predominantly CO, CO_2, H_2 and CH_4 and this mixture can be used directly as a source of combustion energy, e.g. in agricultural and horticultural settings where energy is required for heating purposes (animal sheds, greenhouses, etc.); using suitable anaerobic digestion conditions, the product can consist predominantly of 'bio-methane'. There is currently much activity in many countries in which anaerobic digestion is

---

[h] Human and agricultural waste are also a consequence of food production and the methods available for the treatment of such wastes are also discussed briefly below.

[i] Aerobic digestion is also widely practised as in sewerage works. However, such systems are used largely for waste disposal rather than for the generation of energy and chemicals, and any solid waste is used as a fertiliser. An exception to this practice would be the use of the solid organic residues from an aerobic sewerage disposal plant in anaerobic digestion or gasification processes as described here.

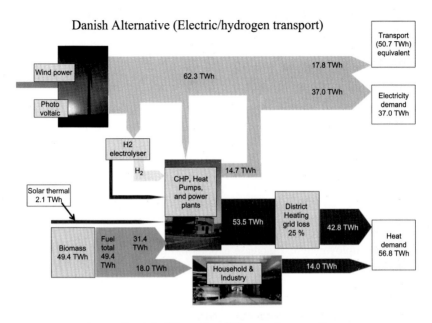

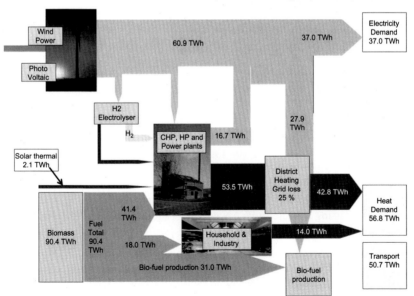

Fig. 7.9 The energy flow in a Danish 100 percent renewable energy system. The top diagram is based on electric and hydrogen fuel cell vehicles referring to Fig. 7.8, and the bottom diagram refers to the biofuel alternative.

power, which will involve a large share of hydrogen or similar energy carriers leading to certain inefficiencies in the system design.

In the project of the Danish Society of Engineers, the method applied to the design of a future energy system in Denmark was a combination of two phases: a creative phase involving the inputs of a number of experts and a detailed analytical phase involving technical and economic analyses of the overall system and feedback on each individual proposal. In a back-and-forth process, each proposal was formed in such a way that it combined the best of the detailed expert knowledge with the ability of the proposal to fit well into the overall system in terms of technical innovation, efficient energy supply, and socioeconomic feasibility.

First, the Danish Society of Engineers appointed 2006 as the "Energy Year" in which the organization aimed at making specific proposals to advocate an active energy policy in Denmark. The targets formulated for the future Danish energy system, 2030 (IDA, 2030), were the following: (1) to maintain the security of energy supply, (2) to cut CO_2 emissions by 50 percent by 2030 compared with the 1990 level, and (3) to create employment and increase exports in the energy industry by a factor of four. The target of maintaining the security of supply referred to the fact that Denmark, at that time, was a net exporter of energy from the production of oil and natural gas in the North Sea. However, the reserves are expected to last for only a few more decades. Consequently, Denmark will soon either have to start importing energy or develop domestic renewable energy alternatives.

Based on these targets, the work of the Danish Society of Engineers was divided into seven themes under which three types of seminars were held: a status and knowledge seminar, a future scenario seminar, and a roadmap seminar. The process resulted in a number of suggestions and proposals on how each theme could contribute to the national targets.

The contributions involved a number of energy demand-side management and efficiency measures within households, industry, and transportation, together with a wide range of improved energy conversion technologies and renewable energy sources, emphasizing energy efficiency, CO_2 reduction, and industrial development. All proposals were described in relation to a Danish 2030 business-as-usual reference (Ref 2030). These descriptions involved technical consequences as well as investment and operation and maintenance costs.

In a parallel process, all proposals were analyzed technically in an overall energy system analysis using the EnergyPLAN computer model. The energy system analysis was conducted in the following steps:

1. First, the Danish Energy Agency's official business-as-usual scenario for 2030 (Ref 2030) was replicated and recalculated on an hourly basis using the EnergyPLAN model. It was possible, on the basis of the same inputs, to come to the same conclusions regarding annual energy balances, fuel consumption, and CO_2 emissions. Consequently, a common understanding of Ref 2030 was established.

2. Second, each of the proposals for year 2030 was defined as a change of the reference system, and a first rough alternative was calculated including all changes. The creation of this system led to a number of technical and economic imbalances, and, consequently, proposals of negative feasibility were reconsidered and suitable investments in flexibility were added to the system.

In the EnergyPLAN model, the analysis was done by basing the operation of the system on a business economic optimization of each production unit. This optimization included taxes and involved electricity prices on the international electricity market. The calculation of the socioeconomic consequences for Danish society did not include taxes. This calculation was based on the following basic assumptions:

- World market fuel costs equaled an oil price of 68 USD/barrel (with a sensitivity of 40 USD and 98 USD/barrel).
- Investment and operation costs were based on official Danish technology data, if available, and if not, on the input from the "Energy Year" experts.
- An interest real rate of 3 percent was used (with a sensitivity of 6 percent).
- Environmental costs were not included in the calculation, apart from CO_2 emission trade prices of 20 EUR/ton (with a sensitivity of 40 EUR/ton).

A technical analysis and a feasibility study were conducted of each individual proposal. Since many of the proposals were not independent in nature, the analysis was conducted for each proposal, in both the reference business-as-usual system (Ref 2030) and the alternative system (IDA, 2030). One proposal, the insulation of houses, may be feasible in the reference but not in the alternative system: for instance, if solar thermal was applied to the same houses or if the share of CHP was increased as part of the overall strategy. Consequently, several of the contributions and proposals had to be reconsidered and coordinated with other contributions.

The proposed alternatives of the Danish Society of Engineers (IDA, 2030, 2050) were compared to both the reference year (2004) and to a business-as-usual reference scenario for 2030 (Ref 2030), assuming that the gross energy consumption (primary energy supply) would rise from 850 PJ in 2004 to 970 PJ in 2030. The IDA 2030 and 2050 alternatives were defined as a series of changes to the business-as-usual reference in 2030. IDA 2030 was an alternative for the year 2030, and IDA 2050 was a 100 percent renewable energy system alternative for 2050. The different energy systems were comprehensive, also including natural gas consumption on the drilling platforms in the North Sea and jet fuel for international air transportation.

After completing the back-and-forth process of comparison and discussion among experts and the overall systems analysis, the proposals of IDA 2030 ended up as follows:

- Reduce space heating demand in buildings by 50 percent
- Reduce fuel consumption in industry by 40 percent
- Reduce electricity demand in private households by 50 percent and in industry by 30 percent
- Supply 15 percent of individual and district heating demands by solar thermal
- Increase electricity production from industrial CHP by 20 percent
- Reduce fuel consumption in the North Sea by 45 percent through savings, CHP, and efficiency measures
- Slow down the increase in transportation demand through tax reforms
- Replace 20 percent of road transportation with ships and trains
- Replace 20 percent of the fuel for road transportation with biofuels and 20 percent with electricity
- Replace natural gas boilers by micro–fuel cell CHP, equal to 10 percent of house heating

- Replace individual house heating by district heating CHP, equal to 10 percent
- Replace power stations planned for construction after 2015 by fuel cell CHP stations, equal to 35–40 percent of the total amount of power stations in 2030
- Increase the total amount of biomass resources (including waste) from 90 to 180 PJ in 2030
- Increase wind power from 3000 to 6000 MW in 2030
- Introduce 500 MW wave power and 700 MW PV power
- Introduce 450 MW_e heat pumps in combination with existing CHP systems and flexible electricity demand to achieve a better integration of wind power and CHP into the energy system

It should be emphasized that the proposal of adding heat pumps and flexible demand was an outcome of the overall energy systems analysis process, which pointed out that the potential of flexible production should be exploited in the best possible way to overcome balancing problems in electricity and district heating supplies. Especially regarding CHP stations based on solid oxide fuel cell technology, the stations should exploit the potential for changing production quickly without losing efficiency and within the full range of loads.

The results of the socioeconomic feasibility study and the export potential are shown in Figs. 7.14 and 7.15. Fig. 7.10 illustrates the economic costs related to Denmark's energy consumption and production in Ref 2030 and in IDA 2030, respectively. In Fig. 7.11, the business potential of IDA 2030 is shown, calculated as expected exports in 2030 and compared to the data of 2004.

Socioeconomic feasibility was calculated as annual costs, including fuel and operation, and annual investment costs based on a certain lifetime and interest rate. The feasibility study was carried out with three different oil prices (as mentioned previously), and the IDA 2030 alternative was compared with Ref 2030, assuming that

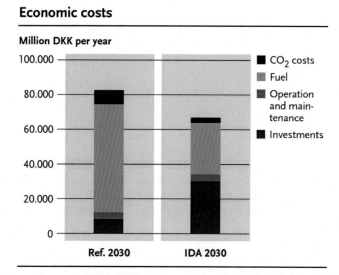

Fig. 7.10 Economic costs of IDA 2030, the energy plan for 2030 of the Danish Society of Engineers.

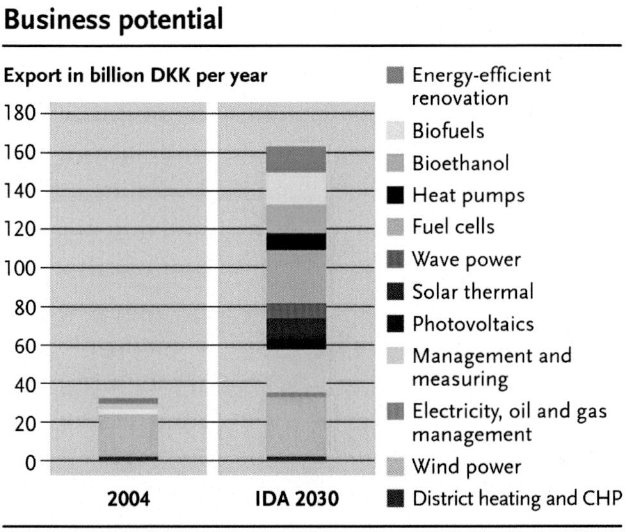

Business potential

Fig. 7.11 Business potential of IDA 2030.

the average oil price was applicable 40 percent of the time, and the low and high oil prices each were applicable 30 percent of the time.

Compared to Ref 2030, the IDA 2030 alternative converted fuel costs into investment costs and also had lower total annual costs. Such a shift in cost structure is very sensitive to two factors: the interest rate and the estimation of the size of total investment costs. Consequently, sensitivity analyses were made. In the first analysis, the interest rate was raised from 3 to 6 percent, and in the other, all investment costs were raised by 50 percent. In both cases, the IDA 2030 alternative was competitive to the reference.

Fig. 7.11 gives an indication of the export potential of IDA 2030. This potential was estimated on the basis of the Danish development of wind turbine manufacturing and is considered a very rough estimate. However, the estimate provides valuable information on both the different relevant technologies and the size of the total potential. The socioeconomic feasibility and the CO_2 emissions of the two energy systems, Ref 2030 and IDA 2030, are shown in Fig. 7.12. All measures were evaluated marginally in both Ref 2030 and IDA 2030. As can be seen, the back-and-forth process led to the identification of measures that would be predominantly feasible. However, some proposals with negative feasibility results were included in the overall plan for other reasons. Some had good export potential, whereas others were important to be able to reach the final target of 100 percent renewable energy in the next step, and yet others had important environmental benefits.

The socioeconomic feasibility shown in Figs. 7.10 and 7.12 was calculated for a closed system without any exchange of electricity on international markets. On the basis of these calculations, a separate study was conducted of the potential benefits of electricity exchange to assess whether the IDA 2030 energy system in this respect differed from the reference system Ref 2030. The evaluation was done for the three

Economic savings achieved through individual measures estimated in relation to the energy systems of the Danish Reference and the Danish Society of Engineers' Energy Plan

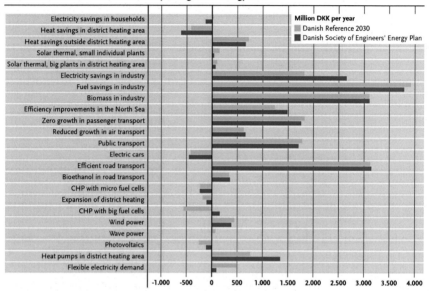

CO_2 reduction achieved through individual measures estimated in relation to the energy systems of the Danish Reference and the Danish Society of Engineers' Energy Plan

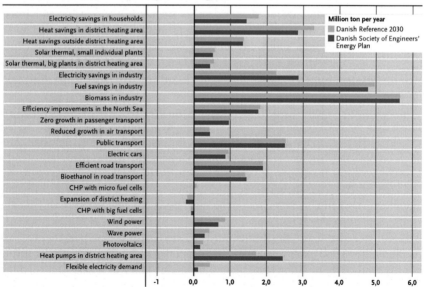

Fig. 7.12 Feasibility and CO_2 emission reduction of each of the individual measures.

different fuel price levels and the two CO_2 emission trade cost levels, as well as for the three Nordic hydropower circumstances: wet, normal, and dry years. The results are shown in Fig. 7.13 in terms of socioeconomic net revenues for Danish society. Moreover, the diagram shows the import and export of each system: Ref 2030 and IDA 2030.

The net revenue from exchange was calculated in the EnergyPLAN model by comparing the results of a reference calculation of a closed system to the results of a calculation of an open system. The closed system had no exchange, while the open system benefited from exchange by selling electricity when the price exceeded the marginal production costs of the Danish energy system and buying electricity when

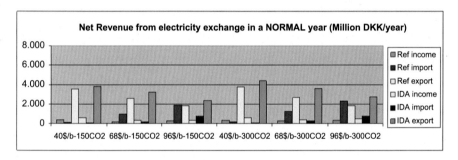

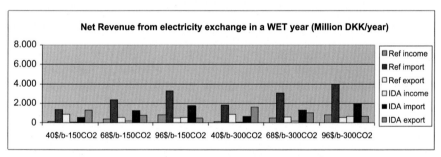

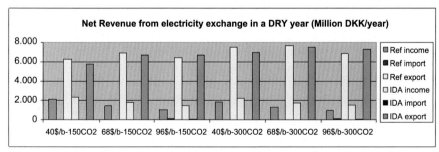

Fig. 7.13 Net revenue from electricity exchange for Ref 2030 and IDA 2030 at different fuel prices, CO_2 emission trading prices, and under different hydro power circumstances: wet, normal, and dry years, which are determined by the hydropower reservoirs of the Nordic electricity supply system.

the price was lower than the marginal costs. The modeling took into consideration bottlenecks among the countries. The whole procedure of calculation was based on the assumption that each of the electricity production units optimized its business economic revenues.

As can be seen in Fig. 7.13, Denmark would be able to profit from the exchange of electricity on the Nordic Nord Pool market in all situations. The net revenue was typically on the order of 500–1000 million DKK/year. In years with low fuel prices and high electricity market prices, revenues were primarily earned from exporting, while in years with high fuel prices and low electricity prices, revenues were earned from importing electricity. It should be mentioned that not all combinations are equally probable. The electricity market price will, to some extent, follow the changes in the fuel price levels.

A comparison of the reference Ref 2030 and IDA 2030 was made by calculating the average net revenues during a period of years in which the following conditions occurred:

- Wet, normal, and dry years were 3:3:1.
- Low, medium, and high fuel prices were 3:4:3.
- Low and high CO_2 emission trading prices were 1:1.

Based on these ratios, the average net revenues of the IDA 2030 system were 585 million DKK/year compared to 542 million DKK/year of Ref 2030. Based on this analysis, it is only fair to say that the two systems can benefit equally from the exchange of electricity on the Nord Pool market. However, compared to the total annual cost of 60–80 billion DKK/year, the net revenue gained from the exchange of electricity is only marginal. The important economic benefits come from the fuel savings achieved by changing the system from Ref 2030 to IDA 2030.

To achieve a 100 percent renewable energy supply, the following additional initiatives were proposed by the steering committee, thus extending the IDA 2030 energy system and creating the IDA 2050 system:

- Reduce heat demand in buildings and district heating systems by another 20 percent compared to the year 2030.
- Reduce fuel demand in industry by another 20 percent.
- Reduce electricity demand by another 10 percent.
- Stabilize transportation demand at the 2030 level.
- Expand district heating by 10 percent.
- Convert micro-CHP systems from natural gas to hydrogen.
- Replace oil and natural gas boilers by heat pumps and biomass boilers in individual houses.
- Replace 50 percent of road goods transportation with trains.
- Replace remaining fuel demand for transportation equally with electricity, biofuels, and hydrogen.
- Supply 3 TWh of industrial heat production from heat pumps.
- Replace all CHP and power stations with fuel cell–based, biogas, or biomass gasification.
- Supply 40 percent of the heating demand of individual houses by solar thermal.
- Increase wave power from 500 to 1000 MW.
- Increase PV power from 700 to 1500 MW.

The necessary wind power and/or biomass resources were calculated as residual resources and had to be increased, as described in the following.

The 100 percent renewable energy system for 2050 (IDA, 2050) was calculated in more than one version. First, all the proposals just mentioned were simply implemented in the modeling, which led to a primary energy supply consisting of 19 PJ of solar thermal, 23 PJ of electricity from RES (wind, wave, and PV), and 333 PJ of biomass fuels. In this scenario, wind power was equal to the figure of the year 2030: 6000 MW installed capacity. A figure of 333 PJ of biomass fuels, however, may be too high. According to the latest official estimate when the study was conducted, Denmark had approximately 165 PJ of residual biomass resources, including waste. Residual resources consist of straw that is not needed for livestock, biogas from manure, organic waste, and waste from wood industries. However, the potential of biomass fuels from the change of crops was considered to be huge. Denmark, for example, grows a lot of wheat that can be replaced by other crops such as corn, leading to a much higher biomass production while maintaining the same output for food. This reorganization of the farming areas together with a few other options may lead to total biomass fuel potential as high as 400 PJ (Mathiesen et al., 2008).

The analyses proposed a compromise with 10,000 MW of wind power and 270 PJ of biomass fuels. All three versions are shown in Fig. 7.14. The energy flow of the system is illustrated in Fig. 7.15, and the primary energy supply and the resulting CO_2 emissions are shown in Fig. 7.16. However, as described later 270 PJ of biomass is hardly within the limits of sustainable biomass resources in Denmark.

On the basis of the first version of the 100 percent renewable energy scenario, it was analyzed how much the need for biomass fuels would decrease if more wind power

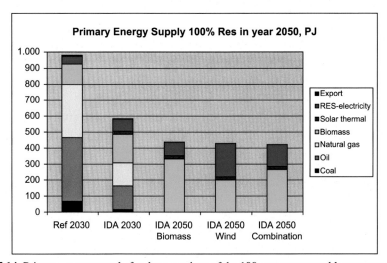

Fig. 7.14 Primary energy supply for three versions of the 100 percent renewable energy system, IDA 2050, compared to the reference, Ref 2030, and the proposal for the year 2030, IDA 2030.

100 PERCENT RENEWABLE ENERGY

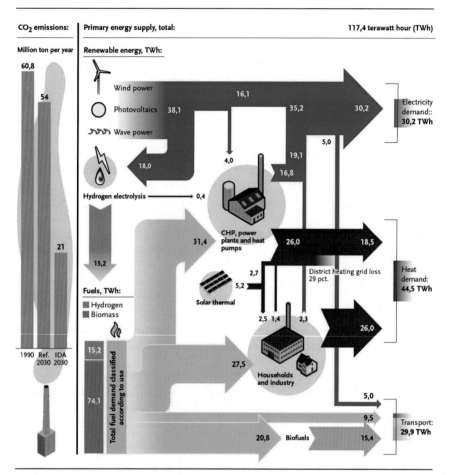

Fig. 7.15 Flow diagram of the 100 percent renewable energy system (IDA, 2050).

was added. If wind power was raised from 10,000 to 15,000 MW, then a rise in electricity to 200 PJ would lead to a decrease in biomass fuel consumption to 200 PJ. It should, however, be emphasized that this replacement led to a significant increase in the demand for hydrogen as an energy carrier, which resulted in considerable efficiency losses.

The primary energy supply was expected to increase from approximately 800 PJ in 2004 to nearly 1000 PJ in the business-as-usual reference (Ref 2030). If the proposed IDA 2030 plan was implemented, the primary energy supply would fall to below 600 PJ and CO_2 emissions would decrease by 60 percent compared to the year 1990.

If the 100 percent renewable energy system proposed for 2050 (IDA, 2050) is implemented, the primary energy supply will fall to approximately 400 PJ, and the

Primary energy supply

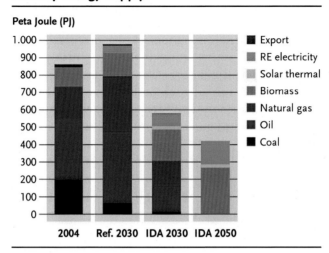

CO_2 emissions

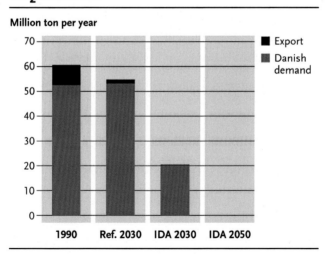

Fig. 7.16 Primary energy supply and CO_2 emissions. CO_2 emissions are divided into domestic electricity demand and electricity net exports.

CO_2 emission will, in principle, be equal to zero. However, some waste is included in the biomass resources, of which some will result in a minor CO_2 emission. Moreover, it should be mentioned that Denmark will still contribute to greenhouse gas emissions from gasses other than CO_2. In total, however, the Danish greenhouse gas emissions will decrease by approximately 80 percent.

The IDA Climate Plan

In 2009, three years after creating the IDA Energy Plan, the Danish Society of Engineers made a follow-up and an extension of the study in terms of an IDA Climate Plan. The follow-up was part of a joint effort of engineering societies from various countries to contribute to the COP15 meeting in Copenhagen. As described by Mathiesen et al. (2011), the IDA Climate Plan added to the previous study among others by including not only the energy sector but all sectors contributing to greenhouse gas emissions. Further, the study involved additional efforts regarding transportation and biomass as well as an estimate of the influence on health costs and on job creation based on applied concrete institutional economics.

Again, the conclusion is that a 100 percent renewable energy system is possible, but the balance between a large consumption of biomass and a large amount of electricity for direct use or for the production of synthetic fuels appears to be a challenge. In combination with changes in the agricultural sector and including the extra contribution from aviation, the emission of greenhouse gasses can be reduced to 10 percent in 2050 compared to 2000 levels. In Chapter 8, it is shown how Denmark can become fully carbon neutral.

The change from a conventional energy system to a renewable energy system provides socioeconomic savings, because the savings achieved in fuel costs are higher than the annual cost to be paid for the additional investments. On top of this, there is a benefit for the health of the population, which increases the socioeconomic savings. Further, if the changes are implemented relatively early in the period, there is a potential for exports and related jobs. In all cases, the changes will increase the number of jobs, even if the commercial potential for increased exports is not met. These results indicate the possibility of continuing economic growth while implementing climate mitigation strategies.

3   The CEESA coherent 100 percent renewable energy scenario[c]

This section is based on a part of a chapter in Lund et al. (2011a), "Coherent Energy and Environmental System Analysis." The report is the result of the CEESA project, which was partly financed by the Danish Council for Strategic Research. Beyond the IDA energy and climate plans, the CEESA project applied and implemented several of the smart energy system points of Chapter 6 to the analysis of a 100 percent renewable energy scenario. This involved, among others, the fuel pathways described and the hour-by-hour analysis of the gas grid as described and emphasized in Chapter 6.

As in the IDA plans, the aim of the CEESA scenarios was to design a 100 percent renewable energy system by the year 2050. A focus point was that this transition should highly rely on the technologies that were assumed to be available within

[c] Excerpts reprinted from Lund, Hvelplund et al. (2011), "Coherent Energy and Environmental System Analysis", Department of Development and Planning, Aalborg University, Aalborg 2011.

the specified time horizon and which would have different effects on the biomass consumption. To highlight this, the CEESA project identified scenarios based on three different assumptions regarding the available technologies. This methodology allowed a better optimization and understanding of the energy systems. To enable a thorough analysis of the different key elements in 100 percent renewable energy systems, two very different 100 percent renewable energy scenarios and one recommendable scenario were designed:

- *CEESA 2050 Conservative*: The conservative scenario was created using mostly known technologies and technologies that were available at the time of the CEESA study in 2010. This scenario assumed that the market could develop and improve existing technologies. In this scenario, the costs of undeveloped renewable energy technologies were high. Very little effort was made to push the technological development of new renewable energy technologies in Denmark or at a global level. However, the scenario did include certain energy efficiency improvements of existing technologies, such as improved electricity efficiencies of power plants; more efficient cars, trucks, and planes; and better wind turbines. Moreover, the scenario assumed further technological development of electric cars, hybrid vehicles, and bio-DME/methanol production technology (including biomass gasification technology).
- *CEESA 2050 Ideal*: In the ideal scenario, technologies still in the development phase were included on a larger scale. The costs of undeveloped renewable energy technologies were low, due to significant efforts to develop, demonstrate, and create markets for new technologies. For example, the ideal scenario assumed that fuel cells were available for power plants, and biomass conversion technologies (such as gasification) were available for most biomass types and on different scales. Co-electrolysis was also developed and the transportation sector moved further toward electrification compared with the conservative scenario.
- *CEESA 2050*: This scenario aimed to be a "realistic and recommendable" scenario based on a balanced assessment of realistic and achievable technology improvements. Less co-electrolysis was used and a balance was implemented between bio-DME/methanol and syn-DME/methanol in the transportation sector. This was the main CEESA scenario.

In all scenarios, energy savings and direct electricity consumption were given a high priority, and all scenarios relied on a holistic smart energy system approach as explained in Chapter 6. This included the use of heat storages, district heating with CHP plants, and large heat pumps as well as the integration of transportation fuel pathways with the use of gas storage. These smart energy systems enable a flexible and efficient integration of large amounts of variable electricity production from wind turbines and PVs. The gas grids and liquid fuels allow long-term storage, while the electric vehicles and heat pumps allow shorter-term storage and flexibility.

Transportation fuel pathway

The transportation sector poses two main challenges in the transition to renewable energy: First, the obvious easily accessible source, biomass, is limited, and second, the increase in the transportation demands is historically high. The scenarios included a suggestion for a new transportation system, with a medium increase in demands (except goods) and more rail transportation. To replace oil and keep the biomass consumption at a low level, the following strategy was applied: focus was placed on maximizing the use of electricity in the transportation sector, and, where liquid fuels were needed in some cars, vans, trucks, and aviation, priority was given to DME/methanol.

With DME/methanol fuels, conventional cars can be used in the short term (up to 3 percent blend), and with minor changes in vehicles, the share can be increased. In the CEESA scenarios, bio-DME/methanol was produced from a combination of gasified biomass and hydrogen from electrolyzers and not from waste products. In the longer term, land-use effects can be lowered further by replacing bio-DME/methanol with syn-DME/methanol, which requires co-electrolyzers and carbon sequestration, as explained in Chapter 6 regarding the fuel pathways. This strategy reduces the biomass use and allows the integration of more wind and PV power into the energy system in general; i.e., the transportation sector becomes an important part of the smart energy system.

As explained in Chapter 6, it is too early to know if the liquid fuel DME/methanol solution is the best alternative as other options exist such as a gas methane solution or even a combination. However, approximately the same energy balances can also be achieved with a gas solution as considered here for the liquid solution. DME/methanol was used in the scenarios for the concrete calculations to illustrate the principle of using biomass resources in combination with electrolyzers to replace fossil fuels in the transportation sector in the short term. In the longer term, carbon from other sources than biomass was used to replace larger amounts of fossil fuels without putting further strain on the biomass resource. Other types of fuels that fulfill this principle could also be relevant in the future, but the scenarios showed that this principle could reduce the biomass consumption significantly.

All three technology scenarios above were designed in a way in which renewable energy sources, such as wind power and PV, were prioritized, taking into account the technological development in the scenarios and the total costs of the system. Moreover, they were all based on decreases in the demand for electricity and heat as well as medium increases in transportation demands. Consequently, none of the scenarios could be implemented without an active energy and transportation policy. However, sensitivity analyses were conducted in terms of both a high energy demand scenario as well as the unsuccessful implementation of energy saving measures. These analyses pointed in the direction of higher costs, higher biomass consumption, and/or an increased demand for wind turbines. The important differences between the scenarios are highlighted in Table 7.4.

In the *conservative* technology scenario, wave power, PV, and fuel cell power plants were not included and emphasis was put on bio-DME/methanol and on direct electricity consumption in the transportation sector. The electrolyzers were based on known technology in this scenario. Smart energy systems and cross-sector system integration were required between the electricity systems and district heating sectors, as well as in the transportation system and gas grid in all scenarios. The integration of the transportation system and gas grids was, however, not as extensive in the conservative scenario as in the ideal scenario. In the *ideal* scenario, wave power, PV, fuel cells and a number of other technologies were used to their full potential, while in the *recommendable* scenario, the technologies were assumed to be developed to a degree in which they could make a substantial contribution.

Table 7.4 Main differences among the 100 percent renewable energy scenarios in CEESA.

	CEESA 2050 conservative	CEESA 2050 Ideal	CEESA 2050
Renewable energy and conversion technologies			
Wind power	12,100 MW	16,340 MW	14,150 MW
PV	–	7500 MW	5000 MW
Wave power		1000 MW	300 MW
Small CHP	Engines	Small fuel cell CHP	Engines/fuel cells gas turbine
Large CHP and power plants	Gas turbine combined cycle/combustion	Large fuel cell combined cycle CHP/PP	Combined cycle/large fuel cell combined cycle CHP/PP
Gasification for electricity and power production	Yes partly	Yes	Yes
Transportation			
Direct electricity	13 percent	23 percent	22 percent
Bio-DME/methanol	87 percent	0 percent	44 percent
Syn-DME/methanol	–	77 percent	34 percent
Bio-DME/methanol plants	Yes	No	Yes
Electrolyzers for bio-DME/methanol plants	Yes	No	Yes
Co-electrolyzers for syn-DME/methanol plants	No	Yes	Yes

Primary energy and biomass resources

The levels of primary energy consumption for 2050 for the three scenarios and the reference energy system are compared in Fig. 7.17. Compared to the reference energy system, all the scenarios are able to reduce the primary energy supply to a level of approximately 500 PJ. There are, however, large differences among the scenarios regarding the use of biomass as illustrated in Fig. 7.18. In the conservative technology scenario, a 100 percent renewable energy system was possible with a total biomass consumption of 331 PJ. The ideal technology scenario could decrease this consumption to 206 PJ of biomass. In the CEESA 2050 recommendable scenario, the biomass consumption was 237 PJ and thus 30 PJ higher than in the ideal and 96 PJ lower than in the conservative scenario, respectively.

The CEESA project included a careful examination of the pathways to provide biomass resources. The starting point was an overview of the number of residual resources in terms of straw, wood, and biogas from manure, etc., summing up to approximately 180 PJ/year. A shift in forest management practices and cereal cultivars could increase the potential further to approximately 240 PJ/year by 2050.

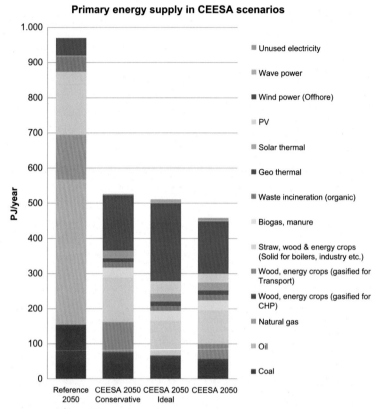

Fig. 7.17 Primary energy supply in the 2050 reference energy system and the three CEESA 100 percent renewable energy scenarios.

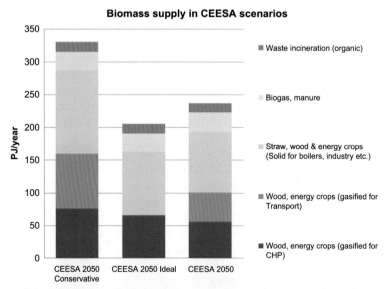

Fig. 7.18 Biomass supply in the three CEESA 100 percent renewable energy scenarios.

The 180 PJ/year could also be increased to 200 PJ by enacting dietary changes. This potential represented the use of residual resources only. This means that the CEESA 2050 recommendable scenario was kept within the boundaries of residual resources, and the CEESA 2050 conservative scenario illustrated that an active energy and transportation policy would be required to stay within these limits. It should be noted that a target of 240 PJ/year by 2050 would imply a number of potential conflicts due to many different demands and expectations from ecosystem services; it would require the conversion of agricultural land otherwise allocated to food crop production to energy crop production, potentially reducing food and feed production. All crop residues must be harvested, potentially reducing the carbon pool in soils. A way to reduce these potential conflicts would be to reduce the demand for biomass for energy or to further develop agriculture and forestry to increase the biomass production per unit of land.

If biomass in a future non-fossil society should cover the production of materials currently based on petrochemical products, even more pressure would be put on the biomass sector. To meet these demands, 40–50 PJ would have to be allocated to that purpose. It should be noted that, in addition to the 240 PJ of residual biomass resources, waste resources are also available amounting to 33–45 PJ or a total of approximately 280 PJ. In this respect, CEESA 2050 recommendable and ideal scenarios would enable the allocation of biomass resources to the materials currently based on petrochemical products.

Based on the "realistic and recommendable" scenario, a roadmap toward 2050 was designed and compared to a business-as-usual reference development. It should be noted that this reference was designed before the Danish Parliament first decided to aim for a 100 percent renewable energy system by 2050 and later, for a carbon neutral society by 2045. This context will be described further in Chapter 8. The primary energy supply in Denmark (fuel consumption and renewable energy production of electricity and heat for households, transportation, and industry) was approximately 850 PJ at the time of the CEESA study. This takes into account the boundary conditions applied to transportation in this study, in which all transportation is accounted for, i.e., national/international demands for both passenger and freight transportation. If new initiatives were not taken, the energy consumption was expected to decrease marginally until 2020, but then increase gradually until 2050 to about 970 PJ. The reference energy system followed the projections from the Danish Energy Agency from 2010 until 2030, and the same methodology was applied to create a 2050 reference energy system. The measures relating to savings, transportation, renewable energy, and the integration of the electricity, heat, transportation, and gas sectors could reduce the primary energy supply to 473 PJ in CEESA 2050. The primary energy supply is illustrated in Fig. 7.19. The energy flows in the CEESA 2050 recommendable 100 percent energy system are illustrated in a Sankey diagram in Fig. 7.20.

In CEESA, the greenhouse gas emissions from fossil fuels were reduced significantly in the energy system. In Fig. 7.21, the greenhouse gas emissions from the energy system in the CEESA scenario are illustrated in relation to the reference energy system, including an extra contribution from aircraft due to discharges at high altitudes. In 2050, the emissions would not be zero due to aviation, but the emissions from these sources were reduced to 2 percent compared to the level of the year 2000.

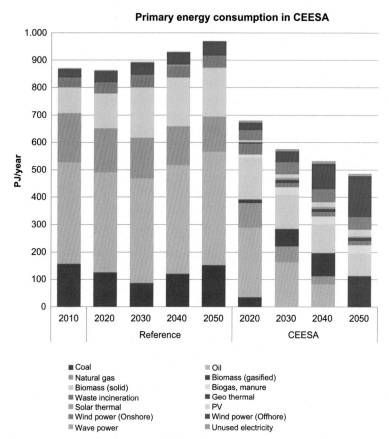

Fig. 7.19 Primary energy supply in CEESA.

Greenhouse gas emissions from industrial processes and from agriculture or land use changes were not included in this figure.

The CEESA scenarios document that it is possible to find technical solutions for a 100 percent renewable energy system. However, a certain technological development becomes essential, notably in enabling the efficient direct use of electricity in the transportation sector with better electric, hybrid electric, and plug-in-hybrid electric vehicles and in biomass gasification technologies (small and large scale). The results also show that if these technologies are not developed sufficiently, the biomass consumption could be larger than in the CEESA 2050 conservative scenario.

In CEESA, a 100 percent renewable energy system was designed, which may potentially be supplied by domestic residual biomass resources. It must, however, be emphasized that there were no objectives in the CEESA project against international trade with biomass. The scenario recommended in CEESA, however, would ensure that Denmark did not merely become dependent on imports of biomass, replacing the dependence on imports of oil, natural gas, and coal, which was the case

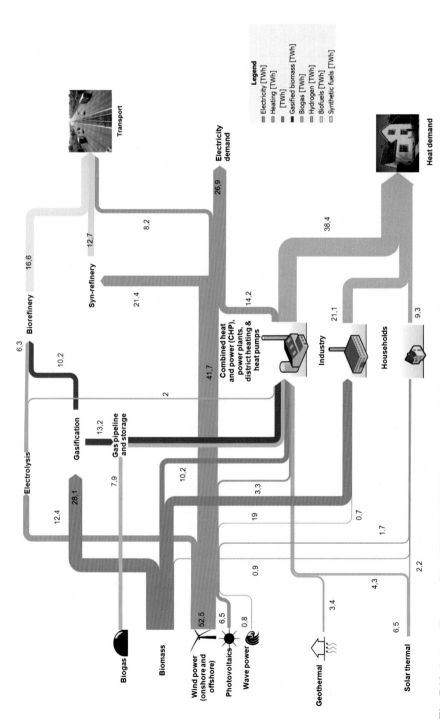

Fig. 7.20 Sankey diagram of the CEESA 2050 100 percent renewable energy scenario.

Legend

- Electricity [TWh]
- Heating [TWh]
- [TWh]
- Gasified biomass [TWh]
- Biogas [TWh]
- Hydrogen [TWh]
- Biofuels [TWh]
- Synthetic fuels [TWh]

Transport

Electricity demand

Heat demand

Biorefinery 6,3

16,6

Syn-refinery 12,7

8,2

26,9

38,4

21,4

14,2

21,1

9,3

Combined heat and power (CHP), power plants, district heating & heat pumps

41,7

Industry

Households

10,2

2

Electrolysis

Gasification 13,2

Gas pipeline and storage

12,4

28,1

7,9

10,2

3,3

19

0,7

1,7

0,9

4,3

2,2

Biogas

Biomass

Wind power (onshore and offshore) 52,5

Photovoltaics 6,5

Wave power 0,8

Geothermal 3,4

Solar thermal 6,5

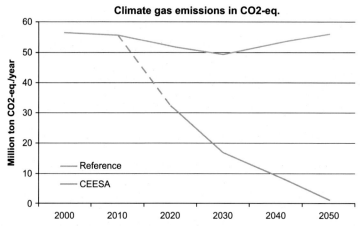

Fig. 7.21 Emissions of greenhouse gasses in CEESA.

in the reference scenario (once Denmark would not have any resources left in the North Sea). It would also ensure that the biomass consumption was within the limits of the Danish residual biomass resources. Looking at the global residual resources, the residual biomass potential is higher per person in Denmark, however, implying that Denmark should go even lower than the domestic residual biomass resources available.

Smart energy systems and cross-sector integration

In all three scenarios, hour-by-hour energy system analyses were used to increase the share of wind turbines to an amount ensuring that the unused electricity consumption, also referred to as excess electricity, was lower than 0.5 TWh (1.8 PJ). These analyses also ensured that the heat supply and gas supply were balanced. To achieve this balance, a smart energy systems approach was applied in the following way.

The integration of sectors is very important in 100 percent renewable energy systems to increase fuel efficiency and decrease costs. The first, and most important, step is the integration between the heating and the electricity sectors. In Denmark, this is already implemented to a large extent, and approximately 50 percent of the electricity demand is produced by CHP plants. This integration requires thermal storages of today's sizes (about 8 hours of average production), a boiler, and district heating networks to enable the flexible operation of the CHP plants, as already implemented in the Danish energy system. This can reduce the fuel consumption and help integrate variable wind power efficiently. As concluded in Chapter 5, around 20–25 percent of the wind power can normally be integrated without significant changes in the energy system. With more than 20–25 percent wind power, the next step in the integration is to install large heat pumps. In the CEESA scenarios, a significant amount of onshore and offshore wind power was installed by 2020. By then, approximately 40 percent of the electricity demand could be covered by these sources.

This resulted in some imbalance in the electricity grid, and heat pumps alone were not able to ensure the balance. The transportation sector had to be integrated into the energy system with more than 40–45 percent wind power. As a consequence, some electric vehicles were implemented and flexible demand was included in households and industry. This, however, was not sufficient. Thus, small amounts of electrolyzers based on known alkaline technology were also implemented to facilitate wind power integration and for the production of bio-DME/methanol in combination with gasified biomass. This also enabled the integration of larger amounts of renewable energy into the transportation sector.

By 2030, a larger proportion of electric vehicles would be included and it was assumed that they would be able to charge according to a price mechanism. To make sure that electric vehicles could fulfill this function, the low-voltage grid needed to be enforced in some areas. The electricity production from onshore and offshore wind power in combination with PVs would then be approximately 60 percent in 2030.

In CEESA 2050, more and new technologies would be necessary to make sure that the renewable energy was integrated efficiently into the system and that fossil fuels were being replaced totally. Hence after 2030, the share of electrolyzers for hydrogen production for bio-DME/methanol would be gradually increased to provide larger amounts of liquid fuels to the transportation sector, while the electrolyzers themselves would also be more efficient. Also, carbon capture would be utilized to produce syn-DME/methanol without using biomass.

In the CEESA 2050 energy system, gasified biomass and gas grid storages would also be utilized in combination with electric vehicles, fuel production in the transportation sector, and district heating systems. This would create an energy system in which smart energy systems would be integrated and the storage options would be used in combination to enable the final scenario.

The CEESA project took a closer look at the balancing of gas supply and demand. The hourly activities of all gas-consuming units, such as boilers, CHP, and power plants, as well as production, such as biogas and gasification (syngas) units (including hydrogenation), were calculated and analyzed regarding the need for import/export, gas storage, or flexibility and extra capacities in the gas-producing units. This analysis is reported for the CEESA 2050 scenario in the following.

First, the annual need for import/export was calculated in the case of no gas storage and no extra capacities in the production units. Then, similar analyses were made with storage capacities gradually being increased from zero to 4000 GWh. In all scenarios, the need for import was equal to the needed export on an annual basis, since the systems were designed to have a net import of zero. However, the need for import/export decreased along with increases in the domestic storage capacity. The results are shown in Fig. 7.22, which indicate that a storage capacity of about 3000 GWh is able to completely remove the need for import/export.

The current Danish natural gas storage facilities have a gas content of 17,000 GWh in Stenlille in Jutland and 7600 GWh in Lille Torup in Zealand. The work content of the storages is smaller, approximately 6500 and 4800 GWh, respectively. This means that the total current storage capacity assuming natural gas quality is 11,350 GWh. If we assume that the gas quality in the entire grid is lowered to biogas standard, the

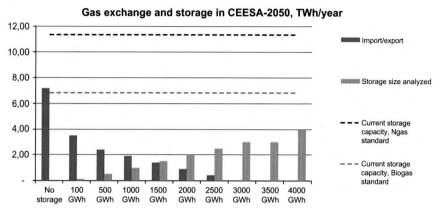

Fig. 7.22 Annual gas exchange and storage analyses in CEESA 2050 in TWh/year.

storage would be reduced to approximately 6800 GWh, as the capacity is reduced by 40 percent. As illustrated in Fig. 7.22, this indicates that the current storage capacity is more than twice as large as required in the CEESA 2050 scenario, even when assuming no extra capacity at the gasification plant, i.e., no flexibility in the production of syn-gas.

Next, to analyze the influence of adding flexibility and extra capacity to the gas-producing units, the analysis of gas storage was repeated while gradually increasing the production capacities as illustrated in Fig. 7.23. As can be seen, an increase in production flexibility and capacity decreases the need for storage capacities. In the CEESA 2050 scenario, it was chosen to include 25 percent overcapacity in combination with a storage capacity above 3000 GWh.

One important learning outcome from the hourly analysis of the complete system including both electricity and gas balances is that relatively cheap gas storage capacities (which in the Danish case already exist) can be used to balance the integration of wind power into the electricity grid. Consequently, in the CEESA 2050 scenario, it was possible to decrease excess electricity production to nearly zero at the same time as high fuel efficiencies were achieved by using heat and gas storages. No electricity storage was included and, as already explained in Chapter 5, investments in electricity storage would not be profitable due to the very limited hours of utilization. Furthermore, it would not be efficient, either. The reason why heat and gas storage is more effective than electricity storage is that electricity has to be converted into heat and gas anyway because of the demand for heat and fuels in the heating and transportation sectors.

Cost and job estimates based on concrete institutional economics

CEESA would be implemented over a period from 2010 until 2050 by continuously replacing technologies, buildings, and vehicles when their lifetime expired. Hence, many of the elements of the current energy and transportation systems in society

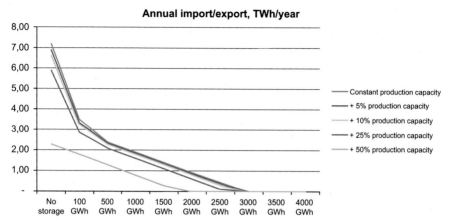

Fig. 7.23 Annual gas exchange and storage analyses in CEESA 2050 in TWh/year and with different levels of surplus capacity.

would need to be replaced even if the scenarios in CEESA were not implemented. Therefore, as a point of departure of the study, the expenses included were calculated as the extra costs generated through investment in better facilities in comparison to the reference energy system. However, exceptions to this might be seen.

The socioeconomic costs were calculated as annual expenses for each of the years 2020, 2030, and 2050, including an interpolated approximation for 2040. The annual costs in CEESA's energy systems were compared with the costs of the reference in each of the applicable years. The costs were categorized as fuel costs, operation and maintenance costs, and investment costs. The investment costs were transformed to an annual cost using a real interest rate of 3 percent. Furthermore, the investment costs were further divided into investment costs in the energy sector and extra investment costs in the transportation system. The transportation investment costs included were additional to the annual investment already made in the 2010 system (approximately 28 billion DKK in road and rail). The economic analyses were based on the assumptions regarding fuel prices and CO_2 quota costs which were defined by the Danish Energy Agency (2011). Three fuel price levels were used. The middle price level was based on fuel price projections for 2030, which corresponded to an oil price of 113 USD/barrel according to the Danish Energy Agency (2010 prices). The high fuel price was based on the prices in the spring/summer of 2008 and corresponded to an oil price of 159 USD/barrel (2010 prices). The low price level was based on assumptions which the Danish Energy Agency used in its forecast in July 2008 and corresponded to an oil price of 70 USD/barrel (2010 prices). Calculations were also done with long-term CO_2 quota costs of 35 and 70 EUR/ton for 2030 and 2050, respectively. The energy systems were analyzed with higher biomass costs than in these assumptions, which did not, however, change the overall results. The CO_2 quota costs did not include all potential costs, such as flooding, for example, but were only anticipated quota costs. If these types of effects were included in the calculation, the energy

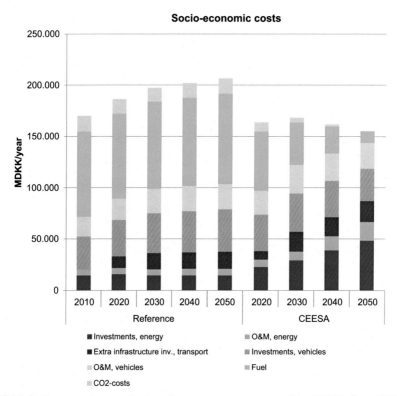

Fig. 7.24 Socioeconomic costs in the reference energy system and in CEESA from 2010 to 2050. Extra infrastructure costs are relative to 2010 total costs.

systems in CEESA would have an economic advantage compared to the reference energy system.

As illustrated in Fig. 7.24, a first result of the work in CEESA was that the total annual energy and transportation system costs could be quantified to approximately 170 billion DKK/year. As illustrated, the costs of the reference increased gradually. This is mainly due to increased costs in the energy system and also due to the investments required in road infrastructure and new vehicles, which are necessary to meet the high increase in the transportation demand. In CEESA, the investment costs increased significantly; however, for the energy and transportation systems, these infrastructure investments were necessary to improve the efficiency of the technologies utilized, to reduce demands, and to consequently reduce the demand for fuel. Since the savings associated with these investments were larger than the initial investments, there was an overall decrease in costs in the CEESA scenarios. In conclusion, the 2020 energy and transportation systems in CEESA were more than 20 billion DKK less costly than in the reference energy system. This means that, by implementing known technologies, it was possible to implement changes that provided lower socio-economic costs than the traditional energy and transportation systems. In the longer term, the costs of the CEESA scenario were rather stable; however, large investments were made to meet a 100 percent renewable energy system. The CEESA 2020 energy and transportation systems even had lower costs than the traditional systems (approximately 7 billion DKK). Continuing the business-as-usual path would only enhance the socioeconomic savings achieved by changing to a renewable energy system combined with energy savings.

The CEESA scenarios also proposed energy systems that were more robust to fluctuations in fuel prices. It is worth noting that Danish society at the time of the study in 2010 spent 50–100 billion DKK/year (7–15 billion EUR/year) depending on whether the fuel costs were low or high (the high costs represented real costs experienced in 2008). In the future, one must expect that the world will continue to experience fluctuating fuel prices—neither constantly high nor constantly low. Hence, energy and transportation systems less dependent on fuels such as the systems proposed in CEESA are less vulnerable. In addition to the potential economic savings in the renewable energy system scenarios mentioned previously, society can benefit from savings related to health costs, commercial potential, and extra employment effects.

As a starting point for the estimation of the employment effect, the annual costs in the reference energy system and the CEESA scenarios were divided into investments and operations. An implementation of the CEESA scenarios would include increasing investment costs and increasing costs for biomass, but also lower costs for fossil fuels. These changes would enable higher Danish employment levels while also improving the balance of payments. The effects on the amount of jobs in the energy sector and the effects on the balance of payment could be further increased if the commercial potential in the form of increased exports would also be realized.

In the CEESA scenarios, expenditures on fuels were reduced while expenditures on operation and maintenance were increased. CEESA also involved a heavy shift from the imports of fuel to investments including an increase of more than a quarter of a trillion, which was dispersed between 2010 and 2050. For each cost type, an import share was estimated based on experiences from previous collections of foreign exchange and employment data for investments in energy facilities, infrastructure, and buildings. In relation to the previous data, a general upward adjustment of the import share was done, as this, from experience, is known to increase. The data sources and methodology are the same as in Chapter 6 regarding the economic crisis and infrastructure investment.

Employment effects were estimated on the basis that two jobs would be created for each million DKK based on the share that was left after removing the import share. This included derived jobs in the finance and service sectors. It should be emphasized that these estimates are subject to uncertainties and, again, it is emphasized that they are based on adjusted figures from previously collected data. The extra employment created in Denmark by the implementation of the CEESA scenarios compared with the reference was estimated by the use of these methods and was assumed to correspond to approximately 20,000 jobs. Jobs would be lost in the handling of fossil fuels, but jobs would be created through larger investments in energy technology than in the reference, as well as larger investments in energy savings. In the reference energy systems, large investments were made in roads, while in the CEESA scenarios, these were replaced by jobs and investments in rail infrastructure.

It was important for a number of reasons to place the large employment effort as early as possible in the period. The first reason was that the labor force as a share of the total population would be falling in the entire period to the year 2040 and, therefore, the largest labor capacity to undertake a change of the energy system would be present in the beginning of the period. The second reason is that the Danish North Sea resources would run out during the next 20 years. Hence, it was important to develop the energy systems and changes as early as possible in the period.

The above-mentioned effects on employment did not include job creation as a result of increased exports of energy technology, i.e., the commercial potential. These advantages would be an additional benefit of implementing the CEESA scenarios. With the assumption of a 50 percent import share, an annual export of 200 billion DKK would generate up to 170,000 jobs, depending on the location of the exports without an ambitious implementation of the scenarios, the extent of unemployment, and the potential employment of these people in other export trades. In relation to this, it should be noted that all other things being equal, a share of Danish labor will be made available as the oil and gas extraction in the North Sea comes to an end. In addition, the energy system is more effective and also less vulnerable to fluctuations in energy prices. Hence, this can increase the competitiveness of Danish society and of Danish businesses.

4   Smart energy Aalborg[d]

This section is based on Thellufsen et al.'s (2020) article "Smart energy cities in a 100 percent renewable energy context." It presents a methodology to design Smart Energy Cities within the context of 100 percent renewable energy at a national level. Cities and municipalities should act locally to local demands but acknowledge the national and global context when addressing resources, industry and transportation. The method was applied to the case of transitioning the municipality of Aalborg in Denmark to a 100 percent renewable smart energy system within the context of a Danish and European energy system. The case demonstrates how it is possible to transition to a Smart Energy City that fits within a 100 percent renewable energy context of Denmark and Europe. The suggested methodology is framed in a way that makes it applicable to other cases globally.

Methodology and guidelines

The overarching guiding principle is that local action should be balanced to match national or global action. This requires that a region, city and municipality situate themselves within national and international contexts in order to make a coherent strategy for the transition to for instance 100 percent renewable energy. In the case of the municipality of Aalborg, the goal is to transition to a 100 percent renewable

[d] Excerpts reprinted from *Renewable and Sustainable Energy Reviews* 129 J.Z. Thellufsen, H. Lund, P. Sorknæs, P.A. Østergaard, M. Chang, D. Drysdale, S. Nielsen, S.R. Djørup and K. Sperling (2020), "Smart energy cities in a 100% renewable energy context" with permission from Elsevier.

energy supply in 2050 in such a way that Aalborg forms an integrated part of the Danish national goals. To achieve this integration, one must define national goals and establish a context of a national strategy. In this case, the context was the national strategy *IDA Smart Energy Denmark* (Mathiesen et al., 2015). The strategy was an update from 2015 of the IDA scenarios of 2006 and 2009 as described in the previous sections of this chapter. The national strategy *IDA Smart Energy Denmark* was aligned with a similar strategy for Europe expressed in the paper "Smart Energy Europe: The technical and economic impact of one potential 100 percent renewable energy scenario for the European Union" (Connolly et al., 2016). Thus, by being an integrated part of the national Danish strategy, Aalborg was also an integrated part of a similar European strategy.

Next step of the guiding principle is to identify how to deal with issues that are not logically embedded in neither local nor global action. The study pointed specifically to five elements that should be dealt with and highlighted specific guidelines for including these in a local 100 percent renewable energy strategy, namely:

1. *To identify available sustainable biomass resources.* When using biomass as a fuel, it is important to identify the amount of biomass that can be used due to resource availability. To maintain a sustainable resource availability, only residual biomass should be used to keep the environmental impacts low in terms of CO_2 emissions. According to this principle, the local region should only use its share of sustainably available biomass resources on a national, European and global scale. This can be defined using the share of population. Later in Chapter 8, we will go more into detail with the issue of sustainable biomass.
2. *To define and include transportation needs.* Transportation demand should include international aviation and shipping. This can be defined using the share of population. Thus, this also includes demands for aviation and freight.
3. *To define and include the industry.* Since energy demands for industry result in products available nationally and globally, the principle is that the local region should cover its share of the national, European and global industrial use of energy.
4. *To define available wind and solar resources taking into account the national offshore potential and similar.* The principle is to identify the local renewable energy sources but also investigate the total national sources. Some regions will have more renewable energy available, while others will have less. The goal is that the intake of renewable energy should fit the local energy demands, including points 1–3 in these guidelines. In principle, some regions will be net exporters of renewable energy while others will be net importers. For instance, good onshore wind resources might be utilized locally to leave offshore wind resources for the major cities.
5. *To define and establish electricity balancing.* The import and export of electricity should be balanced to take the city's share of national and international storage and flexible production technologies into account.

In the following, these guidelines are applied to the case of Smart Energy Aalborg.

The Smart Energy Aalborg 100 percent RES scenario

Based on the general guidelines and methodology described above, it is possible to design a Smart Energy Aalborg vision for a sustainable future of Aalborg Municipality. The analyses were carried out in EnergyPLAN, in which a reference model of Aalborg for 2018 was made, based on the actual energy demands in Aalborg, combined with the guiding principle to determine industrial and transportation demands. The 2018 reference scenario was extrapolated to a 2050 reference scenario, the 2050 Business as Usual (BAU) scenario. From the 2050 reference scenario, a 2050 smart energy system was designed, based on the principles of smart energy systems and the guiding principle described above. The starting point for the design of the vision was the energy supply for the year 2018 illustrated in the Sankey diagram of Fig. 7.25. The figure is based on the inputs and outputs representing one year simulated in EnergyPLAN. As one can see, the traditional system was largely based on the burning of fossil fuels, even though wind, biomass and waste made certain contributions.

The energy balance in the 100 percent renewable smart energy city of Aalborg can be seen in the Sankey diagram in Fig. 7.26, which shows the energy demands and the results from EnergyPLAN in terms of fulfilling these demands in 2050. In accordance with the guiding principle, the smart energy city would be able to fulfill the local energy demands for heating while utilizing local resources. However, the local resources were used in a way that allowed other Danish cities, Denmark as a country and Europe to transition to 100 percent renewable smart energy systems. Due to the methodology suggested, the excess heat was accounted for as a local energy source, instead of being tied to the industry demand. This is due to the fact that the industrial energy demand related to Aalborg's share of the national industrial energy demand.

The Sankey diagram makes it possible to identify the usage of electricity from renewable energy in terms of wind, solar and biomass. The wind resources were predominantly used for covering the electricity demand, while biomass was gasified to a large extent and some of the waste resources were burned directly in an incinerator. The biomass, combined with the electricity used in electrolyzers, was used to produce gas for power stations and fuels for the transportation and industrial demands which were determined based on the guiding principle. The biomass usage was equal to the local resources in Aalborg Municipality, as this amount coincidentally corresponded to the nationally allocated biomass resource.

Electric heat pumps covered the heating demand for individual households. The individual heating demand was only 0.21 TWh out of a total demand of 1.86 TWh. For the district heating demand of 1.65 TWh, excess heat from the local industry provided 0.86 TWh of heat, which included low-temperature excess heat sources that needed boosting with a heat pump. Furthermore, heat pumps, geothermal sources and heat from the CHP plants (gas turbines and gas engines) supplied the remaining district heating demands.

As already described above, the system fulfills the basic principles mentioned in the guiding principle. However, some remarks should be made regarding the analysis of the balancing of the electricity exchange. The Sankey diagram of the Smart Energy City of Aalborg shows a balanced system that uses electrolyzers in combination with the gas demand to reduce the amount of excess electricity production from wind turbines and solar power. However, other options such as batteries and high-temperature steam storage are available. A comparison was made to show the differences between the different technologies and its results are presented in the following section. Additionally, it is important to mention that with the continuous decrease in prices of renewable energy technology, it might become economically viable to simply install more turbines and shut them off during hours of excess production.

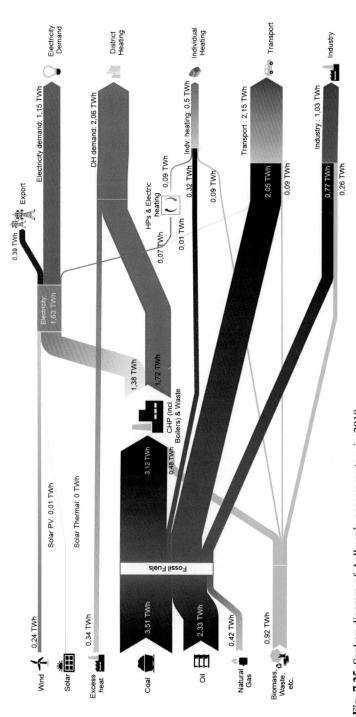

Fig. 7.25 Sankey diagram of Aalborg's energy system in 2018.

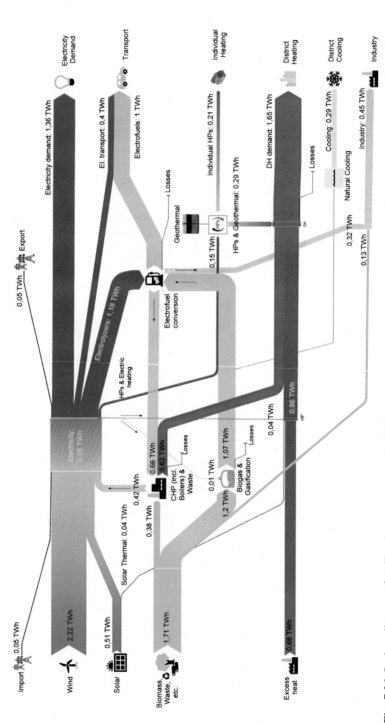

Fig. 7.26 Sankey diagram of Aalborg's energy system in 2050.

If no action was taken to deal with the potential excess production from renewables, the imbalances would be as high as an export of more than 500 GWh/year and an import of 300 GWh/year. This situation was used as a starting point for the analysis on how Aalborg could make its contribution in the most affordable way. As already mentioned above, the target was set at a level at which Aalborg would be able to decrease the import and export to 40–50 GWh/year, thus exploiting mutual benefits of exchange and utilizing common European hydropower options. To deal with the import, an investment in a gas-fired power plant operated on green gas was chosen. In principle, such a plant represented Aalborg's share of European needs for power capacities for the hours of insufficient wind power and PV production. In the Aalborg case, this corresponded to a power plant of 120 MW and green gas use of 450 GWh/year.

To evaluate the different options of reducing excess export, a reference situation was defined as a benchmark. The reference involved building slightly more wind power and PV and then waste the electricity by curtailment or similar. To calculate the cost and benefits, a simple calculation was made of the amount of saved wind power and PV and the corresponding reduction in investments, compared to the investments in balancing technology. The feasibility was estimated based on this calculation.

To minimize the export (excess power production), the most obvious and affordable measures would be (1) to operate heat pumps flexibly and make use of thermal storage capacities and (2) to smart charge the electric vehicles instead of using dump charge. These measures alone would reduce the excess production from more than 500 GWh/year to approximately 250 GWh/year, as illustrated in Fig. 7.27. Furthermore, they would reduce the necessary wind and PV from 1277 to 1099 MW. As illustrated in the figure, these efforts are clearly feasible and align well with the general findings expressed in Chapter 6.

The next step was to evaluate the mix of variable renewable energy production. The combination of wind and solar power could affect the utilization rate of the variable

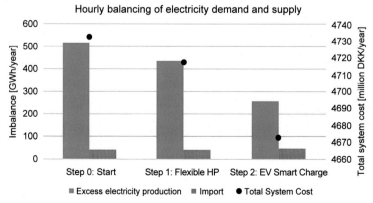

Fig. 7.27 The influence of flexible operation of heat pumps and smart charge of electric vehicles on reducing the excess electricity production of the Aalborg Smart Energy Vision.

renewable energy, since these sources were able to complement each other. Therefore, the study assessed different mixes of renewable energy. In the first mix, 10 percent of the variable renewable energy production came from PV; in the second, 20 percent came from PV, and in the third, 30 percent came from PV. The result of the analysis showed that from both an economic perspective and in terms of imbalances, the mix between PV and wind power where PV delivered 20 percent of the production would result in the best system. This result aligns well with the findings of Chapter 5 regarding optimal combinations of wind and solar.

The final step in bringing down excess electricity production was more difficult. The analysis evaluated three different alternatives, as shown in Fig. 7.28.

The first alternative was to invest in batteries. Since the cost of batteries was similar to the cost of many other electricity storage technologies, this option also represented pump hydro, compressed energy storage, fuel cell hydrogen systems and similar (Lund et al., 2016). The additional benefit of these technologies was their ability to reduce the investments in wind and PV, while avoiding investment in the power plant. However, as illustrated in the figure, the cost of the batteries was extremely high compared to the benefits and the expected investment costs. It would be much cheaper to curtail the excess production.

The next technology option was to invest in a high-temperature steam storage heating by excess electricity and utilize it for electricity production in the power plant. The advantage of this option was the much lower storage cost. However, investments would have to be made in a steam boiler and turbine power station, raising the cost of the power plant. This option was more or less in balance, and given the uncertainties of expected investment costs, it could not be ruled out but would need further investigation.

The third option was to invest in additional electrolyzer capacity in the system. The electrolyzers should not produce more hydrogen for electrofuels, but the additional capacities would make it possible to produce the same amount in a more flexible way. This was the option with the lowest costs. Thus, it was used as part of the final

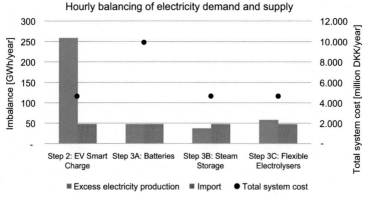

Fig. 7.28 Comparison of batteries, steam storage and flexible electrolyzers to regulate excess electricity.

energy system in the Aalborg vision. However, it should be emphasized that the feasibility compared to simply curtailing more wind was low and the sensitivity to additional needs for hydrogen storage capacity was high. Therefore, this option should also be subject to further investigation in the future.

In total, the energy transition resulted in significant fuel efficiency increases, due to electrification and other efficiency measures such as energy savings and conversion to district heating.

5   The potential of renewable energy systems in China[e]

This section is based on Liu et al.'s (2011a) article "Potential of Renewable Energy Systems in China," which discusses the perspectives of renewable energy systems in China and analyzes whether the methodologies described in previous parts of this chapter, or similar methodologies, can be applied to China to create a future renewable energy system.

China's energy consumption has influenced the energy demand on a global scale significantly, since China has become both the largest energy consumer and CO_2-emitting country in the world. This dramatically increasing energy consumption means that the domestic energy production cannot meet the demand. Already by 1993, China became a net crude oil importer, and only 19 years later, in 2012, the Chinese net oil imports reached 285 Mtoe. Thereby, China ranked as the world's second largest oil importer after the United States and accounted for about 59 percent of the global oil demand. China became a net importer of primary energy in 1997 and, since then, energy security and the maintenance of the balance of energy production and consumption have been vital problems in the country. The country's long-lasting fossil fuel-dominated energy structure, which is mainly based on coal, has brought severe challenges to the goal of maintaining environmental protection and decreasing greenhouse gas emissions.

Given the fact that the amount of fossil resources is finite, and focusing on the aims to fill the gap between domestic energy production and consumption as well as to maintain high economic growth, it is essential and meaningful for China to integrate renewable energy into future sustainable energy development strategies. China is endowed with an abundant reserve of renewable energy sources, which are currently underexploited and which offer a significant potential for renewable energy system development.

Through a review of the Chinese energy supply and consumption in the past three decades, it can be concluded that the share of renewable energy has increased steadily and has begun to play a role in the energy structure. However, the energy supply and consumption structure, which is dominated by fossil fuels, especially coal, has basically remained the same. Although China has made some achievements in decreasing the energy intensity, the energy demand in China is expected to continue to grow,

[e] Excerpts reprinted from *Applied Energy* 88, Liu, Lund, Mathiesen, and Zhang, "Potential of Renewable Energy Systems in China", pp. 518–525 (2011a), with permission from Elsevier.

driven by the country's highly energy-intensive economy and strong GDP growth. Meanwhile, energy conservation is an absolutely necessary strategy for China to stabilize its energy demand.

In Liu et al. (2011a), a review was made of potential renewable energy sources in China and, as a follow-up, Liu et al. (2011b) highlighted some of the barriers to and potential for the large-scale integration of wind power into the Chinese energy system. Similar to the case of Denmark shown in Table 7.1, the potential for renewable energy sources in China was categorized into electricity, heat, and biomass fuel (see Table 7.5). It should be noticed that the potential listed here is shown as a range. With the expected future technological, economic, and social development, the potential may increase, as, e.g., geothermal energy and biomass fuels are not considered here.

In Fig. 7.29, the potential for renewable energy sources is compared with the gross energy consumption of 2011 as well as the energy demand prediction for China for 2015 and 2030. In Fig. 7.30, the potential for electricity supplied by renewable energy is compared with the electricity consumption in 2011 and 2030. The minimum potential for renewable energy sources in China is smaller than the country's current energy consumption, while the maximum potential for renewable energy sources is larger than the estimated energy demand in 2030. A more optimistic tendency appears in the comparison between the potential renewable energy sources of electricity and the future electricity demand. The minimum and maximum potentials for electricity supplied by renewable energy sources exceed both the current electricity demand and the demand in 2030, respectively. This shows the possibility for renewable energy sources to cover the future energy consumption in China.

The total amount of potential renewable energy sources per capita in Denmark and China were 137 and 103 GJ/capita, respectively. Fig. 7.31 presents the result of the comparison. As shown, although the territories, population, and renewable energy

Table 7.5 Potential renewable energy sources in China.

Renewable energy sources	Unit	Potential
Wind	TWh/year	7644–24,700
Photovoltaic	TWh/year	1296–6480
Tidal energy	TWh/year	>620
Wave	TWh/year	>1500
Hydro power	TWh/year	2474–6083
Total electricity	TWh/year	13,434–39,383
Solar thermal	PJ/year	6000–30,000
Geothermal	PJ/year	1000
Total heat	PJ/year	7000–31,000
Straw	PJ/year	5561–6440
Wood	PJ/year	4332–5210
Waste (combustible)	PJ/year	1170–3454
Biogas	PJ/year	1258–2517
Energy crops	PJ/year	3660–10,500
Total biomass fuel	PJ/year	15,981–28,121

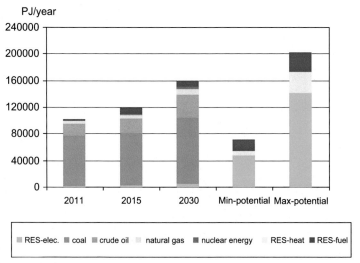

Fig. 7.29 Potential for renewable energy sources in China compared to the primary energy consumption.

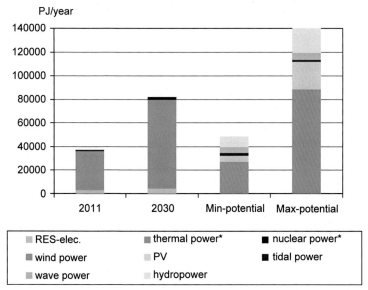

Fig. 7.30 Potential for renewable energy electricity in China compared to the electricity consumption (*fuel equivalent: electricity multiplied by 3).

sources differed significantly, the total amount of potential renewable energy sources per capita was less than twice as high in Denmark as in China. The climate situation in China was more suitable for agriculture, but Denmark owned more potential biomass fuel per capita. The Danish cultivated land area was about 4900 square meters per capita, corresponding to four times the cultivated land area per capita in China.

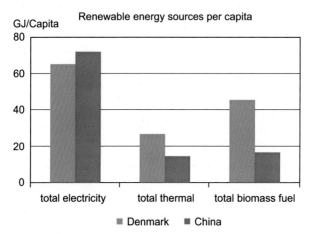

Fig. 7.31 Comparisons of renewable energy sources per capita and per area in Denmark and China.

Three kinds of indexes are compared between China and Denmark in Table 7.6. As illustrated, the energy demand per capita was distinctly lower in China than in Denmark in 2011. The electricity demand per capita in China was about half of that in Denmark, and the heat demand per capita in China was only about one-eighth of that in Denmark. However, in relation to energy intensity, one GDP unit in China required more energy consumption than in Denmark. The gross energy consumption per GDP in China was about five times higher than in Denmark. In terms of the replacement of

Table 7.6 Comparison of energy demand and energy intensity in China and Denmark.

	Index (2011)	Unit	China	Denmark
Energy demand	Primary energy supply per capita	GJ/capita	65.39	159.40
	Gross energy consumption per capita	GJ/capita	75.16	144
	Electricity demand per capita	*GJ/capita*	*10.23*	*20.30*
	Heat demand per capita	*GJ/capita*	*2.35*	*18.82*
Energy intensity	Primary energy supply per GDP[a]	TJ/million USD	12.75	2.65
	Gross energy consumption per GDP[a]	TJ/million USD	14.55	3.03
Renewable energy	Share of total gross energy consumption	Percent	8.4	17.0
	Share of total domestic electricity supply	Percent	19.3	29.3

[a] 2011 exchange rate.

fossil fuel by renewable energy sources, the shares of renewable energy in the energy consumption and the domestic energy supply in China were both lower than the shares in Denmark.

The substantial renewable energy supply potential and the low energy demand per capita, as well as distinct differences in energy efficiency and renewable energy deployment, would give China the opportunity to design and approach future renewable energy systems.

In conclusion, China is facing two severe challenges of maintaining the energy balance as well as handling environmental pollution. Both challenges are rooted in the inappropriate economic and energy structures that have changed little in the past three decades. Energy demands in China will continue to grow driven by the country's highly energy-intensive economy and strong GDP growth. In order to fill the increasing gap between domestic energy production and consumption and to change the inappropriate energy structure, it is essential and meaningful to integrate renewable energy into future energy systems in China.

Today, renewable energy is at a rapid development stage; however, compared with the potential, renewable energy is presently underexploited in China. There is much room and promise for developing large-scale renewable sources in the future energy system, especially along with technological and economic development.

There is no obvious difference in the renewable energy resources per capita between China and Denmark. However, the energy demand per capita in China is distinctly lower compared with Denmark and the energy efficiency gap between the two countries is distinct. The methodologies used to design a 100 percent renewable energy system for Denmark and to analyze typical technological changes in terms of energy conservation, energy efficiency, and renewable energy integration are applicable to China. On this basis, it is reasonable to propose the issue for discussion and conduct further analyses of a 100 percent renewable energy system in China.

6   Reflections

Reflections on the analyses of 100 percent renewable energy systems in this chapter are made regarding principles and methodologies, as well as the implementation of renewable energy systems in Denmark and other countries.

Principles and methodologies

From a methodology point of view, it can be concluded that the design of future 100 percent renewable energy systems is a very complex process. On the one hand, a broad variety of measures must be combined to reach the target, and on the other, each individual measure has to be evaluated and coordinated with the new overall system. In the IDA Energy Plan case, this process was implemented by combining a creative phase involving the inputs of a number of experts and a detailed analytical phase with technical and economic analyses of the overall system, providing feedback on the individual proposals. In a back-and-forth process, each proposal was formed in such a way that it combined the best of the detailed expert knowledge with the ability of the proposal to fit well into the overall system, in terms of technical innovation, efficient energy supply, and socioeconomic feasibility. In the CEESA case, the process was carried out in collaboration with approximately 25 researchers from five Danish universities and companies with quite different backgrounds and fields of expertise. A key focus in the project was the biomass resources available for Denmark, in combination with the identification of transportation fuel pathways to add to the design of 100 percent renewable energy systems with a smart energy systems approach; i.e., trying to identify and make use of the synergies between the different subsystems to achieve an optimal solution for the overall system as well as for the individual sectors.

Next, this chapter formulated a guiding principle and a methodology to be applied to cities and local communities for the design of sustainable energy futures in a way that makes such a transition compatible with a national and global transition to renewable energy. The argument is that local action is needed to identify and implement the common best solutions to the green transition at the national and international levels. However, no isolated solutions should be defined at the local level. Instead, it is vital that the local communities aim at identifying their role to play in the best strategy for the overall system. Such an aim poses several essential questions and challenges as well as practical issues to be dealt with in the design of strategies for implementing the green transition in future sustainable energy solutions.

7   Conclusions and recommendations

Based on the cases of Aalborg and Denmark, this chapter presented a series of studies on the challenges and perspectives of converting present energy systems into a 100 percent renewable energy system. Moreover, the perspective of applying the methodologies to China was discussed. The three studies of Denmark each include two or three alternatives that are based on either biomass or wind power. The main data of all eight alternatives are listed in Table 7.7. For the Danish case, the conclusion is that a 100 percent renewable energy supply based on domestic resources is physically possible and that the first step toward 2030 is feasible for Danish society.

All three studies show that, when reaching a high share of variable resources in combination with CHP and savings, the development of renewable energy strategies becomes a matter of introducing and adding flexible energy conversion and storage technologies and designing integrated energy system solutions. The first study identifies specific improvements of system flexibility as essential to the conversion of the energy system into a 100 percent renewable energy system. First, oil for transportation must be replaced by other sources. Given the limitations on the Danish biomass resource, solutions based on electricity become key technologies. Moreover, these technologies increase the potential for including wind power in the ancillary services to maintain the voltage and frequency of the electricity supply.

The next improvement involves including small CHP stations in the regulation as well as adding heat pumps to the system. These technologies are of particular importance because they provide the possibility of changing the ratio between electricity and heat demand while maintaining the high fuel efficiency of CHP. The third key

Table 7.7 Main data of the eight alternative 100 percent renewable energy systems.

	First approach		IDA Energy Plan (IDA 2050)			CEESA 100 percent RES scenarios		
	EV/H₂FC	Biofuels	Biomass	Main	Wind	Conserv.	Ideal	CEESA
Demands (TWh/year)								
Electricity	37.0	37.0	30.2	30.2	30.2	26.9	26.9	26.9
Heating (including process)	56.8	56.8	44.5	44.5	44.5	68.8	68.8	68.8
Transportation (electricity)	17.8	–	5.0	5.0	5.0	27.7	57.6	39.9
Transportation (biomass)		50.7	24.9	24.9	24.9	31.1	0	14.3
Primary energy supply								
Biomass (PJ/year)	180	325	333	270	200	331	206	237
Solar thermal (PJ/year)	8	8	19	19	19	23.3	23.3	23.3
Geothermal (PJ/year)	–	–	–	–	–	9.8	12.5	12.4
PV (GW installed)	5	5	1.5	1.5	1.5	–	7.5	5
Wave (GW installed)	–	–	1	1	1	–	1	0.3
Wind (GW installed)	15	15	6	10	15	12	16	14

point is to add electrolyzers to the system and, at the same time, create the basis for further inclusion of wind turbines in the voltage and frequency regulation of the electricity supply.

Based on the implementation of these three key technological changes, the analyses of the first study show that the Danish energy system can be converted into a 100 percent renewable energy system, by combining 180 TJ/year of biomass with 5000 MW of PV and 15–27 GW of wind power. In the reference, 27 GW of wind power is necessary, while in combination with savings and efficiency improvements, the required capacity is reduced to around 15 GW. Thus, the first study emphasizes the importance of implementing energy conservation as well as efficiency improvements in the supply sector.

In the second study (IDA Energy Plan), electric or hydrogen fuel cell vehicles are introduced in the entire transportation sector. If this solution is replaced with biofuel-based transportation technologies, the need for biomass resources may be nearly doubled. Consequently, the study also emphasizes the importance of further developing electric vehicle technologies. Moreover, it indicates that biofuel transportation technologies should be reserved for the areas of transportation in which the electricity/hydrogen solution proves insufficient.

The next part of the second study (IDA Climate Plan) goes a step further, especially regarding energy conservation and the design of a coherent transportation solution. The study implements energy conservation measures at a high level, and, as a consequence, energy demands will decrease compared to the first study. On the other hand, the transportation technologies applied in the second study are much more differentiated and include the combination of electric vehicles and biofuel technologies, which increases the demand compared to the first study. Regarding the design of a suitable transportation solution, it must, however, be emphasized that both studies are far from coherent or optimized. Nevertheless, the studies do provide a sufficient overview of the principal possibilities.

The second study shows that Denmark can convert into a supply of 100 percent renewable energy constituted by 280 PJ/year of biomass, 19 PJ of solar thermal, 2500 MW of wave and PV, and 10,000 MW of wind power. It should be emphasized that the 280 PJ/year of biomass does not include all conversion losses. However, the study shows how biomass resources can be replaced by more wind power, and vice versa, and points out that Denmark will have to consider to which degree the country should rely mostly on biomass resources or on wind power. The solution based on biomass will involve the use of present farming areas, while the wind power solution will involve a large share of hydrogen or similar energy carriers leading to certain inefficiencies in the system design.

The third study (CEESA) adds to the two previous studies as it goes into more detail with the discussions and analyses of several important issues. First, a further step has been taken in the discussion and identification of suitable transportation fuel pathways, and some of the potential pathways have been quantified and integrated into the overall identification of a roadmap toward a 100 percent renewable energy system in 2050. A DME/methanol-based pathway has been used for the concrete calculations of the scenarios to illustrate the principle of using biomass resources in combination with electrolyzers to replace fossil fuels in the transportation sector in the short term.

In the longer term, carbon from other sources than biomass is used to replace larger amounts of fossil fuels, without putting further strain on the biomass resource. It is too early to know if a liquid fuel DME/methanol solution is better than, for instance, a gas methane solution or a combination. However, approximately the same energy balances can be achieved with a gas solution.

Next, the study furthers the discussion on the influence of technological development. Three different scenarios, each representing different assumptions regarding the availability of technologies, are compared. The analyses indicate that especially technological development in electric cars, hybrid vehicles, and vehicles to utilize either DM/methanol and/or methane gas becomes essential. Moreover, bio-DME/methanol or methane production technologies (including biomass gasification and electrolyzers) are important in the short or medium term and should be supplemented by carbon capture technologies in the longer term.

Finally, the study emphasizes and illustrates the importance of applying a smart energy systems approach to the identification of a suitable 100 percent renewable energy systems design. In particular, the study combines the analysis of gasified biomass and gas grid storages, in combination with electric vehicles and fuel production in the transportation sector, as well as district heating systems. This creates an energy system into which smart energy systems are integrated and the storage options are used in combination to enable the final scenario. The analyses ensure that there is an hourly balance in the gas supply and demand, while the results indicate that the capacity of the current Danish salt cavern storage facilities is more than sufficient to facilitate this balancing.

All three studies apply the EnergyPLAN energy systems analysis tool and together they illustrate how such a tool can be used to design 100 percent renewable energy solutions, as well as form the basis for an evaluation of systems based on the use of concrete institutional economics.

The case of Aalborg Municipality in Denmark has shown a way to deal with the issues of designing a local strategy which fits into the national context. A Smart Energy System vision for Aalborg to become 100 percent renewable in 2050 was designed in such a way that it fit into a common best solution in Denmark as well as in Europe. At the same time, this vision was used to conduct detailed analyses of the options of balancing electricity supply and demand on an hourly basis with a focus on how to deal with excess electricity production. The result indicates that least-cost options are flexible operations of power to heat in terms of heat pumps in combination with thermal storage and district heating and smart charging of electric vehicles. Batteries and similar electricity storage options (except for electric vehicles) are far from feasible compared to curtailment. Instead, steam storage and especially overcapacity in electrolyzers are promising. These conclusions confirm, quantify and further similar conclusions presented in Chapters 5 and 6.

Finally, the analyses also confirm previous conclusions of Chapter 5 with respect to the best combination of PV and wind power (under Danish circumstances). This means that, on an annual basis, 20 percent of the electricity should come from PV and the rest from wind in order to minimize excess electricity problems. However, this study takes these previous conclusions a step further by also introducing this combination as the economically least-cost solution.

Carbon neutral societies and smart energy systems

8

In recent years, political goals of green transitions tend to focus on *carbon neutral societies* rather than renewable or decarbonized *energy systems*. The Paris Agreement (United Nations, 2015) constitutes the global framework for such aims and goals. The European Commission (2018) report "A Clean Planet for All" put forward a strategic vision for a *climate neutral* economy and several countries around the world have engaged in similar political aims and goals. In 2020, the Danish government passed a Climate Act through Parliament (Danish Ministry of Climate, Energy and Utilities, 2020). The Danish Climate Act sets a near-term target of reducing Denmark's total greenhouse gas emissions by 70 percent by 2030 compared to the 1990-level and sets a long-term target of achieving *climate neutrality* by 2050 at the latest. The International Energy Agency (2021a) published a Roadmap for the Global Energy Sector to reach *net-zero* emissions by 2050 in consistence with the Paris Agreement's goal of limiting the long-term increase in average global temperatures to 1.5 °C.

Political goals of carbon neutral societies add to the challenge of designing renewable smart energy systems such as described in the previous chapters. The energy and transportation sectors—in sum called the energy system—account for around three-quarters of today's greenhouse gas emissions and, therefore, hold the key to any strategy for achieving a fully decarbonized society. However, strategies for reducing CO_2 emissions from fossil fuels in the energy and transportation sectors would have to align with reducing climate gas emissions in other sectors as well. This coordinated action calls for an integrated approach in which the interactions between the sectors in question are taken into consideration. Moreover, the global goals cannot be reached without active contributions from every region and country. The individual countries would also have to coordinate their efforts to properly address common issues such as the decarbonization of international shipping and aviation, the global use of sustainable biomass resources, and the international exchange of electricity as well as—potentially—hydrogen and other green fuels.

This chapter presents and calculates in detail first a Danish national and then a European strategy for decarbonizing the energy sector within the context of achieving a carbon neutral society in all sectors. Thus, the design of the energy sector solution is made while considering the need for coordinating with other greenhouse gas sectors, as well as coordinating internationally with neighboring countries, and in the end, with the rest of the world. The chapter builds on the concept of Smart Energy Systems by focusing significantly more in detail on bioenergy and biomass pathways as well as including carbon capture and storage (CCS) in combination with carbon capture and utilization (CCU). Moreover, a section details the aspects of achieving hourly balance in the future power supply as well as stabilization in the power grid.

Renewable Energy Systems. https://doi.org/10.1016/B978-0-443-14137-9.00008-5

The focus in this chapter is on the general and fundamental guidelines for a transition strategy for a country to become a fully decarbonized society. To demonstrate such general guidelines, they have been applied first to the case of Denmark by the year 2045 and then to the case of Europe by the year 2050. The study also includes an analysis on how to reach the first important steps already in 2030. This makes it possible to discuss how ambitious 10-year goals could work as a stepping stone toward eliminating fossil fuel and becoming carbon neutral by the mid-century. The design of suitable energy systems must consider both conversion and storage technologies. Renewable energy will have to be compared not to fossil fuels, but to other sorts of renewable energy system technologies, including conservation, efficiency improvements, and storage and conversion technologies—for example, wind turbines versus the need for biomass resources.

This chapter begins by presenting the overall Smart Energy Denmark 2045 scenario with a focus on how it fits into achieving a carbon neutral society in all sectors. Then, the chapter focuses on three important elements of the efforts involved: First, the issue of the limits of sustainable biomass resources available at the national as well as the global level; next, the issue of how to transform and integrate the transportation sector into the overall solution; and third, the issue of securing stabilization and balancing of the power supply.

Finally, this chapter presents a European case based on applying the smart energy systems principles to the European Commission (2018) report "A Clean Planet for All."

1 Smart energy systems in the context of a carbon neutral society[a]

This section is based on Lund et al. (2022b) article "Smart Energy Denmark. A consistent and detailed strategy for a fully decarbonized society." The article presents a strategy for achieving a fully decarbonized Danish energy system in 2045. The input for the strategy is based on the results of a collaboration with the Danish Society of Engineers (Lund et al., 2021a) as a follow up on the previous work already presented in Chapter 7. However, here the smart energy system is coordinated with other sectors to achieve a fully decarbonized society. The article highlights a set of guidelines for the design of such a strategy for a single country, considering the global context and character of the green transition. Thus, the strategy could also be relevant for most countries at a global level.

Like the strategies presented in the previous chapters, this energy system analysis also included hour-by-hour computer simulations on the EnergyPLAN model leading to the design of a smart energy system with a focus on sector integration and the ability

[a] Excerpts reprinted from *Renewable end Sustainable Energy Reviews 168*, Henrik Lund, Jakob Zinck Thellufsen, Peter Sorknæs, Brian Vad Mathiesen, Miguel Chang, Poul Thøis Madsen, Mikkel Strunge Kany and Iva Ridjan Skov "Smart Energy Denmark. A consistent and detailed strategy for a fully decarbonized society", October 2022, with permission from Elsevier.

to balance all sectors of the complete energy system. In the analysis, issues such as international shipping and aviation, the sustainable use of biomass, and the exchange of electricity and gas with neighboring countries were all considered. Moreover, the size of the employment impact of investing in decarbonizing the Danish economy was discussed.

Overall governing guidelines

A country may convert to 100 percent renewable energy and/or fully decarbonize in different ways. E.g., a country such as Denmark could replace its fossil fuels by imported bioenergy, or Denmark could build a lot of wind turbines and export the electricity based on the viewpoint that the CO_2 saved in other countries would compensate for oil in the Danish transportation. However, if the efforts of a single country should make sense in terms of combating greenhouse gas emissions, it must be seen in the global context of all neighboring countries following similar strategies. Therefore, the transition to a fully-fledged renewable society requires the fulfillment of the two following criteria or governing principles:

Firstly, *Denmark would have to fulfill its objectives of increased renewable energy and CO_2 reductions in a way that would fit well into a context in which the rest of Europe—and the world—realistically would be able to do the same.* This overall criterion would imply several choices and issues. Some of the most important are that:

- Denmark should include the Danish share of *international aviation and shipping* even though it is not yet included in the UN method for calculating national CO_2 emissions.
- Denmark should not exceed the Danish share of *sustainable use of biomass* in the world.
- Denmark should make its contribution in terms of *flexibility and reserve capacity* that would allow for the increased integration of variable RES, mainly wind and solar, into the *European electricity supply.*

Secondly, Denmark should fulfill its short-term objective of a 70 percent CO_2 reduction in 2030 in a way that would fit well into a fully decarbonized society by 2045. To fulfill this principle, existing technologies should be combined with paving the way for the right technologies in the long term. Therefore, before 2030, politicians, the public administration, and private firms would need to prioritize:

- Energy technologies which fit well into the future zero carbon solutions, as expanded on in the following.
- Innovation efforts for technologies that would be needed after 2030, even if they are not likely to play a significant role before 2030.

The Smart Energy Denmark 2045 scenario

The concrete proposals of Smart Energy Denmark 2045 have been divided into *four sectors* and designed with a focus on *five crosscutting areas.* The *four sectors* are:

- *Heating and cooling*: Supply of heat for all buildings.
- *Industry*: Service, commercial and heavy industry, including gas and oil in the North Sea.
- *Transportation*: Persons and goods including aviation and shipping.

- *Electricity*: Primarily, the classic electricity consumption including the overall balancing of the electricity supply.

Five crosscutting focus areas following from the two main criteria are:

- *Energy efficiency*: Including the fulfillment of the EU Energy Efficiency Directive.
- *Sector integration*: Including energy storage, conversion and electrification, as well as the integration between electricity, gas, district heating and cooling grids.
- *Biomass*: With a focus on the issue of sustainable biomass.
- *Renewable energy*: Which type and how much of it?
- *Technological challenges*: Identifying the most important technologies as well as their potential role in 2030 and 2045 based on their expected development.

The Smart Energy Denmark 2045 scenario is described in the following with a focus on the four sectors and five focus areas. As a benchmark, a Sankey diagram illustrates the starting point situation in Denmark (2020) in Fig. 8.1. As can be seen, Denmark already has a high input of wind power equivalent to approx. 50 percent of the electricity supply as well as a high share of biomass. However, the share of fossil fuels in terms of coal, oil and natural gas is still substantial. Especially, within the transportation sector and manufacturing, the share of fossil fuels is relatively high compared to the other sectors.

With the Danish energy system of 2020 as a starting point and by using the beforementioned five cross-cutting focus areas, a smart energy systems scenario was designed to include the following elements:

Energy efficiency (including fulfillment of EU Energy Efficiency Directive):

- *Heat savings* in all existing buildings (resulting in average savings of 12 percent in 2030 and 30 percent in 2045 as compared to 2020).
- Gradual transition to *4th generation district heating* (half in 2030, full in 2045), which will both reduce the grid loss of the district heating system while also improving the efficiency of the energy conversion units and allowing for increased utilization of waste heat from industrial processes.
- 12 percent *savings and efficiency* improvements in the industrial energy consumption in 2030 and 32 percent in 2045.
- More *energy efficient technologies in data centers*, resulting in a reduction in electricity consumption of 5 percent.
- *Curbing of growth in private car* mileage as compared with growth in the baseline projection (1.6 percent/year instead of 2 percent/year) and similar savings within modular shifts from car and aviation to trains and public transportation.
- 10 percent *savings in the "classic electricity consumption"* in 2030 (20 percent in 2045), with more energy efficient energy technologies, such as more energy efficient appliances.

Sector coupling (energy conversion, storage and electrification):

- Oil and gas boilers phased out and replaced with a mix of *district heating and individual heat pumps*, depending on the heat density of the given area.
- *District heating* expanded to 63 percent of heating requirement using heat pumps, excess heat from industries, data centers and Power-to-X technologies in combination with large-scale thermal storage. The remaining heating demand is supplied by individual heat pumps supplemented by solar thermal to reduce electricity demand, both connected to heat

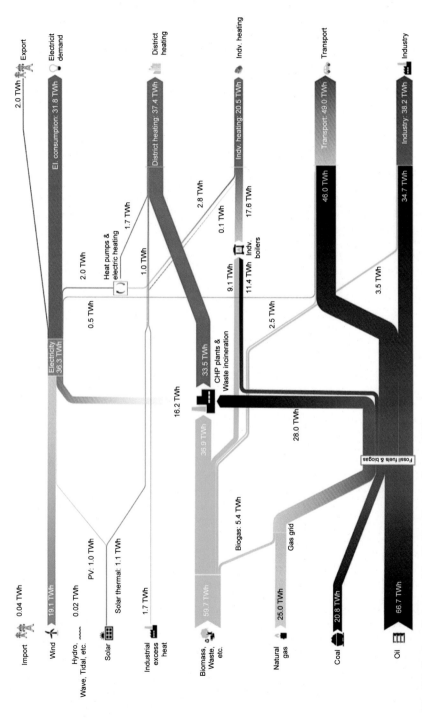

Fig. 8.1 Sankey diagram of the Danish Energy System 2020.

Import 0.04 TWh

Wind 19.1 TWh

Hydro, 0.02 TWh
Wave, Tidal, etc.

Solar
PV: 1.0 TWh
Solar thermal: 1.1 TWh

Industrial 1.7 TWh
excess
heat

Biomass, 59.7 TWh
Waste,
etc.

Natural 25.0 TWh
gas

Coal 20.8 TWh

Oil 66.7 TWh

Export 2.0 TWh

Electricit demand

El. consumption: 31.8 TWh

Electricity 36.3 TWh

District heating: 37.4 TWh
District heating

Indv. heating: 20.5 TWh
Indv. heating

Transport: 49.0 TWh
Transport

Industry: 38.2 TWh
Industry

Heat pumps & electric heating

2.0 TWh
0.5 TWh
1.7 TWh
1.0 TWh
2.8 TWh
0.1 TWh
17.6 TWh

Indv. boilers
11.4 TWh
9.1 TWh
2.5 TWh

CHP plants & Waste incineration
33.5 TWh
16.2 TWh
36.9 TWh

28.0 TWh

Biogas: 5.4 TWh

Gas grid

Fossil fuels & biogas

46.0 TWh
3.5 TWh
34.7 TWh

storages. The solar thermal corresponds to 16 percent of the annual heating demand and the
heat storage capacity corresponds to the heating demand of 1 average day.
- Balancing of wind, solar and wave power with *flexible electricity consumption* units. This
  mainly refers to smart charging electric vehicles that charge when it is suitable for the energy
  system, heat pumps connected to heat storages in individual households and district heating,
  and electrolysis plants from the other sectors, including the use of existing natural gas
  storage and the establishment of hydrogen storage.
- *Electrolysis capacity* is set at about 60–65 percent to allow for a flexible operation of the
  electrolysis with the utilization of H_2 storage, which is set to allow for about 4 days of storage
  of average production. This enables the increased utilization of variable renewable
  electricity sources for Power-to-X.

Biomass (kept at a sustainable level):

- Wood, waste and upgraded biogas to be used in *CHP plants*. In 2045, waste for incineration
  baseload CHP is reduced to a minimum to allow for more recycling, while upgraded biogas
  are used for flexible gas-fired CHP plants for backup and periods with low wind and PV
  potential.
- 200 MW woodchip gasification (syngas and CO_2 for electrofuels) in 2030 and approx.
  2000 MW in 2045, divided between *thermal gasification, HTL and pyrolysis*.
- *Biogas* production to be increased to 35 PJ in 2030 and 60 PJ in 2045. The biogas will be
  upgraded by removing the CO_2 part that is then utilized for Power-to-X processes.

Renewable energy:

- *Solar heating* in the district heating system and in individual dwellings to supplement
  heat pumps.
- *Geothermal heat* for district heating; 500 MW in 2030 increasing to 1000 MW in 2045.
- *PV systems* to be expanded from approx. 1000 MW in 2020 to 5000 MW in 2030, rising to
  10,000 MW in 2045. The PV systems should primarily be on large roofs to reduce the
  land-use for PV systems.
- *Onshore wind power* to be expanded from approx. 4200 MW in 2020 to at least 4800 MW in
  2030 and 5000 MW in 2045.
- *Offshore wind power* to be expanded from approx. 2000 MW in 2020 to 6630 MW in 2030,
  rising to nearly 14,000 MW in 2045.
- *Wave power* (132 MW) (but with wind as an alternative) in both 2030 and 2045.

The underlying principles of these proposals are:

- To make it possible to exploit technological positions of strength and to maximize the number of jobs created.
- To promote and facilitate a technological development after 2030 bringing the Danish
  energy system closer to the goal of decarbonization in 2045.
- To ensure a degree of energy efficiency in accordance with the EU directive on this issue.
- To aim at attaining the 70 percent CO_2 reduction objective by 2030 in a socio-economically
  efficient way.
- And finally, to ensure a broad popular involvement in the implementation of these
  political goals.

Additional to the Danish share of global sustainable biomass resources, the Smart
Energy Denmark 2045 scenario is based on the use of substantial wind power and

PV. This corresponds to similar scenarios for Europe (Rasmussen et al., 2012) and other parts of the world such as, e.g., Central and South America (De Barbose et al., 2017). Since Denmark has good wind potential, it may be a future exporter of wind to other European countries with less wind potential.

In Fig. 8.2, the 2030 scenario for the Danish energy system is summarized. As can be seen, the input of wind power, and—to some extent—also solar energy, will increase substantially. However, to reach a sustainable level, the amount of biomass would have to decrease. In 2030, the remaining fossil fuels would primarily be used in the transportation sector, while in 2045, there would be no fossil fuels at all. In transportation, priority will be given to the direct use of electricity in electric vehicles. However, liquid fuel would still be required to cover all the needs. Especially in 2045, when the Danish shares of international shipping and aviation are included, green fuels based on biomass and hydrogenation in combination with CCU are to become important technologies in a full decarbonization of the energy system.

In Fig. 8.3, the Danish energy system has moved on to 2045. The most important changes compared to 2030 are that fossil fuels are completely phased out and, therefore, the direct electrification of the heating and transportation sectors plays a major role in combination with electrofuels.

In both 2030 and 2045, the Smart Energy Denmark scenario will exchange electricity with neighboring countries based on the principle of mutual benefits, e.g., by providing electricity from wind power to Norway to reduce the use of water in the relatively large dammed hydro power capacity in Norway. Export occurs when there is an excess of variable renewable electricity production in the modeled energy system, and import occurs when there would be a need for power plant operation. The scenario does not include scenarios for the rest of Europe, but instead limits these import and export potentials based on an assumed limit of the relative population of Denmark in relation to the total European population, modeled as an external storage capacity of 820 GWh. The scenario assumes a capacity of 6000 MW of interconnectors.

With the implementation of the Smart Energy Denmark 2045 scenario, it is possible to achieve a fully renewable energy system. Fig. 8.4 synthesizes Figs. 8.1–8.3 into an overview of the primary fuel consumption in the system, while Fig. 8.5 details the heating demand and production and Fig. 8.6 details the electricity demand and production units implemented in the systems, both in 2030 and 2045.

The results reveal that through the implementation of a smart energy system, it is possible to achieve a fuel-efficient transition to a 100 percent renewable, decarbonized energy system in 2045. Furthermore, it is demonstrated that it would be possible to design an energy system in 2030 that not only fulfills the 70 percent reduction target, but also serves as a stepping stone toward the fully decarbonized energy system in 2045.

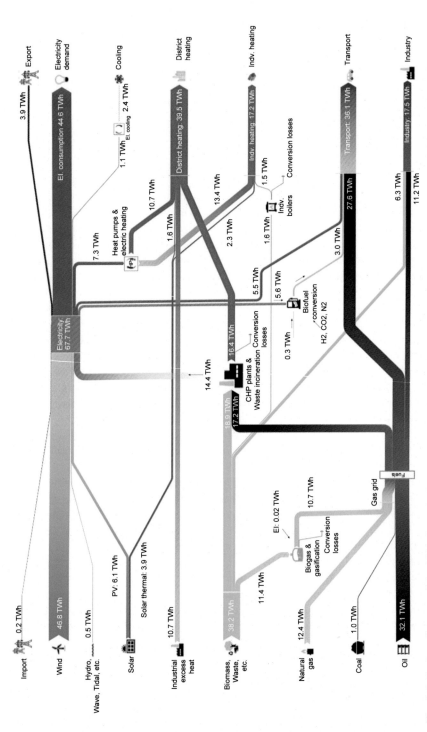

Fig. 8.2 Sankey diagram of the 2030 Scenario for the Danish energy system in accordance with the Smart Energy Denmark proposal.

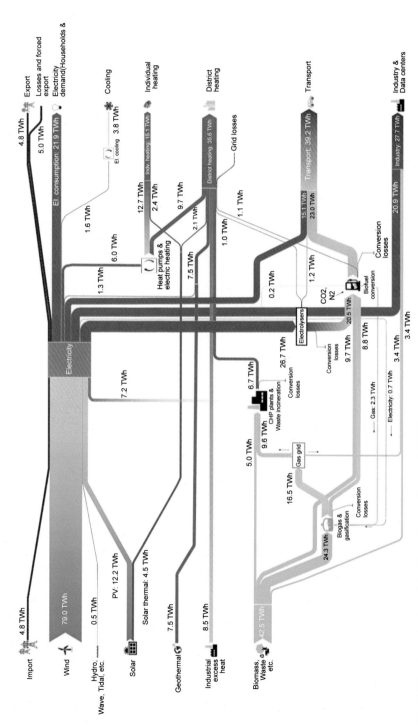

Fig. 8.3 Sankey diagram of the Smart Energy Denmark 2045 energy system.

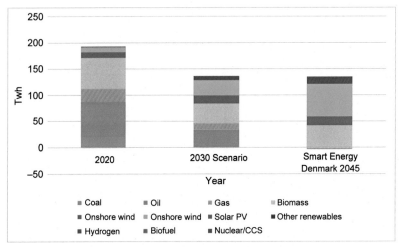

Fig. 8.4 Primary sources of energy in 2020, 2030, and 2045.

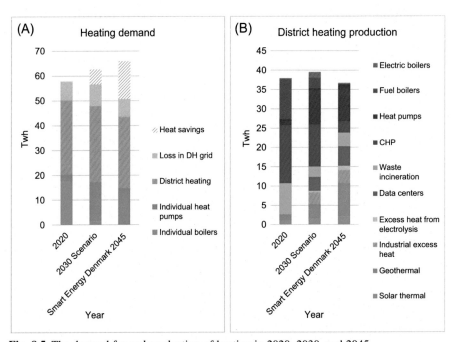

Fig. 8.5 The demand for and production of heating in 2020, 2030, and 2045.

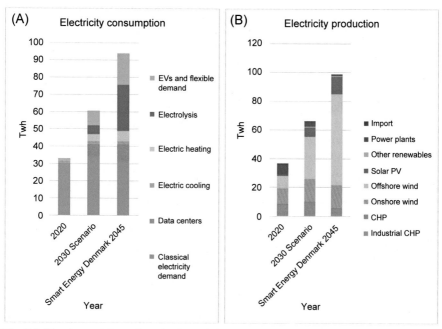

Fig. 8.6 The demand for and production of electricity in 2020, 2030, and 2045.

The context of a carbon neutral society

The Danish political target of a 70 percent reduction in greenhouse gasses is defined in accordance with the UN and IPCC accounting methodology, using the year 1990 as the point of departure. Fig. 8.7 illustrates the development in Danish emissions since 1990 divided into the different sectors: energy and transportation, agriculture, process,

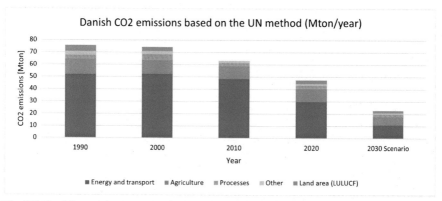

Fig. 8.7 Danish greenhouse gas emissions in CO_2 equivalent since 1990. The Smart Energy Denmark scenario will reduce emissions by 70 percent in the 2030 Scenario. The numbers do not include international shipping and aviation.

additional and land area. In 1990, emissions were equivalent to 75.7 Mt CO_2, and these had been reduced to a little less than 50 Mt CO_2 in 2020. As in the rest of the world, the highest share comes from the energy and transportation sectors. An investigation of other sectors shows that the emissions of the energy and transportation sectors should be reduced to approx. 11.5 Mton CO_2 by 2030 for the 70 percent reduction target to be achieved. The energy system achieves this goal by 2030 as illustrated in Fig. 8.7. Fig. 8.8 illustrates the Danish greenhouse gas emissions of the energy and transportation sectors divided into the four sectors of the proposal. As can be seen, the major challenge is to be found within the transportation sector.

The Danish emissions shown in Figs. 8.7 and 8.8 are accounted for with reference to the criteria defined by the UN International Panel of Climate Change (IPCC). For each member state, the accounting is made with reference to the Kyoto and UNFCCC obligations originating directly from activities in the land in question. This includes stationary plants and domestic transportation, but not international transportation. Fossil fuel emissions are attributed to the country in which the fuel is being burned, while emissions in relation to the production of biomass, on the other hand, are accounted for in the country of origin. Biomass emissions are part of the LULUCF sector, i.e., Land Use, Land Use Change and Forestry.

In the Smart Energy Denmark 2045 scenario, not only domestic transportation should be included. To achieve a fully decarbonized society, Denmark would have to include the Danish share of international shipping and aviation. In the UN system, the Danish share in 2020 is identified to be 11.6 TWh for international aviation and 6.9 TWh for international shipping (Lund et al., 2021a).

If the greenhouse gasses from these sectors are included, the situation changes quite a bit, as illustrated in Fig. 8.9.

Thus, the Smart Energy Denmark 2045 scenario allows for a full decarbonization of the energy sector, but further decarbonization of society is necessary. To achieve complete climate neutrality, several emissions are expected to remain in 2045.

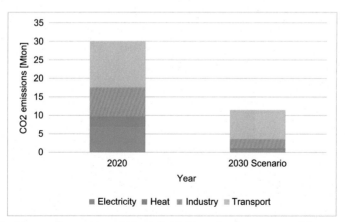

Fig. 8.8 Danish greenhouse gas emissions divided into the four sectors of the proposal highlighting the 2030 Scenario.

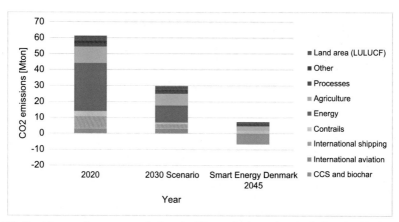

Fig. 8.9 Danish greenhouse gas emissions in CO_2 equivalent when including the Danish share of international shipping and aviation. To achieve a fully decarbonized society in Smart Energy Denmark 2045, the proposal includes CCS and Biochar to compensate for the remaining emissions in the other sectors than energy and transportation, as well as greenhouse gas equivalents arising from contrails in aviation.

Primarily, these come from agriculture, industrial processes, flight contrails, and from the burning of bio and e-fuels. To limit these emissions, the Smart Energy Denmark 2045 scenario suggests carbon capture, both through biochar from pyrolysis and by point capture from industry and power stations. The implementation of CCS and biochar is carried out based on the principle that carbon capture should be used as a final abatement step and not for mitigating fossil fuel use in industry, transportation or power production. The purpose of CCS and bio sequestrations is to offset non-avoidable carbon emissions from industry and transportation, for instance from the chemical process of producing cement or contrails from shipping. A detailed description of the biomass and CCUS part of the proposal is given in part 2 of this chapter.

Cost assessment

So far, focus has been on technological possibilities. The next elements to be investigated here are the resulting costs of the 2030 Scenario and Smart Energy Denmark 2045 systems. With these changes to the energy system, many investments would have to be made. Table 8.1 shows the main investments, targeted for 2030 and 2045, respectively. The calculation of cost is based on a comprehensive cost database, primarily using consensus investment data from the Danish Technology catalog (Danish Energy Agency, 2022). For a few technologies—not included in the catalog—costs are based on relevant literature.

When combining the increase in investments in the energy sector with lower fuel costs, it is possible to assess the total energy system costs. These are represented in Fig. 8.10. When comparing the costs in Table 8.1 and in Fig. 8.10, it is important to mention that only the 2045 scenario includes international shipping and aviation.

Table 8.1 Investments to be made to implement the Smart Energy Denmark 2045 scenario from 2020 to 2030 and 2030–2045.

	From 2020 to 2030		From 2030 to 2045	
	Investments	Annual costs	Investments	Annual costs
	Billion DKK	Million DKK/ year	Billion DKK	Million DKK/ year
Building renovations	124	5360	185	7986
Onshore and offshore wind	78	4173	102	5150
EVs and e-roads	73	6896	52	4947
Individual heat pumps	70	5114	7	946
Industry (savings and electrification)	36	2570	28	2079
District heating expansion and 4GDH	30	1467	7	462
Solar PV	21	937	22	969
Biogas plants	18	1223	12	857
New gas power plants	16	897	1	18
EV charging, distribution grid and ITS	14	825	25	1463
Large-scale heat pumps	9	499	28	1594
Electrolysis and H₂ storage	8	501	78	3531
Geothermal energy	8	440	8	410
Wave power	5	303	5	303
Gasification, pyrolysis and e-fuels	5	316	25	1579
Intelligent flexible electricity demand	3	235	1	93
Solar thermal, excess heat and thermal storage	3	176	2	97
District cooling	2	89	0	0
Gas and hydrogen grid	2	89	10	390
Sum	**525**	**32,110**	**598**	**32,874**

From Fig. 8.10 it can be seen that a 100 percent renewable decarbonized system is possible to achieve, and not at an increased cost compared to the reference scenario in 2045. In terms of cost, the 2030 scenario is also like a 2030 baseline. However, due to expected population and demand growth, the overall annual cost for the energy system will increase when looking at the period from 2020 to 2045, no matter if the Smart Energy Denmark scenario or the reference is implemented. While the direct cost is

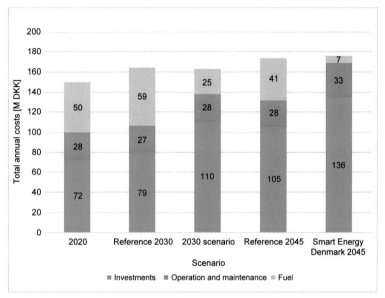

Fig. 8.10 Socio-economic cost of transforming the economy in 2020, 2030, and 2045.

the same, the cost structure makes an important difference. The Smart Energy Denmark 2045 scenario replaces fuel costs by investment costs. Such a change constitutes an important possibility to increase domestic economic growth and job creation as further detailed in the following.

2   Sustainable bioenergy in a carbon neutral society[b]

This section is based on Lund et al. (2022a) article "The role of sustainable bioenergy in a fully decarbonised society." The article is based on the same Smart Energy Denmark 2045 scenario as presented in the previous section of this chapter. However, this article aims to identify the role of sustainable bioenergy in the context of a carbon neutral society. The focus is on strategies to further develop sustainable biomass resources and conversion technologies within energy and transportation paired with CCUS (Carbon Capture Utilization and Storage) to coordinate with other sectors and achieve a fully decarbonized society. By using hourly energy system modeling and a smart energy systems approach, it is possible to create a robust multiple technology strategy and keep the use of bioenergy at a sustainable level. The results are presented as principles and guidelines on how to include the use of sustainable biomass in the individual country as an integrated part of global decarbonization.

[b] Excerpts reprinted from *Renewable Energy 196*, Henrik Lund, Iva Ridjan Skov, Jakob Zinck Thellufsen, Peter Sorknæs, Andrei David Korberg, Miguel Chang, Brian Vad Mathiesen and Mikkel Strunge Kany "The role of sustainable bioenergy in a fully decarbonised society", August 2022, with permission from Elsevier.

Overall governing guidelines

As already described in the previous part of this chapter, the design of the Smart Energy Denmark 2045 scenario fulfills the overall principles to achieve a carbon neutral solution within the energy and transportation sector at the same time as being coordinated with the rest of the greenhouse gas emitting sectors. To remain within the limits of the Danish share of global sustainable biomass resources, there is a strong need to convert both biomass resources and CO_2 together with electrolytic hydrogen to different end-fuels via Power-to-X pathways or biomass resources by advanced biofuel pathways. As already elaborated in Chapter 7, these fuels can then be used for parts of the transportation demand not suitable for electrification due to the requirements for high energy density fuels such as heavy-duty road transportation (long-haul), marine and aviation. Even though the overall demands for liquid and gaseous fuels will decrease in the future, due to the high levels of electrification, the gas demands for the industry and electricity production are expected to increase, highlighting the need for aligning the production of green gaseous products with the available biomass resources.

Several biomass conversion technologies are relevant to minimize biomass use and generate needed fuels. In addition, some Power-to-X pathways with CCU are needed to relieve the biomass demand. However, many of these technologies are not yet commercially ready at the relevant scales. Furthermore, the biomass demand for different fuel productions poses competition, as the same biomass types are used to produce electricity.

Fuel pathways that can use feedstocks, like waste biomass that would otherwise not be utilized, bring new possibilities to the system. Biomass resources with low net CO_2 emissions should be prioritized. Such resources are, e.g., biogas produced on manure having additional greenhouse gas emission advantages to the agricultural sector. The production implies the need to transform diverse bio-resources into liquid or gaseous fuels, which poses challenges in process development to improve the conversion efficiency and overall sustainability, while decreasing production costs. For these reasons, the Smart Energy Denmark 2045 scenario promotes multiple technology solutions, focusing on using different biomass technologies and Power-to-X technologies by enabling synergies between them and the energy system. The aim is to design a robust multiple technology strategy in which one technology can replace another in case some of these are not commercially ready at the proper scale in time. In general, if more pathways are available options, this will be better to design a suitable, efficient, and effective system.

Consequently, the following overall guidelines for the use of sustainable biomass have been identified:

- Keeping the biomass within the limits of the Danish share of global sustainable biomass resources.
- Biomass resources and technologies with low net CO_2 emissions are given priority.
- Solutions integrated with hydrogen from flexible electrolysis plants are given priority.
- Due to uncertainties related to future technological development, priority is given to robust multiple technology pathways.
- Priority is given to technical solutions that provide negative CO_2 emissions, i.e., both CCS and biochar.

Sustainable bioenergy scenario

The limits for how much biomass can be used sustainably for energy purposes on a global scale are still being debated (Mai-Moulin et al., 2021). Following the Danish Energy Agency (2014, 2020b), the future amount of sustainable biomass is estimated at somewhere between 100 and 300 EJ/year, equal to 10–30 GJ/capita annually in a scenario with 10 billion global inhabitants. However, other estimates (International Energy Agency, 2016; Creutzig et al., 2015) point out that, in the long run, it will only be around 100 EJ/year. It is forecasted (Connolly et al., 2016) that approximately 14–46 EJ of bioenergy will be available in the European Union, and in their Smart Energy Europe study, the authors assumed that a future 100 percent renewable energy system might consume a maximum of approximately 14 EJ/year of bioenergy, which is the minimum forecast from all studies identified. 14 EJ/year for the EU28 corresponds to 27 GJ/capita annually, while the global bioenergy resources for 2050 are expected to be 33 GJ/capita annually. By limiting the EU28 bioenergy consumption to a similar level as the global availability, the EU28 contributes to a sustainable global solution. It should be emphasized that the amount of sustainable biomass in a specific country and at the global level depends on whether an active policy is conducted or not. This is also illustrated by the wide range in the estimates.

In the case of Denmark, it is possible to increase the harvesting of sustainable wood by increasing the share of forest (as elaborated in Table 8.2). Similarly, there is also an option of coordinating with biochar production and other types of CO_2 sinks. In IDA's Climate Response 2045, the Danish contribution to global sustainable biomass resources is estimated, as illustrated in Table 8.2.

Energy crops are not included in Table 8.2 and can be an additional resource. The estimate of wood chips depends on increases in forestry in Denmark. In this study, it was chosen to base the numbers on an increase of 100,000 acres of productive forest, which is aligned with the current political plans. Given such an assumption, the potential in 2045 is 40 PJ of wood chips. Including an additional resource of firewood, wood pellets, and waste wood of approximately 15 PJ in 2045, the sum of potential sustainable wood becomes 55 PJ in 2045. The 45 PJ of straw can be reached either by supplying 75 PJ to a biogas plant with a return of 30 PJ to the fields or by supplying 45 PJ directly to other technologies such as combustion or thermal gasification.

The 158 PJ represent the potential Danish input to the global amount of sustainable biomass resources (see Fig. 8.11). However, in its strategies for achieving a zero-carbon solution, Denmark should still limit itself to the Danish share of sustainable biomass. 158 PJ Biomass correspond to approximately 26 GJ/capita, which is within the range of potential sustainable biomass on the global level, yet, likely in the high end. Therefore, a Danish strategy should try to limit its use of biomass for energy purposes and leave some of it for potential export to other countries, where sustainable biomass resources are less plentiful.

To make the best use of the available sustainable biomass, the Smart Energy Denmark 2045 scenario proposes a combination of the following conversion technologies, also described below: biogas plants, thermal gasification, pyrolysis, hydrothermal liquefaction,

Table 8.2 An estimate of the Danish potential for sustainable biomass in 2045.

Biomass type	PJ/year	Notes and source
Biogas based on manure, deep bedding, industry and residual waste, discarded crops and green agricultural waste	49	The estimate is based on (Wenzel et al., 2020). In this assessment, the potential change over time and the shown numbers are valid for the year 2040. This potential should be regarded as an upper limit
Straw	45–75	The estimate is based on (Wenzel et al., 2020). The potential is an upper limit, however, without straw use for cows. Historically, typical 20–25 PJ has been used for energy purposes. The potential of 75 PJ could be converted into approx. 45 PJ biogas with a return of 30 PJ to the fields. The net CO_2 emissions are more or less the same as if all 75 PJ were put directly on the fields
Wood such as wood chips, firewood, wood pellets and wood waste	40–80	The potential for wood chips is based on (Danish Forest Industry, 2020) and is an estimate of waste product that has to be removed from the forest. The potential includes power-cultures, i.e. fast-growing trees such as, e.g., larch trees. A potential of 35 PJ is based on the current forest area, while 60 PJ include an increase in forest areas of 200,000 acres. 90,000 m^3 of wood is allocated to improving biodiversity. Additional to wood chips, there will also be a potential for firewood equal to the current level of 3–4 PJ (Nord-Larsen et al., 2020) plus a similar contribution from rural areas and a potential of 2.7 PJ of wood pellets and wood waste 9 PJ
Waste	13	Waste resources are estimated based on the current level of 40 PJ/year being reduced due to the recycling and subtraction of organic waste
Total	160–220	Estimated maximum potential

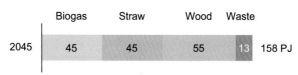

Fig. 8.11 Sustainable biomass resources for 2045.

combustion and CCUS. All investments in biomass technologies are decided as inputs to the EnergyPLAN model. The EnergyPLAN model is used to calculate the operation and consequences in the context of a fully sector-integrated smart energy system. Each of the biomass conversion technologies are elaborated on in the following:

Biogas plants are a well-established technology in Denmark, with its first biogas plant inaugurated in 1920. In 2022, around 40 percent of the Danish natural gas demands distributed in the Danish natural gas grid were green gas based on upgraded biogas from biogas plants. This makes Denmark the country in Europe with the highest share of biogas in gas consumption. In Denmark, biogas production is primarily based on manure and various types of organic waste. In the future, straw might be a promising additional resource to boost biogas production. The inclusion of straw calls for a relatively small increase in reactor volume and, consequently, involves only minor additional investments (Wenzel et al., 2020). The production of biogas with the addition of straw results in equal shares of methane and CO_2, while for traditional manure-based biogas, the relation is typically 35 percent CO_2 and 65 percent methane. The cleaning and upgrading of biogas with physical or chemical upgrading methods are energy and chemicals intense, but necessary to increase the calorific value of the gas, as the energy content of gas is in the methane part. Separated CO_2 can be captured and utilized in CO_2 intensive industries, reacted with the addition of electrolytic hydrogen in methanation units or stored with CCS.

Thermal gasification is one of the biomass conversion technologies that can decompose dry biomass as woody products and straw into syngas in the presence of controlled amounts of oxygen or steam at relatively high efficiencies. Ren et al. (2020) give an overview of different gasification technologies and see methane or synthetic natural gas (SNG) production via biomass gasification as an economical technology in the future. Furthermore, output gas generated by gasification is rich in carbon and is well suited as a Power-to-X pathway with hydrogenation and further conversion into liquid electrofuels such as methanol, DME or jet fuel. However, thermal gasification is still on the demonstration level. Demonstration of the coupling of biomass gasification and electrolytic H_2 has not yet been realized, but has been modeled for methanol production with high potential for flexible operation and carbon conversion. Hannula (2016) finds that the efficiency of biomass gasification to methanol or methane is highly sensitive to the type of gasifier, even with the use of the same biomass resources. This may add to the uncertainty of the technology if some gasifier designs do not deliver on the expected technical performances. Nevertheless, previous energy system analyses showed that the production of electrofuels via gasification could lower the energy system costs and improve the system efficiency (Connolly et al., 2014; Korberg et al., 2021).

Pyrolysis is another thermochemical process that decomposes organic matter in the absence of oxygen into bio-oil and a solid residual coproduct named biochar. Slow pyrolysis is employed to maximize the solid product yield, while fast pyrolysis is used to maximize the liquid product yield. Both of these use the same type of dry feedstocks as gasification, i.e., primarily woody biomass and straw. The amount and distribution of products generated depend on the pyrolysis temperature, resource composition and heating rate, but the main product is bio-oil, which can directly replace fossil oil in regular refining processes and is also the main attractiveness of this technology. However, the pyrolysis bio-oil has a high oxygen and low energy content; thus, it requires integration with deoxygenation processes based on hydrogen addition to upgrade the fuel to higher quality. The pyrolysis off-gasses, a combination of CO and CO_2 may also be used for permanent storage or for producing additional fuels via Power-to-X, in which case, more hydrogen will be needed than for the regular deoxygenation process.

Biochar from pyrolysis may be used as carbon sink, a technical solution that has been discussed for decades. Studies report different global carbon sequestration potentials of biochar at 0.7–1.8 Gt CO_2-$C_{(eq)}$ per year (Hrbek, 2019). The most commonly applied utilization of biochar is soil amendment; however, there are other potential utilization pathways that can be considered and eventually improve both the economic viability and environmental sustainability. In general, pyrolysis is likely to involve lower investments and costs than thermal gasification, but the output of liquid and gaseous fuels is lower.

Hydrothermal liquefaction (HTL) is a thermochemical conversion of biomass into liquid fuels in a hot pressurized environment. A literature review (Gollakota et al., 2018) looked into various aspects of HTL, such as the types of processes, different feedstock, advantages/disadvantages and energy efficiency. The results show that HTL is best used for processing high-moisture biomass, as it does not have any moisture-level requirements and is able to process a wide variety of biomass and waste feedstock from industrial, food and forest industry, swine manure, algae, arboreous crops and sewage sludge. The process can also convert plastic to oil, which is particularly useful for the removal of plastic and similar waste products from organic waste, before this is returned to the fields. One of the byproducts of HTL is CO_2, which can either be sequestrated or reacted together with process gas and electrolytic H_2 via a Power-to-X unit to methanol. In this way, the additional demand for hydrogen is higher than in the regular upgrading process, which also uses hydrogen to remove oxygen. This concept, therefore, brings the benefit of cross-sectoral integration. The HTL process can be designed for various combinations of outputs.

Biomass incineration has typically been used in Denmark for processing waste feedstock in incineration plants, which are in most cases combined heat and power (CHP) plants connected to district heating networks. Combustion is also likely to be a part of future solutions, especially to convert those types of biomass which are hard to utilize using other technologies. These processes can be combined with CCUS, either for storage purposes or by utilizing CO_2 for fuel production.

Carbon capture, utilization and storage (CC, CCU, CCS or CCUS) come in various combinations. Carbon can be captured from different biomass conversion processes, combustion plants or industrial processes such as cement industries. While the CCU concept is aligned with circular economy principles and treats CO_2 as a resource rather than waste, most CCU options do not bring the permanent sequestration of carbon. The use of CCU via Power-to-X pathways to generate various electrofuels, such as methane, methanol or even jet fuels, could also bring economic benefits and support the electricity grid balancing, thereby reducing the need for direct electricity storage. On the other hand, CCS as a technology has been criticized as an enabler for the prolonged use of fossil fuels and not fitting into renewable energy systems.

The assessment of the potential for carbon capture in the Smart Energy Denmark 2045 Scenario is based on a calculation of the gross CO_2 output combined with assuming utilization shares, as shown in Table 8.3. The low shares of CHP and power plants in the system illustrate that these plants, in a future renewable energy system, are likely

Table 8.3 Assessment of the Danish carbon capture potential in the Smart Energy Denmark 2045 scenario.

Input	Gross potential (Mt CO$_2$/year)	Share of utilization	Net potential (Mt CO$_2$/year)
Biogas	1.75	0.9	1.57
Combustion in industry	2.03	0.8	1.63
Waste incineration	1.32	0.8	1.06
Green gas CHP and power plants	1.93	0.3	0.58
Industrial processes (cement industry)	1.00	0.8	0.80
Total	**8.04**		**5.64**

to have very few operating hours. They will mostly produce in peak hours or hours with little to no production from variable renewable electricity sources. The gross potential also includes an output from industrial processes from the Danish cement industry as an industrial emitter with continuous emissions. This results in a gross potential of approximately 8 Mt CO$_2$/year and a net potential of approximately 6 Mt/year.

Bioenergy and CCUS in a carbon neutral society

In the Smart Energy Denmark 2045 scenario, biomass resources are converted into green gaseous and liquid fuels using the previously presented conversion processes. The multi-technology approach is implemented in close connection to the biomass potential and availability identified in the previous sections, and the mix of technologies is therefore aligned with the resource availability and demands that need to be supplied with different fuel products. Two scenarios were constructed, one with the biomass consumption of 23 GJ/capita (see Fig. 8.12) and one with 20 GJ/capita (see Fig. 8.13). These figures also illustrate the potential for CCUS in these scenarios.

Both scenarios include biomass and its relation to the use of hydrogen and CO$_2$ sink potentials from CCS and biochar. As will be elaborated in the next section of this chapter, the gaseous and liquid fuel outputs cover most of the needs of the Danish transportation sector, including the Danish share of international shipping and aviation. In the scenarios, approximately 5 percent of the total transportation demand is met by direct use of hydrogen and parts of the marine demand are covered by ammonia; however, this is not included in the two figures as they focus on the biomass and CCUS energy flows.

In the Smart Energy 2045 scenario, Denmark will contribute to the global sustainable biomass resources with 153 PJ of biomass, of which 140 PJ is used to cover Danish energy demands, including the Danish share of international shipping and aviation, and the rest is exported in the form of green methane. Moreover, the strategy provides a potential for CCS and biochar equivalent to 4–5 Mt/year that can be used to

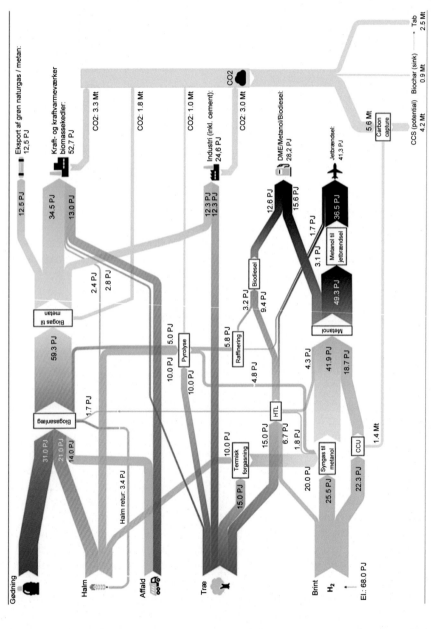

Fig. 8.12 Overview of the use of biomass in the Smart Energy Denmark 2045 scenario (153 PJ minus export 13 PJ = 140 PJ equal to 23 GJ/capita).

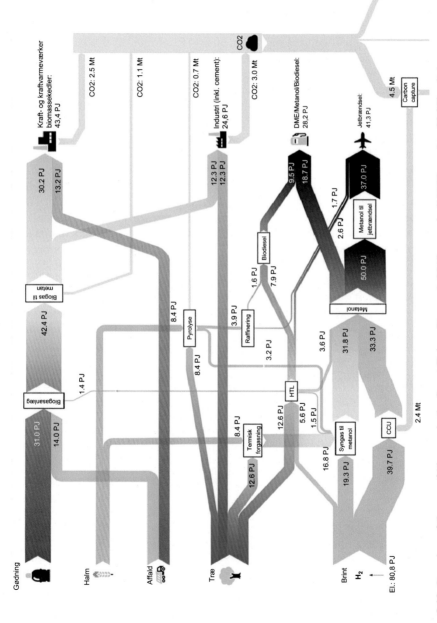

Fig. 8.13 Overview of the use of biomass in the Smart Energy Denmark 2045 scenario (117 PJ equal to 20 GJ/capita).

compensate for the flight contrails and greenhouse gasses from other sectors than energy and transportation, such as agriculture, Land Use, Land-Use Change and Forestry (LULUFC) and industrial processes. Due to the high agricultural activities in Denmark, it is assumed that the biochar is applied as soil amendment to improve soil health.

In 2045, the use of straw and wood chips in boilers and CHP plants will be reduced as the existing biomass-fired plants cease; however, smaller amounts of waste and solid biomass will still be used in CHP plants. 14 PJ of municipal waste is used as an input for biogas production, while the rest is supplied by organic waste, including manure supplemented with straw. There is an increased demand for gas in the peak and reserve load power and heating plants, which are supplied by upgraded biogas. In industry, part of the existing consumption of fossil fuels is displaced by biomass and upgraded biogas. The transportation demand is met by bio-based electrofuels and CCU electrofuels paired with electrolytic hydrogen together with Hydrothermal liquefaction (HTL) and pyrolysis. Aviation fuel demands are met either by methanol-to-jet fuel pathways or by refining bio-oil from pyrolysis and HTL.

However, as illustrated in Fig. 8.14, the scenario can also be accomplished with less biomass input, with a reduction of the total biomass consumption of 36 PJ, which as a consequence, provides a lower potential for carbon capture and biochar production. In the first scenario, biomass consumption is based on the current Danish agricultural and animal production structures with regard to the cultivation of fields and the number of livestock, etc. If the agriculture sector changes direction to more organic farming and reduces animal husbandry, this may reduce the potential for sustainable biomass for energy purposes. It should also be mentioned that the transition in Fig. 8.12 can—to a certain extent—be adapted if, e.g., one technology should become commercially available on large scale sooner than another.

3   Energy for transportation in a carbon neutral society[c]

This section is based on Kany et al.'s (2022) article "Energy efficient decarbonisation strategy for the Danish transport sector by 2045." The article is based on the same Smart Energy Denmark 2045 scenario as presented in the previous two sections. However, this article aims to identify the role of energy for transportation in the context of a carbon neutral society. The transportation sector contributes to approximately one third of Danish greenhouse gas emissions and almost half of the emissions from the energy sector. This section elaborates on the potential for reducing the national transportation sector greenhouse gas emissions in 2030 and proposes a pathway to integrate the transportation sector in the aim of achieving a carbon neutral society in 2045 using a complex set of measures.

[c] Excerpts reprinted from *Smart Energy 5*, Mikkel Strunge Kany, Brian Vad Mathiesen, Iva Ridjan Skov, Andrei David Korberg, Jakob Zinck Thellufsen, Henrik Lund, Peter Sorknæs and Miguel Chang "Energy efficient decarbonisation strategy for the Danish transport sector by 2045", February 2022, with permission from Elsevier.

TransportPLAN

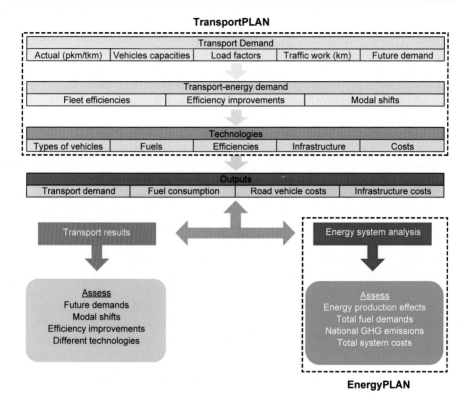

Fig. 8.14 Methodology applied to analyze the transportation sector in TransportPLAN and the outputs from the model that are used in EnergyPLAN.

Toward 2030, the major focus is on an extensive electrification of passenger cars, alongside the implementation of significant measures to achieve lower growth rates for kilometers traveled by car and aircraft. From 2030 onwards, a decisive focus is set on sector integration. The production of electrofuels proves to be a key measure to decarbonize aviation, shipping and long-distance road freight transportation. The results show a reduction of greenhouse gas emissions of 41 percent in 2030 and full decarbonization in 2045. The reduction is achieved without a significant increase of socio-economic costs. From 2030 to 2045, a substantial electrification of road transportation and a focus on moving the need for mobility from roads toward rail and bicycles drive the full decarbonization together with the replacement of fossil fuels with electrofuels for aviation, shipping and heavy-duty road transportation.

The TransportPLAN tool and methodology

The design of a strategy toward decarbonizing the Danish transportation sector is modeled using the TransportPLAN and EnergyPLAN tools. TransportPLAN goes into detail with the energy needs for the transportation sector and provides inputs for the energy systems analysis of the Smart Energy Denmark 2045 scenario in the

EnergyPLAN model. In TransportPLAN, it is possible to analyze transportation developments in detail and assess the effects of implementation of policy measures or alternative fuels and propulsion technologies. System effects of implementations in the transportation sector are then analyzed in the hour-by-hour energy system analysis tool EnergyPLAN. TransportPLAN is developed as a stand-alone tool, but with the interaction with EnergyPLAN in mind. Combining a detailed bottom-up transportation modeling tool with an hour-by-hour energy systems analysis tool allows for very detailed and specific analyses and results. In Fig. 8.14, the general outputs from TransportPLAN and the outputs for EnergyPLAN are highlighted.

TransportPLAN is a back-casting modeling tool, which allows for detailed scenario analysis of transportation systems. The detailed resolution of input data provides the possibility of adjusting the development of the transportation demand precisely and in-depth. The tool requires inputs regarding annual transportation demand, vehicle fleet composition, utilization rates and fuel distribution in the first modeling year. Here, data for all modes of transportation allocated into subcategories of trip lengths are gathered. Transportation demand is split between passenger and freight transportation measured in passenger-kilometers for passenger transportation and tons-kilometers for freight transportation. Transportation demand data in combination with capacity utilization factors enable the calculation of total kilometers traveled for all modes of transportation (in the following referred to as either traffic work or vehicle-kilometers). Additionally, an average energy consumption for the fleet of vehicles for all modes of transportation is necessary to calculate the final energy consumption.

TransportPLAN is built to enable the development of transportation system strategies and scenarios that follow the Avoid-Shift-Improve approach (International Energy Agency, 2021b). The tool provides options to increase/decrease annual growth in transportation demand and annual modal shifts across all modes of transportation. Moreover, the tool allows for creating pathways to implement alternative fuels and propulsion technologies as well as introducing different trajectories for the development of energy efficiency improvements in existing and new engine technologies and improvements in capacity utilization rates. The model results are annual transportation demand and traffic work for each modeled year, annual final energy consumption, greenhouse gas emissions, and transportation system costs.

These features facilitate the implementation of policy measures or urban development strategies directly into the TransportPLAN tool and visualize the effects in terms of future traffic work, energy consumption, greenhouse gas emissions, and transportation system costs. The transportation system costs include the costs of road vehicles, fuel production costs, and infrastructure costs. Infrastructure costs include investment and maintenance costs for road and railway infrastructure, charging infrastructure for electric road vehicles, and infrastructure costs related to public transportation systems and walking and biking infrastructure. All infrastructure costs are calculated based on historical investments and maintenance expenses in relation to historical growth rates. Vehicle costs and costs related to fueling infrastructure for trains, ships and aircrafts are not included in TransportPLAN. The costs of fueling infrastructure for fossil fuels are not included either.

A transportation decarbonization scenario

By use of the TransportPLAN model, this section elaborates on the design of a strategy toward a zero-emission Danish transportation sector in 2045 (in the following referred to as IDA2045). The scenario is an integrated part of the Smart Energy Denmark 2045 scenario presented in the previous two parts of this chapter. Thus, unlike the general UN methodology, the scenario includes energy consumption and greenhouse gas emissions of international aviation and shipping. Like in the previous analyses, the fully decarbonized scenario is compared to a Business-As-Usual (Frozen Policy) scenario. The Frozen Policy scenario is based on the frozen policy scenario published by The Danish Energy Agency (2019). The IDA2045 scenario presents a pathway toward decarbonization in 2045, using a combination of measures such as:

- lower growth in traffic work while still accommodating a growing demand for transportation and mobility
- comprehensive modal shifts from energy intensive modes of transportation toward active modes (i.e. walking or bicycling) and public transportation (i.e. buses or railway)
- limiting transportation energy demand and greenhouse gas emissions by implementing an extensive electrification replacing internal combustion engines with energy efficient electrical engines and substituting energy-dense liquid fossil fuels for aviation and shipping with renewable electrofuels produced by Power-to-X pathways

The two scenarios will be compared in terms of annual final energy demand, annual transportation system costs, as well as annual CO_2 emissions. Figs. 8.15 and 8.16 show the details of the composition of the Danish transportation sector, for both passenger and freight, in 2020. For passenger transportation, the majority of passenger-km (pkm) are covered by car and air traffic, even though >90 percent of all km traveled are traveled in cars. This is, among other factors, explained by the poor capacity utilization of cars, which on average is only 1.49 passengers/vehicle. The remaining 15 percent of the transportation demand and <10 percent of traffic work comprise traveling by bus, rail, bicycling and walking or by sea.

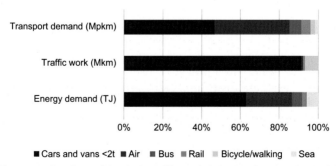

Fig. 8.15 Transportation demand (mpkm), traffic work (mkm) and energy demand (TJ) divided by modes of Danish passenger transportation in 2020 (Danish Energy Agency, 2020a). Data from Danish Energy Agency, 2020a. Data, tabeller, statistikker og kort. Energistatistik 2019. Danish Energy Agency, Copenhagen. Figure is original content by the author.

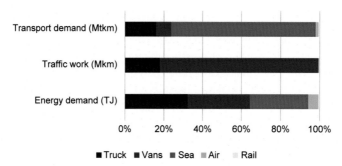

Fig. 8.16 Transportation demand (mtkm), traffic work (mkm) and energy demand (TJ) divided by modes of Danish freight transportation in 2020 (Danish Energy Agency, 2020a).
Data from Danish Energy Agency, 2020a. Data, tabeller, statistikker og kort. Energistatistik 2019. Danish Energy Agency, Copenhagen. Figure is original content by the author.

In freight transportation, the majority of goods are transported by sea; hence maritime transportation covers >75 percent of the freight transportation demand. Trucks and vans cover 24 percent while rail and aviation only cover approximately 2 percent of the freight transportation demand. The capacity of maritime freight vessels compared to road freight vehicles entails that trucks and vans are responsible for the vast majority of total km traveled. The relatively low utilization capacity of trucks and vans of approximately 50 percent on average also contributes to the dominance of traffic work. While the freight transportation demand and traffic work covered by aviation and rail are insignificant, they cover 6 percent of the energy consumption. The remaining energy demand is split evenly between trucks, vans, and sea transportation.

For both passenger and freight transportation, a lot of passengers and goods are moved in vehicles with relatively low utilization of capacity, thus highlighting the potentials of energy efficiency improvements in the transportation sector as a whole. Reducing the traffic work while accommodating the transportation demand is a key factor in successfully decarbonizing the transportation sector. This will also ease the transition from fossil fuels to renewables, as fewer renewable fuels such as electricity, hydrogen and electrofuels are necessary.

Fossil fuels are predominant in the final energy demand in the Danish transportation sector (see Fig. 8.17). Almost all road transportation consumes diesel or gasoline, and the small share of renewable biofuels are admixtures. Biofuels cover 4 percent of the fuel consumption, while only 1 percent is covered by electricity. This is primarily electric trains that receive electricity from overhead wires. In 2020, only 15,500 BEVs and 9800 PHEVs were registered in Denmark out of a total of 2,650,000 registered cars. A small share (>1 percent) of electric and biofuel buses are operational and no trucks or vans are fueled by renewable alternatives. Jet-fuel for aviation covers approximately 17 percent of the final fuel consumption.

For both the Reference scenario and the IDA2045, the annual transportation demand (mpkm) increases from 2020 to 2045 (Fig. 8.18). In the Reference scenario, this increase is driven primarily by a significant increase in the transportation demand for cars. From 2000 to 2018, the annual vehicle-km for Danish cars increased on

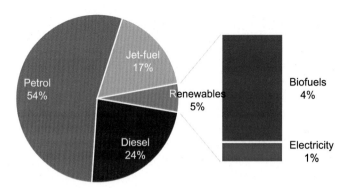

Fig. 8.17 Share of fuels in the Danish transportation sector in 2020 (Danish Energy Agency, 2020a).
Data from Danish Energy Agency, 2020a. Data, tabeller, statistikker og kort. Energistatistik 2019. Danish Energy Agency, Copenhagen. Figure is original content by the author.

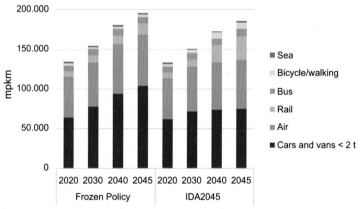

Fig. 8.18 Development of the annual Danish passenger transportation demand from 2020 to 2045 in the Frozen Policy scenario and Smart Energy Denmark 2045 (IDA2045).

average 1.6 percent annually and since 2010, this increase has been 2.5 percent annually. A contentious increase of 2 percent annually toward 2045 is assumed in the Reference scenario.

The Smart Energy Denmark 2045 (IDA2045) scenario presents a slight decoupling of growth in mobility and growth in vehicle-km for cars. While the general mobility demand still increases from 2020 to 2045, much of the growth is shifted from cars to other, more energy-efficient modes of transportation, e.g., railways, buses, and bicycles. Moreover, the expectations to the annual increase in vehicle-km for cars are reduced. Instead of an annual increase of 2 percent, a 1.6 percent annual increase is assumed. The shift away from cars toward public transportation and bicycles along with a general reduction of annual increase in vehicle-km can be achieved with several measures. A key measure is the implementation of road pricing with varying prices in different zones depending on land area and usage. A new taxation scheme like this aims to restructure the expenditures of owning vehicles, in such a way that the purchase and registration of the vehicle is less cost-intensive and instead the usage and driving of vehicles are taxed more.

Creating a pathway toward a transportation sector supplied by 100 percent renewable energy requires a present focus on infrastructure investment that enables energy efficient mobility. In the Smart Energy Denmark 2045 (IDA2045) scenario, a guiding principle of shifting investments from road infrastructure toward, e.g., railways, public transportation systems and bicycle infrastructure is followed. This entails that the increase in demand for mobility is not met by an equal increase in car traffic but energy efficient modes of transportation such as trains, subways, buses, bicycles, etc.

For aviation, Smart Energy Denmark 2045 (IDA2045) proposes a pathway to reduce transportation demand and thus reduce fuel consumption. Airborne travel has allowed for convenient interconnection via long distances that would otherwise not be possible. To reduce the demand, instruments such as increased fuel and passenger taxes could be utilized. A more expensive ticket price could, in combination with the measures proposed above to enhance public transportation and railways, contribute to reducing demand and fuel consumption. The demand for domestic aviation in Denmark is negligible, and due to the geographic scale, most routes are less than 300 km. This ensures a potential to shift a lot of these trips toward rail or buses. An improved railway system would provide an attractive alternative if the journey costs are appropriate. To succeed with reducing the impact from aviation, a joint strategy from the EU or a group of European countries is needed.

All of the above measures do not limit the development and accessibility for mobility but assist in the transition toward a 100 percent renewable transportation sector. The transportation demand in Smart Energy Denmark 2045 (IDA2045) still increases by 45 percent from 2020 to 2045, but the increase is met by more energy-efficient modes of transportation. The measures elaborated translate into these specific alterations in the transportation modeling tool:

- Attenuation of the growth in pkm for cars from 2 percent annually to 1.6 percent in the entire modeling period from 2020 to 2045.
- A shift from cars (2 percent) and national aviation (10 percent) toward rail and public transportation by 2030. In 2045, additionally 25 percent of pkm of car transportation and 87 percent of national aviation are shifted toward rail and public transportation.
- A shift from cars (2 percent) toward bicycles in 2030. Additional 5 percent from cars toward bicycles in 2045.
- A shift from international aviation (17 percent) toward international rail in 2045.
- Attenuation of the growth in pkm for national and international aviation by 10 percent in 2030.

In the IDA2045 scenario, alternate developments for the freight transportation demand have not been explored. In *White Paper on Transport*, the European Commission (2011) formulated targets to shift 30 percent of the freight transportation demand from roads toward rail and seaborne transportation by 2030 and 50 percent by 2050. It is uncertain how this will affect the Danish freight transportation, which is primarily travels for shorter distances. Instead, a single scenario for the development of the freight transportation demand is investigated in this work. Shipping covers the majority of the international freight transportation demand, while road transportation is accountable for the bulk of national transportation. In Fig. 8.19, the development of the freight transportation demand is outlined. An increase of 10 percent is assumed, where the majority comes from growth in national road freight transportation.

To successfully implement a transition toward a fully decarbonized transportation sector in 2045, the IDA2045 scenario proposes a combination of alternative propulsion technologies along with a pathway for renewable production of electrofuels to substitute energy-dense fuels for primarily aviation and maritime transportation. A reduction in the growth in VKT for cars decreases the annual fuel consumption substantially, but an implementation of energy efficiency improvements and the replacement of fossil fuels with renewable alternatives will still be necessary.

IDA2045 suggests that all internal combustion engine (ICE) cars and vans are replaced with battery electric vehicles (BEV). The substantial developments in battery technologies in recent years have allowed for a realistic implementation of large shares of BEVs. The well-to-wheel (WtW) efficiencies of BEVs compared to ICE vehicles are approximately three times higher. Hence, replacing all ICE cars and vans with BEVs significantly reduces the total energy consumption. The implementation of electricity in IDA2045 reduces the consumption of gasoline and diesel from 102 PJ in 2020 to 59 PJ in 2030 for Danish passenger transportation. To accommodate the transportation demand in 2030, approximately 1.3 million BEVs and plug-in hybrid electric vehicles (PHEV) are needed (Fig. 8.20). The PHEVs are not suggested as a long-term solution toward 2045, but they are expected to play a role toward 2030.

The strategy of electrification is also followed for Danish railways and buses. An extensive electrification of the railway system is already being rolled out with a combination of overhead wires and battery electric trains. For buses, it is estimated that it is possible to replace 75 percent of the transportation demand with battery electric buses. This covers all the urban buses and 50 percent of the regional and inter-city buses.

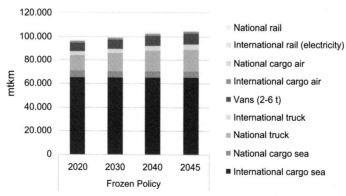

Fig. 8.19 Development of the annual Danish freight transportation demand from 2020 to 2045 in the Frozen Policy scenario.

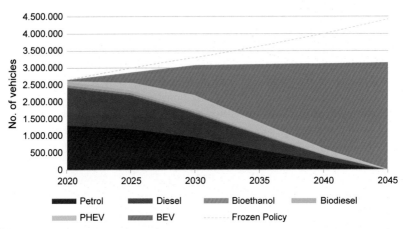

Fig. 8.20 No. of cars in IDA2045 divided by fuel consumption. Yellow line is the no. of cars in the Frozen Policy scenario.

A less significant shift toward BEVs is suggested for heavy-duty road transportation. It is estimated that all trips less than <150 km can be directly replaced by battery electric trucks (BET). To assist the electrification, IDA2045 proposes an implementation of Electric Road Systems (ERS) to allow for in-motion charging. This technology is still only on the demonstration stage for road transportation, but it has been used and developed for decades for railway transportation. ERS represents a large infrastructure investment but could offer an energy-efficient solution to electrify heavy-duty road transportation.

The additional investment costs for electric vehicles in IDA2045 compared to the Frozen Policy scenario from 2020 to 2030 are 9.8 billion Euro. These cover the investment in 1.3 million PHEV and BEV cars as well as the electrification of buses and trucks including minor investments in pilot projects for ERS. From 2030 to 2045, the corresponding costs are 7 billion Euro. The costs include approximately 5.6 billion Euro to replace the remaining ICE passenger cars with BEVs, 0.5 billion Euro to substitute ICE buses and trucks with BEVs, and 1 billion Euro in investment costs for approximately 400 km of ERS. The total investment costs are lower in the period from 2030 to 2045, primarily due to an expected decrease in the purchase prices for BEVs concurrent with the development of battery technology. For the transportation system, infrastructure investments are required for the expansion of the railway network and improved charging infrastructure for BEVs. The historical costs of the increased transportation demand on rail in Denmark have been almost identical to the costs of increased transportation demand on road. Hence, the proposed modal shift from road to rail does not entail any additional infrastructure costs. The costs of expanding a reliant network of charging stations for BEVs amount to 0.4 billion Euro in 2030 and additionally 0.7 billion Euro from 2030 to 2045.

For aviation and shipping, electrification is currently only applicable to a few specific usages. In the Smart Energy Denmark 2045 scenario, all domestic ferries are electrified in 2045. The trip length of Danish ferries, except the route to the island of Bornholm, are short and it will be possible to equip a battery and charging stations at each port. For aviation, battery electrification is proposed only to have an insignificant role.

To accommodate the remaining transportation demand, that is not fit for electrification, the scenario proposes a significant upscale of the electrofuel production. Electrofuels produced from renewable electrolytic hydrogen combined with a source of carbon or nitrogen in fuel synthesis allow for sustainable, energy-dense hydrocarbons or ammonia that can directly replace fossil fuels like jet-fuel, heavy fuel oil (HFO) for shipping, etc. Electrofuels fill the direct purpose of replacing fossil fuels in sectors that are otherwise hard to decarbonize, while providing a flexibility service to integrate large penetrations of renewable energy into the energy system.

The Smart Energy Denmark 2045 scenario proposes that 55 percent of heavy-duty road transportation is replaced with electrofuels and 10 percent with hydrogen in 2045. Hydrogen is expected to be utilized in niche markets and only take up a relatively small market share. Building a hydrogen infrastructure to support a full transition is not believed to be feasible; instead electrofuels like methanol or DME are proposed to replace diesel. Hydrogen is combined with nitrogen in fuel synthesis to produce ammonia that should replace 50 percent of fossil fuels in the maritime freight transportation. The remaining share is replaced with other types of electrofuels, which are not specified here. Many different pathways are possible, but it would not make a significant difference for the total energy consumption. For aviation, the proposal suggests that electric aircrafts should cover 25 percent of the shorter inter-European routes of less than 1000 km and hydrogen-powered aircrafts should cover 10 percent of the longer routes of more than 1000 km. The integration of electric and hydrogen aircrafts on these routes will cover approximately 14 percent of the aviation transportation demand in 2045, after the implementations of measures to reduce the transportation demand by air. The remaining transportation demand is proposed to be covered by electrofuels. In Fig. 8.21, the annual fuel consumption and CO_2 emissions for the Frozen Policy scenario and Smart Energy Denmark (IDA2045) scenario are compared.

Cost assessment

Total transportation system costs in IDA2045 are reduced by 4.4 percent in 2030 compared to the Frozen Policy scenario (Fig. 8.22). The lower growth in transportation demand for cars is the main reason for the reduction. The lower growth results in fewer cars, and this counterbalances the higher purchase price of BEVs compared to ICE vehicles. The lower growth significantly reduces annual road infrastructure investment and maintenance costs, and the energy efficient BEVs also reduce annual fuel costs. Additional costs for railway infrastructure, charging infrastructure and bicycling and public transportation infrastructure are needed in IDA2045.

IDA2045 reduces the annual transportation system costs by 10 percent compared to the Frozen Policy scenario. The lower growth in the transportation demand for cars along with notable modal shifts toward public transportation and bicycling are still significantly important. The investment in vehicles is reduced by 20 percent compared to the Frozen Policy scenario. The anticipated reduction in the costs of BEVs ensures

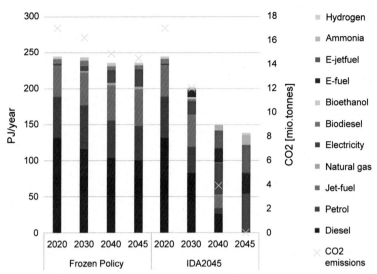

Fig. 8.21 The fuel consumption in the Danish transportation system in the Frozen Policy scenario and the Smart Energy Denmark 2045 scenario. Annual CO_2 emissions are indicated by the yellow *x* on the secondary *y*-axis.

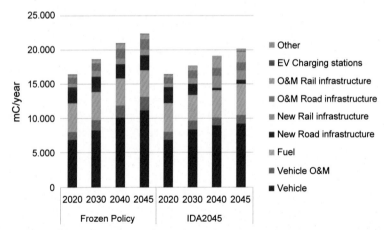

Fig. 8.22 The annual transportation system costs in the Frozen Policy scenario and the Smart Energy Denmark scenario IDA2045.

that the full economic effects of reducing growth and enforcing modal shifts are achieved. Investment and maintenance in road infrastructure are reduced by half compared to Frozen Policy, while investments in rail are doubled and hence counterbalance any potential savings. Fuel costs increase in IDA2045 as the production of electrofuels is scaled up to decarbonize the heavy-duty road transportation, aviation, and shipping sectors. The development of production costs of electrofuels is expected to be 3–4 times higher than fossil alternatives in 2045 (Brynolf et al., 2017). Hence, the fuel savings from replacing fossil fuels with electricity in most of passenger road transportation is evened out by the electrofuel production costs.

4   Electricity balancing and grid stabilization

As already described in the previous chapters, the issue of electricity balancing and grid stabilization of the power grid becomes essential to address as part of the green transition. In the traditional power system based on fossil fuels, frequency and voltage control are sustained by either large-scale hydro power or steam turbines based on nuclear or fossil fuels. In a 100 percent renewable energy system—no matter if it is in the context of a carbon neutral society or not—these tasks will have to come from other technologies. On one hand, this is a challenge that should not be neglected; on the other hand, this is also fully doable and many of the technologies are already implemented. In the following, it is illustrated how to deal with the matter using the Smart Energy Denmark 2045 scenario as a case. The discussion is divided into two important aspects. The one is the balancing of the electricity demand and supply and the other is grid stabilization, i.e., to secure the voltage and frequency control on the power grid.

Balancing electricity demand and supply

The traditional electricity demand varies throughout the day, the week, and the year. In general, it peaks in the morning and in the late afternoon. It is higher in the weekdays than in the weekend, and typically in a country such as Denmark, it is higher in the winter than in the summer. If the traditional Danish electricity demand was to be covered solely by wind and PV without building more wind and PV than needed on an annual basis, it would generate large imbalances in the demand and supply. Typically, only around 60 percent of the supply would meet the demand. The rest would be excess production and corresponding lack of supply when needed. However, it is important to understand that such terms will undergo radical changes in the green transition to a 100 percent renewable energy system in the context of a carbon neutral society. As explained in the previous sections, such a transition involves a substantial electrification introducing large amounts of new flexible demands. The most important are electric heat pumps and electric boilers for the heating supply in individual as well as district heating and electric vehicles, electrolysis, and Power-to-X. Such new demands are essential to fully decarbonize the heating, the industry, and the transportation sectors.

Typically, green transition scenarios will result in a doubling or more of the electricity demand due to these new flexible demands. Fig. 8.23 illustrates the changes in the electricity demand in the case of the Smart Energy Denmark 2045 scenario. As can be seen, there is a small increase in the "classical" demand, here referred to as the nonflexible demand. This is constituted by the traditional demand together with parts of the electricity demands for industry and for the share of transportation which is not expected to be flexible in the same degree as the charging of electric vehicles and

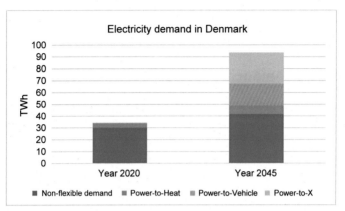

Fig. 8.23 Illustration of the changes in the amount and the composition of today's electricity demand in the Smart Energy Denmark 2045 scenario. In the future, electricity demands will increase substantially by new flexible demands for Power-to-Heat, Power-to-Vehicle and Power-to-X. The diagram corresponds to Fig. 8.6—only the electricity demand is divided into different categories.

similar. As illustrated, the electricity demand increases substantially in the future and therefore a green transition requires many more wind turbines and PV installations than needed to cover the traditional electricity demand. As already mentioned, in the Smart Energy Denmark 2045 scenario, there is a need for 19 GW of wind and 10 GW of PV. However, a consequence of such an expansion of wind and PV is that the ability for RES to directly meet the non-flexible demand increases, as illustrated in Fig. 8.24.

Fig. 8.24 compares hourly productions from RES in 2045 to the non-flexible demands in the future Smart Energy Denmark 2045 scenario. The production from RES is mostly wind and PV but also includes a small share of electricity production from waste incineration. The latter, however, will be small in the future due to a high degree of recycling. As can be seen, the share of non-flexible demand will be low compared to the RES production and there is only a small deficit. To secure a fully balance, the small deficit will have to be produced at some other units. However, on an annual basis, this deficit will be as small as in the order of 5–10 percent. In the case of the Smart Energy Denmark 2045 scenario, the number is only 7 percent. A number of different measures exist to cover a small deficit of only 7 percent such as:

- To exchange wind and PV with neighboring countries and benefit from the differences in weather between the geographical areas. On a European scale, the deficits will become lower than on a Danish scale.
- To exchange wind and PV with countries with hydro power, for instance between Denmark and Norway.
- To make use of V2G as explained in Chapter 6.
- To produce the small remaining deficit on CHP plants of gas turbines or similar based on green gas. Such plants have low investment costs and high energy costs which correspond well with the task of peak production in a limited number of hours.

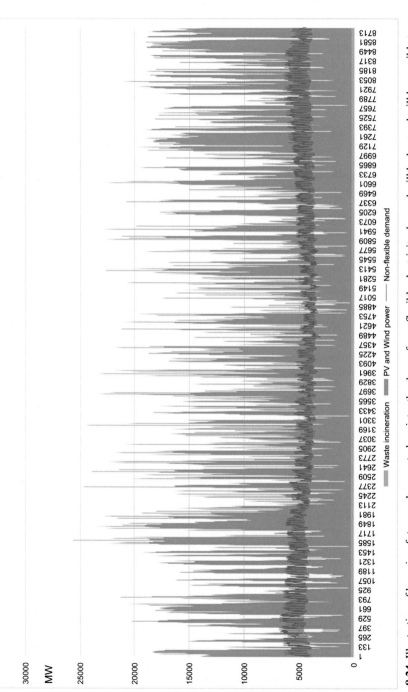

Fig. 8.24 Illustration of how, in a future carbon neutral society, the share of non-flexible electricity demand will be low and will be possible to cover almost directly by electricity production from wind, PV and to a small extent waste incineration. In the case of the Smart Energy Denmark 2045 scenario, 93 percent of the non-flexible demand can be covered directly from RES without any storage. Only 7 percent will have to be covered in another way.

When the non-flexible demand has been covered, most of the flexible demand will be able to adjust to the variations in production. However, it will be subject to technical and economical limitations such as capacities of heat pumps and chargers, electrolyzers and associated storage systems as well as driving patterns. When hour-by-hour calculations are carried out in the EnergyPLAN model, such limitations are taken into account. In the case of the Smart Energy Denmark 2045 scenario as described above, it has been documented how a fully balanced system can be achieved with only a small share of excess electricity production. Such excess production or curtailment could in theory be brought to zero, but that would not be economically feasible. It arises from an economic optimization in which it is more cost efficient to overinvest in wind turbines rather than increasing capacities in electrolyzers and other technologies.

Voltage and frequency control

As already highlighted in Chapter 6, an AC power system requires a delicate balancing of demand and supply not only to the hour-by-hour level but also below. This is to secure voltage and frequency in the power supply. Traditionally, such tasks have been fulfilled by hydro power and steam turbines with synchronous generators and rotating mass to secure the spinning reserve. Along with the replacement of these plants by other units in the future, the grid stabilization will also have to be replaced. As described in Chapter 6, already today other units have been included in the grid stabilization market—also known as the Frequency Containment Reserves (FCR) market. Thus, already today, small CHP plants, electric boilers, batteries, synchronous compensators, and wind turbines have been equipped with and have participated in the grid stabilization tasks. Along with replacing the steam turbines, these new units will gradually have to take over. In the case of the Smart Energy Denmark 2045 scenario, grid stabilization could be secured in the following way:

- In most hours (80–85 percent as shown in Fig. 8.24), the production from wind and PV will exceed the non-flexible demand. In such hours, grid stabilization should come from flexible demand in the same way as today, e.g., from electric boilers in the district heating supply. However, in the future, Power-to-X such as electrolyzers should also have this ability.
- In 15–20 percent of the hours, the non-flexible demand will exceed the RES production. In such hours, import and/or production on gas turbines or the like are required. During these hours, grid stabilization can simply be supplied from thermal units as today. However, as highlighted in Chapter 6, it is important that small CHP plants are activated on the primary automatic reserve markets—as they already are in Denmark. In such hours, one may also activate V2G in the grid stabilization, if they are already involved in the hourly balancing as explained above.
- In very few hours, it may happen that productions from wind and PV match the non-flexible demand. In such hours, one must either secure grid stabilization from wind turbines in combination with batteries (as has also been proven) or one can have a flexible demand together with a production on the thermal units.

In some of the situations, there will not be a rotating mass as in traditional steam turbines and therefore it will in some cases be necessary to apply power electronic solutions.

5 A smart energy systems approach to a carbon neutral Europe[d]

This section is based on Thellufsen et al.'s (2023) article "Beyond sector coupling: Utilizing energy grids in sector coupling to improve the European energy transition." This section applies the same smart energy systems principles as in the previous sections; however, this time to the whole of the Europe Union instead of just one country. The European Commission (2018) report "A Clean Planet for All" is used as a starting point. Essential scenarios from this report are implemented and replicated by use of the EnergyPLAN model to achieve the appropriate level of modeling details for the design of a smart energy Europe scenario which fits into a fully decarbonized Europe.

Recreating "A Clean Planet for All" scenarios in EnergyPLAN

The scenarios from "A Clean Planet for All" are initially developed in the PRIMES model. However, the PRIMES model does not go into detail with the hourly operation of the energy system. Therefore, the PRIMES scenarios were converted and implemented into a model using EnergyPLAN. As part of this replication, it was necessary to make a few adjustments to the Baseline and 1.5 TECH scenarios to ensure the same technology costs and efficiencies across the systems, especially for electrolyzers and e-fuel production. Moreover, power plant capacities had to be raised as the detailed hourly modeling revealed that, otherwise, the European model would have to import electricity from energy systems outside Europe. The idea behind the adjustments above is to ensure that the differences between the systems in terms of costs, primary energy consumption and CO_2 emissions are due to the energy system configuration and not due to different assumptions for the same technologies. To validate the EnergyPLAN replication, the primary energy consumption was compared between the two models as illustrated in Fig. 8.25. As can be seen, there is a good match between the results of the two different models.

Smart Energy Europe scenario

Next step is to apply the smart energy systems principle to the EnergyPLAN model of the European 2050 baseline scenario and create an energy system which fits into a carbon neutral European society—here named a Smart Energy Europe scenario. Following an analytical approach inspired by Connolly et al. (2016), the steps are as follows:

1. *Energy efficiency first.* Based on the energy efficiency first principle, the potential for energy savings is implemented. Specifically, the end-use demands for the heating, industry and

---

[d] Excerpts reprinted from *Submitted to Smart Energy 5*, Jakob Zinck Thellufsen, Henrik Lund, Andrei David Korberg, Peter Sorknæs, Steffen Nielsen, Miguel Chang and Brian Vad Mathiesen "Beyond sector coupling: Utilizing energy grids in sector coupling to improve the European energy transition", June 2023, with permission from Elsevier.

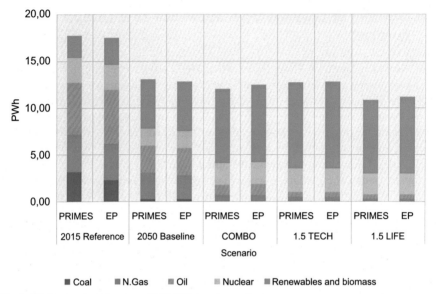

Fig. 8.25 Comparison between PRIMES documented primary energy outputs and EnergyPLAN (EP) modeled outputs from the "A Clean Planet for All" replication (European Commission, 2018).

transportation sectors are included. The Smart Energy Europe scenario uses knowledge from the Smart Energy Denmark and similar studies to determine levels of energy consumption in industry, transportation, and heating.

2. *Updating heating systems.* Based on Heat Roadmap Europe, and in line with the energy efficiency principle, district heating is implemented in areas identified as feasible for district heating. Within the district heating system, it is necessary to utilize the potential for thermal storage, waste heat and large-scale heat pumps, to make the supply system as efficient and well-integrated with renewables as possible. Outside areas feasible for district heating, an efficient use of electricity is key, which means that most of the individual heating demands are covered by individual heat pumps. This ensures that biomass can be used later in the transition, since only a limited biomass resource is available (see step 5).

3. *Electrification of the transportation sector.* The third step is the use of batteries and direct electrification of the transportation sector. This has high potentials in terms of the system's energy efficiency, and the more it is possible to electrify, the easier will the final steps of carbon neutrality become. All personal vehicles are expected to be electrified as well as most of the light-duty vehicles, with heavy transportation only being electrified to some extent. Shipping and aviation are expected to have a high demand for fuels still.

4. *E-fuels in transportation.* To supply the demand for fuels in the heavy part of transportation as well as for shipping and aviation, e-fuels are produced from gasified biomass and point source carbon capture in combination with hydrogen produced with the use of renewable electricity from wind turbines and photovoltaics.

5. *Replace the remaining fuel demands in industry and power production.* In the case of Smart Energy Europe, the choice is to use biogas and biomass to the extent that this is sustainable, as these provide the most cost-efficient solution. The biomass limit is the same as in the 1.5 TECH scenario, approximately 20 GJ/person.

For the design of the Smart Energy Europe scenario, the following constraints have been made to facilitate results comparable to the PRIMES model scenarios of "A Clean Planet for All":

- Biomass is limited to 2.71 PWh/year, similar to the biomass consumption in the 1.5 TECH scenario.
- No electricity import should be needed from outside the modeled area.
- Curtailment of electricity production from renewables should be a maximum of 5 percent of the total electricity consumption to avoid installing over capacity of wind and solar with low utilization rates.

In each design step, the variable renewable energy capacity is increased to cover the demands and fulfill the design principle above. The European energy system is ensured not to rely on imported electricity by having sufficient power plant capacity in the system.

Comparing results

To illustrate the potential benefits of a smart energy systems approach to a fully decarbonized society, the Smart Energy Europe scenario has been compared to the Baseline 2050 scenario and the 1.5 TECH scenario of "A Clean Planet for All." Fig. 8.26 shows the primary energy consumption for the three scenarios, highlighting the points that the Smart Energy Europe scenario has the lowest total primary energy consumption and achieves the largest reduction in the fuel consumption. This is due to energy efficiency gains obtained by utilizing both the heating, gas and electricity grid as well as energy storages across all three grids. By utilizing all grids, waste heat becomes a key resource for district heating systems. Furthermore, a heavier focus is on electrification, eliminating fossil fuel demands in transportation and industry. Both the baseline and 1.5 TECH scenarios include fossil fuels, which in the 1.5 TECH are offset by

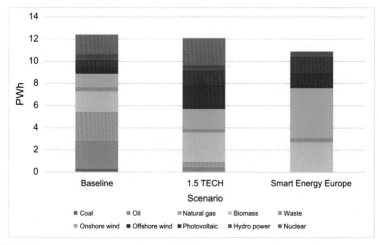

Fig. 8.26 Primary energy consumption in the three energy systems.

carbon capture and storage. Nuclear is omitted in the Smart Energy System as it is more expensive than implementing more variable renewable energy with flexibility from electrolyzers and hydrogen storage. Thus, by using the smart energy systems approach with a focus on sector coupling and utilizing smart grids in all energy sectors, it is possible to achieve a more fuel-efficient energy system that would be less sensitive to changes in fuel prices due to the lower fuel demands. The biomass consumption is the same between the 1.5 TECH and the Smart Energy scenario, with a consumption of around 20 GJ/person, which is within a sustainable use of biomass as discussed in the previous sections.

With regard to CO_2 emissions, the baseline 2050 system is not CO_2 neutral and, therefore, cannot fulfill the target of reaching a carbon neutral society. In the 1.5 TECH scenario, there is a demand for offsetting fossil carbon emissions from using natural gas and oil in the transportation and industrial sectors. Therefore, the scenario includes a need for CCS technology which, besides offsetting fossil carbon emissions, is needed to account for missing carbon reductions in other sectors. The Smart Energy Europe energy system is completely carbon neutral; but still includes some CCS. However, this is not to offset fossil fuels in the energy system, but solely to allow for easier mitigation strategies in hard-to-abate sectors such as agriculture and certain industries. Therefore, as illustrated in Fig. 8.27, the Smart Energy Europe scenario with a net carbon budget of 0.131 Gton is less dependent on CCS than the 1.5 TECH scenario.

The final comparison regards the total annual costs of the three scenarios. The costs include investments, calculated as annuity payments based on lifetimes and a discount rate of 3 percent, fixed and variable operation and maintenance costs, fuel costs and CO_2 costs. The calculations have been made both with and without the transportation sector. As seen in Fig. 8.28, the systems have similar total annual costs, around 2000 billion Euros. However, the Smart Energy Europe scenario has the lowest costs no

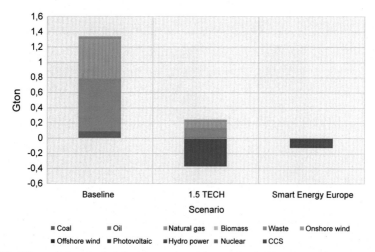

Fig. 8.27 CO_2 emissions and carbon capture and storage in the three energy systems.

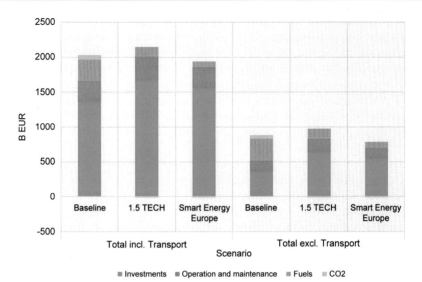

Fig. 8.28 Total annual costs in the three energy systems, one including transportation costs and one excluding transportation costs.

matter if transportation costs are included or not. The reason is the increased energy efficiency as a result of smart utilization of all energy grids. By applying the smart energy systems approach, utilizing the potentials of all energy grids, and not focusing on individual solutions requiring either power-to-heat or direct use of hydrogen, it is possible to identify least-cost solutions. The implementation of large-scale heat pumps in district heating, waste heat use from electrolysis and fuel storage, which are all possible to store using more affordable thermal storages, offsets the need for more expensive individual solutions and massive renovations in the building sector. By making the supply system cheaper, it is possible to not only lower investment costs but also achieve a system with lower fuel demands and, therefore, lower fuel costs.

6   Reflections

Reflections on the design of smart energy systems in the context of carbon neutral societies in this chapter are made regarding the principles and methodologies, as well as conclusions and recommendations related to the design of feasible solutions in Denmark and similar countries.

Principles and methodologies

From a methodology point of view, it can be concluded that the design of smart energy systems in the context of carbon neutral societies is a very complex process. This chapter has discussed and presented a set of ***guidelines*** for the design of a transition strategy for a country to become a fully decarbonized society.

As a core element of the guidelines, a country would have to fulfill its objectives of increased renewable energy and CO_2 reductions in a way that would fit well into a context in which the rest of the world realistically would be able to do the same. Moreover, a country should fulfill its short-term objective of CO_2 reduction in a way that would be able to lead to a fully decarbonized society as a next step. These overall criteria imply several choices and issues. Some of the most important are to include the country share of *international aviation and shipping* and not exceed the country share of a global *sustainable use of biomass*, as well as enabling the country's contribution in terms of *flexibility and reserve capacity* to increase the integration of non-continuous RES such as wind and solar into the electricity supply. Finally, countries should include innovation efforts for technologies that would be needed in the long-term, even if these are not likely to play a significant role in the short term.

An additional and essential part of the guidelines is to base the design of the strategy on the smart energy systems principles of taking a cross-sectoral approach to identify a cost-efficient, technically reliable, and politically doable solution.

Based on these overall guidelines, this chapter has also detailed a set of guidelines for the important issues of sustainable biomass and transportation. A comprehensive transportation modeling tool has been presented and applied to the case of decarbonizing the Danish transportation sector by 2045. The modeling tool TransportPLAN is a bottom-up back-casting transportation model that allows for the user to assess the implementation of measures to lower the growth of the transportation demand, promote modal shifts and introduce novel and alternative propulsion technologies and fuels. Designing a pathway for decarbonizing the Danish transportation sector is complex, especially when accounting for the Danish share of international aviation and maritime transportation. This significantly increases the demand for bioenergy and the production of electrofuels, which inflate transition costs and presents the issue of sustainable use of bioenergy resources.

The methodology of identifying the role of sustainable bioenergy in a fully decarbonized society is based on the basic principle that if the world should decarbonize within the limits of global sustainable biomass resources, each country should fulfill its objective of renewable energy and CO_2 reductions in a way that fits well into a context where the rest of the world will be able to do the same. The current UN greenhouse gas inventories do not include international shipping or aviation when accounting for total emissions. This reduces the focus on the emissions from these parts of the transportation sector. Moreover, it also eliminates the consideration of the biomass demand for fuel production for these transportation modes. Furthermore, the decarbonization of the energy and transportation sectors will have to align with the emission of climate gasses in the other sectors to achieve a climate neutral society. Since it may not be possible to reduce climate gasses in the other sectors to zero, such coordinated actions involve sinks in terms of CCS and Biochar.

7 Conclusions and recommendations

The above guidelines have been applied to *the case of Denmark* in the year 2045 as well as *the case of Europe* in 2050. With regard to Denmark, the key findings of the case are that a fully decarbonized energy system in the context of a carbon neutral society by 2045 is technically doable within the limitations of Denmark's share of the global sustainable biomass resources. Such a solution involves (1) that the Danish share of international aviation and shipping is considered even though it is not included yet in the UN way of calculating CO_2 emissions; (2) that the Danish share of worldwide sustainable biomass is not exceeded; and (3) that the use of bioenergy provides both sinks to compensate for other sectors and as an integrated part of a broad spectrum of renewable energy sources to fully replace the use of fossil fuels in the entire system. Compared to a business-as-usual reference, the societal costs do not have to increase. However, the costs of importing fossil fuels are replaced with investment costs and thus lead to increasing employment and economic growth. In the present situation, there is, however, a labor shortage in the Danish economy in several sectors. But in many other countries, the situation is different. In the longer run, the benefits of new types of employment and a greener industrial development would also be crucial to the general Danish economic development.

Hence, the case illustrates that a green transition of the Danish economy and society is doable, economically responsible, but also realistic within a sufficiently ambitious deadline. However, this does not necessarily imply that it will also happen. It would require decisive long-term public investments in a green transition of the entire energy system and that the private sector is encouraged to follow suit and decides to do so. The challenge is also to coordinate the many larger and smaller private and public initiatives to go green, thus enabling the decarbonization of not only the Danish but also the European energy system.

To achieve the best role for bioenergy, it is recommended to conduct a multiple technology solution with a focus on the use of different biomass technologies by enabling synergies between them and the energy system. This results in a robust multiple technology strategy in which one technology can replace another, in the case that some technologies are not commercially ready at the proper scale in time. In order to make the best use of the available sustainable biomass, as a result of energy system analysis, this paper proposes a combination of the following conversion technologies: biogas plants, thermal gasification, pyrolysis, hydrothermal liquefaction, combustion and CCUS.

With this in mind, the specific balance between these and other relevant technologies may differ from country to country and should be adjusted on an ongoing basis as time passes and the technologies are subject to further developments.

With regard to the transportation sector, a series of measures are necessary to implement. The main focus should be on lowering the growth of the transportation demand in inefficient modes of transportation while not compromising the development and growth of mobility. In the scenario, several measures are implemented to reduce growth in car travel and create modal shifts toward public transportation and active modes of transportation as cycling and walking. A way of achieving this is with political instruments, i.e., the introduction of a road price scheme and urban planning measures such as building extensive infrastructure for both rail and public transportation systems and bicycle pathways. Lowering the growth of transportation demand for aviation is proposed to be achieved with the introduction of elevated passenger taxations to increase the ticket purchase price and the improvement of railway infrastructure to enhance transportation alternatives.

Also, the issues of balancing and stabilizing the power grid become essential to address as part of the green transition. In the traditional fossil fuel-based system, frequency and voltage control are sustained by either large-scale hydro power, nuclear or fossil fuel-based steam turbines. In a 100 percent renewable energy system—no matter if it is in the context of a carbon neutral society or not—these tasks will have to come from other technologies. On one hand, this is a challenge that should not be neglected; on the other hand, this is also fully doable and many of the technologies are already implemented. In this chapter, it is illustrated how to deal with the matter using the Smart Energy Denmark scenario as a case.

In Table 8.4, the case of a smart energy system in the context of a carbon neutral society is compared to the 100 percent renewable energy systems presented in Chapter 7. As can be seen, biomass consumption is lower in order to align with the sustainability goals, and therefore, the need for wind and PV are higher. The latter, however, is also caused by the inclusion of international aviation and shipping as well as the need for carbon capture and biochar to compensate for greenhouse gas emissions from other sectors.

With regard to Europe, this Chapter has presented a case of taking a Smart Energy Systems approach to the design of a European energy transition into a fully decarbonized society and comparing the results to the 2050 baseline and the 1.5 TECH scenario of the European Commission's report "A Clean Planet for All." The study emphasizes the need to fully investigate all energy grids and the synergies between the grids. It is seen that, by utilizing all system benefits including waste heat from industry, e-fuel production and power-to-heat in combination with cheap thermal and fuel storages, it is possible to find a decarbonized energy system with lower costs and higher fuel efficiency than the comparable 1.5 TECH scenario, which mostly uses electrification as a mean of sector coupling.

The reason behind achieving this result is that increased sector coupling, i.e., the smart energy system, changes energy efficiency from focusing only on the end-use heating demand to being a holistic principle, both in the supply and demand sides. End-use heating demand is not reduced to the same extent as the 1.5 TECH scenario, but by implementing sector coupling in all grids and utilizing the energy more efficiently in the supply system, the Smart Energy Europe scenario achieves an overall more efficient energy system.

Table 8.4 Main data of the eight alternative 100 percent renewable energy systems.

	First approach	IDA (2050) IDA Energy Plan	CEESA	Smart Energy Denmark 2045
Demands (TWh/year)				
Electricity (including cooling and industry)	37.0	30.2	26.9	46.6
Heating (including process)	56.8	44.5	68.8	57.5
Transportation (electricity)	17.8	5.0	39.9	15.1
Transportation (biomass)		24.9	14.3	24.2
Primary energy supply				
Biomass (PJ/year)	180	270	237	140
Solar thermal (PJ/year)	8	19	23.3	16
Geothermal (PJ/year)	–	–	12.4	27
PV (GW installed)	5	1.5	5	10
Wave (GW installed)	–	1	0.3	0.13
Wind (GW installed)	15	10	14	19
Carbon capture and biochar for compensation in other greenhouse gas emission sectors				
CO_2 equivalent (Mt/year)	–	–	–	5.1

Consequently, for modeling pathways for future energy systems, this highlights the point that current scenarios may be missing crucial sector coupling benefits simply due to a modeling scope that captures many but not all potential sector coupling potentials between electricity, heating, transportation, and industry. It is crucial to include the entire energy system, not only to include all demands but also to enable an analysis of the links within the energy system, between the grids. Furthermore, it is important to utilize different types of cheap energy storages in the different sectors to create the necessary flexibility in an energy system relying on large shares of renewable energy.

Choice Awareness cases

9

This chapter returns to the discussion of the theoretical framework by presenting a number of cases of energy investments that have occurred since 1982. Choice Awareness strategies have been applied to the specific decision-making processes of these cases. Typically, the cases involve the design and introduction of concrete technical alternatives and/or other Choice Awareness strategies. The cases refer to a large series of publications and documentation mentioned in each section. The overall purpose of the chapter is to deduce what can be learned from the cases regarding the Choice Awareness theses and strategies formulated in Chapters 2 and 3.

The cases use the research method described in Chapter 3. Most cases are based on my personal involvement and that of my colleagues at Aalborg University. Basically, our involvement has had a twofold purpose. First, we intended to raise the awareness of choice in specific situations and thereby help society make better decisions. Second, we wished to observe and learn how different actors react to the existence of alternatives. In that way, the description and promotion of concrete technical alternatives, as well as institutional alternatives in specific decision-making processes, can be regarded as our way of applying a "questionnaire" to the complex system of actors involved in the process. From their reactions, we can observe and learn. Among other aspects, we are able to identify institutional barriers to new energy technologies and thereby form a platform for the design of concrete public regulation measures and institutional alternatives.

The cases are listed chronologically, and they consist of the application of mainly the two first Choice Awareness strategies: the description and promotion of concrete technical alternatives and the use of feasibility studies based on concrete institutional economics. However, as mentioned, the same cases also form the basis for applying the two other Choice Awareness strategies: the design of concrete public regulation measures, including institutional changes, and proposals to improve the democratic infrastructure, as indicated in the coming sections.

The cases in this chapter focus on the descriptions of technical alternatives and socioeconomic evaluations. In the debates, I contributed to this aspect, while my colleague, Frede Hvelplund, contributed to the design of institutional alternatives. However, it should be emphasized that technical alternatives and institutional alternatives create an important synergy and should be seen together. In most of the cases, the description of technical alternatives led to the proposal of some sort of public regulation measures.

In some cases, the institutional proposals are directly related to the case issue, such as in the Aalborg heat planning case in which specific changes in energy taxation were proposed, or in the Biogas case in which a comprehensive series of public regulation measures was designed to implement a scheme of large-scale

Renewable Energy Systems. https://doi.org/10.1016/B978-0-443-14137-9.00009-7

biogas stations. In other situations, the information from several cases forms the input to the design of comprehensive institutional alternatives. One such institutional alternative has been described in the book by Hvelplund et al. (1995) titled *Democracy and Change*.[a]

1   Case I: Nordkraft power station (1982–1983)

The case of Nordkraft power station is basically the story of a decision-making process that in the beginning was based on only one project proposal. It is also the story of how the introduction of a concrete technical and radically different alternative reveals the extent to which the main proposal is linked to existing organizations. It shows the severe institutional barriers that the radically different alternative meets, even though this alternative may provide an environmentally better solution at the same costs. The case description is based on the book *When ELSAM Makes Plans. Aalborg, Brønderslev … Pieces of the Puzzle* (Lund and Bundgaard, 1983)[b] and the article "When ELSAM Taught Aalborg All about Planning" (Lund, 1984).[c]

Nordkraft (North Power) was the name of both the power station and the utility company located in the center of Aalborg until approximately 2000. Nordkraft was one of seven similar power stations and companies constituting the joint electricity supply of West Denmark: Jutland and Funen. The cooperative was named ELSAM and it managed the joint financing of the power companies and made the decisions on which company should be the next to implement a new unit.

In 1967, ELSAM decided to build a new unit at Nordkraft. That unit ended up having the most unfortunate life that any power station unit could possibly have. It was built as an oil-fired steam turbine of ~250 MW of electric power, and it started to produce in August 1973, only a few months before the beginning of the first oil crisis. It continued its production during the time of the two oil crises but was given relatively low priority in the ELSAM cooperation because of high oil prices. In the early 1980s, it was decided to convert the power unit into a coal-fired unit, which meant a substantial extra investment, since the boiler had to be rebuilt and expanded to approximately twice its size. Moreover, coal storage and harbor facilities had to be adjusted. The conversion into coal was completed in the mid-1980s, exactly when oil prices dropped again. The unit was then operated on coal during periods of low oil prices up to around the turn of the century. Then the unit was demolished only a few years before oil prices rose again. The life of this power station unit clearly demonstrates the difficulty of adjusting to shifting oil prices.

The following case is based on the discussion that took place in the early 1980s on whether to convert the unit from oil- to coal-based production.

---

[a] Translated from Danish: *Demokrati og forandring*.

[b] Translated from Danish: *Når ELSAM Planlægger. Aalborg, Brønderslev … brikker i spillet*.

[c] Translated from Danish: *Da ELSAM lærte Aalborg om planlægning*.

The "no alternative" situation

The first coal unit plans were introduced to the public by ELSAM in 1980, and, consequently, the issue became part of the debate during the city council elections in Aalborg in 1981. The obvious alternative to coal was natural gas from the Danish North Sea, and several city council candidates expressed their preference for natural gas. The national natural gas grid was not yet completed; however, in June 1982, the Danish government and Parliament decided to accelerate the project. Consequently, natural gas became a realistic alternative when the project was discussed in late 1982 and early 1983.

Nordkraft supplied not only electricity, but the power station also supplied heat produced by combined heat and power (CHP) to the district heating of Aalborg city, which the municipality of Aalborg managed. It is important to note that it was a joint ELSAM decision that Nordkraft was an oil-fired station, whereas most of the other power stations in the ELSAM area were coal-fired. Consequently, the electricity and heat consumers of Nordkraft did not pay a higher price for fuel than the other power station consumers. Internally, within ELSAM, the power stations defined common average prices of coal and oil, which were used at all power stations. The Nordkraft consumers, consequently, did not personally suffer from the mistake of opening an oil-based unit only a few months before the first oil crisis. The extra costs were shared among all electricity consumers within the ELSAM area.

During the decision-making process, in 1982, ELSAM suggested a change in the contract between the power station and the municipality. The aim of this change was to raise the amount paid by heat consumers if the authorities did not approve the coal project. The political chair of the technical committee of Aalborg, who was also a member of the board of representatives of Nordkraft, suggested to the board of representatives that the price of heating was likely to increase if the authorities did not examine the project proposal (in its present version) in a satisfactory way—presumably implying satisfactory to ELSAM.[d]

Confronted with the threat of increased heating prices, the city council approved the ELSAM coal project (see Fig. 9.1). This approval was made without analyzing or describing the natural gas alternative. However, in accordance with Danish Planning law, the physical planning procedure included a public participation phase. In this phase, the city council received 700 written objections from local citizens, mostly arguing against the environmental problems arising from coal and the lack of proper analyses and suggestions of alternatives.

The local radio station asked the managing director of Nordkraft, P. E. Nielsen,

> *Could you imagine that the project, which has now really met a lot of resistance, will not be approved—that it will not be implemented?[e]*

[d] Translated from Danish: ... *Nordkrafts formand antydede overfor repræsentantskabet, at en forhøjelse af varmeprisen var sandsynlig, hvis ikke behandlingen af forslaget (i dets nuværende form) hos plan—og miljømyndigheder forløb tilfredsstillende—formentlig underforstået for elsammenslutningen (ELSAM).* Aalborg Stiftstidende, December 16, 1982.

[e] Translated from Danish: *Har du fantasi til at forestille dig, at* det *projekt, som nu altså har mødt en del modstand, at* det *ikke kommer igennem—at* det *ikke bliver gennemført? Nordjyllands radio, April 27, 1983.*

Fig. 9.1 Drawing from 1983 illustrating the threat from ELSAM against Aalborg to raise the heat prices if the city council did not approve the plans to convert into coal.

Nielsen replied,

I cannot imagine which other alternative one would suggest instead.[f]

What the managing director could not do—suggest alternatives—was instead done by local citizens, as illustrated in Fig. 9.2.

The concrete alternative proposal

As a member of a local nongovernmental organization (NGO),[g] I participated in the design and promotion of a concrete alternative representing radical technological change. Our motivation was based on our firm belief that the coal project would

---

[f] Translated from Danish: *Jeg har ikke fantasi til at forestille mig, hvad man så vil stille i stedet for.* Nordjyllands radio, April 27, 1983.

[g] The NGO, called Aalborg Energy Office, was the local group of three national energy organizations and movements: Cooperating Energy Offices, the Danish Antinuclear movement (OOA), and the Danish Organization for Renewable Energy.

Fig. 9.2 Drawing from 1983, illustrating the situation in which the managing director of Nordkraft power company needed the help of local citizens to be able to imagine any alternatives to coal.

increase pollution, anticipate the upcoming heat planning in Aalborg, and deteriorate the possibilities of introducing renewable energy (Lund and Bundgaard, 1983).

The alternative, which is illustrated in Fig. 9.3 in a sketch from 1983, consisted of three elements. The first was to convert Nordkraft into a natural gas-fired station. This conversion would cost only 20 million DKK compared to the estimated cost of 640 million DKK of the conversion to coal. The coal project required a completely new boiler double the size of the old one, while the natural gas solution could use the existing boiler. The next step was to insulate approximately 8000 houses and thereby save 1600 TJ of expensive gas oil in houses outside the district heating area. Finally, the plan was to invest in small CHP stations on existing district heating grids in small towns and villages in Northern Jutland. This investment would make it possible to expand the CHP production. Not all electricity production at Nordkraft was used to the benefit of CHP, simply because of the lack of district heating demand compared to the capacity of Nordkraft. Consequently, by replacing part of the production by small CHP units, one could expand the amount of heat produced by CHP, thereby reducing fuel consumption.

The investments of the alternatives amounted to only 320 million DKK compared to 640 million if Nordkraft was to be converted to coal. Moreover, the alternative would *reduce* the primary energy consumption by 3300 TJ/year, partly by energy conversion and partly by CHP expansion, while the coal project would *replace* almost 6000 TJ/year. Whether the 3300 TJ savings would prove cost-effective compared to

Fig. 9.3 Drawing from 1983 illustrating the alternative to the investment in a coal-fired Nordkraft power station.

the replacement of 6000 TJ/year of oil by coal naturally depended on fuel prices. In the promotion of the alternative, the actual fuel prices of the previous 12 years (from 1970 to 1982) were applied to the coming future period, and in this calculation, the alternative came out with an economically better result than the coal project, even when environmental benefits were not included.

Later, I made a calculation on the basis of actual oil and coal prices between 1985 and 2000, both years included (see Fig. 9.4). The cost of converting Nordkraft into a

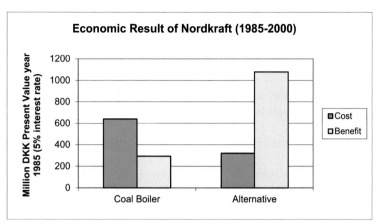

Fig. 9.4 Feasibility of the conversion of Nordkraft to coal compared to the alternative of CHP and energy conservation. Investment costs were compared to a net present value of annual savings between 1985 and 2000 based on actual coal and oil prices.

coal-based production, 640 million DKK, has been compared to the economic benefit of replacing an annual amount of 6000 TJ of oil by coal. The benefit has simply been identified as the actual price difference between coal and fuel oil when delivered to a Danish power station. To compare investment costs and savings, the annual savings were converted into the present value of 1985 using an interest rate of 5 percent. In the same way, the benefits of saving an annual amount of 1700 TJ of fuel oil and 1600 TJ of gas oil were identified for the alternative.

The calculation was not based on actual production data of Nordkraft, and the differences in operation costs between coal and fuel oil were not included. In the alternative, the price of natural gas was determined at the level of the fuel oil price. Thus, the calculation represents an estimate. Nevertheless, the result clearly indicates that the project of converting Nordkraft to coal was hardly beneficial to society, the power companies or the electricity consumers. Meanwhile, if it would have been possible to implement the alternative, the saved fuel would have made such a project cost-effective.

Conclusions and reflections

In the end, Nordkraft was converted into a coal-fired station. The following observations can be made from the case: the initial proposal put forward by the power company was a *one, and only one, alternative* proposal. The proposed technology fit well into the existing organizations of the power companies. No radically different alternatives were presented to the public when the proposal was to be approved.

City council members expressed their preference for an alternative based on natural gas, but such an alternative did not even form part of the basis for the decision-making process. Local citizens seemed to be the only ones who were free to describe and promote a concrete technical alternative. The citizens analyzed this alternative, indicating that it would prove cost-effective compared to the project of conversion to coal. Since then, a calculation based on the exact fuel prices of the lifetime of the Nordkraft coal power station has shown that the local citizens were right when they claimed that their alternative was competitive. Actually, it was preferable both in terms of environment and economics.

The main point to be deduced from the preceding observations is that, in this case, the institutional set-up could not identify and implement the best alternative by itself. The alternative entailed a radical technological change; that is, it could not be implemented without introducing changes in institutions, including the existing organizations.

The discourse of the power companies focused on the optimization of the fuel use within the existing technical and organizational set-up. The identification of alternatives that represented radical technological change was not a part of their interests or perception of reality, and even if it was, the implementation of such alternatives would be out of their reach, since it would involve investments in the insulation of private houses as well as CHP units in district heating companies owned by others.

The discourse of the city council focused on maintaining low district heating prices. Moreover, the council also had to manage urban and environmental concerns in the physical planning process. Again, the implementation of insulation and CHP outside the municipality was out of their reach. On the other hand, natural gas was an option within the reach and perception of the city council. However, the city council did not have the power or the resources to ensure a proper analysis and description of such an alternative when faced with the risk of substantially increasing district heating prices.

The proposal of radically different technological alternatives had to come from citizens outside the power companies and the city council. The existence of alternatives could raise public awareness of the fact that, from a technoeconomic point of view, choice *did* exist. As a result, 700 citizens claimed that alternatives should be discussed and included in the debate. However, given the institutional set-up, such radically different technological alternatives could not be implemented. Institutional changes had to be implemented at a higher level.

2 Case II: Aalborg heat planning (1984–1987)

The case of Aalborg Heat Planning is the story of how the municipality by law was forced to choose an inconvenient solution and how the municipality sought to exclude this choice from the decision-making process. The inconvenient solution represented a radical technological change in the direction of small CHP plants and renewable energy as opposed to coal. The solution was "inconvenient" because it would mean higher heating prices for the consumers. However, it would also mean a better environment, and, in an overall socioeconomic evaluation (as defined by the authorities), it proved more cost-effective to Danish society than the other alternatives. The law stated that the municipality had to choose the alternative with the best socioeconomic feasibility. The case description is based on the books *Low Taxes on Coal Spoil the*

Heat Planning—A Commented Collection of Documents from the Heat Planning Process in Aalborg (Hvelplund and Lund, 1988)[h] and *Energy Taxation and Small CHP Plants* (Lund, 1988).[i]

The case illustrates some of the basic mechanisms of excluding technical alternatives from the public discussion. However, it also shows how the description and promotion of concrete technical alternatives can lead to the identification of institutional barriers and promote the design of institutional alternatives.

In 1979, the Danish Parliament passed a law on heat supply. According to the law, all municipalities had to conduct a heat planning procedure in which different heat supply options were described, analyzed, and compared. The overall objective of the law was to

> *promote the best socioeconomic use of energy for the heating and hot water supply to houses and to reduce the energy supply's dependency on oil.[j]*

When asked how to include externalities such as environmental considerations in the socioeconomic feasibility studies, the Danish Ministry of Energy answered, "The socioeconomic analysis has to describe and compare all costs and benefits."[k] In practice, this is done by combining an economic calculation and a description of relevant externalities, typically including the environment, energy security, balance of payment, and job creation. However, the assessment and identification of the best solution should include all relevant considerations.

The alternatives in question

The heat planning procedure included a public discussion phase, and, consequently, in the late summer of 1984, the municipality distributed to all households a written invitation to participate in the discussions. The invitation put forward three alternatives as illustrated in Fig. 9.5 (left). In all of the alternatives, the main city area of Aalborg would continue to be supplied with district heating from the Nordkraft CHP station, which, as mentioned in the previous section, was now based on coal. The issue in question was how to supply the small suburban areas, towns, and villages around Aalborg with heat. The municipality proposed three alternatives. One was to expand the district heating system of Aalborg to most of the small urban areas. This alternative would mean that all heating would be based on coal, but it would be an efficient use of coal,

---

[h] Translated from Danish: *De lave kulafgifter ødelægger varmeplanlægningen—en kommenteret aktsamling fra varmeplanlægningen i Aalborg.*

[i] Translated from Danish: *Energiafgifterne og de decentrale kraft/varme-værker.*

[j] Translated from Danish: *A fremme den mest samfundsøkonomiske anvendelse af energi til bygningers opvarmning og forsyning med varmt vand og at formindske energiforsyningens afhængighed af olie.* Law on heat supply included in Hvelplund and Lund (1988).

[k] Translated from Danish: *Den samfundsøkonomiske analyse skal beskrive og sammenholde de samlede fordele og ulemper.* Letter from the Danish Ministry of Energy, February 29, 1984. In Hvelplund and Lund (1988).

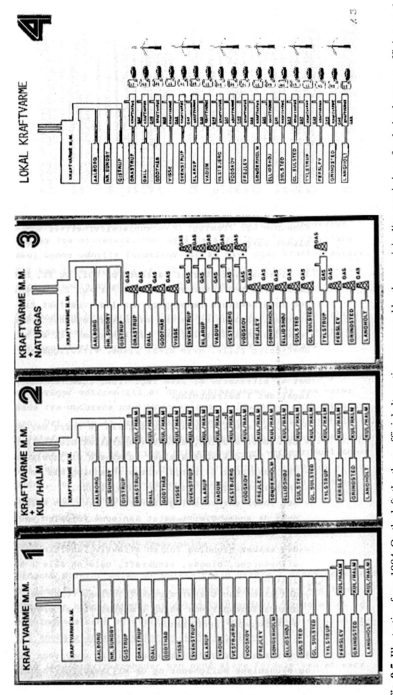

Fig. 9.5 Illustration from 1984. On the left, the three official alternatives proposed by the municipality consisting of a choice between efficient coal in CHP or inefficient boiler solutions. On the right, the citizens' alternative 4, proposing CHP based on natural gas to pave the way for renewable energy.

since it would all be CHP production. The two other alternatives suggested using boilers for heat-alone production either in small district heating systems and/or as individual boilers using natural gas. The fuel would be coal, straw, or natural gas. Both alternatives would involve the replacement of coal, but these productions would not be as efficient, since they would not benefit from CHP.

As part of the public discussion phase, I and six other students and teachers from Aalborg University proposed a fourth alternative that was simply to build small CHP stations based on natural gas in all areas. We illustrated the alternative as shown in Fig. 9.5 (right). Our intent was to pave the way for renewable energy in the forms of wind power and biogas. The main idea was to have both an efficient CHP production and, at the same time, avoid the use of coal. We presented three advantages of alternative 4 compared to the others. In terms of *energy efficiency*, alternative 4 was substantially better than alternatives 2 and 3. For two reasons, it was also slightly better than alternative 1. First, one could save district heating losses in huge pipelines of 10–15 km from Aalborg to the small urban areas. Second, the efficiency of the small natural gas units was slightly better than the efficiency of the Nordkraft coal unit.

In terms of *environment*, alternative 4 was the best solution, since it would either replace coal by natural gas when compared to alternative 1 or save substantial amounts of fuel when compared to alternatives 2 and 3. Alternative 4 would pave the way for renewable energy, since a system of small gas engine units would ease the introduction of biogas and, at the same time, provide a better integration of wind power than a coal-fired steam turbine.

Often such an environmentally better solution would be more costly than other alternatives. However, if the costs of cleaning SO_2 emissions from coal were included in the calculations, alternatives 1 and 4 came out equally cost-effective from an economic point of view. And since the natural gas/CHP solution had further benefits with regard to energy security, balance of payment, and CO_2 emissions, and these externalities according to the authorities were to be included in the overall assessment, it was concluded that alternative 4 was the best solution.

Though alternative 4 was preferable in terms of socioeconomics as defined by the authorities according to the law, the implementation would still mean increasing prices for the consumers compared to the coal alternative. The reasons for this were to be found in the institutional set-up of the Danish energy taxation system. The energy tax on coal was only 27 DKK/GJ compared to a tax of 51 DKK/GJ on natural gas. Moreover, this tax was only to be paid for the fuel used for heat and not the fuel used for electricity production. Also, an administrative practice had developed in which coal-fired steam turbines were only taxed a small part of the fuel (typically around 40 percent), while natural gas engines were taxed a major part (typically 60 percent). A detailed description and analysis of the administrative practice and the consequences are presented in Lund (1988).

All in all, in accordance with the law, the alternative involving natural gas and CHP should be chosen, since it proved to have the best socioeconomic feasibility. However, the municipality wanted to implement the CHP and coal alternative, since it proved to have the best consumer heat prices.

Choice-eliminating strategies

As mentioned, the group from Aalborg University described and promoted alternative 4: small CHP stations based on natural gas paving the way for future renewable energy systems. The proposal was discussed in the newspaper, and we mailed it to the municipality as part of the public discussion phase in 1984. The municipality answered that no decision had been made and that they would closely follow the development within natural gas-based CHP stations.

In the following years, two occurrences took place that helped the promotion of the natural gas-based CHP alternative. First, an agreement was made between the government and the major opposition party to expand small CHP stations on domestic resources by 450 MW. Second, the Danish Energy Agency issued a report that concluded that natural gas-based CHPs proved socioeconomically cost-effective, even in small urban areas.

Nevertheless, in February 1987, the newspaper stated that the Municipality of Aalborg would now decide on the matter. References were made to calculations showing that the best heat prices would be achieved by implementing alternative 1; that is, coal-based district heating from Nordkraft was to be supplied to 10 surrounding urban areas. We asked for the report and could see that alternative 4 had not even been analyzed.

We protested in the local newspaper. The city council postponed the decision for 2 weeks and invited my colleague, Frede Hvelplund, and me to analyze the alternatives together with the municipality administration. We soon agreed on the premises and chose one of the small towns, Frejlev, as an example. Both parties made a series of calculations and met on April 6, 1987, to compare the results.

Our calculations showed that natural gas-based CHP was the best solution in a socioeconomic perspective, whereas, regarding heating prices, the result depended somewhat on the expectations of future fuel prices. The calculations of the municipality administration confirmed that natural gas-based CHP was the best solution in terms of socioeconomic feasibility. However, their calculations showed that coal-fired central CHP would provide the lowest consumer heating prices.

The fact that the two calculations came to the same result regarding socioeconomic feasibility caused a major problem for the city council. According to the law, they were supposed to choose natural gas CHP, but they wanted coal. The city council had its next meeting on April 13, and the committee on heat supply discussed the matter at a meeting on April 9.

Then something happened: when sending out documents prior to the city council meeting, the two calculations of April 6 were not included. Instead, the municipality administration made a new calculation dated April 9, which was distributed to the committee at the meeting. This calculation showed, in terms of socioeconomics, that coal was even slightly better than small natural gas-fired CHPs. However, these calculations did not include all the environmental and energy security benefits of the natural gas alternative. Consequently, even based on the new calculation, natural gas would still be best in an overall socioeconomic assessment as defined by the law, but this issue was not mentioned.

We responded by sending our calculations of April 6 directly to the city council members before the city council meeting on April 13. At the meeting, some city

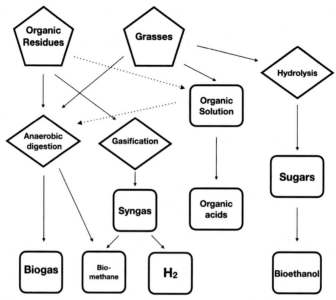

Fig. 5.7 The conversion of organic residues and grasses to valuable products. *(Based on data from the reference of footnote 'a'.)*

practised that is aimed at adding methane to existing natural gas supply networks. As the bio-methane is generally produced at or near atmospheric pressure, it can either be pressurised prior to admitting it to a high-pressure gas pipeline or it can be supplied at lower pressure directly to the consumer.

Grass is one of the most common biomass species and it is most commonly used as a feed for cattle and other ruminants, this ultimately giving rise to some of the organic residues discussed above. However, grass can also be treated to produce a solution from which organic acids can be extracted, the remaining solid residue also being used in anaerobic digestion systems. Further, grass can be hydrolysed to give a solution containing a variety of sugars and these can then be further processed to form products such as bioethanol as discussed in more detail in the following section.

Ethanol and bioethanol production

Ethanol is an important chemical that is used in many industrial applications. It is an important product of the petrochemical industry in which it is formed by the hydration of the ethylene produced from oil in catalytic crackers.

$$C_nH_m \rightarrow CH_2 = CH_2 \, (+ \, CH_3 - CH = CH_2)$$
$$CH_2 = CH_2 + H_2O \rightarrow C_2H_5OH$$

This hydration reaction requires the use of an acidic catalyst such as supported phosphoric acid at temperatures about 250°C. The majority of the ethanol formed is subsequently used in other petrochemical processes. The annual world production of ethanol by the petrochemical route, about 0.4 million tonnes per year, is approximately 4% of the total world production.

The vast proportion of the remaining ethanol now used worldwide (Table 5.2) is produced by various fermentation routes. In consequence, the product is generally described as 'bioethanol'. These methods in many cases differ little from those that have been used for centuries for beverage production. The total annual world production of ethanol by both routes is around 40 million tonnes per year, of which 77% was used as a fuel (largely for transport, see below), 8% as a component of beverages and 15% in industrial applications. As can be seen from the data in Table 5.2, the predominant producers of bioethanol are the United States and Brazil.

Bioethanol can be derived from the starch or sugars contained in many different biomaterials but it is currently mostly produced in commercial quantities from either the starch content of maize (which is the predominant feed in the US) or the sucrose content of sugar cane (the predominant feed in Brazil). Corn

Table 5.2 The major global producers of bioethanol.

Country	Production in 2019/million gallons	% of world production
United States	15,788	54
Brazil	8590	30
European Union	1370	5
China	1000	3
Canada	520	2
India	510	2
Thailand	430	1
Argentina	280	1
Rest of World	522	2
Total Production	29,000	100

Adapted from data provided by the Renewable Fuels Association (https://ethanolrfa/statistics/annual-ethanol-production/).

contains 60%–70% of starch and the remainder consists mainly of proteins (8%–12%) and water (10%–15%). (Other predominantly starch-containing plants which can also be used for bioethanol production include cereals—such as wheat, rye, barley and sorghum—or root crops such as sugar beet.) The sugar from these sources consists largely of two polymeric species: amylose and amylopectin. The former is a water-insoluble straight-chain polymer composed of alpha-glucose subunits while amylopectin is a branched-chain polymer of the same alpha-glucose subunits. Prior to fermentation of a feed such as corn, these polymeric species have to be broken down to give fermentable sugar units and this is generally carried out in an acid-catalysed hydrolysis process occurring in two steps: initial breakdown to oligomers and subsequent reaction to form the sugar monomers. (A process involving enzymatic hydrolysis is also nowadays increasingly practised.) The all-over process of obtaining alcohol from a starch crop such as corn is outlined in Fig. 5.8.[j]

In the case of sugar cane and other sugar-containing crops such as sugar beet, the hydrolysis step is not necessary and so the operational costs associated with bioethanol production from these crops are significantly lower. The sucrose contained in the cane or beet is extracted by crushing them in water

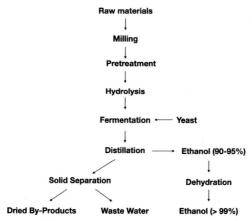

Fig. 5.8 The production of bioethanol from starch-containing plants. *(Diagram modified from 'Bioethanol: Market and Production Processes', M.J. Taherzadeh and K. Karami, 'Biofuels Refining and Performance', Ed. A. Nag, McGraw Hill, (2008) 69–123.)*

---

[j] This and many other related topics are covered in some relatively recent textbooks and collections of reviews. An example of the latter type is 'Biomass to Biofuels-Strategies for Global Industries' edited by A.A. Vertès, N. Qureshi, H.P. Blaschek and Y. Yukawa, 2010, John Wiley and Sons, Chichester.

and it is then purified prior to the enzymatic fermentation step.[k] Following the fermentation stage, after which the maximum alcohol content is about 12 wt%, the product is subjected to distillation, this resulting in an azeotropic mixture of alcohol and water containing 95.57 wt% alcohol. This product can be used as such in most applications and the solids remaining after the distillation process can be further processed, used as animal food or combusted for energy production. However, in order to obtain pure alcohol ('absolute alcohol') for use, for example, as an additive for petrol, further purification is needed. This involves either further distillation of a ternary solution containing an additive such as benzene or cyclohexane or the extraction of the water from the azeotropic mixture by the use of molecular sieve drying technology or by selective membrane separation methods.

One of the primary current uses of bioethanol, now also produced from lignocellulose (see below), is its dehydration to produce ethylene (by the reverse of the reaction shown above for the formation of ethanol from the ethylene produced by cracking crude oil). This ethylene is then used for the preparation of a variety of important chemicals and polymers as shown in Fig. 5.9. Table 5.3 shows the quantities of some of the products

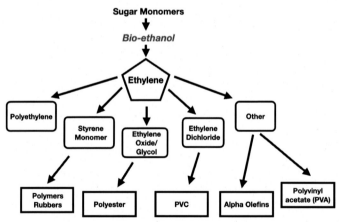

Fig. 5.9 The conversion of bioethanol to a variety of products.

[k] Sugars suitable for fermentation can also be produced from oil crops and from materials comprising predominantly of lignocellulose; however, the preprocessing in these cases is more complex. As we will see, the lignocellulose components of the residues from other crops such as sugar bagasse (the material remaining after sugar extraction) or various other stover materials can also be used as feedstocks for further processing.

Table 5.3 Worldwide production of some important industrial commodities from bioethanol.

Product	Saving in greenhouse gas emission/tonne CO$_2$ per tonne product	World capacity/ mtonne per year	Annual decrease in greenhouse emissions/m tonne per year
Acetic acid	1.2	8.3	9.6
Acrylic acid	1.5	2.9	4.4
Adipic acid	3.3	2.4	7.9
Butanol	3.9	2.5	9.6
Caprolactam	5.2	3.9	20.0
Ethanol	2.7	2.6	7.1
Ethylene	2.5	100.0	246
PHA	2.8	57.0	15.0
PLA	3.3	11.1	35.6

PHA, polyhydroxyalkanoate (an easily biodegradable polymer); *PLA*, polylactic acid (less easily biodegradable polymer formed from lactic acid).
Table constructed from data of reference in footnote 'a'.

currently produced globally from bioethanol and the savings in greenhouse gas emissions achieved by adopting a biomass-based route. The predominant commodity in this listing is ethylene, the main use of which is polyethylene manufacture.

Conversion of oil crops and oil-based residues to biodiesel and chemicals

Just as the fermentation of sugar-based crops to produce alcohol has been practised for many centuries, naturally occurring oils such as palm oil, olive oil and rapeseed oil that are easily extracted from the associated biomass have been used since early times for lighting, medicinal and cooking purposes. Oil for such uses can also be extracted from many other types of seeds and berries and there are also many uses for the residual oil-containing materials. More recently, there has been a significant interest in the use of crops for the production of energy carriers, particularly biodiesel, and also of chemicals. The crops now most commonly used for biodiesel production are rapeseed (*Brassica napus*) and soya bean (*Glycine max*).[1] The processes that will now be discussed are in direct competition

[1] The methods used for biodiesel production can also be applied to the treatment of used cooking oil, this making it cost-effective to collect and treat this material rather than disposing of it as waste.

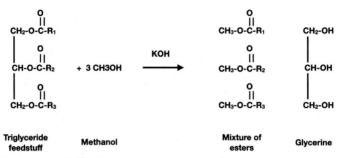

Fig. 5.10 The transesterification process.

with food production and are therefore regarded as being first-generation methods.

The oils extracted from the various sources mentioned above consist of triglycerides. As shown on the left-hand side of Fig. 5.10, these are esters of glycerine containing three different $RC(O)O$ groupings derived from the corresponding acids. The simplest of the possible acids is acetic acid, CH_3COOH, in which case R is the CH_3 group, but in practice R is generally much larger, containing between 14 and 22 carbon atoms.[m] Hydrolysis of a glyceride in the presence of simple alcohol such as methanol results in the formation of the methyl esters of the acid groupings.[n] This 'transesterification' process, catalysed by a base such as KOH, is shown in the equation of Fig. 5.10. The mixture of esters formed in the reaction can be used as diesel fuel ('biodiesel') as its combustion properties are similar to those of conventional diesel fuel as prepared from crude oil.[o] Biodiesel can be used undiluted as fuel for conventional diesel engines but it is more frequently blended with diesel produced from crude oil sources. As is discussed later, bio-diesel is now also produced from the bio-oil formed in fast pyrolysis processes. A fuller description of the methods of production and use of biodiesel is beyond the scope of this book and the reader wishing more detail should refer to one of a wide variety of comprehensive texts currently available (see footnote 'j').

[m] Triglycerides also exist in animal fats; they can also be extracted from oil residues of the type referred to earlier.

[n] Ethyl esters are also produced but this occurs much less frequently.

[o] Some quantities of residual methanol in the resultant fuel can be tolerated but its concentration must not be too high since methanol addition lowers the flash-point of the diesel.

Fuels and chemicals from lignocellulosic crops

All growing matter, ranging from the wood from trees to the stems and leaves of annual crops, contains cellulose, hemicellulose and lignin (see Box 5.2) in various proportions and many of these crops are distinct from those grown for food. This section therefore discusses the non-food uses of a number of chemicals that can be generated from lignocellulosic crops in 'second generation bio-refineries'. These ligocellulosic crops include not only the wood from trees that are not used for fuel or pulping but also forest residues together with some of the residues resulting from the use of food crops such as straw and corn stover. Additionally, there is increased use of specially grown crops such as miscanthus grasses. This genus includes *Miscanthus × Gigantica*, a perennial species grown in Europe that flourishes on fertile soils but can also be grown on set-aside land unsuitable for the growth of food crops. It requires minimal use of fertiliser and can be harvested in the springtime rather than in the winter months.[P]

Some of the processes that can be used for the conversion of lignocellulosic crops and their residues (i.e. lignin plus other components of the crops) are shown schematically in Fig. 5.11 and a selection of these routes will now be discussed.

As discussed in Box 5.2, cellulose is made up of linear polymeric chains comprising exclusively of the C_6 sugar, D-glucose. In contrast, hemicellulose contains a number of different sugar varieties; the main polymeric chain of hemicellulose comprises mostly of the C_5 sugar xylose but the polymer also has side branches made up of several other different types of sugar rings. The lignin that makes up the majority of the remainder of all the fibrous parts of lignocellulosic crops acts as a 'glue', holding the cellulose and hemicellulose together, the consequence being that these structures generally resist hydrolysis ('saccharification') by dilute acids or bases. It is well established that different hydrolysis procedures can result in various different degrees of depolymerisation of each of these three components. In consequence, many different hydrolysis methods are used commercially to depolymerise lingocellulose materials, the details dependent on the desired product. As an example, the depolymerisation of the lignin component of wood used for paper-making is achieved, as already discussed, using SH^- anions

[P] D.J. Hayes and M.H.B. Hayes, 'The role that lignocellulosic feedstocks and various biorefining technologies can play in meeting Ireland's biofuel targets', Biofuels, Bioproducts and Biorefining, 3 (2009) 500–520.

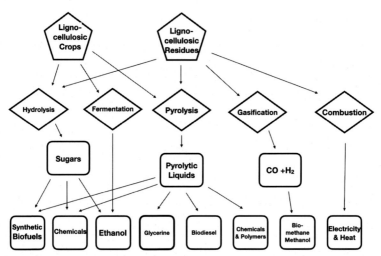

Fig. 5.11 The conversion of lignocellulosic crops and residues to useful products.

(Fig. 5.6), and the cellulose and the hemicellulose structures are relatively unaffected. Once the lignin structure is broken down, the polymer chains of the cellulose and hemicellulose components become more accessible to further hydrolysis by bases or acids and this additional hydrolysis then results in solutions containing a predominance of the constituent sugars; see Fig. 5.11. However, as the rates of hydrolysis of the chains of the polymers of cellulose and hemicellulose are significantly different, the hemicellulose structure is broken down more easily and the cyclic sugars obtained are able to break down further to form linear molecules before the hydrolysis of the cellulose is complete. Hence, many hydrolysis procedures involve two steps: the first is carried out under relatively mild conditions in order to achieve the liberation of both the lignin and hemicellulose contents; then, after separa-tion of the sugars formed in the first step, a more severe hydrolysis process is carried out to break down the cellulose chains.[q]

Once saccharification and separation of the sugars have been carried out, the constituent saccharides can be reacted further. In particular, as they are now accessible to conventional fermentation routes, bioethanol is the pre-dominant end product. The route for the production of bioethanol from

---

[q] When timber is the raw material being treated, the process is known as 'wood saccharification'. For more details of some of the work that has been carried out on hydrolysis procedures, the reader might consult the reference given in footnote d.

lignocellulosic materials contributes a growing part of the worldwide supply as discussed in a section above (see Table 5.2), resulting in a year-by-year increase in the use of non-food feedstocks such as corn stover and sugarcane bagasse. The ethanol can then be used for the production of ethylene and hence of many other important products such as polyethylene and ethylene glycol, also discussed above (see Fig. 5.9 and Table 5.3). All these products had conventionally been made from ethylene produced by the petrochemical industry. Hence, their production from biomass has the potential to make significant reductions in the emissions of CO_2 that would otherwise have occurred.

Careful control of the hydrolysis procedure applied to lignocellulosic feedstocks can also give rise to various other useful products. One example of such a procedure is the Biofine Process[r,s] shown in Fig. 5.12 that is used to produce levulinic acid (4-oxopentanoic acid, CH_3-CO-CH_2-CH_2COOH) and formic acid (HCOOH) by reaction of the D-glucose component of the cellulose polymer and of furfural (C_4H_3OCHO) by reaction of the xylose component of the hemicellulose polymer. The levulinic acid formed can then be converted directly to a wide range of useful chemical compounds (see Fig. 5.13 and the references of footnotes 'r' and 's'). Alternatively, it can be esterified with ethanol to form ethyl levulinate, a valuable bio-diesel substitute. The formic acid, initially seen as a waste by-product, can easily be decomposed to give pure hydrogen (we will return to this subject in a later section) while the furfural produced in parallel (from the hemicellulose content of the lignocellulose) also has many uses, including its conversion to the important industrial solvent, tetrahydrofuran. The lignin fraction ends up as a char that is burnt to provide the energy needed to operate the plant.

A full-scale Biofine plant was constructed in Caserta (Italy) and this started production in about 2005 using material such as vine trimmings as feed. The plant was then acquired by GF Biochemicals (founded in 2008 and based in Geleen, The Netherlands) who later acquired the American company Segetis, a firm specialising in the downstream uses

[r] D.J. Hayes, J.R.H. Ross, M.H.B. Hayes and S.W. Fitzpatrick, 'The Biofine Process: Production of Levulinic Acid, Furfural and Formic Acid from Lignocellulosic Feedstocks', in Biorefineries Industrial Processes and products, B. Kamm, P.R. Gruber and M. Kamm (Eds.), Wiley-VCH, 2006.

[s] J.J. Bozell, L. Moens, D.C. Elliott, Y. Yang, G.G. Neuenscwander, S.W. Fitzpatrick, R.J. Bilski and J. L. Jarnefeld, 'Production of Levulinic Acid and Use as a Platform Chemical for Derived Products', Resources, Conservation and Recycling, 28 (2000) 227–239.

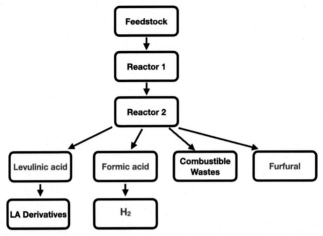

Fig. 5.12 The biofine process for the conversion of lignocellulose to various products. *(Adapted from the references of footnote 'r'.)*

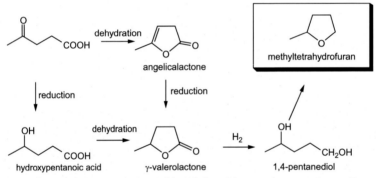

Fig. 5.13 The catalytic conversion of levulinic acid to various products. *(From 'The Biofine Process: Production of Levulinic Acid, Furfural and Formic Acid from Lignocellulosic Feedstocks', in Biorefineries Industrial Processes and products, B. Kamm, P.R. Gruber and M. Kamm (Eds), Wiley-VCH, 2006.)*

of levulinic acid. In 2020, GF Biochemicals announced a new joint venture with Towell Engineering Group of Oman for the production and marketing of various of their products, including ethyl and butyl levulinates and several ketal esters,[t] all of these intended for use as solvents.

---

[t] A ketal is derived from a ketone by replacement of the C=O group by two OH groups.

Gasification and pyrolysis of biomass

Gasification

Virtually all types of biomass can be gasified by a process similar to that used for many years for the gasification of coal. The gasification process, producing a mixture of CO, H_2 and CO_2, is carried out at a temperature of at least 700°C in the presence of oxygen and/or steam. Air can also be used as the source of oxygen but the product gas mixture will then also contain nitrogen. The syngas produced without any further purification contains additional condensable organic molecules plus methane and this makes it unsuitable for further direct processing. The removal of these contaminant species using catalytic methods has been reviewed elsewhere.[u] Syngas can also be produced directly from biomass by catalytic steam reforming, a topic also reviewed elsewhere.[v] The syngas produced in these gasification processes can be used to produce methane or methanol or used in the Fischer Tropsch process to produce motor fuels. However, the most common use of the unpurified syngas is in direct combustion for the production of electricity. The advantage of using this syngas rather than using direct combustion of the raw biomass material is that the all-over efficiency of the combustion process is greater at the higher temperature produced by syngas combustion. One of the most common uses of biomass gasification is in the treatment of biomass residues that would be otherwise difficult to process due to the inhomogeneity of these residues.

Pyrolysis

So-called 'fast pyrolysis' is now commonly used to convert both specially grown biomass and also the many forms of residue available to give a range of different products; see Fig. 5.14. The pyrolysis process involves heating the biomass or residue to a relatively high temperature (about 500°C) in the absence of and air and collecting the so-called 'pyrolysis liquid' formed. The pyrolysis liquid can then be treated chemically to give a very large range of chemical products. One of the great advantages associated with such processing is that the pyrolysis stage can be carried out on sites close to the source of biomass and the bio-oil product can then be transferred to a central processing facility for further upgrading, this being much more economical than transporting larger amounts of low-density biomass.

[u] D. Sutton, B. Kelleher and J.R.H. Ross, Review of Literature on Catalysts for Biomass gasification', Fuel Processing Technology, 73 (2001) 155–173. D.A. Bulushev and J.R.H. Ross, 'Catalysis for Conversion to Fuels via Pyrolysis and Gasification: A Review', Catalysis Today, 271 (2011) 1–13.

[v] D.A. Bulushev and J.R.H. Ross, 'Catalysis for Conversion to Fuels via Pyrolysis and Gasification: A Review', Catalysis Today, 271 (2011) 1–13.

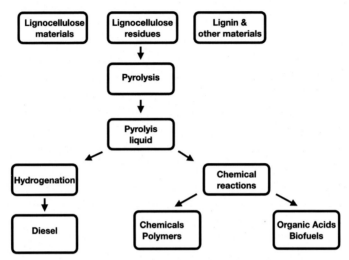

Fig. 5.14 The pyrolysis process.

One of the earliest versions of the pyrolysis process was the production of charcoal from wood, a process that has been used for many centuries. In the charcoal-burning process, the wood is heated slowly in the absence of air to produce what is essentially pure carbon, the remaining components of the wood being liberated as smoke from the kiln. However, in a modern pyrolysis system, the process is carried out using a fluidised bed reactor (see Fig. 5.15) in which the residence time is very short and from which the organic vapours emitted are collected as a 'pyrolysis liquid'. The pyrolysis process can be either non-catalytic, in which case the fluidisation is carried out using fine quartz sand as a heat transfer medium, or catalytically, in which case a suitable catalyst is included in the system (see below). The reactors used are similar to the fluid catalytic converter (FCC) units used by the petrochemical industry to convert crude oil to lower molecular weight products. Char is formed as a significant product during the pyrolysis process and is removed from the reactor together with the sand/catalyst flow; it is then combusted to provide the energy necessary for the operation of the pyrolysis process and to recirculate the sand/catalyst (Fig. 5.15).

The composition of the pyrolysis liquid obtained in the pyrolysis process depends on a very large range of factors. Of particular importance is the composition and particle size of the material being pyrolysed. However, secondary factors include the temperature of the reactor column and the residence time of the components in the reaction zone. Early versions of the fast pyrolysis process did not include a catalyst but there has relatively recently been an upsurge in the use of catalytic systems. The presence of a catalyst,

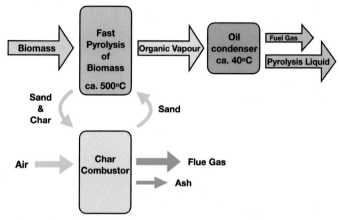

Fig. 5.15 The combustor setup of a pyrolysis unit.

either included with the sand fed to the riser system or contained in a separate catalyst bed (placed after the riser unit but before the condenser), has a very major influence on the product composition. Catalysts used in the riser reactor typically include zeolites such as ZSM-5, Y-zeolite and beta-zeolite. ZSM-5 encourages the formation of aromatic hydrocarbons during the processing of biomass materials from a wide variety of sources. Catalytic treatment after the riser reactor frequently involves hydrodeoxygenation in order to increase the H/C ratio of the products. Fig. 5.16 illustrates some of the changes that can occur in such processing.[w]

Pyrolysis oil is also a source of biodiesel by transesterification as described in a previous section and the production of biodiesel by this route is now very significant. A detailed description of all the processes that are applied to the upgrading of pyrolysis oil is beyond the scope of this book and the interested reader should consult one of a number of useful reviews of biomass pyrolysis.[x]

[w] A.O. Oyedun, M. Patel, M. Kumar and A. Kumar, The Upgrading of Bio-Oil via Hydrodeoxygenation, in Chemical Catalysts for Biomass Upgrading, M. Crocker and E. Santillan-Jiminez, eds., Wiley-VCH (2020) 35–60.

[x] Examples of the many reviews available on biomass pyrolysis reactions include the following: A.V. Bridgewater, Biomass Fast Pyrolysis, Thermal Science, 8 (2004) 21–49. A.V. Bridgewater, Review of Fast Pyrolysis of Biomass and Product Upgrading, Biomass and Bioenergy, 38 (2012) 68–94. N. Dahmen, E. Henrich, A. Kruse and K. Raffelt, Biomass Liquefaction and Gasification, in Biomass to Biofuels: Strategies for Global Industries, A. Vertès, N. Qureshi, H. Blaschek and H. Yukawa, ads., John Wiley and Sons (2010) 91–122. C.A. Mullen, Upgrading of Biomass via Catalytic Fast Pyrolysis (CFP), in Chemical Catalysts for Biomass Upgrading, M. Crocker and E. Santillan-Jiminez, eds., Wiley-VCH (2020) 1–33. A number of other reviews in the last two listed books also contain relevant information.

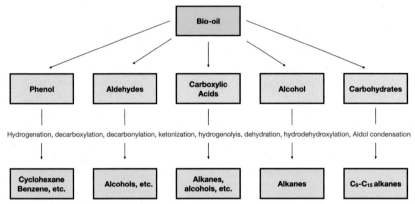

Fig. 5.16 Hydrodeoxygenation pathways for some of the compounds found in pyrolysis oil. *(Modified from the review by Oyedun et al. (Footnote w).)*

Other sources of biomass

Seaweed and algae

Seaweed and algae have long been used as a source of food, chemicals and energy. There are many different forms of both, ranging from kelps to micro-algae. The world production of seaweed exceeds 2 million metric tons a year, with the predominant usage being in China and France. The total global annual production of all aquatic plants exceeded 30 million tonnes in 2016.

Seaweed (macro-algae) has been used for centuries as a fertiliser and soil-enhancing component. For example, in rocky areas of the West of Ireland it was used in combination with sea-shore sand to form an artificial soil in which potatoes could be grown; it is now sold in powdered form as a soil enhancer to promote plant growth. Seaweed 'burning' under oxygen lean conditions similar to those used for charcoal production was a means of producing a cement for construction purposes; straightforward combustion was also used for heating and cooking purposes. Seaweed is used in many countries as a food and a series of hydrocolloids used in food products such as alginates, agar and carrageen can be produced from it. Alginates are also used in medicinal products. The growth of seaweed, in common with that of all plants, consumes CO_2 and also produces oxygen, thus contributing significantly to the reduction of greenhouse emissions. During its growth, seaweed also consumes nutrients such as ammonia, nitrate and phosphate anions as well as iron and copper cations and as a result it provides a filtering action on reefs. In China, seaweed is used to purify phosphate effluents.

It has been proposed that 'sea afforestation' could be used as a method of removing carbon emissions. The harvested material would then be treated in anaerobic digesters, producing biogas containing 60% CH_4 and 40% CO_2. The CH_4 can be used as a fuel and the CO_2 collected and stored. It is claimed that 9% afforestation of the world's oceans would remove more than the annual global emissions of CO_2. Such an approach would clearly require efficient methods to cultivate and then harvest the seaweed and this would be a serious barrier to its adoption.[y]

Microalgae

Microalgae are microscopic species with sizes ranging from several micro-metres to several hundred micrometres. Unlike normal plants, they do not have roots, stems or leaves. A significant proportion of the world's micro-algae are found in the oceans, rivers, lakes and ponds; a well-known exam-ple is the buildup of algal blooms in relatively static areas of water. These algae are particularly effective in photosynthesis and are responsible for the consumption of a very large proportion of the CO_2 used in global pho-tosynthesis processes, producing much of our atmospheric oxygen. Micro-algae contain a very wide range of potentially important components and are already used for the production of low-volume high–value products such as pharmaceuticals and cosmetics. There is an increased interest in cultivating them as a source of oils for use in the production of bio-fuels. Cultivation can be carried out in either marine or freshwater systems. As with seaweed, one of the main problems is associated with harvesting and the main sources are in specially-constructed ponds in which the move-ment of the water is carefully controlled. They can also be grown in the equivalent of greenhouse conditions (see Fig. 5.17) in which the tem-perature, the amount of light and the water circulation can be carefully controlled.

Chitin

The final source of biomass to be considered is *chitin*, one of the main com-ponents of the cell walls of fungi and is also found in sea shells, fish scales and

---

[y] Many web entries are available to describe the use of seaweed, e.g. https://en.m.wikipedia.org/wiki/Seaweed; https://seaweed.ie/uses_general/.

Fig. 5.17 A photo-bioreactor used for the growth of micro-algae. *(From https://commons.wikimedia.org/wiki/File:Photobioreactor_PBR_4000_G_IGV_Biotech.jpg)*

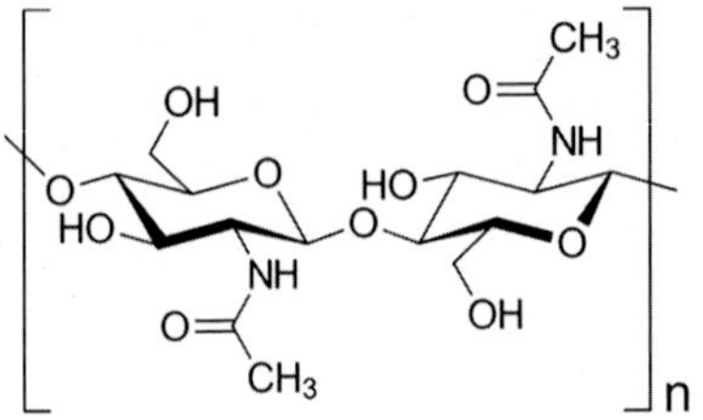

Fig. 5.18 The polymeric chain of chitin. *(From https://commons.wikimedia.org/wiki/File: Chitin.svg.)*

insect shells. It is a long–chain polysaccharide with a structure similar to that of cellulose but with units made up of *N*–acetylglucosamine. This polymer forms microfibrils that provide strength to the shells in which it is present (Fig. 5.18). It has a range of applications, from acting as a fertiliser and plant conditioner to being used in paper–sizing applications and food processing.

This chapter would be incomplete without a discussion of the organic matter contained within soil: 'soil organic matter'. Its importance in relation to the growth of all types of biomass is discussed in Box 5.4.

BOX 5.4 Soil organic matter

Soil organic matter (SOM) contains organic species such as the microorganisms, bacteria and fungi, these being stored in sub-surface soil. Although SOM does not provide a source of energy, our crops would not grow without its presence. The global quantity of this organic material exceeds by a factor of three the total amount of organic material above the surface of the soil. SOM is the 'cement' that binds the inorganic content of the soil (clays and various hydroxides, especially those of iron and aluminium) and this combination of colloidal organic and inorganic particles is the origin of soil fertility. It is important to recognise that when soils are in long-term (especially monoculture) cultivation, SOM degradation occurs. The most fertile soils of the world are being so over cultivated and has been estimated that after a further 50 to 100 crop cycles from now, the SOM of these soils will be so depleted that the soil structure will be degraded and its fertility lost. It is therefore essential that attention is also given over the next century to ensuring that SOM levels are replenished so that it will remain possible to supply food for the expanding world population.

The composition of SOM is quite complex and research on the subject is quite extensive. Extraction and fractionation of the components have been a major challenge. Techniques developed recently involve exhaustive extractions with the aqueous base at increasing pH values, followed by further extraction using an aqueous base containing urea (6 M) and finally extraction of the dried residues with a solution of dimethyl sulfoxide (DMSO) and concentrated sulphuric acid (6%). This sequence allows the isolation of up to 95% of the components of SOM, each with different polarities. The more polar components isolated in the aqueous systems are enriched in organic acid functionalities and include polysaccharides, peptides and humic substances (these predominating). The generally dominant low polarity humin materials, composed of long-chain fatty acids, waxes, cuticular materials, cutin, cutan, suberin and suberan are isolated in the DMSO-acid medium.

Concluding remarks

It is clear that while biomass has been used for many centuries for a wide range of applications, there are many novel applications that are still in development which have the potential to contribute very significantly to reductions in the all-over emissions of greenhouse gases. In so doing, they can contribute to greatly improved sustainability. Many of the applications currently available are relatively little used. Hence, in common with many of the other topics discussed in this book, they require much more attention, needing substantial investment from both governments and industrial operators. As just one example, the oil industry has the potential to diversify much more extensively into the provision of biofuels. However, that would require that both political support and legislative pressure are given. Many of the areas discussed in this chapter are also in need of increased levels of basic research.

CHAPTER 6

Transport

Introduction

Transport is one of the main contributors to global greenhouse gas emissions (see Fig. 1.6). Much attention is therefore currently being given to methods by which these emissions can be reduced. This chapter starts by tracing some of the history of the development of various forms of transport currently used, almost all of which being based on the use of internal combustion engines.[a] Not only is road transport for both personal and commercial applications considered but the use of air and sea transport is also discussed. The chapter then discusses each type of transport in more detail, considering methods of improving operational efficiencies that are currently being introduced or have been proposed for future application. As it is likely that conventional systems will continue to be used for some years to come in at least in some applications, the important topic of the control of emissions from conventional automobiles is also discussed.

Historical development of mechanically driven transport

Electrical vehicles. Our ancestors relied for many centuries on horse-driven transport. However, in the late nineteenth century, following the development of the steam engine and the widespread establishment of the railways, the first efforts to produce motorised vehicles took place. Some efforts were made to create steam-driven transport. However, the first successful vehicles were electrically driven. Electricity had been known to exist for many centuries but it was the experiments by Franklin around 1752 that it began to be understood. Michael Faraday then showed in 1831 how electricity could be generated and this was followed by the introduction of the light bulb and the gradual electrification of our cities and towns. It was therefore to electricity that the early innovators in transportation turned.

---

[a] An internal combustion engine depends on the combustion of a fuel within the engine. This is distinct from the external combustion that occurs in steam engines used for railway traction and also for the operation of machinery (Chapter 1).

Sustainable Energy
https://doi.org/10.1016/B978-0-12-823375-7.00009-3

Fig. 6.1 The Groß-Lichterfeld tram (1882). *(https://upload.wikimedia.org/wikipedia/commons/9/90/First_electric_tram-_Siemens_1881_in_Lichterfelde.jpg.)*

The first operational electrical vehicles were trams. The first public tramway was opened in Lichterfeld near Berlin (Germany) in 1881, thus preceding the introduction of the first automobile in 1885 (see below). Fig. 6.1 shows a photograph of a tram from this system taken in 1882. It was constructed by Siemens and was 5 m long, weighing 4.8 t; it travelled at up to $40 \, \mathrm{km \, h^{-1}}$ and carried 20 people at a time. This tramline initially operated using electrical current supplied through the rails but the system was modified in 1883 to use overhead wires. Similar tram systems were installed in many countries and there still exist a number of such tramways, one of the oldest of these being the Volk's Electric Railway in Brighton (UK) that was first opened in 1883. Tram systems are limited by having to have permanent track installed and such tracks can cause significant problems, as encountered for example by the large cycling population in Amsterdam (The Netherlands), a city that is still serviced by a significant and very efficient tram network; this system first operated in 1875 using horse-drawn trams but it was converted to electrical operation between 1900 and 1906.

The problem of having to have rails was circumvented in many cities by introducing trolley buses drawing the required current through moveable poles as illustrated in Fig. 6.2. The first trolley buses were introduced in Berlin in 1882 and this was soon followed by services in other German cities. The first trolley-bus service in the UK was opened in Bradford in 1911 where it operated until 1972. Belfast (Northern Ireland) opened its system in 1936, this operating until 1968; a typical trolley bus from the Belfast operation is shown in Fig. 6.3[b] Trolley buses are still in operation in many cities

---

[b] Many of the road surfaces in Belfast were very smooth and the trolley buses were very silent; one of the reasons given for their disbandment was based on safety grounds as pedestrians often did not hear their approach.

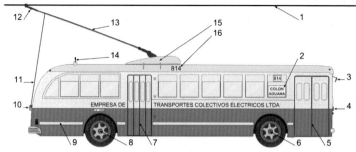

Fig. 6.2 The electrical connection of a trolley bus. 1: parallel overhead wires; 2: destination sign; 3: rear-view mirror; 4: headlights; 5: entry doors; 6: direction (turning) wheels; 7: exit doors; 8: traction wheels; 9: decoration; 10: trolley retractors; 11: pole rope; 12: contact shoes; 13: trolley poles (current collector); 14: pole storage hooks; 15: trolley pole housing; 16: bus number. *(https://en.wikipedia.org/wiki/Trolleybus.)*

Fig. 6.3 The last Belfast trolley bus to operate in 1968. *(https://en.wikipedia.org/wiki/Trolleybuses_in_Belfast.)*

throughout the world, particularly in the USSR, and there is currently a movement to reintroduce them in others.

To avoid the need for power cables, battery-operated vehicles were also developed but these had only limited success in the early attempts. Early electric buses operated using rechargeable batteries stored under the front seats and had limited ranges. Fig. 6.4 shows an electric bus developed by Edison that dates from 1915; this operated with Ni-Fe batteries. Other types of electric vehicle developed included the battery-driven milk float used for the delivery of milk and bread. Milk floats, which first appeared in use in the UK in 1889, could travel between 60 and 80 miles on one charge. Other uses of electric vehicles included road-sweepers and dust carts. We will return to

Fig. 6.4 Edison electric bus from 1915. *(https://upload.wikimedia.org/wikipedia/ commons/6/6e/Edison_electric_bus_from_1915.jpg.)*

the topic of electrical traction below when discussing modern electrically operated vehicles.

Vehicles with internal combustion engines. Although there had previously been several attempts to build engines relying on combustion, for example using mixtures of hydrogen and oxygen (Francois Rivaz, 1807) or burning towns gas (Samuel Brown, 1824), the first successful spark-ignition internal combustion engine, fuelled by coal gas, was invented and patented by Jean Lenoir in 1858.[c] However, it was not until 1876 that Nikolaus Otto invented and later patented a four-stroke engine operating with gasoline that he incorporated in a motor cycle. This design became the basis of all four-stroke engines. In 1885, Gottlieb Daimler built a two-wheeled vehicle ('reitwagen') with a four-stroke engine comprising of a vertical cylinder and a carburettor, the prototype for the modern engine; the following year, he adapted a stage-coach to be the first four-wheeled vehicle, this having the same type of internal combustion engine. That same year (1886), Karl Benz received a patent for the first gasoline-fuelled vehicle, a tricycle (see Fig. 6.5); he went on in 1889 to establish Benz et Cie, this company becoming the world's largest company producing automobiles by 1900. In the same period, Daimler founded the company Daimler Motoren-Gesellschaft to build and market his designs and in 1901 this company launched the Mercedes, this having been designed by Daimler's partner, Wilhelm Maybach.

The production of automobiles in the United States was dominated in the early years by Ford and an early Ford car is shown in Fig. 6.6. First

[c] http://www.energybc.ca/cache/oil2/inventors.about.com/library/weekly/aacarsgasa6fc5.htm.

Fig. 6.5 The original Benz Patent Motorwagen (1885). *(https://en.wikipedia.org/wiki/Benz_Patent-Motorwagen.)*

Fig. 6.6 A Ford Model-T in Geelong, Australia for the launch in 1915. *(http://www.slv.vic.gov.au/pictures/0/0/0/doc/pi000357.shtml.)*

launched as early as 1908, it was the most widely used four–seater car of its era and is considered to have been one of the most influential designs of the twentieth century. Automobile production in the US tended towards large vehicles with very high petrol consumptions, this being related to the low cost and easy availability of gasoline. A typical American car from mid-twentieth century is shown in Fig. 6.7[d]

In Europe, there were equivalent developments of large comfortable cars, examples being those produced by Mercedes Benz and Daimler mentioned above. However, the general preference was for smaller vehicles. The engines of these cars typically had higher compression ratios and this gave greater manoeuvrability on what were generally more congested roads. Following the Suez crisis and the resultant petrol shortages, the preference for

[d] The author was once a part-owner of a Buick of this vintage. It was roomy and comfortable but relatively unmanoeuverable.

Fig. 6.7 A Buick Super from 1957. *(https://upload.wikimedia.org/wikipedia/commons/5/50/Buick_Super_1957.jpg.)*

Fig. 6.8 The first Morris Mini-Minor, now housed in the British Motor Museum. *(https://upload.wikimedia.org/wikipedia/commons/2/2f/Morris_Mini-Minor_1959_%28621_AOK%29.jpg.)*

smaller more compact vehicles became more emphasised. This resulted in the development of even smaller and more efficient vehicles such as the Mini series, first introduced in the UK in 1959 by Morris and Austin; see Fig. 6.8.

Vehicles with diesel engines. The first diesel engines were used in stationary applications and were only later used in vehicle propulsion. Rudolf Diesel first patented the concept of the diesel engine in a US patent application lodged in 1895 (granted 1898, No. 608845), having previously lodged another application in 1892 based on an incorrect description of the principal of operation.[e] Fig. 6.9 shows his successful prototype that ran for a total of 88 revolutions in February 1894. Diesel worked with both

---

[e] There was considerable controversy at the time as to whether or not Diesel's ideas were novel but it is now generally accepted that he was the originator of the modern diesel engine.

Fig. 6.9 Experimental diesel engine from 1894. MAN-Museum, Augsburg. *(https:// upload.wikimedia.org/wikipedia/commons/c/c7/Experimental_Diesel_Engine.jpg.)*

Krupp in Essen and Machinenfabrik Augsberg and this collaboration led to the development of diesel–powered machines that were used in many larger applications in manufacturing facilities.

The first fully functional diesel engine, completed in 1896, is shown in Fig. 6.10. This engine was rated at 113.1 kW and had an efficiency of 26%. Subsequent developments that occurred quite rapidly included the construction of

Fig. 6.10 Historical diesel engine in Deutsches Museum. *(https://commons.wikimedia. org/wiki/File:Historical_Diesel_engine_in_Deutsches_Museum.jpg.)*

Fig. 6.11 The Mercedes-Benz 260D released in 1936. The earliest diesel was launched in 1933 by Citroen. *(https://en.wikipedia.org/wiki/History_of_the_diesel_car.)*

ocean-going ships and submarines powered by diesel engines and this was followed by the construction of tractors and lorries.

The first car with a diesel engine appeared in 1929 and various companies in Europe and in the United States started to produce different models in the 1930s. As a typical example, Fig. 6.11 shows a Mercedes Benz from 1936. Improvements in the design of the diesel engine led to the high-speed engines launched by Perkins and Chapman in 1932[f] and these were soon being used in racing cars and other applications such as tractors and farm machinery. Four diesel engines were used to power the Hindenburg airship. As will be discussed further below, a diesel engine operates with greater efficiency than does an Otto engine. It can also work efficiently with a variety of fuels, these including biofuels of the type described in Chapter 5. Because the ignition of the fuel occurs at relatively low temperature, the risk of catching fire is much lower than for an engine operating with spark ignition. As they have no ignition systems, the reliability of diesel engines is also markedly greater than of Otto engines.

The efficiency of internal combustion engines. Both Otto and diesel engines are classed as being internal combustion engines since the fuel is ignited within the cylinders of both types of engine and work is done by the expanded gaseous combustion products. (This contrasts with the operation of the *external* combustion engines of the type discussed in Chapter 1.) The operation of the Otto engine is now described and this is followed by a brief discussion of the operation of the diesel engine.

---

[f] See: https://en.wikipedia.org/wiki/Perkins_Engines.

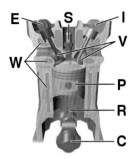

Fig. 6.12 The structure of an internal combustion engine. *(C—crankshaft; E—exhaust crankshaft; I—inlet crankshaft; P—piston; R—connecting rod; S—spark plug; V—valves; W—water-cooling jacket. https://en.wikipedia.org/wiki/Internal_combustion_engine.)*

The Otto engine. The operation of a typical four-stroke internal combustion engine is illustrated in Fig. 6.12, the form having changed little in the period since the original development of the Otto engine.

Each cylinder of the engine, which is water-cooled, contains a piston (P) attached to the crankshaft (C) of the engine by a connecting rod (R). As the piston first descends, it draws fuel (normally gasoline, although it could also be a gaseous molecule such as hydrogen or methane) into the cylinder through an inlet valve operated by the inlet crankshaft (I). When the fuel is gasoline, typically a non-aromatic hydrocarbon (C_nH_{2n+2}), it is vaporised due to the high temperature achieved in the cylinder and remains in vapour form while it is compressed by the piston while rising to the top of the cylinder once more. At that moment, the spark plug is fired, this causing the combustion of the gasoline vapour and resulting in a significant increase in the pressure in the cylinder due to the formation of water and CO_2:

$$C_nH_{2n+2} + O_2 \rightarrow nCO_2 + (n+1)H_2O$$

as well as from the temperature increase due to the evolution of the heat of combustion. The product gas forces the piston to descend once more. Finally, the piston rises again, this time with the exhaust valve open so that the gases pass to the exhaust system. The timing of the inlet and outlet valves as well as of the ignition spark is critical to efficient operation.

The efficiency of an Otto engine cannot exceed that predicted by the theoretically ideal Carnot cycle discussed in more detail in Box 6.1. The efficiency of such a cycle is given by:

$$\eta = W/Q_2 = (Q_2 - Q_1)/Q_2 = 1 - T_1/T_2$$

The energy to drive the motor arises from the work done by the expansion of the combustion products during the second cycle described above. The maximum efficiency achieved during this cycle is dependent on the upper and lower temperatures of the cycle. In practice, there are substantial losses of energy to the surroundings and the maximum theoretical efficiency is never achieved. Hence, on average, only 40%–45% of the energy supplied to such an engine is converted into mechanical work and used to drive the vehicle. A large proportion of the energy lost from the system is associated with heat released to the environment through the cooling system. Various

BOX 6.1 The Carnot cycle

Fig. 6.13 shows the idealised form of the reversible Carnot cycle. This cycle determines the upper limit of the efficiency of the cylinder of a thermodynamic engine such as that of the second cycle of the Otto design described in the main text, this process depending on the conversion of heat into work.[9] It can equally well apply to an engine working in the opposite direction as in a refrigeration system.

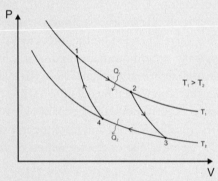

Fig. 6.13 The Carnot cycle for an Otto engine. *(https://upload.wikimedia.org/ wikipedia/commons/0/06/Carnot_cycle_p-V_diagram.svg.)*

The steps in can be described as follows:

Step 1. At this stage, which corresponds to that when ignition has occurred and the product gas is compressed at the top of the cylinder at high pressure, the gas expands isothermally (at constant temperature, T_1) by moving the piston and therefore does work against the surroundings. During this stage, heat (Q_1) is being transferred to the gas from the hot surroundings. The entropy of the gas, ΔS_1, is thus increased by Q_1/T_1.

Step 2. The gas is now insulated against loss or gain of energy to or from the surroundings and expands adiabatically, moving the piston further and losing

Continued

BOX 6.1 The Carnot cycle—cont'd

internal energy equivalent to the additional work done on the surroundings. There is no entropy change in this step.

Step 3. The gas is now in contact with the surroundings at the lower temperature T_2. The surroundings do work on the piston, pushing it back towards the top of the cylinder. This causes the transfer of heat (Q_2) out of the cylinder and there is a corresponding decrease of the entropy, ΔS_2, given by Q_2/T_2.

Step 4. The gas is again insulated from loss or gain of energy and the surroundings do work on the piston, pushing it back to the starting point of the cycle. Again, there is no entropy change associated with this step.

Hence, when the cycle is complete,

$$\Delta S_1 = \Delta S_2 \text{ or } Q_1/T_1 = Q_2/T_2$$

The efficiency, η, of the cycle is given by:

$$\begin{aligned} \eta &= \text{work done}/Q_H \\ &= (Q_H - Q_C)/Q_H \\ &= 1 - T_1/T_2 \end{aligned}$$

The efficiency therefore depends on the temperature difference achieved: the higher that T_2 is compared with T_1, the higher is the efficiency. In practice, complete isolation of the cylinder containing the piston from its surroundings can never be achieved and there will also inevitably be losses of energy due to frictional forces.

[g] For a full description of the Carnot cycle and it applications, the reader should consult web sources such as https://en.wikipedia.org/wiki/Carnot_cycle; https://en.wikipedia.org/wiki/Perkins_Engines.

ways to recover some of the waste heat have been developed, such as the use of a supercharger to introduce compressed air into the cylinder, thereby increasing the efficiency of the process. Efficiency is also improved if the compression ratio is relatively high, thus causing increased engine temperatures, but that requires higher octane fuel to avoid pre-ignition or engine 'knocking'. For this reason, the high compression ratio cars used in Europe and running on relatively high octane fuel generally have better petrol consumption behaviour than the larger vehicles with lower compression ratios and more sluggish performance that have been favoured until relatively recently in the United States.

The diesel engine. The diesel engine operates on a completely different principle to that of a four-stroke Otto engine. Ignition in the cylinder of a diesel engine is brought about by using highly compressed hot air rather than a spark and uses a very different fuel. The cycle which occurs is shown in Fig. 6.14.

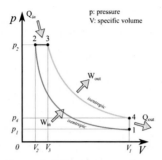

Fig. 6.14 The Diesel Cycle. The cycle follows the numbers 1–4 in a clockwise direction. The horizontal axis is volume of the cylinder. In the diesel cycle, the combustion occurs at almost constant pressure. In this diagram, the work that is generated for each cycle corresponds to the area within the loop. *(https://en.wikipedia.org/wiki/Diesel_engine#/media/File:DieselCycle_PV.svg.)*

The cycle begins at *point 1* where air is introduced into the cylinder of volume V_1. It is then compressed to volume V_2 when the pressure is now p_2, this compression step heating the air and providing energy Q_{in}. At that stage (*point 2*), the diesel fuel is injected into a space at the top of the cylinder in such a way that the liquid is broken down into small droplets and distributed evenly throughout the void space. Molecules of the fuel evaporated from the droplets then ignite spontaneously, the rate at which this occurs depending on the taste of evaporation, until all the fuel is combusted at *point 3*. (The combustion step occurs suddenly and causes the knocking sound that is characteristic of the diesel engine.) When combustion is complete at what is essentially constant pressure, the product gas expands and the piston descends once more to *point 4*.[h] The exhaust valve is opened at 4 and the product exhaust gas is then expelled, there being a decrease of pressure upon returning to *point 1*. An additional cycle that is not shown then follows during which the exhaust gases are expelled from the cylinder and a fresh charge of air is introduced before the combustion cycle is repeated. The valve arrangements of the diesel engine are similar to those in the Otto engine shown in Fig. 6.12.

The high level of compression used in a diesel engine gives a very significant improvement in efficiency when compared with the Otto engine. For this reason, the fuel consumption figures for diesel engines always exceed those for spark ignition engines of equal capacities; see Box 6.2. However, although the diesel engine has a theoretical efficiency of 75%, this

[h] The term isentropic shown in Fig. 6.14 refers to an idealised thermodynamic process that is both adiabatic and reversible; there is no transfer of heat or matter.

BOX 6.2 Fuel consumption figures and CO_2 emissions

In Europe, the most efficient petroleum-fuelled cars are claimed to give petrol consumptions approaching 60 miles per imperial gallon while the most efficient diesel-powered cars can give significantly higher efficiencies: greater than 70 miles per imperial gallon. In contrast, the larger cars favoured in the US generally give much worse fuel consumptions; a recent report shows that the average vehicle produced by the Ford Motor Company has a consumption of 22.5 miles per US gallon (27 miles per imperial gallon). To convert a figure in miles per gallon (mpg) to actual CO_2 emissions (g/km), for a *petrol fuelled car*, divide 6760 by the mpg figure and for a *diesel car*, divide 7440 by the mpg value; for two cars each doing 40 mpg, the values are 169 g/km (petrol) and 189 g/km (diesel). A simple rule of thumb is the following: the combustion of one US gallon of gasoline (this containing about 87% carbon) causes the emission of 20 pounds of CO_2. (www.climatekids.nasa.gov).

is never achieved as a result of heat losses through the cylinder walls and elsewhere in the system. A modern diesel car in practice has an efficiency up to about 43% while larger engines such as those in trucks and buses have slightly higher efficiencies (45%). The diesel engine has a number of operational advantages over the spark–ignition engine, including the fact that they tend to be more efficient at low loads and in consequence are often used for vehicles making short journeys with frequent stops. As is shown in later sections, their emissions are also in principle more easily controlled than those from Otto engines.[i] Further, they can be operated using a variety of different types of fuel, these including biofuels (Chapter 5). For further details of the operation of cars with diesel engines, the reader should consult articles such as https://en.wikipedia.org/wiki/Diesel_engine.

Exhaust emission control

Fig. 6.15 shows a typical layout of a modern car, this being largely self-explanatory. Apart from the internal combustion engine, essential features include a fuel injection system and an electronic control module that provides full control of all aspects of the operation of the vehicle. The fuel tanks are mostly sited under the rear seats and there is therefore a fuel line running

[i] Although, as discussed in a previous section, European cars tend to operate at higher compression ratios that their US counterparts, the level of compression is still lower than that in the diesel engine. This is because the use of any higher compression ratios would lead to pre-ignition and uneven operation.

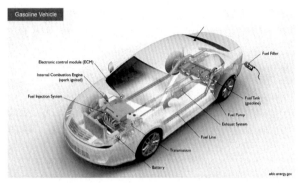

Fig. 6.15 A gasoline vehicle. *(https://afdc.energy.gov/vehicles/how-do-gasoline-cars-work.)*

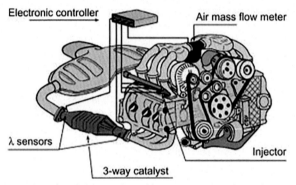

Fig. 6.16 The positioning of the catalytic filter in an internal combustion automobile engine. *(From (2003). Catal. Today, 77, 419–449, reproduced with kind permission of Elsevier.)*

under the main passenger compartment. The other very important feature of all vehicles is the exhaust system that is also found under the vehicle. This section describes in outline the methods that are used to control engine emissions.[j]

Modern vehicles with internal combustion engines always include a catalytic emission control system and this is placed between the exhaust manifolds of the engine and the silencer unit as shown in Fig. 6.16. An electronic controller is included in the system that is linked to two sensors, one before and one after the catalyst unit, and a control unit to adjust the flow of air to

---

[j] This section only gives a rudimentary description of the operation of a catalytic exhaust converter. A more detailed description of the catalysts used is to be found in Contemporary Catalysis - Fundamentals and Applications, Julian R.H. Ross, Elsevier (2019).

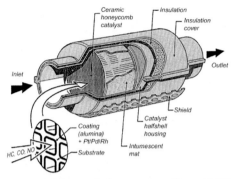

Fig. 6.17 The arrangement of the catalytic converter. *(From (2001). Appl. Catal. A, 221, 443–457, reproduced with kind permission of Elsevier.)*

the engine. The structure of a catalytic converter unit is shown schematically in Fig. 6.17. The catalyst support material, which may be in the form of either a metallic foil or ceramic honeycomb structure such as those illustrated in Fig. 6.18, is surrounded by insulation material held within a cylindrical container.

The chemical reaction occurring within an internal combustion engine of either the Otto or diesel type is the total oxidation of the hydrocarbon fuel to produce CO_2 and water. However, the product gases inevitably contain traces of unburnt hydrocarbon fuel (or lower molecular weight fragments formed from the fuel) plus carbon monoxide and the oxides of nitrogen, N_2O, NO and NO_2, the latter mixture described for convenience as NO_x. (The proportion of N_2O is generally very low.) The quantities of each of these emissions depend on the operating conditions. Fig. 6.19 shows very qualitatively how their concentrations depend on the air/fuel ratio in the

Fig. 6.18 Typical honeycomb supports used in catalytic converters; *left*: metal foil; *right*: ceramic monolith. *(From (2011). Catal. Today, 163, 33–41, reproduced with kind permission of Elsevier.)*

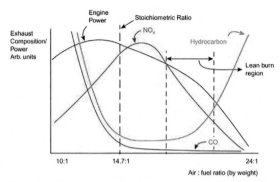

Fig. 6.19 The effect of air-fuel ratio on the operation of a four stroke Otto engine. *(From Julian R.H. Ross, Contemporary Catalysis, 2019, reproduced with kind permission of Elsevier.)*

feed to the engine; also shown in this figure is the engine power as a function of the air/fuel ratio. The stoichiometric ratio value of 14.7:1 corresponds to the optimum combustion conditions. (Below this value, the atmosphere is effectively a reducing one and above it is an oxidising one.) CO is formed at lower air/fuel ratios when there is insufficient oxygen to provide complete combustion but drops off to a relatively low level at higher ratios. The hydrocarbon component in the exhaust follows a similar path to that of CO at low air/fuel ratios but rises again at values well above the stoichiometric ratio due to there being uneven combustion of the fuel under these conditions, this being generally associated with 'knocking' of the engine. The NO_x is formed by gas–phase reactions between the nitrogen and oxygen of the feed gas at the high temperatures in the ignition stage of the cycle, the reaction being most pronounced when the temperature is at its highest near the stoichiometric composition. The engine power produced is also shown, this being at its maximum slightly below the stoichiometric air/fuel ratio; the power drops off significantly at higher values, particularly in the so-called 'lean-burn' region.

Strict air-purity control regulations were introduced in many countries in the second half of the last century. Prior to the introduction of these environmental controls, many car manufacturers, particularly in the US, produced engines that operated in the lean-burn region of Fig. 6.19 with which the NO_x emissions are relatively low and the formation of CO is negligible. However, with the introduction of the new strict regulations, the car manufacturers moved towards the production of engines operating with an air/fuel ratio slightly below the stoichiometric ratio so that the exhaust gas contains a mixture of the CO, hydrocarbon and NO_x. (This value was still above that corresponding to maximum engine power in Fig. 6.19 where the emission of CO and unburnt hydrocarbon is very high.) So-called three-way catalysts, generally containing the precious metals Pt, Pd and Rh as discussed above, were developed to oxidise the unburnt hydrocarbon and CO while at the same time reducing the NO_x component to nitrogen gas. For further details, see for example the reference of footnote 'j'.

With increasingly exacting environmental requirements, a number of car manufacturers returned to the concept of using lean-burn engines, these operating under oxidising conditions corresponding to the region with the higher air–fuel ratios shown in Fig. 6.19. In this region, both the CO and any unburnt hydrocarbon can easily be oxidised using a simple Pt catalyst. However, the NO_x remains unaffected. One solution to the control of these NO_x emissions in the oxygen-containing atmosphere of the lean-burn engine was introduced by Toyota who developed the use of NO_x traps. These traps contain BaO to adsorb the NO_x content of the exhaust gas by forming $Ba(NO_3)_2$. A schematic representation of these traps is shown in Fig. 6.20. The traps are regenerated periodically by cycling the system briefly to fuel-rich conditions in order to convert the nitrate species formed back to the original BaO.

The diesel engine also operates in the lean-burn region of but has the advantage that combustion occurs at a lower temperature than that in the Otto engine. The main gaseous emissions from a diesel engine are therefore low concentrations of unburnt fuel as well as low NO_x concentrations. However, these emissions have associated with them particulate matter that result from incomplete combustion of the fuel. The low hydrocarbon concentrations in the exhaust gases can be controlled by using a simple oxidation catalyst. However, while the particulate matter can be collected in simple filter assemblies of the type illustrated in Fig. 6.21, these filters will become gradually blocked, this resulting in deterioration of the engine performance. Regeneration of the filters can, at least in theory, be achieved reasonably simply by heating them to a sufficiently high temperature in an oxygen-containing stream, this being termed 'active regeneration'. However, such combustion is difficult to control without damaging the filter and so another method of controlling the build-up, termed 'passive regeneration', is generally used. For this, a catalyst is included within the filter (see Fig. 6.22) that oxidises any NO in the exhaust stream to NO_2. The NO_2 then catalyses the oxidation of the soot that is collected in the filter. The latter passive regeneration system operates without the need for active control involving periodic changes of the reaction conditions.

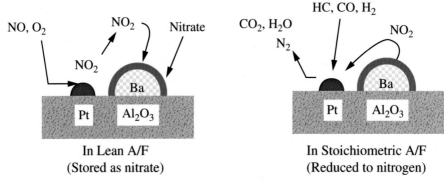

In Lean A/F
(Stored as nitrate)

In Stoichiometric A/F
(Reduced to nitrogen)

Fig. 6.20 The Toyota model for the NOx storage catalyst. *(S. Matsumoto, Catal. Today, 90 (2004) 183-190, reproduced with kind permission of Elsevier.)*

Fig. 6.21 Schematic representation of a ceramic filter for the removal of particulates from a diesel exhaust stream. *(M.V. Twigg, Catal. Today, 163 (2011) 33–41, reproduced with kind permission of Elsevier.)*

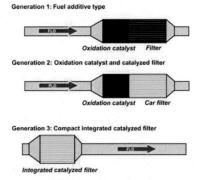

Fig. 6.22 Three different catalysed filter systems for the removal of particulates from a diesel exhaust stream. *(M.V. Twigg, Catal. Today, 163 (2011) 33–41, reproduced with kind permission of Elsevier.)*

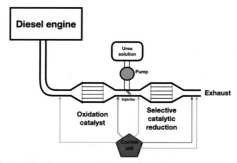

Fig. 6.23 A typical diesel selective catalytic reduction system combined with a diesel oxidation catalyst.

With increasingly strict emission regulations, there is now also a need to control the NO_x emissions from diesel engines. Under the oxidising conditions of the diesel exhaust, selective catalytic reduction using three-way catalysts as used in Otto engines cannot be used. In consequence, NO_x trapping of the type described above can be used. An alternative method of removing the NO_x is to carry out selective reduction using ammonia as reductant.[k] This process is similar to that used in the selective reduction of NO_x in the oxidising atmospheres of the exhaust gases from power stations. Fig. 6.23 shows such a selective reduction system for use in diesel systems.

The diesel exhaust gas is first passed through an oxidising catalyst to remove any unburnt hydrocarbons and CO from the stream and also to convert any NO to NO_2. The reductant is then introduced, generally as a solution of urea. The urea decomposes thermally to produce ammonia according to the equation:

$$(NH_2)_2CO \rightarrow NH_3 + HNCO$$

The isocyanic acid formed as a byproduct is then being hydrolysed to form additional ammonia by the reaction:

$$HCNO + H_2O \rightarrow NH_3 + CO_2.$$

The reaction occurring in the selective catalytic reduction bed, which contains either a zeolite or a vanadia-containing catalyst similar to those used in the selective reduction of NOx emissions from power stations, is:

$$6\,NO_2 + 8\,NH_3 \rightarrow 7\,N_2 + 12\,H_2O$$

---

[k] See for example, 'The pollutant emissions from diesel-engine vehicles and exhaust after treatment systems', I.A. Resitogglu, K. Altinisik and A. Keskin, Clean Techn. Environ. Policy 17 (2015) 1715–27. (This extensive review is open access.)

The all–over performance of the selective reduction unit is monitored and controlled by a series of temperature sensors and a pair of NO_x sensors. Because of the need to have a supply of urea in order to carry out the selective reduction, such systems are most commonly found only in larger engines such as those of lorries and other heavy-duty vehicles. However, some automobile manufacturers are currently beginning to introduce similar systems into light-duty vehicles. It should be noted that many of the catalytic systems discussed above encounter problems if the fuel used contains any significant quantity of sulphur-containing impurities as sulphur is a non–reversible poison for most catalysts. The introduction of the technologies discussed above have therefore been associated with a move by the petroleum refining industry towards the production of low–sulphur containing fuels with the aim of obviating catalyst poisoning problems. The introduction of stricter and stricter regulations have however led to some serious difficulties in the control of diesel emissions that have resulted in the reduction of sales of diesel automobiles: see Box 6.3.[1]

Hybrid vehicles

With an increased awareness of the need to decrease global greenhouse gas emissions and strengthening regulations, car manufacturers started to examine methods of improving the fuel consumption behaviour of their vehicles. While many started to concentrate on the production of lean-burn petroleum- and diesel-fuelled vehicles, Toyota introduced the world's first hybrid vehicle, their Prius model, at the Tokyo Motor Show of 1995. The first production models of the Prius went on sale in December 1997 but international sales only started in 2000. Shortly after the Toyota development, Honda produced their Insight, this being launched at the Tokyo Motor Show in 1997; the Insight became generally available in Japan in 1999 and it was launched in the US in 2000, becoming the first hybrid vehicle available there.

A hybrid vehicle such as the Prius has two batteries: the 'auxiliary battery', with low voltage (12 V) which is used for starting the car and providing initial power for the on-board computer; and the second 'traction battery' (with a voltage over 250 V) of relatively low capacity which is used to power the electric traction motor. This battery is charged during braking of the car ('regenerative braking') and the generation process at the same time supplies

[1] See for example: https://en.wikipedia.org/wiki/Volkswagen_emissions_scandal.

BOX 6.3 Dieselgate

In 1999, the US introduced new Tier 2 rules decreasing over a period of several years the limit for NOx emissions from 1.0 g/mile to 0.07 g/mile. It transpired in 2015, by which time the new US limits applied, that Volkswagen (who had adopted a lean-NO_x trap approach rather than selective reduction methods) had been unable to meet the new standards for their turbocharged direct injection (TDI) diesel engines under road conditions. In consequence, they had introduced a 'defeat device' linked to the control system of the engine that allowed it to operate with acceptable limits under bench-test conditions but bypassed the system for the remainder of the time (see Fig. 6.24). The ensuing scandal caused an immediate drop in the Volkswagen share value (almost 40% drop over a period of 3 weeks) and a series of dismissals of personnel and court cases. It later transpired that very few brands of diesel engines were able to meet the Tier 2 standards under road conditions. Manufacturers affected included Volvo, Renault, Mercedes, Jeep, Hyundai, Citroen, BMW, Mazda, Fiat, Ford and Peugot. It was also shown that many brands of diesel cars failed to meet similar European road-test standards.

Fig. 6.24 A 2010 Volkswagen Golf TD1 fitted with a 'defeat device' exhibited in the Detroit Auto Show. *(https://commons.wikimedia.org/wiki/File:VW_Golf_TDI_Clean_Diesel_WAS_2010_8983.JPG.)*

Although it is possible using the selective reduction approach described able to produce diesel engines that emit very low concentrations of NOx, the attitude of the general public to diesel cars has been seriously affected. The consequence of this is that there has been a very concerted move by a number of the car manufacturers to produce battery-powered vehicles, a topic discussed later in this chapter.

a proportion of the braking capacity required, giving a further energy saving. The traction battery is used to power a secondary electric motor which is engaged at startup and also at lower speeds of up to 25 mph (40 km/h). The secondary motor is also used as a source of additional power for acceleration or climbing, benefitting from the higher efficiency of the electric motor compared with that of the main petrol engine. The electric motor automatically takes over as the vehicle slows down to stop and it shuts off when the vehicle is stationary, thus helping to economise on fuel consumption. The hybrid concept gives significantly improved all-over fuel economy. The Honda Insight, which operates with a primary lean-burn engine,[m] was rated as the most efficient gasoline-fuelled vehicle in the US in 2014 by having a road fuel usage of 61 miles per US gallon (73 miles per imperial gallon or 3.9 L/100 km).

The battery of a hybrid vehicle such as the Prius or Insight (or their equivalents from other manufacturers) is never completely charged, nor is it fully discharged: the range of charge is normally in the range 40%–60%; a computer-controlled operation system ensures that the level of charge never gets outside the range 38% to 82% and this effectively extends the useful life of the battery well beyond the operating age of a normal vehicle. With the Toyota Prius, the battery is of the nickel metal hydride type (see Chapter 7), made by the Panasonic EV Energy Company. If the petrol motor is not operated and the speed is kept below 25 mph, the vehicle can be driven using only the battery for very short journeys but such use would be very uncommon. It is worth noting that a number of other types of hybrid systems have also been designed, for example using compressed air as an energy source. However, such concepts are not discussed here as they do not appear to have been put into production.

Plug-in hybrid vehicles

A marked improvement on the simple hybrid design of the Toyota Prius and other equivalent vehicles such as the Honda Insight discussed above was the introduction of the 'Plug–in Hybrid' shown schematically in Fig. 6.25. Such a vehicle has a somewhat larger traction battery, normally of the Li–hydride variety (Chapter 7), and this is coupled to a relatively small electric traction

---

[m] Honda and a range of other manufacturers are now using so-called Atkinson cycle engines in their hybrid vehicles. This cycle operates in a similar way to that of the Otto engine but with a shorter compression stroke and a longer expansion stroke, this giving improved fuel economy. (See, for example: https://en.wikipedia.org/wiki/Atkinson_cycle.)

Plug-in Hybrid Electric Vehicle

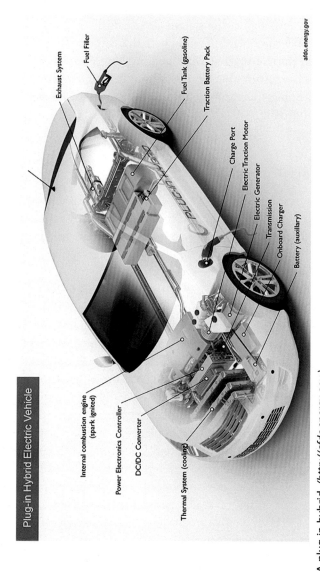

Exhaust System

Fuel Filler

Fuel Tank (gasoline)

Traction Battery Pack

Charge Port

Electric Traction Motor

Electric Generator

Transmission

Onboard Charger

Battery (auxiliary)

Internal combustion engine (spark ignited)

Power Electronics Controller

DC/DC Converter

Thermal System (cooling)

afdc.energy.gov

Fig. 6.25 A plug-in hybrid. (*http://afdc.energy.gov.*)

motor. The great advantage of this arrangement is that it is possible to travel significantly further when using only the stored electrical energy fed to the car from an external source. This means that it is not necessary to recharge the battery before driving further when the charge runs low since the system then changes automatically to use of the petroleum-fuelled engine. When operating with this engine, the system also gradually recharges the battery. The first Toyota Prius Plug-In Hybrid with this conformation had a range of only 18 km of normal driving without using any petroleum fuel but this figure has been significantly improved in subsequent models so that currently available plug-in hybrids do not require the use of any petroleum fuel for most local driving.[n]

Battery electrical vehicles

Automobiles powered solely by batteries are very recent developments although battery power was used in specialised vehicles such as milk floats and fork-lift trucks during the previous century, as discussed earlier in the chapter. The sudden rise in the popularity of battery-powered cars in the last 15–20 years is such that the global share of such vehicles is expected to rise to more than 20% by 2030, having only been 2% as recently as 2016. Some countries have announced that they intend to ban the sale of petroleum-powered vehicles in the coming years and many have introduced incentives for the purchase of battery-driven cars. The increased importance of electric vehicles is associated with the very rapid developments of relatively inexpensive and reliable batteries such as those that are currently used in the plug-in hybrid vehicles discussed above. These improvements have been coupled with rapidly increasing battery storage capacities and corresponding increases in driving distances before recharging is necessary. Also contributing to the surge in popularity has been the wide-ranging installation of recharging facilities and the development of equipment to give much more rapid charging. The use of modern rechargeable batteries has been extended to other forms of road transport such as trucks, buses, trains, scooters and bicycles as well as aircraft. For a full discussion of the construction and operation of such batteries, see Chapter 7.

[n] Another variant that does not appear to have operated until now is a battery powered vehicle with a small low-powered petrol engine that is used only to recharge the battery as the electrical charge is used up. If such an engine was fuelled with a biomass-derived fuel and the electricity for the battery was provided from totally renewable sources, this type of vehicle would be completely 'zero emission'.

A number of companies manufacture all-electric vehicles. The best-selling models come from Tesla and Nissan: in March 2020, 500,000 units of the Tesla Model 3 had been delivered and the Nissan leaf passed the 500,000 mark in December 2020. Other successful manufacturers include Hyundai, Jaguar, Kia, a Renault-Nissan-Mitsubishi alliance, BMW and Volkswagen. Electric cars are also manufactured in China by the BAIC Group (Beijing Automotive Industry Holding Company Ltd.), SAIC-GM-Wuling Automobile (a joint venture including General Motors) and Chery (Chery Automobile Co. Ltd.). Fig. 6.26 shows the layout of such a vehicle while Fig. 6.27 is a photograph of the chassis of a Tesla model (Model S) showing the traction engine (nearest camera) and the battery compartment. There has been a very steady increase in the range of battery-powered electrical vehicles. Table 6.1 lists the ranges of some of the models that were available commercially in the UK in early 2021. The Tesla models have the highest ranges in the listing, with values that are comparable with those of the most efficient petrol cars. However, the other lower priced models still have significant ranges that begin to make them very competitive with internal combustion engines. Short recharging times are also important and Tesla has been at the forefront in introducing rapid charging stations for its models, particularly in California; with the fast-charging units, it is claimed that the batteries will reach 80% power in 30 min. Nevertheless, charging times for most models are anything between 4 and 8 h, these values being quoted for the BMWi3 and the Nissan Leaf.[o]

Another very important factor in relation to the use of electric vehicles is whether or not the electricity used for charging the vehicle is from a fully renewable source or not. For this reason, countries or regions in which the proportion of renewable electricity is high are more favourable locations for the use of such vehicles. A recently published report[p] shows that if only 50% of the electricity available in the grid system is renewable, the use of plug-in hybrid vehicles gives a greater reduction in CO_2 emissions than does the use of fully electrical vehicles. Table 6.2 summarises some of these conclusions; it should be noted that these data also take into account the

---

[o] These data were taken from the following site: https://youmatter.world/en/hydrogen-electric-cars-sustainability-28156/ but similar results are to be found on other equivalent sites. The 'youmatter.world' site is also a useful source of additional material relevant to the content of this book.

[p] 'The European Environment - State and Outlook 2020', European Environment Agency (2019), ISBN 978-92-9480-090-9. The full report is downloadable from http://europa.eu; https://www.forbes.com/sites/davidrvetter/2021/01/25/its-official-in-2020-renewable-energy-beat-fossil-fuels-across-europe/?sh=fa3872d60e83.

All-Electric Vehicle

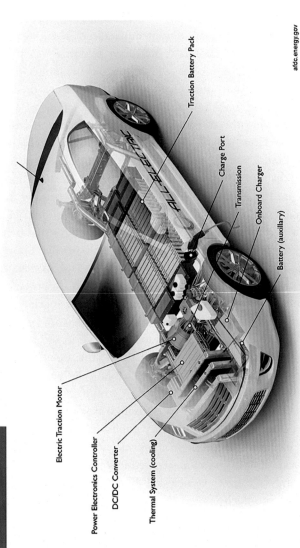

Electric Traction Motor

Power Electronics Controller

DC/DC Converter

Thermal System (cooling)

ALL-ELECTRIC

Traction Battery Pack

Charge Port

Transmission

Onboard Charger

Battery (auxillary)

afdc.energy.gov

Fig. 6.26 Schematic representation of an electric vehicle. *(http://afdc.energy.gov.)*

Fig. 6.27 A Tesla Motors Model S base. *(https://upload.wikimedia.org/wikipedia/commons/f/f3/Tesla_Motors_Model_S_base.JPG)*

Table 6.1 A comparison of the ranges of some of the models of electric cars now available showing the top ten best ranges.

Manufacturer/Model	Maximum range/miles	Price/£
Tesla Model S Long Range	379	77,980
Tesla Model 3 Long Range	348	46,990
Tesla Model X Long Range	314	82,980
Jaguar i-Pace	292	64,495
Kia e-Niro	282	36,495
Hyundai Kona Electric	278	3890
Mercedes-Benz EQC	259	65,720
Audi e-tron	239	58,900
Nissan Leaf e$^+$	239	35,895
BMW i3	193	37,480

(Source of data: Car Magazine, 4 January 2021. https://www.carmagazine.co.uk/electric/longest-range-electric-cars-ev/.)

green-house gas emissions associated with the production of both the fuel used and the vehicle as well as that expected for its eventual recycling at end of life. Only if the vehicle is used in regions where the percentage of renewable electricity production exceeds the average value for the EU (about 58% in 2020[q]) does a fully electric vehicle offer an advantage over the plug-in hybrid vehicle when all the factors regarding its life-time operation are taken into account. As an example of a country where the percentage of renewable electricity is high is Norway which produces 98% of its

[q] https://www.forbes.com/sites/davidrvetter/2021/01/25/its-official-in-2020-renewable-energy-beat-fossil-fuels-across-europe/?sh=fa3872d60e83.

Table 6.2 Life-cycle emissions of CO_2 for a series of different vehicles and fuel types.

Vehicle type	Vehicle production and disposal/CO_2 emissions (g/km)	Fuel production / CO_2 emissions (g/km)	CO_2 exhaust emission/ CO_2 emissions (g/km)	Total life-cycle CO_2 emissions /(g/km)
Petrol	40	25	170	235
Diesel	40	30	130	200
Plug-in Hybrid	55	25	90	170
Battery Electric/100% renewable	60	20	0	80
Battery Electric/EU Average Renewable Electricity	60	115	0	175
Battery Electric/100% Coal Generation	60	240	0	300

(Data from 'The European environment – state and outlook 2020', European Environment Agency (2019), ISBN 978-92-9480-090-9 (http://europa.eu).)

electricity using hydropower; as a result, the proportion of new electric vehicles there is very high, reaching 54% of the total sales in 2020.

Fuel cell vehicles

As discussed above, the relatively long recharging time of an electric vehicle is a drawback in many instances since a longer journey has to be well planned to allow relatively long stop-overs to achieve reasonable levels of recharge. An alternative concept, now being marketed in California and in a limited number of other locations, is the fuel-cell vehicle. For these, the fuel is hydrogen gas and the range of a typical model is roughly equivalent to that of a conventional internal combustion engine powered vehicle. The topic of fuel cells is discussed more fully in Chapter 7. With pure hydrogen as fuel,[r] the fuel cell has the great advantage that it produces only water as product and so the problem of green-house gas creation relates back to the method of production of the hydrogen: whether or not it is 'green'. Important factors of relevance to the operation of fuel cell vehicles are therefore the availability of green hydrogen and its cost. As shown in Fig. 6.28, very few countries or regions have yet established a suitable infrastructure for the widespread sale of hydrogen and relatively few filling stations have been created; only a proportion of these are available for public use, the remainder being for the use of private fleet vehicles. The figure also shows the planned introduction of further refuelling stations, the majority of the future installations being in the EU and Asia (especially China and Korea).

Fig. 6.29 shows the layout of a typical fuel cell automobile. The fuel cell stack is relatively small compared with the battery storage compartment shown in Fig. 6.26. The hydrogen used as fuel is stored as a gas at a pressure of about 700 atm (70 MPa) and this requires an especially strong fuel tank. The tank used in the Toyota Mirai, one of the currently available commercial vehicles, has a three-layer structure made of plastic that is reinforced by carbon fibre and various 'other materials'. Toyota say that they have introduced a high-capacity converter to boost the voltage of the system to 650 V, thereby enabling a decrease of the size of the fuel cell stack compared with a previous model that had previously been available for leasing only in California.[s] The advantage of the fuel-cell vehicles compared with battery

[r] As discussed in this chapter, fuel cells can also be operated using hydrogen generated in-situ from a fuel such as methanol. Although prototypes of a wide variety of vehicles using methanol fuelled fuel cells have been produced and operated, they are not in general operation.
[s] https://www.toyota.ie/world-of-toyota/articles-news-events/2014/mirai-fuel-cell.json.

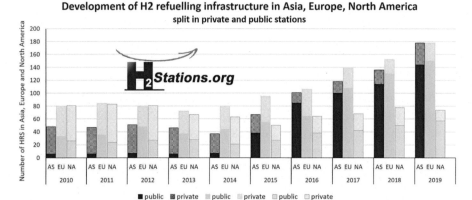

Fig. 6.28 The development of the numbers of hydrogen filling stations in Asia, the European Union and North America from 2011 to 2019. *(www.H2stations.org. Reproducedwith kind permission of Ludwig-Bölkow-Systemtechnik GmbH (LBST).)*

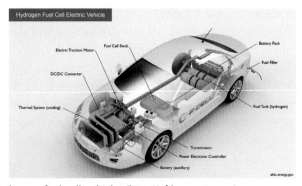

Fig. 6.29 Hydrogen fuel cell vehicle. *(http://afdc.energy.gov.)*

electric vehicles is that the tank can be refilled in less than 5 min and the range is of the order of 300 miles, a value comparable to that of a petroleum-fuelled vehicle.

Concluding remarks

This chapter has shown the development of different means of transport, with particular attention being given to the development of the personal automobile and of low-emission technologies used for them. In most situations it would appear that plug-in hybrid cars are currently among the most environmentally friendly in terms of both contaminants and CO_2 emissions although all-electric battery vehicles win out in regions where a significant proportion of the electric supply is produced from renewable sources. Battery vehicles will also be preferable if the home-owner is equipped to generate the required electricity, for example by using a photovoltaic system, and if the potential range of the vehicle between recharging is adequate for the owners requirements. Despite much publicity regarding the use of hydrogen as a fuel for automobiles and the relatively long history of the development of fuel cells suitable for such use, the lack of an adequate distribution network for hydrogen is a major drawback to the acceptance of hydrogen powered vehicles. Sales of fuel-cell powered cars are therefore likely to be restricted to the very small number of locations where public hydrogen-fuelling stations are situated. Hydrogen is more likely to be used as a fuel in larger vehicles operated from a privately operated hydrogen supply depot and it is probable that there will be significant developments of such usage. Batteries and fuel cells are also very important in relation to energy generation and also to the storage of electrical power. The following chapter (Chapter 7) deals in more detail with fundamental aspects of the development of batteries and fuel cells and discusses their uses for transportation and other purposes.

CHAPTER 7

Batteries, fuel cells and electrolysis

Introduction

With the increasing availability of renewable electrical energy from a wide variety of sources such as wind and photovoltaic arrays, there is an increasing demand for methods of storing this energy in such a way that excess generated in periods of peak production can be stored and then used at other times. Both batteries and fuel cells have important contributions to make in such storage and usage strategies. This chapter addresses some aspects of these two related technologies. It starts with a brief history of their discoveries and developments, proceeds to an outline of some of the background electrochemical science of various battery systems and fuel cell devices, and finally discusses some related electrolysis processes. As some of the uses of batteries and fuel cells in transport applications, particularly in automobiles, were discussed in Chapter 6, the focus here is on other applications.

The Volta pile, Faraday and the electrochemical series

The Italian Physicist, Alessandro Volta, was the first to demonstrate (in 1799) that an electric current could be produced in an external circuit when alternating discs of copper and zinc were placed one on top of another, each layer being separated by either a cloth or a piece of cardboard soaked in brine. The structure of this 'voltaic pile' is illustrated in Fig. 7.1. Volta believed that this current was due to a voltage difference between the metals but Humphry Davy showed that the electromotive force (EMF) that was produced was due to the occurrence of a chemical reaction. When a current flows through the cell, metallic zinc at the surface of a zinc layer (the 'anode') is oxidised to Zn^{2+} and two free electrons are liberated into the metal:

$$Zn \rightarrow Zn^{2+} + 2\,e^-$$

Sustainable Energy
https://doi.org/10.1016/B978-0-12-823375-7.00003-2

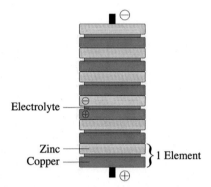

Fig. 7.1 A voltaic pile. *(https://upload.wikimedia.org/wikipedia/commons/0/06/Voltaic_pile.svg.)*

These electrons pass through the external circuit to the copper ('cathode') where they react at the surface with H+ ions from the brine 'electrolyte':

$$2H + 2e^- \rightarrow H_2$$

The allover reaction can thus be written as:

$$Zn + 2H^+ \rightarrow Zn^{2+} + H_2$$

When no current is drawn from the pile, each cell of the stack generates a voltage of 0.76 V with a brine electrolyte. With the stacking shown in Fig. 7.1, six cells in series, the total EMF produced by the stack is $0.76 \times 6 = 4.56$ V.

In this example, the copper of the stack does not take part in the allover chemical reaction but only acts as a catalyst for the proton transfer reaction since hydrogen is evolved at the Cu electrode which remains inert. It was subsequently shown that many other materials could be used as the cathode as long as they are inert under these conditions, examples being silver, platinum, stainless steel or graphite. If a current continues to be drawn, the zinc electrodes are completely consumed as long as sufficient electrolyte is present to provide the H^+ ions needed for the reaction and the Zn^{2+} ions can remain in solution without fouling up the electrodes. In principle, it would also be possible to reverse the reaction as long as gaseous hydrogen was provided to the copper electrode compartment and a suitable voltage of greater than 0.76 V was applied to the cell in question.

Volta's work caused a great deal of excitement in the scientific community and many well-known scientists of the period became involved in carrying out research on related topics. One of these was Michael Faraday, then working as an assistant to Humphry Davy at the Royal Institution in London, who showed that all types of electricity that had been studied up until that time, including voltaic, thermal and magnetic, were equivalent. Faraday went on to carry out experiments on many aspects of electricity, introducing in 1834 the term *electrolysis* and the laws governing the process. In his lectures given to the Royal Institution, he also popularised the use of what are now familiar names: anode, cathode, electrode and ion.

Faraday's Laws of Electrolysis recognised that there is a direct relationship between the voltage generated in any electrochemical cell (such as the Zn-Cu cell of the Volta pile) and the chemistry of the oxidation of the metal involved (the Zn in the Volta pile). Simply stated, the laws are as follows:

First Law. The mass of an element deposited at an electrode (m) is directly proportional to the charge, Q, passed through the electrode (in ampere seconds or Coulombs (C)):

$$m/Q = Z$$

where Z is the electrochemical equivalent of the substance involved.

Second Law. The mass of any substance liberated or deposited at an electrode is directly proportional to their chemical equivalent weight (W):

$$W = \text{Molar mass}/\text{Valence}$$

As long as the electrochemical process in question is carried out reversibly at a constant temperature and pressure, the consequence of these relationships is that the EMF of any cell reaction, E, is directly related to the free energy change associated with that reaction (ΔG) by the equation:

$$\Delta G = -nFE$$

where n is the number of electrons required per ion reacted and F is the Faraday constant. (The value of F is 96,485.332 $C\,mol^{-1}$, this generally being rounded to 96,485 $C\,mol^{-1}$.) Hence, the all-over EMF of any combination of electrodes measured under reversible conditions is directly related to the chemical thermodynamics of the chemical reaction concerned. For a more detailed discussion of the thermodynamics of electrochemical reactions see Box 7.1.

BOX 7.1 The thermodynamics of a reaction in an electrochemical cell

In order to discuss the operation of modern batteries, fuel cells and related systems, it is necessary first to consider briefly some thermodynamic and kinetic aspects of the operation of electrochemical cells. For any reaction:

$$aA + bB \Leftrightarrow cC + dD$$

the value of the free energy change ΔG is given by Eq. (7.1):

$$\Delta G = \Delta G^\circ + RT \ \ln\left(aC^c.aD^d/aA^a.aB^b\right) \qquad (7.1)$$

where ΔG° is the standard free energy change for the all-over reaction. If the reaction is at equilibrium, ΔG is zero and

$$\Delta G^\circ = -RT \ln K \qquad (7.2)$$

where K is the equilibrium constant for the reaction. The quantity ΔG in Eq. (7.1) is an indication of how far the reaction is from equilibrium; a positive value corresponds to composition to the left of the equilibrium position, a negative value to composition to the right. If a random mixture of A, B, C and D is made up and there are no constraints to the reaction occurring, the value of ΔG will tend to zero and an equilibrium composition will be attained. As discussed above, the free energy change for any electrochemical half-cell reaction is given by:

$$\Delta G = -nFE \qquad (7.3)$$

Substituting the values of ΔG and ΔG° in Eq. (7.1) by $-nFE$ and $-nFE^\circ$, where n is again the number of electrons transferred in the reaction and E° is the standard EMF of the cell, we obtain:

$$nFE = nFE^\circ - RT \ \ln\left(aC^c.aD^d/aA^a.aB^b\right)$$

or

$$E = E^\circ - RT/nF. \ \ln\left(aC^c.aD^d/aA^a.aB^b\right) \qquad (7.4)$$

It should be recognised that E° is therefore directly related to the standard free energy of the reaction occurring and the equilibrium constant for the reaction, K, by the following relationship:

$$E^\circ = -\Delta G^\circ/nF = +RT\ln K/nF \qquad (7.5)$$

While a normal chemical reaction will proceed spontaneously towards its equilibrium position with the corresponding value of ΔG tending to zero, the position of an electrochemical reaction can be controlled by the value of the potential E applied to the electrodes. In other words, it is possible by the application of a potential to bring about a reaction that is not thermodynamically permitted in the absence of this electrochemical potential. A battery is an example of a combination of half-cell reactions involving an allover spontaneous chemical reaction. In contrast, electrolysis is an example of a reaction which would not be possible without the application of a potential.

Half-cell EMF's and the electrochemical series

An important consequence of the recognition by Faraday that the EMF associated with an electrochemical process is thermodynamically controlled and that the EMF of any electrochemical process can be determined from the thermodynamics of that process led to the establishment of the so-called 'electrochemical series'. This gives the standard half-cell EMF of any single electrode process relative to that of a standard hydrogen half-cell electrode of the type shown in Fig. 7.2.

This standard hydrogen half-cell involves the reaction of hydrogen ions at an inert electrode composed of a metal such as platinum immersed in a molar acid solution at 298 K and with a hydrogen pressure of 1 atm. (101.325 Pa, 1.01325 bar):

$$2H^+ + 2e^- \Leftrightarrow H_2$$

Table 7.1 gives some selected values of standard half-cell potentials measured under reversible conditions at 298 K.[a] Each value given corresponds to the potential related to a single electron transfer. All species are at concentrations of 1 mol/L, these corresponding to activities of unity for each pure solid, pure liquid or water (when used as a solvent).

As an example of the use of the data in Table 7.1, consider the Zn–Cu cell of the Volta pile shown in Fig. 7.1. The half-cell potential of the Zn electrode $(Zn^{2+} \rightarrow Zn)$ is given as -0.7618 V while that of the Cu electrode $(Cu^{2+} \rightarrow Cu)$ is given as $+0.337$ V. The reaction at the Zn electrode will

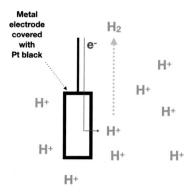

Fig.7.2 A hydrogen electrode consists of a metal coated with Pt black and immersed in an acidic solution. Electrons from the external circuit react with hydrogen ions from the solution to form hydrogen gas.

---

[a] A more complete listing of the electrics-chemical series is given at: https://en.wikipedia.org/wiki/Standard_electrode_potential_(data_page). This listing includes values for many other metals as well as for some commonly encountered chemical compounds that can be used in electrode reactions.

Table 7.1 Standard electrode potentials.

Element	Reactant	No. of electrons involved in the reaction	Product	$E°$/Volts
Li	Li^+	1	Li(s)	−3.040
K	K^+	1	K(s)	−2.931
Ba	Ba^{2+}	2	Ba(s)	−2.912
Sr	Sr^{2+}	2	Sr(s)	−2.899
Ca	Ca^{2+}	2	Ca(s)	−2.868
Na	Na^+	1	Na(s)	−2.71
Mg	Mg^{2+}	2	Mg(s)	−2.70
La	La^{3+}	3	La(s)	−2.379
Y	Y^{3+}	3	Y(s)	−2.372
Mg	Mg^{2+}	2	Mg(s)	−2.372
Ce	Ce^{3+}	3	Ce(s)	−2.336
Sr	Sr^{2+}	2	Sr(s)	−1.793
Al	Al^{3+}	3	Al(s)	−1.662
Zr	Zr^{4+}	4	Zr(s)	−1.45
H	$2 H_2O$	2	$H_2 + 2 OH^-$	−0.828
Zn	Zn^{2+}	2	Zn(s)	−0.7618
Cr	Cr^{3+}	3	Cr(s)	−0.74
Ta	Ta^{3+}	3	Ta(s)	−0.6
Co	Co^{2+}	2	Co(s)	−0.28
Ni	Ni^{2+}	2	Ni(s)	−0.25
C	$CO_2(g) + 2H^+$	2	$CO_2 + H_2O$	−0.11
C	$CO_2(g) + 2H^+$	2	HCOOH (aq)	−0.11
H	$2H^+$	2	H_2	0.00
Cu	Cu^{2+}	1	Cu^+	+0.159
Cu	Cu^{2+}	2	Cu(s)	+0.337
Cu	Cu^+	1	Cu(s)	+0.520
O_2	$O_2(g) + 2H_2O(l)$	4	$4 OH^-$	+0.401
Ag	Ag^+	1	Ag(s)	+0.7996
Hg	Hg_2^{2+}	2	2Hg(l)	+0.80
Pd	Pd^{2+}	2	Pd(s)	+0.915
Pt	Pt^{2+}	2	Pt(s)	+1.188
Cl_2	$Cl_2(g)$	2	$2Cl^-$	+1.36
Au	Au^+	1	Au(s)	+1.83

thus proceed through the oxidation of Zn metal to Zn^{2+} ions in solution. However, as the half-cell potential of the Cu electrode is positive and thus above the value for the hydrogen electrode, hydrogen evolution according to $2H^+ \rightarrow H_2$ occurs at the Cu electrode; even if Cu^{2+} ions had been available in solution so that Cu deposition might have occurred, the hydrogen evolution reaction is thermodynamically preferred. It is possible to construct a large number of different electrochemical cells using various combinations of half cells such as those listed in Table 7.1. These can consist simply of a container in which a single electrolyte exists between the two electrodes, as in the Volta pile discussed above. However, more commonly, they involve two separate half-cell assemblies linked to one another in such a way that a current can pass between the two electrodes. Fig. 7.3 illustrates one such combination in which the anode and cathode compartments are separated by a salt bridge. (A salt bridge is a tube that typically contains a highly concentrated solution of KNO_3—often maintained in an agar gel—the ions of which do not react with either the anode or cathode; see Box 7.2.) The charge balance of the system is then maintained by the movement of K^+ ions to the cathode compartment to compensate for the creation thereof cations by the reaction of the cathode material. In parallel to that, the Cl^- ions perform a similar function in the anode compartment by compensating for the removal of anions at the anode. As an alternative to the salt bridge, the two compartments can be separated by a porous membrane which allows the transport of cations and anions between the two cells but without significant mixing of the solutions in the separate

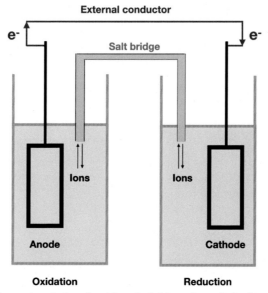

Fig. 7.3 An electrochemical cell with salt bridge. A typical electrochemical cell construction with a salt bridge separating the anode and cathode compartments. In the case of a Daniel cell (see below), the anode is Zn and this is immersed in a solution of $ZnSO_4$ while the cathode is Cu immersed in a solution of $CuSO_4$. The salt bridge can be replaced by a membrane separator.

BOX 7.2 The salt bridge

A salt bridge such as that which can be used in a Daniell cell is a tube that typically contains a highly concentrated solution of KNO_3, often held in an agar gel, the ions of which do not react with either the anode or the cathode. The charge balance of the system is then maintained by the movement of K^+ ions to the cathode compartment to compensate for the removal of Cu^{2+} ions and of NO_3^- ions to the anode compartment to balance the charge of the liberated Zn^{2+} ions. As with the use of a porous membrane, the cell gradually loses potential when used in either a continuous battery mode or a recycling mode due to the mixing of the solutions on either side of the salt bridge.

compartments; see Box 7.2. The Volta pile discussed above functions as a battery, producing electricity until the components responsible for the current are completely used up. The reaction is not reversible as the hydrogen produced in the reaction is released irreversibly to the atmosphere. However, many combinations of half-cell electrodes operate reversibly since the chemicals involved are retained in the containing vessel. Hence, they can in principle be used either for the provision of electrical energy (operating as a battery) or for the storage of electricity supplied to the combination from an external source. Electrolysis is a related process which also involves two half-cell electrodes; the most common example, discussed more fully later in the chapter, is the production of hydrogen and oxygen by the electrolysis of water.

Another important related development that occurred sometime after the work of Volta was the invention of the fuel cell by Sir William Grove in 1838, a topic that is also discussed more fully in a later section of this chapter. The fuel cell involves the production of an electric current by the reaction of suitable chemicals, most frequently hydrogen and oxygen (the reaction being the reverse of the example given for electrolysis), these gases being provided externally to the system.

The kinetics of electrochemical processes

Just as a process occurring at an electrode is thermodynamically controlled, once the conditions deviate from equilibrium, any reaction occurring is also subject to kinetic control. As discussed above, all electrode processes involve electron transfer and the electrons created to travel through an external circuit, the current resulting being a measure of the rate of the electrode process. The current produced in any half-cell reaction is determined by two factors, the *Exchange Current Density*, j_o, for the electrode and the *Overpotential*, η, applied to that electrode. Just as when a chemical reaction is at equilibrium, there is a dynamic situation in which both forward and reverse reactions continue to occur with the forward and reverse reaction rates being equal, an electrochemical reaction is a dynamic situation in which ions are formed and discharged at equal rates. This rate is determined by the *Exchange Current Density*, j_o. Only when an *Overpotential* is applied is there a net current flow to or from the electrode. As the current flowing through each half-cell electrode in an operating cell combination is the same, the overpotentials at the two electrodes will be different. Box 7.3 gives more detail of the theory relating to exchange current density and overpotential, showing the importance of the electrical *double layer* existing at the electrode surface.

BOX 7.3 Exchange current density and overpotential

Consider again the reaction:

$$aA + bB \rightarrow cC + dD$$

but consider now that the rate of this reaction is now kinetically controlled. The conversion rate will now be given by a kinetic expression such as:

$$\text{Rate} = -d[A]/dt = -d[B]/dt = f(A, B).\exp(-E_a/RT)$$

where $f(A, B)$ is some function of the concentrations of A and B and the exponential term includes the 'activation energy' of the reaction, E_a. The all-over reaction may occur in a complex sequence of reaction steps but there is often a single rate-determining step. In an electrochemical process such as that occurring in an electrochemical cell, the rate of that process is in most cases determined by the rate of transfer of ionic species in the electrolyte towards the electrode involved and the all-over reaction rate can be determined by the rate of reaction at the anode or the cathode (or even a combination of both). It is well established that a so-called *double layer* is established at an electrode as depicted in Fig. 7.4.

This model (a slightly refined model of the Gouy-Chapman model due to Bokris, Devanthan and Müllen) shows the surface of the negatively charged cathode as cations approach it. (An equivalent diagram would apply to the movement of anions towards the anode.) The cations first pass through layer 3, the 'Diffuse Boundary Layer', then through a layer containing well-ordered hydrated cations, the 'Outer Helmholtz plane', and finally they encounter the innermost layer of adsorbed solvent molecules, the 'Inner Helmholtz plane'. The double-layer

Continued

BOX 7.3 Exchange current density and overpotential—cont'd

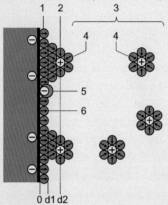

Fig. 7.4 Schematic representation of a double layer. (Bokris/Devanathan/Müllen model.) 1. Inner Helmholtz plane; 2. Outer HP; 3. Diffuse layer; 4. solvated ions (cations); 5. specifically adsorbed ions (redox ion contributes to pseudocapacitance); 6. molecules of electrolyte solvent. (*https://upload. wikimedia.org/wikipedia/commons/7/7e/Electric_double-layer_%28BMD_model %29_NT.PNG*)

provides resistance to the movement of the cations towards the cathode and the current density, j, is given by the Butler-Volmer equation:

$$j = j_o \{ \exp (1 - \alpha) F \eta / RT - \exp \alpha F \eta / RT \}$$

The quantity η is termed the 'over-potential' and is equal to $(E' - E)$, where E is the electrode potential at equilibrium and E' is the electrode potential when a current is being drawn from the cell. When the over-potential is zero, the reaction at the cathode is at equilibrium and the current density is j_o, this being equal to the rate of transfer of ions through the double layer in both directions. The term α is the so-called 'transfer coefficient' and is related to the position of the electron-transfer occurrence within the double layer. The dependence of the ratio j/j_o on the over-potential for different values of the transfer coefficient is shown in Fig. 7.5.

In practice, the value of α is in most cases very close to 0.5. When the over-potential is very small (in practice ca. 0.01 V), the double layer acts like a normal conductor in which the current density is proportional to the applied voltage. However, at higher values of overpotential, there is a logarithmic relationship between the current density and the overpotential that is given by the so-called Tafel equation:

$$\ln j = \ln j_o - a F \eta / RT$$

BOX 7.3 Exchange current density and overpotential—cont'd

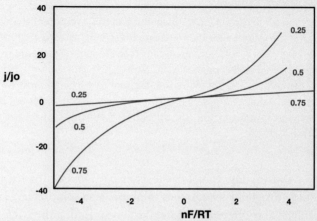

Fig. 7.5 The variation of j/j_o with overpotential for different values of the transfer coefficient.

A plot of the logarithm of current density j versus the overpotential η gives results of the type shown schematically in Fig. 7.6. The intercept on the log j at zero overpotential gives the value of the logarithm of the exchange current density and the gradient gives the value of the transfer coefficient, α, in this case 0.58.

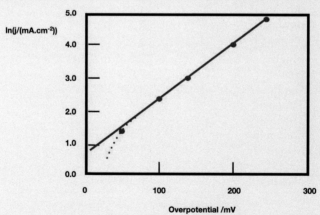

Fig. 7.6 A typical Tafel plot of versus over-potential, η.

In practice, all electrochemical devices when in operation either provide or draw a current. For example, when a rechargeable battery is being charged, a

Continued

BOX 7.3 Exchange current density and overpotential—cont'd

voltage must be applied that is greater than the standard EMF of the electrode assembly in question. However, the magnitude of this overpotential must not be too high or there is a danger that unwanted reactions will occur which shorten the battery life. Hence, it is very important that the rate of recharge is carefully controlled to minimise the possibility of damage. Similarly, when a battery is in use and a current is being drawn from it, the rate of discharge must be carefully controlled, this also avoiding damage to the battery. These topics are discussed further below in relation to the construction and operation of several different electrochemical devices.

Table 7.2 shows some values of the exchange current density, jo, for the evolution of hydrogen from aqueous sulphuric acid solution over a series of transition metals at room temperature. It can be seen that there is a very large range of values, the highest being for the noble metals, palladium, platinum, rhodium and iridium and the lowest being for cadmium, manganese, lead and mercury.[b]

Table 7.2 Exchange current densities for various metals for the hydrogen evolution reaction in an aqueous 1.0 N H_2SO_4 solution at ambient temperature.

Metal	j_o/amp cm^{-2}
Palladium	1.0×10^{-3}
Platinum	8.0×10^{-4}
Rhodium	2.5×10^{-4}
Iridium	2.0×10^{-4}
Nickel	7.0×10^{-6}
Gold	4.0×10^{-6}
Tungsten	1.3×10^{-6}
Titanium	7.0×10^{-8}
Cadmium	1.5×10^{-11}
Manganese	1.3×10^{-11}
Lead	1.0×10^{-12}
Mercury	0.5×10^{-13}

Data from several standard tabulations.

[b] As will be seen later, the high exchange current density for this reaction over noble metals such as platinum is of importance for the construction of electrolytic cells used for the production of hydrogen. The surface area of the platinum is optimised in such systems by coating the electrode with platinum black, a highly dispersed form of the metal.

Byrådskritik af
AUC's brevpapir

Repræsenterer de, som bruger det, AUC?

Af
EIGIL MORTENSEN

Henning Larsen (V): Er det AUC eller...?

AALBORG: AUC-forskerne, som i 11. time såede tvivl om de beregningsmæssige grundlag for Aalborgs varmeplan, blev udsat for en voldsom kritik i Aalborg Byråd i går. Repræsenterer de universitetet eller er de privatpersoner?

- Er det AUC eller blot et par mennesker, som er ansat på universitetet og bruger AUC's brevpapir? spurgte tidligere forsyningsrådmand Henning Larsen (V). Det er altid de samme to underskrifter. Det kan fremstå, som om det er et officiel forskningsprojekt.

- Det er ikke første gang, supplerede Hans Brusgaard (K). Vi har utallige gange fået brev fra to, som tilfældigvis er ansat på AUC. Enten skrives der på universitets vegne med rektors underskrift eller også privat. Med eget brevpapir og egne frimærker. Det er urimeligt med denne metode. Jeg agter fremover at arkivere den slags lodret, såfremt det ikke klart fremgår, hvem der skriver. I magistratens 2.

afdeling ligger der en dynge breve, der er skrevet på AUC brevpapir. Jeg kan da heller ikke skrive privat på kommunens brevpapir og give indtryk af, at det er byrådet. Jeg er glad for AUC og centrets samarbejdet med byens erhvervsliv og kommunen, men dette går ikke.

Også Arne Kristensen (Å) vendte sig skarpt mod brug af AUC's officielle brevpapir - og porto - til private breve, hvorimod både Niels Aage Helenius (S) og forsyningsrådmand Kirsten Hein (SF) gav udtryk for, at det måtte være en intern sag for AUC.

Forsyningsrådmanden lagde dog ikke skjul på, at hun har opfattet brevene som officielle tilkendegivelser fra universitetet.

- Bestemt ikke, fastslog Henning Larsen. Det er et problem for os, at vi ikke ved, hvem der skriver.

Det blev aldrig opklaret, om AUC-forskerne har blandet sig i varmeplanen på universitetets eller egne vegne.

Bortset fra SF-gruppen - minus Kirsten Hein - afviste samtlige byrådsmedlemmer forskernes kritik af embedsmændenes beregninger.

Fig. 9.6 After receiving the results of inviting two university staff members to join the municipality in analyzing heat planning alternatives, the city council criticized the staff members for using the letterhead of the university when proposing an alternative heating plan while disregarding the contents written on the paper.

council members raised a discussion on whether it was appropriate for university employees to write a letter using the university's letterhead (see Fig. 9.6). On the following day, the issue was presented on the front page of the local newspaper: "City Council Criticizes Aalborg University's Use of Letterhead Writing Paper."[1] The contents written on the paper were not discussed. Instead the city council decided to implement the coal-based alternative 1.

[1] Translated from Danish: *Byrådskritik af AUC's brevpapir.* Aalborg Stiftstidende, April 14, 1987. In Hvelplund and Lund (1988).

Soon thereafter, the Danish Minister of Education, Bertel Haarder, who was a member of the same political party as the city council member leading the discussions on the writing paper, made an official inquiry to the university, and we were investigated. However, the rector of Aalborg University, Sven Caspersen, concluded that we acted correctly regarding the use of the official university writing paper.

According to the heat planning procedure, the heat plan of Aalborg had to be approved first by the county and then by the Danish Energy Agency. The process is described in more detail in Hvelplund and Lund (1988). The result was that the Danish Energy Agency attempted to make the municipality conduct further analyses, but in the end, they had to give up, and the coal-based alternative was approved.

Conclusions and reflections

In a way, the case of Aalborg heat planning takes over where the Nordkraft case stops. From the Nordkraft case, it could be learned that given the existing institutional set-up, radical technological alternatives could not be implemented. Institutional changes had to be made at a higher level. In the Aalborg heat planning case, such changes at a higher level had to some extent been made by introducing the heat planning procedures that supported the choice of the socioeconomic least-cost solution in accordance with the purpose of the law.

Some important lessons were learned from this case. Three alternatives were put forward by the municipality, none of which represented the combination of CHP and other fuels than coal. Such solutions did not fit well into the interests and perceptions of the city council and the municipality-owned district heating company and were thus disregarded in the definition of optional alternatives to be discussed in the public participation phase. Alternatives representing radical technological change had to come from the university and local citizens. The case reveals some interesting choice-eliminating mechanisms and strategies:

- When addressing the public discussion phase, the municipality simply left out certain alternatives.
- When such alternatives were proposed by the citizens, the municipality disregarded these in the comparative analyses.
- When comparative analyses showed an inconvenient result, new analyses were made.
- Only the analyses that showed the most convenient results were put forward to the city council.
- When citizens mailed inconvenient results to the city council, the contents of the analysis were disregarded; instead, a discussion of the letterhead used was initiated, presumably to incriminate the senders.

This case confirms what was learned from the Nordkraft case by showing that the existing institutional set-up could not identify and implement the best alternative by itself. The main point of this case is that it is not efficient to tell by law what to do if the institutional market set-up makes it profitable to do something completely different. If the institutional set-up of the energy taxation system favors coal at the cost of, among others, natural gas, the local municipalities are faced with a principal request to choose solutions that are expensive for heat consumers. In this situation, the municipalities will seek to disregard other alternatives than the one with the lowest heating prices. In the case of Aalborg, the county council and the Danish Energy Agency failed to make the municipality live up to the words and the intentions of the law.

However, the case also illustrates how institutional barriers can be identified by the introduction and discussion of concrete technical alternatives. As a follow-up on this case, we promoted concrete proposals on how to change the Danish energy taxation system to create a situation in which the socioeconomic best solution would also generate the best consumer heat prices.

3   Case III: The evaluation of biogas (1990–1992)

The evaluation of the feasibility of large-scale biogas stations in Denmark in 1991 is a story of how traditional cost-benefit studies may provide irrelevant information and, consequently, may draw the wrong conclusion. Traditional cost-benefit studies based on applied neoclassical economics simply do not take into consideration the real-life economic situation and do not refer to the politically decided overall economic goals of the government. In this case, the information was inadequate for making recommendations on how to best achieve the politically defined economic objectives of the Danish Parliament of that time. The case shows how feasibility studies that include such considerations can be done and how such studies can provide information that is relevant to the decision-making process. The case description is based on the books *Socio-Economic Evaluation and Measures. Based on the Case of Biogas* (Lund, 1992a)[m] and *Feasibility Studies and Public Regulation in a Market Economy* (Hvelplund and Lund, 1998a).

In 1990, the Danish Energy Agency was in the process of evaluating the status of large-scale biogas stations to define a new biogas development strategy. Fifteen reports were made, one of which evaluated the socioeconomic impact of biogas stations (Risø, 1991), and another addressed the design of suitable public regulation measures for the implementation of large-scale biogas stations (Lund, 1992a).

The analysis of the socioeconomic impact of biogas stations (the Risø report) concluded that biogas stations were not economically feasible for Danish society. This conclusion created a problem for the report addressing the design of public regulation. It was contradictory to support a massive implementation of biogas stations in Denmark if biogas was not socioeconomically feasible for Danish society. Why, then, write a report on public regulation?

Consequently, Aalborg University conducted another analysis of the socioeconomic feasibility of biogas and described a public regulation strategy (the AAU report). This report concluded that the Risø evaluation method, based on an applied neoclassical cost-benefit study, disregarded the political and economic objectives of

---

[m] Translated from Danish: *Samfundsøkonomisk Projektvurdering og Virkemidler Med biogasfællesanlæg som eksempel.*

Danish society. It was shown that the implementation of a biogas scenario indeed would be of socioeconomic benefit to Danish society when seen in relation to the aims of the Danish Parliament.

The applied neoclassical cost-benefit analysis

The economic evaluation of biogas made by Risø was, according to the report, carried out as a cost-benefit analysis based on neoclassical welfare economic theory and methodologies aiming at comparing socioeconomic costs and benefits of implementing large-scale biogas stations. The analysis was conducted in the form of present value calculations of three different existing biogas stations over 20 years: Fangel, Davinde, and Lintrup. The three stations were then compared to relevant reference stations. The two first biogas stations, Fangel and Davinde, had oil-fired district heating stations as their reference alternatives, whereas the Lintrup biogas station had a natural gas-fired district heating system as its reference alternative. The prices used in the calculations were market prices, excluding value-added tax (VAT) and energy taxes. The price prognoses used for fossil fuel were based on the official prognoses from the Danish Energy Agency. The results of these calculations are shown in Table 9.1, not including the environmental benefits.

As Table 9.1 shows, the projects are assessed as not socioeconomically feasible. The Risø report also made calculations that included the environmental benefits of biogas, based on the following emissions and socioeconomic costs: SO_2 14 DKK/kg; NO_x 8 DKK/kg; and CO_2 100 DKK/ton. The socioeconomic results including these environmental costs are shown in Table 9.2.

Based on the results in Tables 9.1 and 9.2, the Risø report concluded that

> the costs of energy produced by large-scale biogas facilities are approximately twice as high as the costs of energy produced by a reference station, even when certain agricultural and environmental conditions are included in the evaluation,[n]

and the main report concluded that the existing large-scale biogas facilities were not socioeconomically feasible.

Table 9.1 Risø report, 1991 socioeconomic results, not including environmental benefits.

Million DKK	Fangel	Davinde	Lintrup
Investment	26.7	4.3	45.2
Net present value	−15.2	−5.1	−26.8
Annual surplus	−1.4	−0.5	−2.5

[n] Translated from Danish: *Beregningerne viser, at energi produceret på et biogasfællesanlæg er cirka dobbelt så bekosteligt som energi produceret på et reference-energianlæg, selv når visse landbrugs—og miljømæssige forhold er inddraget i beregningerne* (Risø, 1991).

Table 9.2 Risø report, 1991 socioeconomic results, including environmental costs.

Million DKK	Fangel	Davinde	Lintrup
Annual surplus	−0.9	−0.4	−2.2

Feasibility study based on concrete institutional economics

The main problem of the preceding cost-benefit analysis is that it does not provide sufficient information relevant to the specific decision-making process. When the authorities formed the committee, they asked it to

> *provide the basis for deciding whether or not to expand the energy system by implementing large-scale biogas stations in Denmark.*[o]

However, the preceding study did not systematically adapt its analysis to the specific situation. This adaptation could have been performed by systematically asking and answering the questions *what* should be analyzed, for *whom* is the analysis made, and *why* conduct the analysis?

The feasibility study conducted in the AAU report (Lund, 1992a) involved a thorough analysis of those issues and comes to the following conclusions:

Question: What should be studied?
Answer: The socioeconomic feasibility of the biogas scenarios. This means that the study should have a long-term perspective that also integrates the study of technological changes.
Question: For whom and why?
Answer: The study is essentially done for the Danish Parliament, which demands this information be used for deciding the future biogas strategy.

This means that the parameters applied to measure whether biogas is economically good or bad should be relevant to official energy policy objectives as well as to the overall economic objectives of the Danish Parliament. Consequently, a thorough description of these objectives was included in the feasibility study. Regarding the parliamentary energy policy, the analysis showed that both the official energy plan of that time, Energy Plan 81 (Danish Ministry of Energy, 1981), and the Danish law on heat supply declared that their main purpose was to secure and promote the socioeconomically best solutions. When defining the term *socioeconomics*, the official energy plan emphasized the balance of payment and job creation as important considerations to include.

The analysis of the overall parliamentary economic objectives showed that the financial statement of the authority (Danish Ministry of Finance, 1991) emphasized the problem of unemployment. Good results had, in the previous years, been achieved

---

[o] Translated from Danish: ... *at frembringe et grundlag for en stillingtagen til, om der er basis for en bredere udbygning med biogasfællesanlæg i Danmark* (Danish Energy Agency, 1991).

with other economic factors, but the unemployment rate had risen. The authorities targeted unemployment as the important problem to address in the coming period.

All in all, the analysis revealed that energy solutions that would increase employment, improve the balance of payment, decrease pollution, and increase GDP were important measures to fulfill the aims of the Danish Parliament. The analysis also showed that labor was a rather abundant resource, as unemployment mounted to 350,000 in 1990, equal to more than 10 percent of the workforce.

When returning to the applied neoclassical cost-benefit analysis, it is important to note that the study presumes full employment and does not consider foreign debt a problem. In all calculations, positive effects on technological development, balance of payment, state finance, and employment were given no value, although such effects were given high priority by the Danish Parliament.

As a consequence, the AAU study included the preceding effects in its socioeconomic analyses. Calculations were made of a biogas scenario, assuming that 50 percent of all manure in Denmark from cattle, pigs, and poultry was used for biogas production. The outcome of this analysis is shown in Table 9.3.

The analysis revealed that if consumers were to pay the same price for heat and electricity as in the reference, the government had to provide a subsidy equal to 300 million DKK/year. However, Denmark would decrease its net imports (the decrease in import of fuel minus the increase in import of goods to construct the biogas stations) by 450 million DKK/year and increase its GDP by 750 million DKK/year. Such a situation is a token of a positive change. In the study, it is emphasized that the Danish government can choose to benefit from this positive situation *either* by seeking to decrease foreign debts **or** by seeking to raise employment.

In Table 9.3, three examples are presented in which the biogas investment program is supplemented with different degrees of increases in income taxes (or other taxes). The three examples differ from one another. Thus, example 2 includes 1500 million DKK/year in extra tax used to maintain total buying power and production at a

Table 9.3 AAU study: three examples of the consequences of the biogas scenario

	Example 1: (0 million DKK extra tax per year)	Example 2: (1500 million DKK extra tax per year)	Example 3: (750 million DKK extra tax per year)
GDP (million DKK/year)	+1000	0	+500
Employment effects (persons)	+5000	0	+2500
Governmental expenditures (million DKK/year)	+200	+1200	+700
Balance of payment (million DKK/year)	+300	+900	+600

constant level. Example 3 has exactly the same biogas scenario but with 750 million DKK/year in extra tax. Example 1 has no extra tax at all. The three examples show the different possibilities of the biogas scenario.

The positive socioeconomic effects are mainly caused by saved imports due to a decrease in oil and coal imports; increased employment because of the high employment effects linked to building, maintaining, and running biogas stations; and increased incomes and, consequently, increased taxes, which improve the public finances. This increase in income to the governmental budget more than compensates for the subsidies that must be given to motivate the construction of biogas stations.

The resources and environmental effects are shown in Table 9.4, which illustrates that when relating the socioeconomic feasibility of biogas to the goals of the Danish Parliament, the biogas scenario demonstrates higher economic and environmental performance in all important areas. When including employment effects, balance of payment effects, and effects on the state finances, this scenario is, therefore, indeed cost-effective from a socioeconomic point of view.

Conclusions and reflections

When the Danish Parliament wanted to know if biogas was suitable for society, a cost-benefit analysis methodology based on applied neoclassical economics was chosen without further consideration. This type of analysis was not wrong, but it was irrelevant in relation to the specific political context and objectives. The calculations were right when seen isolated from the specific real-life economic situation of Danish society in the 1990s. It was correct to conclude that biogas stations had not yet reached a stage at which they were economically feasible in the existing institutional context and

Table 9.4 AAU study: the resources and environmental effects of the biogas scenario compared to the reference scenario.

	Biogas scenario (50 percent)	Reference: oil for heat and coal for electricity	Environmental advantages of biogas
Primary energy supply in total (TJ/year)	16,196	18,720	–
Fossil fuel consumption (TJ/year)	468	18,720	18,252
CO_2 emissions (1000 ton/year)	34	1494	1460
SO_2 emissions (ton/ year)	470	5840	5370
NO_x emissions (ton/year)	2780	2783	–

when applying existing market prices. But seen from the government's point of view—and this is the relevant standpoint in this case—the biogas scenario just described was socioeconomically feasible, as it pursued and satisfied essential governmental aims better than the references did.

The methods applied here to establish the context and measure the relevance of a certain alternative show the strength of applying the Choice Awareness strategy. When political aims and programs involving elements of radical technological change are in question, it is recommended to conduct feasibility studies based on concrete institutional economics. The main point of the case is that proper attention should be paid to the identification of political and economic objectives when conducting a socioeconomic feasibility study.

4   Case IV: Nordjyllandsværket (1991–1994)

The case of Nordjyllandsværket is the story of how the power company of West Denmark at that time, ELSAM, in 1992, was given permission to construct a 400 MW coal-fired power station, although the Danish Parliament had decided not to build any more coal-fired power stations. This case shows how the creation of a "no alternative" situation played an important role in the decision-making process. Furthermore, it reveals some of the mechanisms applied to the elimination of choice. This case, however, also influenced a subsequent parliamentary decision to change the institutional set-up of market conditions for small CHP stations. Hereby, the Danish Parliament opened up for investments in small CHP stations of more than 1000 MW capacity in the mid- and late 1990s. The case description is based on the report "Danish Energy Policy and the Expansion Plans of ELSAM" (Hvelplund et al., 1991)[p] and the books *An Alternative to ELSAM's Planned Two Power Stations* (Lund, 1992b),[q] *Public Regulation and Technological Change. The Case of Nordjyllandsværket* (Lund and Hvelplund, 1994),[r] *Collection of Documents on the Nordjyllandsværket Case, Volumes I and II* (Lund and Hvelplund, 1993), and *Does Environmental Impact Assessment Really Support Technological Change?* (Lund and Hvelplund, 1997).

In 1990, the Danish government decided on the energy plan Energy 2000 (Danish Ministry of Energy, 1990). The plan is an example of a political wish for radical technological change. At that time, more than 90 percent of the Danish electricity supply was based on large coal-fired power stations. The power companies were organized to operate this exact technology. However, according to Energy 2000, coal-fired power stations were to be phased out and replaced by many small production units based on

[p] Translated from Danish: *Dansk energipolitik og ELSAMs udvidelsesplaner—et oplæg til en offentlig debat* (Hvelplund et al., 1991).

[q] Translated from Danish: *Et Miljø—og Beskæftigelses Alternativ til ELSAMs planer om 2 kraftværker—en viderebearbejdelse af det tidligere fremsatte. Alternativ til et kraftværk i Nordjylland* (Lund, 1992b).

[r] Translated from Danish: *Offentlig Regulering og Teknologisk Kursændring. Sagen om Nordjyllandsværket* (Lund and Hvelplund, 1994).

natural gas and renewable energy, such as CHP and wind power. Moreover, the annual increases in electricity demands were to be slowed down.

In 1991, shortly after the Danish Parliament adopted the CO_2 emission reduction targets of Energy 2000, ELSAM applied for permission to construct a new power station unit. The station, a 400 MW coal-fired power station called Nordjyllandsværket (the North Jutland power station), was to be located outside Aalborg, the main city in Northern Jutland. The application was submitted along with another application for a 400 MW natural gas-fired power station by Fredericia, in southeastern Jutland, Skaerbaekværket. Together they were to form part of a total solution for the whole ELSAM area, that is, Jutland and the island of Funen, constituting approximately half of the electricity supply of Denmark.

The application was sent to the Danish Energy Agency in accordance with the Danish electricity supply law. The Danish Energy Agency had to examine whether new power production capacity was needed before permission could be granted. In this case, the main problem was that the parliamentary energy plan, Energy 2000, had three scenarios identifying how to achieve the objective of CO_2 reduction by 2005, and none of those scenarios included any additional coal-fired power stations.

However, the situation in the Danish government was complex. The coalition government had changed, and the party of the Minister of Energy behind Energy 2000 was replaced by a new minister who was in favor of the new coal-fired power stations. (The details in this complex situation are described in the references at the beginning of this case.) On the one hand, the minority government wanted to approve the new power stations. On the other hand, Parliament still supported the Energy 2000 plan. Consequently, the government and the power companies had to explain to the public how they could back up the Energy 2000 plan while approving the implementation of a completely opposite solution.

The no alternative situation

In the case of Nordjyllandsværket, the political decision makers were asked to choose one, and only one, solution. The board of representatives of the local power company, Nordkraft, which was part of ELSAM, was asked to approve the plans for a new coal-fired power station by voting either for or against the suggestion. The consequences of a *yes* were obvious, but the consequences of a *no* were not suitably described. The year before, some of the representatives had asked for a study of an alternative based on natural gas. Such an alternative involved the replacement of the old facilities of the existing power station in the center of Aalborg, Nordkraft, by a combined cycle CHP power station based on natural gas. Furthermore, the CHP station should be adjusted to the district heating demands of Aalborg and thereby be somewhat smaller than the planned coal-fired unit. Even though this alternative had been examined and indeed was requested by some of the representatives, it was not put forward in the decision-making process, when the decision was made in the power companies, or when the application was examined by the Danish Energy Agency. Everyone was asked to choose a coal-fired power station—or nothing.

In that situation, some of my colleagues and I described and put forward a concrete alterative to the two power stations (Hvelplund et al., 1991; Lund, 1992b). This alternative made it possible to evaluate the many arguments in a relevant context. In principle, the choice was very simple, and everyone had the premises for participating in the discussion. But ELSAM and the government had a problem. The installation of a coal-fired power station was contradictory to the Energy 2000 policy, and that conflict was not to be revealed during the public debate. The situation created a number of arguments, which seemed right but were wrong. Here are some of the statements:

- Because Nordjyllandsværket is a CHP station, it fits well into the strategy of Energy 2000.
- Nordjyllandsværket will replace 30-year-old power stations. Therefore, it is good for the environment.
- The construction of new large power stations will not present any barriers to the construction of small CHP stations.
- The feasibility of the power station is good, even if governmental electricity savings programs are successful.
- ELSAM and the Danish Energy Agency do agree on the prognoses for future electricity demands.
- It is necessary to vote for the two power stations to support the export of Danish know-how.

All of these statements sounded true, and each had arguments to support it. But put into the relevant context, they all proved wrong. For example, it was true that Nordjyllandsværket was constructed as a CHP extraction station (which can be operated as CHP as well as a power-only station). However, due to the size and the location of the station, the project did not allow any increase in the numbers of households being supplied from CHP. Consequently, it did not comply with the Energy 2000 strategy of expanding CHP in Denmark.

The alternative proposal

Two versions of alternatives to the official proposal were designed. The first version was made as part of a discussion paper issued in 1991 in due time before the project was to be approved by the Danish Energy Agency (Hvelplund et al., 1991). The alternative simply consisted of electricity savings in combination with small CHP and the preceding combined cycle CHP unit based on natural gas and located in the town center of Aalborg. The alternative was designed in such a way that it could save or produce the same amount of energy and provide or save the same amount of capacity as the proposed 400 MW coal-fired power station. In concrete terms, it consisted of 100 MW savings, 100 MW small CHP stations, and a 200 MW natural gas-based CHP station in Aalborg.

The construction prices of the alternative were exactly the same as those of the planned coal-fired power station—namely, 2.8 billion DKK—and the annual direct operation costs were also the same. But regarding the environmental impact, as well as the Danish balance of payment and job creation, the alternative came out with much better results than the coal-fired power station. Our conclusion was that the investment in Nordjyllandsværket would increase the problems related to environment, economy, and unemployment in Danish society and also work contradictory to the implementation of the Danish Parliament's energy policy. Better alternatives did exist that could provide or save the same amount of electricity and, at the same time, contribute to solving environmental problems, improving the economics, and creating jobs. We argued that these alternatives should be examined before deciding on the implementation of new coal-fired power stations.

ELSAM responded to the first alternative by criticizing certain factors: a present value had not been calculated, the potential locations of CHP stations had not been specified, investments in district heating pipelines were missing, and the rate of employment was not correct. All in all, ELSAM agreed that the alternative was much better for the environment, but the costs of this alternative were underestimated, and the estimation of job creation was wrong, according to ELSAM, which did not comment on the substantial improvements of the balance of payment. Moreover, ELSAM argued on the discount of building two power stations at the same time: one in Northern Jutland based on coal and one in Southern Jutland based on natural gas.

Consequently, we made a new and more detailed version of the alternative in which both of the two new power stations were included. Moreover, we made present value calculations and included investments in district heating pipelines, and so forth. In Tables 9.5 and 9.6, the alternative is compared to the ELSAM proposal, and present value calculations are given of both alternatives.

As can be seen in Table 9.6, the alternative was composed in such a way that it produced or saved the same amount of energy as the ELSAM proposal. Moreover, the alternative proposal provided or saved even more capacity than the reference. The alternative was identical to the ELSAM proposal in terms of the following points:

Table 9.5 The ELSAM project proposal.

Capacity (MW)	Electricity production (GWh/year)	Cost present value, 30 years, 7 percent (billion DKK)		Sum (billion DKK)
340	1700	**Nordjyllandsværket**		7.9
		Power station	3.1	
		District heating pipeline	0.1	
		New transmission line	0.3	
		Operation and maintenance	1.5	
		Coal (13,600 TJ/year)	2.9	
340	1700	**Skærbæk**		7.1
		Power station	2.1	
		Operation and maintenance	0.5	
		Ngas electricity	4.4	
		(13,000 TJ/year)	0.4	
		Ngas heat (1200 TJ/year)	−0.3	
		Saved coal heat		
		(1200 TJ/year)		
6400	3400	Sum		15.0

Table 9.6 The alternative project proposal.

Capacity (MW)	Electricity production (GWh/year)	Cost present value, 30 years, 7 percent (billion DKK)		Sum (billion DKK)
200	1000	**Electricity savings**		
		Investment	1.3	
		Reinvestment after 15 years	0.5	1.8
100	500	**CHP on natural gas**		
		CHP station	0.6	
		District heating system	0.6	
		Operation and maintenance	0.4	
		Natural gas (net 2000 TJ/year)	0.7	2.3
100	500	**CHP on biomass straw**	2.3	3.9
		CHP station	1.4	
		Operation and maintenance	1.5	
		Straw (6000 TJ/ear)	−1.3	
		Saved natural gas (4000 TJ/year)		
280	1400	**Ngas combined cycle, Aalborg**		
		CHP station	1.6	
		Operation and maintenance	0.4	
		Ngas electricity (10,700 TJ/year)	3.6	
		Ngas heat (500 TJ/year)	0.2	
		Saved coal heat (1700 TJ/year)	−0.4	5.4
0–60		Insulation of houses (during a period of 30 years)		
		Investment	1.7	
		Saved cost (0–1700 TJ/year)	−0.1	1.6
680–740	3400	Sum		15.0

- The present value was 15 billion DDK calculated over 30 years using a discount real rate of 7 percent.
- The annual electricity production and capacity produced or saved were the same.
- The cost of foreign currency was the same in the *construction* phase.
- The annual natural gas consumption was more or less the same. In ELSAM's proposal, the consumption was 14,200 TJ/year, while in the alternative, it was *either* 9200 TJ/year *or* 17,200 TJ/year combined with 8000 TJ/year of saved gas oil in individual boilers.

The alternative differed from the ELSAM proposal in terms of the following points:

- The alternative contributed to a better environment by simultaneously replacing both the two power stations and approximately 60,000 individual oil and gas boilers or district heating

boilers with the same effect. Sulfur and nitrogen emissions were reduced by 84 and 22 percent, respectively, and CO_2 emissions by as much 62 percent.
- The need for foreign currency in the *operation* phase was halved.
- Job creation was increased in the construction phase as well as in the operation phase. Moreover, the alternative provided much more flexibility in the creation of jobs, both in terms of when and where these jobs were created.
- The houses, which had been insulated, would provide a better quality of living. This benefit was not included in the calculation.
- The need for high-voltage transmission lines was reduced (see the following section).
- The construction of large ELSAM power stations was expected to take 6 years, and they had to be built in parallel, while the small units of the alternative could be built within 1–3 years and could easily be built over a period of time. Consequently, Danish society would be able to use this flexibility to wait and see if new capacity would be needed or if electricity saving programs would succeed.

Discussion of the alternative

One of the interesting aspects of the Nordjyllandsværket case is that the alternative was subject to a detailed technical discussion between ELSAM, my colleagues, and me. ELSAM argued that electricity savings could not be considered as part of the alternative, since such savings would be implemented, if feasible, independently from the power station project. We refuted this by proving how the ELSAM project was based on an electricity demand prognosis that presumed that the parliamentary electricity saving program was not implemented. Without this prognosis, the Energy Agency could not approve the application, since the additional capacity would not be necessary.

ELSAM argued that the new power stations would produce during more hours than assumed in our calculations. ELSAM stated 6000 hours/year against our assumption of 4360 hours. We countered by arguing that this would probably be true for the first couple of years, because the power stations were new and more efficient than the old ones. However, for the same reason, production hours would decrease over 30 years as the new stations became old themselves. Seen in relation to an average power station lifetime, the average production hours of the system were approximately 4000 hours.

ELSAM claimed that it would not be possible to implement small CHP stations because they would exceed the technical potentials. In the application to the Energy Agency, ELSAM assumed a maximum potential of 600 MW. It was not possible to implement more, ELSAM claimed. Therefore, the alternative could not be accomplished. We argued that the potential was much higher and referred to an official report identifying a potential of at least 890 MW. Later, in the mid- and late 1990s, a total capacity of more than 2000 MW was built.

Moreover, ELSAM argued that the cost of a new transmission line related to the building of Nordjyllandsværket could not be included in the calculations, since the line would be established whether or not Nordjyllandsværket was built. As described in the following section, the statement afterward made it difficult for ELSAM to explain that the need for the transmission line was not connected to Nordjyllandsværket.

The alternative was later discussed in the public debate on the power station pro-posals. The debate continued for a couple of years and involved letters from ELSAM to local politicians, various radio and newspaper interviews, approval procedures of the local county and the Energy Agency, and even a discussion in the Danish Parlia-ment. In Lund and Hvelplund (1994) and Hvelplund (2005), a detailed description of the many facets of the process is presented, and one of the elements—the environmen-tal assessment—is further examined later in this chapter.

Conclusions and reflections

This case represents a situation in which the parliament decided to implement a radical technological change. According to the official energy plan, Energy 2000, Denmark was to replace large-scale steam turbine power stations based on coal with, among others, small CHP stations based on domestic resources and energy conservation mea-sures. These policies represent a radical technological change as the organizations linked to the big power stations would have to be partly replaced by other organizations.

If parliament policies were implemented, an additional central power capacity would not be necessary. Even in this situation, the power companies proposed to build another new coal-fired steam turbine. The power company ELSAM, together with politicians in favor of this technology, did seek to make the argument in the public debate that no contradiction existed between the coal-fired power station and the energy plan Energy 2000, which did not include any new coal-fired power stations. The description and promotion of a concrete technical alternative in accordance with Energy 2000 provoked responses from ELSAM and revealed that alternatives did exist. The existence of alternatives made ELSAM exercise a series of choice-eliminating mechanisms and strategies.

Danish society was not powerful enough to avoid the central power stations, which fit well into the organizations of existing power companies. However, the Danish Par-liament was powerful enough to promote small CHP stations more or less simulta-neously with the approval of the central power stations. Therefore, the whole matter ended with the choice of both: two big power stations and a whole program of small CHP stations; even though such a large amount of additional power was not necessary. As one politician of the local county put it:

> It is nice to live in a country which is so rich that we can afford to build two power stations even if we only need one.[s]

Building the two power stations created a situation of significant overcapacity. ELSAM argued against the politicians that this would not create any problems. We (Lund and Hvelplund, 1994) argued that this was indeed a problem. Among other problems, the overcapacity would create the basis for implementing poor sales price agreements for electricity sold from wind power and small CHP. This result was

[s] County council member Karl Bornhøft during the discussion of Nordjyllandsværket.

confirmed in 2002, when the Danish Economic Council made an economic evaluation of whether the expansion of small CHP and wind power had been good for society. As we will see later in this chapter, the evaluation was based on the presumption that, because of the overcapacity created by the two big power stations, the capacity of the small CHP stations did not have any value in the calculation.

The above facts from the preceding observations support and add to the lessons learned from the Nordkraft case. The institutional set-up was not able to identify and implement the best alternatives by itself. In this case, a core element was that the plan of the Danish Parliament entailed a radical technological change.

The discourse of the power companies focused on the importance of finding a solution that would satisfy the internal power balance between the seven member power companies of ELSAM. The identification and implementation of small CHPs, wind power, and energy conservation were not part of their interest or perception of reality. And even if they were, the implementation would be out of their reach.

Even in a situation in which the parliament had decided on a plan including the elimination of new coal-fired power stations, the interest of the existing organizations was so strong that they would still promote such an alternative. As in the case of Nordkraft, the proposal representing a radical technological change had to come from outside the organizations linked to the existing technologies of coal-fired power stations. The promotion of a concrete technical alternative was subject to public discussions at the local municipality and county level as well as at the national level, including the parliament. It contributed to the public awareness that alternatives existed that would fulfill parliamentary objectives better than another coal-fired power station.

This case revealed several levels of institutional barriers to the implementation of radical technological change, including the awareness of the democratic infrastructure, as emphasized in Chapter 3. Afterward, this awareness was used to propose changes in public regulation as well as the improvement of the democratic infrastructure, as described in the book *Democracy and Change* by Hvelplund et al. (1995).

The design of the concrete technical alternative was helped and qualified because ELSAM led a policy of complete openness in terms of providing access to all technical and economic data. Consequently, it was possible, to a large degree, to base both the reference and the alternative on the same data and main assumptions. This openness indeed raised the level of debate. However, as we will see in the next case, after introducing a liberalization of the electricity sector, many such data became inaccessible, which created a barrier to a competent public debate.

5   Case V: The transmission line case (1992–1996)

The case of the transmission line is the story of how the power company ELSAM did almost anything to avoid the presence of concrete alternatives to a new 400 kV air-borne transmission line between the cities of Aalborg and Aarhus. ELSAM even ended up withholding technical data on the grounds of "national security," thus claiming that concrete alternatives that had been designed and promoted by local citizens were based on "incorrect" data. The case description is based on the books *Evaluation of the Need for a High-Voltage Transmission Line between Aalborg and Århus* (Andersen et al., 1995),[t] *Public Regulation and Technological Change. The Case of Nordjyllandsværket* (Lund and Hvelplund, 1994),[u] and *Does Environmental Impact Assessment Really Support Technological Change?* (Lund and Hvelplund, 1997).

ELSAM proposed investing in a new transmission line, which was part of a centralized system that was well suited for the ELSAM organization. However, the idea of a centralized system opposed the wishes of a majority of politicians and the ideas of decentralization expressed in the parliamentary energy plan Energy 2000, as described in the previous section.

In November 1991, ELSAM asked for an approval of a high-voltage transmission line between Aalborg and Aarhus, and the Danish Energy Agency granted permission in January 1992. In accordance with the law on electricity supply, the Danish Energy Agency should have evaluated whether the new transmission line proposed was consistent with the overall energy plans and policies. However, the Energy Agency never made a proper review of this question. The Energy Agency never asked if the assumptions behind the ELSAM calculations were in accordance with the parliamentary energy policy of Energy 2000, and it never assessed whether the transmission line was necessary to secure the supply.

After the permission was given, the local county of Northern Jutland had to attend to the physical planning process and make an environmental impact assessment of the project, which involved a public participation phase. In principle, the politicians and the public had to weigh the costs of environmental damage against the benefits of security of supply. The local county carefully described the prospective damage to nature, but ELSAM could not—or would not—provide an equally careful description of how and to which extent the transmission line would improve the security of supply.

Three river valleys in untouched nature had to be used for 50 m high voltage towers. The transmission line was supposed to cross beautiful Mariager Fiord and a huge meadow area between Rold Forest and the famous Lille Vild-mose marshland. Massive interests of nature protection were at stake. But what exactly did Danish society gain from paying such a price? How would the transmission line improve the security of supply? This question was posed again and again by the local citizens, but ELSAM refused to answer it.

Shifting arguments for the need

ELSAM's reasons for constructing a new high-voltage transmission line changed substantially as time went on. First, it was caused by building Nordjyllandsværket. In the expansion plans of ELSAM from 1991 (ELSAM, 1991), Nordjyllandsværket was

---

[t] Translated from Danish: *Vurdering af behovet for en højspændingsledning mellem Aalborg and Århus.*

[u] Translated from Danish: *Offentlig Regulering og Teknologisk Kursændring. Sagen om Nordjyllandsværket* (Lund and Hvelplund, 1994).

named "NEV B3" (unit number 3 at NEFO Vendsysselsværket, the location of the new unit), and the transmission line in question was named "400 kV NEV-TRI." ELSAM stated the following reasons for the need for the new transmission line:

> When NEVB3 is put into operation, limitations on the simultaneous operation of NEVB3 and the transmission line to Sweden will sometimes occur. Therefore, it is recommended that the grid is expanded for the sake of the exploitation of NEV B3 and the line to Sweden.[v]

The citation thus says that the need for the new transmission line was based on the construction of Nordjyllandsværket. Consequently, when analyzing Nordjyllandsværket, we included the cost of the transmission line of 300 million DKK in the calculation, as shown in Table 9.5. However, now ELSAM changed their view and said:

> The transmission line 400 kV NEV-TRI is to be constructed regardless of whether Nordjyllandsværket is built or not. Consequently, Henrik Lund cannot save 300 million in the alternative as calculated.[w]

But if it was not caused by Nordjyllandsværket, why did Denmark then need the transmission line? ELSAM put forward several arguments: security of supply, reserve capacity in the case of renovation of other transmission lines, and electricity transit, as well as the implementation of the old centralization plan from the 1960s. Local politicians expressed the opinion that transit was not a valid reason to compensate for the environmental consequences of locating such high-voltage towers in nature. In 1992, ELSAM stated in an article:

> Transmission lines are always built for the sake of security of supply in Jutland-Funen,[x]

and

> It is the strict policy of ELSAM that transmission lines are never built for the sake of transit.[y]

---

[v] Translated form Danish: *Når NEV B3 er idriftsat, vil der i en række driftssituationer være begrænsninger på samtidig udnyttelse af NEV B3 og Konti-Skan 2. I NUP87 blev det derfor indstillet, at nettet af hensyn til udnyttelsen af NEV B3 og Konti-skan 2 udbygges med … .* ELSAM Netudvidelsesplan 1991, p. 29.

[w] Translated from Danish: *Højspændingsledningen "400 kV NEV-TRI" skal bygges uanset om Nordjyllandsværket etableres. Henrik Lund kan derfor ikke spare de 300 mio.kr., han regner med i den alternative plan.* ELSAM, February 27, 1992. In Aktsamlingen, aktstykke 1i.

[x] Translated from Danish: *Ledningerne bygges altid af hensyn til forsyningssikkerheden i Jylland-Fyn.* ELSAMposten. August 1992. In Aktsamlingen, aktstykke 3i.

[y] Translated from Danish: *Det er ELSAMs helt klare politik, at ledningerne aldrig bygges af hensyn til transit.* ELSAMposten, August 1992. In Aktsamlingen, aktstykke 3i.

The hesitation of ELSAM in presenting any specific calculations to the public, thus documenting the need for the new transmission line, must be viewed in the light of these facts. ELSAM did not want to reveal that the need was a direct consequence of the construction of Nordjyllandsværket. Neither did ELSAM want to reveal that the need was also related to transit. Moreover, ELSAM did not want to show the public that the new transmission plan was one further step in the direction of the centralization plans arising in the 1960s and thereby a contradiction to the political wishes expressed in Energy 2000.

Local citizens made several requirements directly to ELSAM to see the documentation for the improvements in security of supply, but ELSAM would not share any calculations. A national newspaper also tried to access this information, but ELSAM responded:

We have no obligation to let a random civic association evaluate our calculations.[z]

Then the local county made a formal request for further information to the Danish Energy Agency. It answered by returning a calculation made by ELSAM, which still claimed that the new transmission line was needed because of security of supply. However, that calculation did not account for any improvement in the security of supply.

Security of supply

The ELSAM calculation, which was supposed to support the claim that the transmission line was needed to provide security of supply, revealed the following two conditions:

- An existing 150 kV line will be due to overload in the future, in case 1000 MW must be transmitted from Northern Jutland to the south of Denmark. The overload will occur if all the following assumptions are fulfilled simultaneously: all three power stations in Northern Jutland (including Nordkraft and Nordjyllandsværket) are producing full power, both connections to Sweden are importing full power, the consumers in Northern Jutland demand only 60 percent of full load, *and* the existing 400 kV southward transmission line is out of order.
- The line is only due to overload if all assumptions occur simultaneously. Overload only occurs if 1000 MW are to be transmitted, while 700 MW can be handled without any overload. This means that *if* one power station is not producing, *if* one connection to Sweden is not used for import, *if* Northern Jutland has maximum electricity consumption, or *if* the existing 400 kV line is not out of order, then there is no overload.

The decisive question regarding the security of supply, in the preceding situation, was: would there be a need for 1000 MW in the rest of the ELSAM area? That question was never asked by the Danish Energy Agency, and, consequently, it granted an approval without knowing the answer.

---

[z] Translated from Danish: *Vi har ikke pligt til at lade en vilkårlig borgerforening vurdere vore beregninger.* Det fri Aktuelt, July 23, 1992.

Table 9.7 Overview of need for transmission capacity.

Capacity in Northern Jutland	
Sweden I	300 MW
Sweden II	300 MW
Nordkraft power station	240 MW
Vendsysselsværket	295 MW
Nordjyllandsværket	350 MW
Sum	1485 MW
Consumption in Northern Jutland (60 percent load)	−443 MW
Possible transmission to the rest of ELSAM	1042 MW
Capacity in the ELSAM area	
Installed capacity by year 1998	5619 MW
Of which full available power	4852 MW
Of which in Northern Jutland	−1185 MW
Remaining capacity in the rest of ELSAM	3667 MW
Consumption in the rest of ELSAM (60 percent load)	−2247 MW
Full available spare capacity in ELSAM	1420 MW

We, the local citizens, had to evaluate the question ourselves, which we did, as illustrated in Table 9.7. And the answer was clear. There was no need for the transmission line in terms of security of supply. No electricity consumer could be identified who would be out of electricity if the new transmission line was not built and who, in the same situation, would have electricity if the transmission line was built. In fact, no consumer would be out of electricity supply at all, whether or not the transmission line was built. The whole issue had nothing to do with security of supply. It was solely related to the use of the many power stations in Northern Jutland—all too many, one may argue.

As shown in Table 9.7, it is correct that a situation can be identified in which Northern Jutland could export 1042 MW. However, in such a situation, there would be no need for 1042 MW in the rest of the ELSAM area. On the contrary, a full available spare capacity of 1420 MW would be present, even if Northern Jutland was not exporting at all. Consequently, no electricity consumer in ELSAM would be out of electricity.

Given this evidence, ELSAM provided a series of new calculations, which the local citizens asked to see. However, this time ELSAM withheld the calculations by referring to national security matters. If terrorists knew all the technical data of the transmission line system, it would be easy for them to know which tower to blow up, ELSAM argued, and this view was backed up by the Danish Ministry of Defense.

Concrete technical alternatives

Local citizens also proposed concrete technical alternatives to the 400 kV transmission line and asked to have these alternatives properly designed and included in the discussions. The local county of Northern Jutland could not conduct such studies, and they could not make ELSAM do it either. As local citizens, we developed a series of alternatives and presented these to the public several times. Finally, these alternatives were described and carefully evaluated in the report from 1995 mentioned at the beginning of this section.

In the report, a detailed assessment of the transmission line systems was made similar to the situation defined in Table 9.7, and the preceding conclusions were confirmed. The transmission line was not necessary in terms of security of supply. Then, the following technical alternatives were defined:

1. A direct current high voltage cable
2. A 150 kV alternating current high voltage cable instead of a 400 kV airborne transmission line
3. To close down the two old power stations, Nordkraft and Vendsysselsværket, when the new power plant Nordjyllandsværket is put into operation
4. To reinforce the existing 150 kV transmission line system

All of these alternatives were able to remove the overload which ELSAM claimed to be the reason for needing the 400 kV airborne transmission line.

Alternative 4 is of special interest regarding the conflict between ELSAM, which wanted a centralized system, and the Danish Parliament's Energy 2000, which planned for a decentralized system. In the case of a centralized system, the system would benefit from strong 400 kV connections between the big power stations located near the big cities. However, in a decentralized system with a high amount of widely distributed small CHP stations and wind turbines, the system would obtain a greater benefit from a strong local grid. This could be accomplished by reinforcing the existing 150 kV grid.

The report included a calculation of the consequences of *not* putting Nordjyllandsværket into operation. In such a situation, there was no overload at all. Consequently, the first statement of ELSAM was confirmed; namely, that the reason for establishing the transmission line was directly related to Nordjyllandsværket.

Later, ELSAM claimed that the calculations of the local citizens were performed using incorrect data. The calculations were based on the latest technical grid data published in 1991 and 1992. However, ELSAM claimed to have slightly changed the data, but due to national security, the data were no longer publicly accessible.

In the end, the result of the discussions was as a compromise in which the transmission line was approved, however, in a new location and in a new design to try to make the line fit better into the landscape.

Conclusions and reflections

The case of the transmission line illustrates how the power station company ELSAM proposed technical solutions that fit well into their existing organization and how the Energy Agency was not capable of analyzing how such a technical solution contradicted the idea of the Danish Parliament's energy plan Energy 2000. The Energy

Agency was also unable to conduct an analysis of alternatives relevant to the implementation of the parliamentary energy policies. The case also reveals the following choice-eliminating mechanisms:

- ELSAM presented one—and only one—technical alternative.
- ELSAM changed its argument along the way and ended up emphasizing that the need was based solely on security of supply.
- ELSAM withheld the assumptions behind their calculations from the public, even in a situation in which the local politicians and the public were supposed to evaluate the benefit of security of supply against the damage to nature.
- When ELSAM was forced to reveal their calculation, it became evident that there was no need for the transmission line in terms of security of supply.
- ELSAM withheld any further calculation and data by referring to national security.
- ELSAM argued that the alternatives put forward by local citizens were based on incorrect data. ELSAM withheld small changes in the data for national security reasons.

A main point to be learned from this case is that the existing institutional set-up cannot identify or promote alternatives representing radical technological change. Such alternatives cannot be expected to come from organizations linked to existing technologies. In this case, the discourse of the power companies was focused on maintaining and expanding the transmission line infrastructure that would help along an energy supply based on central power stations, such as the existing ones. Changing the strategy by expanding the local grid and thereby helping the introduction of small CHP plants and distributed wind turbines was not within their interest or perception of reality. However, in this case, unlike Nordkraft and Nordjyllandsværket, they would have been able to implement such an alternative. Most of the local 150 kV transmission lines were owned and operated by the same organizations as the proposed 400 kV transmission line systems.

As in the former cases, the proposal representing a radical technological change had to come from outside the organizations linked to the existing technologies of coal-fired power stations. Again, the promotion of a concrete technical alternative was subject to public discussions. It contributed to raising the public awareness that alternatives existed that would fulfill the political wishes of environmental protection and still provide energy security.

The case revealed several levels of institutional barriers to the implementation of radical technological changes, including the awareness of the democratic infrastructure, as emphasized in Chapter 3. This awareness was later used to propose changes in the public regulation as well as improvements in the democratic infrastructure, as described in *Democracy and Change* (Hvelplund et al., 1995).

Regarding the openness of technical data, this case differs from the former case of Nordjyllandsværket. In the Nordjyllandsværket case, the design of a concrete technical alternative was helped and qualified because ELSAM led a policy of complete openness in relation to all technical and economic data. In this case, data were held back and used to incriminate the technical status of the alternatives designed and promoted by the local citizens.

6   Case VI: European environmental impact assessment procedures (1993–1997)[aa]

The case of the European Environmental Impact Assessment (EIA) directive illustrates how planning procedures aimed at promoting cleaner technology alternatives to, among others, coal-fired power stations, failed to even accomplish the description of such alternatives. As shown in the previous cases, alternatives representing radical technological change cannot be expected to come from organizations linked to the existing technologies. Therefore, to implement technological change, it is necessary to consider other proposals. In principle, the EIA procedure proposes an assessment of alternatives, including alternatives representing radical technological change. This case description is based on the book *Does Environmental Impact Assessment Really Support Technological Change?* (Lund and Hvelplund, 1997).

This study is based on the cases of Nordjyllandsværket and the transmission line between Aalborg and Aarhus, as described in previous sections. Moreover, the study includes a similar case of a planned power station in Copenhagen, called Avedøreværket.

In 1985, the European Union (EU) decided on a directive on EIA (European Council, 1985). It was based on the preventive principle: to eliminate the pollution source rather than attempting to counteract it subsequently. According to the Danish implementation of the directive, an EIA must review the main alternatives to the project in question and assess the environmental consequences of each alternative. If this is done properly, EIAs can assist in raising Choice Awareness and help local citizens and upcoming new technology industries with a proper assessment of their cleaner technologies, including public discussions on the alternatives.

Implementation of the EIA principles in Denmark

According to the 1985 EU directive on EIA, all EU Member States must commission an assessment of the environmental consequences of certain types of projects, including power stations, before a building permission will be granted. The EU directive was adopted on the basis of the assumption that the best environmental policy consists in preventing pollution or nuisance, rather than counteracting the effects subsequently.

In Denmark, the EU EIA procedure was implemented in the planning legislation, and the EU directive was translated into Danish law in 1989. Compared to the EU EIA directive, Danish law was more stringent on two counts: Danish law used existing planning procedures for public participation and it clearly stipulated that alternatives should be described and assessed. The aim of the EU EIA directive or the Danish EIA legislation was not to compel the use of a specific solution. Rather, the aim was to promote cleaner technologies by providing information on the environmental

[aa] Excerpts reprinted from *Environmental Impact Assessment Review*, 17/5, Henrik Lund and Frede Hvelplund, "Does Environmental Impact Assessment Really Support Technological Change? Analyzing Alternatives to Coal-Fired Power Stations in Denmark", pp. 357–370 (1997), with permission from Elsevier.

consequences of a solution, before permission was granted for a project with major environmental impact. EU legislation required only that the environmental situation was taken into account, not that the environmentally best solution was chosen. The Danish implementation, on the other hand, contained specific requirements: that the assessment must include an outline of the main alternatives and that public participation was ensured in two phases.

Nonetheless, the Danish implementation of the directive was fragmented and thus weakened. At that time, Denmark was divided into 14 counties (see Fig. 9.7), each with an elected regional council and a professional administration. At the regional level, the Environmental Act and the Planning Act were administered by different departments. In the case of a new power station, the EIA was the responsibility of the authorities of the region in question. The EIA was carried out with public participation according to the Planning Act, which involved the publication and distribution of a proposal followed by public hearings, but the EIA of cleaner technology alternatives was beyond the Planning Act's normal procedure.

Example 1: Nordjyllandsværket

As mentioned previously, my colleagues and I designed and promoted an alternative that could be expected to produce the same amount of electrical power and heat as Nordjyllandsværket but with less fuel and less pollution. The cleaner technology alternatives to Nordjyllandsværket should cover the same geographical area as the existing supply area: Jutland and Funen. Physically, the geographical locations differ. Where Nordjyllandsværket is placed in Northern Jutland, the decentralized cogeneration units and energy conservation would be located in the smaller towns, closer to the consumers.

Table 9.8 shows the most significant alternatives proposed to Nordjyllandsværket. The following describes why two EIAs were presented. Both assessments were subject to our objections to the Nature Protection Appeal Board under the Planning Act. All of the alternatives shown in Table 9.8 were included in the complaints from us to the first and the second Appeal Boards about deficiencies in the EIAs.

In the summer and autumn of 1993, the North Jutland Regional Authority prepared an EIA of Nordjyllandsværket. Prior to its preparation, my colleagues and I, with reference to the EU EIA directive, called the Regional Authority's attention to two of the alternatives presented in Table 9.8. However, the Regional Authority claimed that it was not their responsibility to make an environmental assessment of these alternatives.

Later, in a written reply (North Jutland Regional Authority, 1993), the Regional Authority emphasized that the terminology "description of alternatives" should be understood as alternative locations and designs. According to this definition, the Regional Authority was convinced that it was irrelevant to include alternative modes of production, such as cleaner technologies or a division of Nordjyllandsværket into smaller units.

Consequently, a number of citizens complained about the EIA report to the Nature Protection Appeal Board, the statutory appeals body under the Planning Act. The

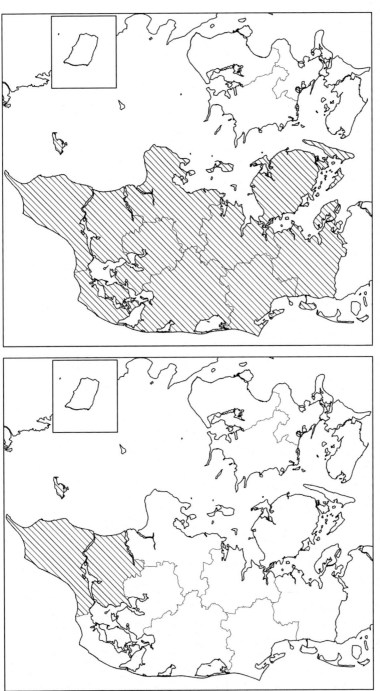

Fig. 9.7 The problems of geographical boundaries. The map of Denmark on the left shows the boundaries of the regional administrations. The region of Northern Jutland is shown by a hatched signature. The map of Denmark on the right indicates ELSAM's supply area. A new coal-fired power station is included in the supply of the whole area but must be placed in one of the regions. Alternatives in the form of reductions in power consumption, decentralized CHP, and renewable energy must be dispersed throughout the whole supply area and, therefore, involve several regions.

Table 9.8 Alternatives to Nordjyllandsværket, with supply areas and geographical locations.

Alternative	Supply area	Geographical location
ELSAM's proposal		
Nordjyllandsværket (400 MW coal-fired power station)	Jutland/Funen	Aalborg
Alternatives		
1. Natural gas-fired combined cycle station in Aalborg	Jutland/Funen	Aalborg
2. Natural gas-fired station at Nordjyllandsværket	Jutland/Funen	Aalborg
3. Environmental alternative (combination)	Jutland/Funen	Jutland/Funen
4. "Only one station," such as Fredericia	Jutland/Funen	Fredericia
5. Location south of Trige (avoids need for high-voltage lines)	Jutland/Funen	Such as Aarhus
6. Brundtland Plan (combination)	Jutland/Funen	Jutland/Funen

Nature Protection Appeal Board's first decision (Nature Protection Appeal Board, 1993) was distinct and precise:

1. An EIA must not only include alternative locations but also other significant alternatives.
2. An EIA must perform a more or less thorough examination of the environmental advantages and disadvantages of each of the alternatives considered as well as their locations. The regional planning authority has the responsibility, more or less, of revealing the environmental consequences of alternatives arising from public participation.
3. An EIA must include information on the background for the desired location regarding environmental impact.
4. An EIA of an electric power station must consider that high-voltage lines are connected from the station. This situation must be dealt with in the impact assessment.
5. The question of cleaner technology must be dealt with if the question is significant for the assessment of whether one can avoid, reduce, and, if possible, neutralize threats to the environment.

The Nature Protection Appeal Board did not state that their decision was valid only within the geographical boundaries of the region. They concluded that the listed conditions were not fulfilled and that the Amendment to the Regional Plan was invalid.

The North Jutland Regional Authority issued a new EIA for Nordjyllandsværket in December 1993. Again, this assessment did not fulfill the EU EIA conditions or the conditions formulated by the Nature Protection Appeal Board in its first decision. The impact assessment was thus unsuitable as a basis for public discussion. It listed a number of alternatives that could not be compared, and two of the alternatives were not assessed regarding their environmental impact. Nothing was said about the background for the desired location in terms of environmental impact; no environmental assessment of any part of the high-voltage transmission lines that would result from Nordjyllandsværket was included, and the analysis did not treat the question of cleaner technology.

Again, a complaint about these deficiencies was submitted to the Nature Protection Appeal Board. The Board's second decision (Nature Protection Appeal Board, 1994) did not comment specifically on any of the points presented in the complaint. Regarding the alternatives, the Board concluded that the EIA conducted by the Regional Authority was satisfactory. The Board agreed with the Regional Council stating that the Council was unable to conduct an EIA of the energy policy alternatives to Nordjyllandsværket and was therefore not compelled to do so.

Nonetheless, the Nature Protection Appeal Board indicated one problem in the impact assessment: it should have included an alternative with decentralized CHP within the Northern Jutland region. However, the Board determined by a vote, with nine for and two against, that this deficiency was insufficient to declare the assessment invalid. In practice, the consequence was that the regional councils were not required to assess environmentally cleaner technology alternatives to large coal-fired power stations at the project level. These power stations have always formed part of a plan to supply the whole Jutland/Funen area, but per se they will always be placed in only one of the regions. Cleaner technology alternatives, which were necessary for the fulfillment of the Danish Parliament's CO_2 goals, would be valid for the same supply area as a large coal-fired power station, but they consist of technologies that are dispersed over several administrative regions. The decision of the Nature Protection Appeal Board stated that the Regional Authority was not compelled to include alternatives that were located outside the region. With this decision, cleaner technology alternatives were typically excluded.

Example 2: High-voltage transmission lines

As mentioned earlier, together with Nordjyllandsværket, ELSAM wished to construct a high-voltage transmission line from the new generating station in Aalborg southward to the city of Aarhus, which is a distance of approximately 100 km. As described, all documentation supporting the need for this line assumed the establishment of Nordjyllandsværket, and the construction of the transmission line is considered a consequence thereof.

According to the EU directive, an EIA must include all direct and indirect environmental impacts of the proposed project. In a written reply to a question from one of the complainants, the EU directorate-general XI (environment) said that a high-voltage transmission line is considered an impact arising from the project that should be assessed, according to Article 3 in the EU directive. But the North Jutland Regional Authorities had not included the high-voltage transmission line in the EIA of Nordjyllandsværket; therefore, it was included in the complaints to the Nature Protection Appeal Board. The Appeal Board's first decision was in general precise, but vague in its reply to the transmission line question. According to the Board, the EIA should have included the first part of the route of the high-voltage transmission line in question to reveal its environmental consequences. The meaning of this formulation was unclear, because a line between two points only has importance if the whole line is considered in the same way that half of a bridge has no importance. In the concept phase of the second EIA, the complaints emphasized that the Regional Authority should obtain the necessary calculations to establish whether the line was justified, also in the event that Nordjyllandsværket was not constructed. The Authority did not do this; instead, it made an environmental assessment of the station's linkage to the high-voltage grid. This is a distance of a few hundred meters and has never been part of the preceding high-voltage connection between the new power station and Aarhus.

The complainant received a letter from the Danish Energy Agency that confirmed that all the submitted documentation for the high-voltage line had assumed the construction of Nordjyllandsværket. Even though the Regional Authority had not obtained the necessary information and even though no documentation had been submitted that substantiated the need for the line without the power generation station, the Nature Protection Appeal Board's only comment was

> No information has been presented which can give cause to a new consideration of this question.

Later, it was decided that an individual EIA should be made of the high-voltage line, but the relationship between Nordjyllandsværket and the line was never subject to an EIA.

Example 3: Avedøreværket

In 1995, a power company on the island of Zealand (SK Energy) applied for permission to build a 460 MW multifuel power station, Avedøreværket unit II, close to Copenhagen. It was a multifuel station because it was designed to use a combination of coal, natural gas, and biomass. In reality, the main fuel was thought to be coal. The Avedøre II station was to be ready for production in 2003.

Seen from an EIA viewpoint, the building of the Avedøre station had the same problem as Nordjyllandsværket: cleaner technology alternatives were not assessed. In this case, the alternative technologies were heat conservation in the Copenhagen area and electricity conservation and cogeneration outside the county of Copenhagen.

In the foreword to the EIA of the Copenhagen County, it is stated:

> In that way, the evaluations are solely performed within the geographical area of the county of Copenhagen (Copenhagen County, 1996).

This was noticed by OOA, a public interest organization on Energy and Environment, which then sent a complaint to the Minister of Environment and Energy, mentioning the need for including alternatives outside the county of Copenhagen in the EIA procedures (OOA, 1996). The Minister did not respond to this letter by changing the EIA procedures. In the following months, a public debate arose regarding the lack of compatibility between the official Danish CO_2 reduction goal and the establishment of a new, mainly coal-fired station. Public resistance to a coal-fired station (caused by the rather obvious lack of a thorough description of alternatives) resulted in the Minister's rejection of the application in the autumn of 1996. The power companies did not give up their plan and quickly changed their application project into a power station based on natural gas. This new application was approved by the Minister on March 31, 1997.

The approval was motivated mainly by the proposed change of fuel. Again, no analyses were presented of the cleaner technology alternatives. This case also shows that Danish EIA procedures cannot ensure that cleaner alternatives from outside the region in question are included in the assessment.

Conclusions and reflections

When implemented into Danish legislation, the EU EIA directive did require the assessment of alternatives to coal-fired power stations. Such a procedure involved a public participation phase, in which alternative proposals could be made, and the procedure was based on the principle of prevention rather than subsequent counteraction against pollution. Consequently, the whole program in principle aligned with the idea of Choice Awareness by securing a proper description of cleaner technology alternatives. However, in practice, the preceding cases reveal a different reality.

The case of Nordjyllandsværket indicates that serious analyses are not conducted as a matter of course of the decentralized alternatives to large coal-fired power stations in Denmark. The main problem is that the regional administrations are responsible for the preparation of the EIA. The law does not require alternatives that extend beyond the boundaries of a regional authority to be assessed. In the example of Nordjyllandsværket, permission was granted without a proper examination of cleaner technology alternatives.

The example of the high-voltage transmission line between Nordjyllandsværket and the city of Aarhus indicates that it cannot be assumed that a serious analysis is conducted of the relationship between a new power station and a new high-voltage transmission line. This was not the case when it was documented that the justification of the transmission line assumed the existence of the new power station.

The example of Avedøre II is similar to the case of Nordjyllandsværket. It shows that the Danish EIA procedures cannot ensure that cleaner alternatives from outside a region are included in an assessment. In this case, permission was granted to a new natural gas-fired power station before a proper assessment of decentralized cleaner technology alternatives was made.

All in all, the cases reveal that because EIA is implemented on a restricted, regional basis, it does not support the radical technological change represented by cleaner technology alternatives. The responsibility for the preparation of the EIA is given to the regional authorities on the basis of a law that does not require the assessment of alternatives if these extend the geographical boundaries of a regional authority. Thus, one cannot be sure that serious analyses are made of cleaner technology alternatives to large coal-fired power stations.

The main point to be learned from these cases is that an existing institutional set-up may hinder the identification and promotion of relevant and better alternatives if such alternatives represent radical institutional change. The discourse of the power companies aims to identify and promote projects that are relevant within their organizational interests and perception of reality. The discourse of the county council and administration focused on job creation and the identification of proper solutions within the borders of the county. It was outside their interest and perception of reality to evaluate and analyze alternatives outside their jurisdiction.

The fact that a cleaner technology alternative that did not form part of the discourse of the power companies and the county council would officially be rejected on a council meeting was a big problem. To avoid this, the alternative was eliminated beforehand. Even though a proper description and evaluation of a cleaner technology alternative were the basic ideas of the EIA legal procedures, the legal authorities were not able to secure that such an alternative was presented and evaluated.

7 Case VII: The German Lausitz case (1993–1994)[ab]

The German Lausitz case illustrates how, in East Germany after the reunification, massive investments were made in rebuilding both the electricity and the heat supply. However, the rebuilding was done without introducing or expanding the use of CHP and thus deriving substantial fuel saving benefits. By the end of 1992, two of my colleagues at Aalborg University and I were invited to design an alternative, which was introduced and discussed in the debate. The discussion of the alternative revealed how energy conservation and CHP did not fit well into the existing organizations. Thus, representatives of the German brown coal industry heavily criticized the alternative. The case description is based on the books *Rebuilding without Restructuring the Energy System in East Germany* (Hvelplund and Lund, 1998b), *Erneuerung der Energiesysteme in den neuen Bundesländern—aber wie?* (Hvelplund et al., 1993), and *Kommentar zur Kritik der Lausitzer Braunkohlen AG an der Aalborg Universität-Studie* (Hvelplund et al., 1994) and the chapter "Energy Planning and the Ability to Change. The East German Example" in *Institutional Change and Industrial Development in Central and Eastern Europe* (Hvelplund and Lund, 1999).

At the end of 1992, two environmental organizations[ac] in Lausitz asked us to design an alternative to the further expansion of brown coal in the electricity supply of East Germany. The study was financed by Bündnis 90, who at that time was represented in the regional parliament (Landtag) of Brandenburg.

The survey included an area from Berlin in the north to the border of the Czech Republic in the south, as shown in Fig. 9.8. However, Berlin was only partly included in the study. The area encompassed approximately 40 percent of the population of the former East Germany. The area had a comprehensive production of brown coal and several large-scale power stations based on brown coal. From 1960 to 1980, the government eliminated 70 villages and rehoused 30,000 citizens to access the coal resources. Moreover, the mining of coal had a severe negative impact on the groundwater of the area.

[ab] Excerpts reprinted from *Energy Policy*, 26/7, Frede Hvelplund and Henrik Lund, "Rebuilding without Restructuring the Energy System in East Germany", pp. 535–546 (1998), with permission from Elsevier.
[ac] Netzwerk Dezentrale Energienutzung e.V. and Grüne Liga e.V., Cottbus.

Fig. 9.8 The Lausitz area covered by the study.

The planned and, at that time, already partly implemented restructuring of the East German energy system was unique. For approximately 10 years, it was planned to spend 60,000 million DM on this restructuring. Such an extensive investment in changing the energy system of a country within such a short period of time was historic. An extraordinary situation like this case offered the opportunity to modernize the energy system—investing in the most appropriate and most developed technologies wherever possible. But the most important factor was to implement the necessary modernization of the institutional structure of the energy organization. In the German case, this could create a situation in which East Germany became an example to follow for other East European countries that needed the same development. Several export possibilities could emerge from such a position.

Unfortunately, what happened at that time was the complete opposite: a completely new system was built, but it was not modernized. Thus, the worst of all situations was created. East Germany had a new but outmoded energy system that could not be expected to fulfill future environmental demands. Many energy installations have a lifetime of 20–40 years. In this case, East Germany would be placed in a worse situation than most other countries in terms of energy supply and environmental impact.

The main problem with the rebuilding plans was that the obvious opportunity of using CHP was not seized. Huge investments were made in new brown coal power stations together with the environmental improvements of old ones, and the heating of houses based on brown coal briquettes was replaced by natural gas in individual boilers.

This situation happened because of a policy to maintain the existing brown coal production in East Germany for a period of time. Two arguments supported this policy. First, there were good reasons to maintain the existing 17,000 jobs in the brown coal areas, as these areas had already experienced high unemployment figures. Second, brown coal, a domestic resource, was considered a good solution in terms of energy security. Both the national government and the local government in Brandenburg were in favor of the policy.

According to the reference strategy, the electricity capacity would be expanded by building new brown coal-fired power stations and by renovating the best of the old ones. In the Lausitz region, it was planned to invest 15,000 million DM in building or renovating a total capacity of 7200 MW. All power stations were placed next to brown coal mines at the two locations, Jänschwalde and Boxberg/Schwarze Pumpe, far from most urban areas. In practice, this made it impossible to fully exploit the potential of CHP. The planned use of CHP only comprised a minor district heating capacity and heat for a plasterboard production. Only 5 percent of the potential would be exploited in Jänschwalde, and only 11 percent would probably be used in Boxberg/Schwarze Pumpe.

The very lack of CHP in the system made it necessary to use large amounts of oil and natural gas for house heating. The expected development was analyzed and is described in the references listed in the beginning of this section. The analyses showed two characteristics of the reference strategy. First, within the house heating sector, East Germany would become dependent on large imports of foreign fuels, such as natural gas and oil. Thus, the argument of using brown coal for electricity production due to reasons of energy security did not involve the entire energy sector. Second, as most of the house heating boilers in the former East Germany were old and would have to be replaced by new gas and oil boilers, the reference strategy would lead to major investments—not only in big power stations but also in central heating systems and gas and oil boilers in detached houses, apartment buildings, and district heating stations. For the whole East German area, these investments were estimated at approximately 35,000 million DM.

The alternative

An alternative to the reference was designed to raise awareness in East Germany of the possibility of modernizing the entire energy system while rebuilding it. The alternative entailed the benefits achieved from energy conservation, CHP, and renewable energy. Regarding energy conservation in the demand system, the alternative strategy proposed a 20 percent reduction in heat demand and a 15 percent reduction in electricity demand compared with the reference. The investments needed to achieve these targets were included in the cost analysis. The 20 percent cut in heat demands could be achieved simply by means of a number of technical improvements of heating system regulation, such as thermostat valves and the change from serial to parallel radiators.

Regarding efficiency improvements in the supply system, the alternative proposed a substantial use of CHP in combination with heat pumps to replace, among others, electric heating. The marginal fuel consumption for electric heating is twice the fuel consumption of gas and oil boilers and 5–10 times the marginal fuel consumption of CHP. Therefore, the alternative was designed with the purpose of avoiding electric heating and promoting district heating based on CHP. All electric heating was substituted by central heating systems, and district heating systems and CHP were proposed in most urban areas.

The house heating net consumption in the Lausitz supply area was calculated as 190 PJ/year, covering hot water and space heating of private households and office buildings as well as industry that could be supplied by district heating (approximately 10 percent). Based on demographic data for all cities, towns, and rural areas, the total energy consumption was categorized according to different town sizes. The survey showed that by connecting 80–90 percent of the potential households in all urban areas to a district heating system, it was possible to produce 131 PJ/year of the total 160 PJ/year from CHP. This corresponded to approximately 80 percent of the market. Here, a district heating loss of 20 percent and a supply of 131 PJ/year were assumed to lead to a CHP heat production of 157 PJ/year.

To meet the heat demand, a total CHP heat capacity of 5000 MW was required. The capacity was divided into 2000 MW heat/1600 MW electric power from natural gas and 3000 MW heat/1900 MW electric power from hard coal. Meanwhile, the CHP units also had to supply a reserve capacity for a certain wind power effect. Therefore, the alternative proposed a total investment in 2000 MW electric power based on natural gas and 2500 MW electric power based on hard coal to establish a suitable technical framework for the integration of wind power.

Regarding renewable energy, the alternative proposed an extensive use of wind power and biomass resources. The resources of the area were estimated, and the alternative strategy was based on the use of 50 percent of the potential resources in the year 2010 with wind turbines producing electricity for the grid, straw substituted for coal in the CHP units, wood substituted for oil in individual boilers, and biogas substituted for natural gas in CHP units. Moreover, solar thermal energy was included in the strategy for the production of hot water in the detached houses that were not supplied with district heating. This measure equaled a 20 percent decrease in the fuel consumption of these houses. Fig. 9.9 shows the fuel flow diagrams.

Fig. 9.9 illustrates how savings, CHP, and renewable energy substantially decreased the use of fossil fuels and thereby CO_2 emissions in the alternative compared to the reference. An economic feasibility study of the reference as well as the alternative was made by comparing the total costs of each solution during a period

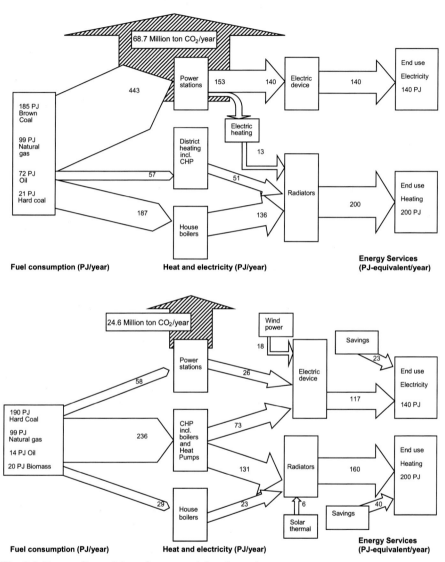

Fig. 9.9 Energy flow of the reference and the alternative.

of 20 years. The study was based on the fuel prices of the early 1990s, including brown coal sales prices from the Lausitz mining company LAUBAG. The prices of biomass resources were based on existing prices in Denmark. Construction and maintenance costs were exact prices for Jänschwalde and Boxberg/Schwarze Pumpe in combination with Danish construction prices for small CHP stations and similar technologies.

A comparison based on the described assumptions showed that the costs of the alternative were slightly lower than those of the reference. However, in this first draft version, the alternative strategy had two political disadvantages: it meant fewer jobs and a higher import of fuels than the reference. These two disadvantages were a direct consequence of the extensive use of imported hard coal. Meanwhile, the higher efficiency of the alternative could also be achieved by using other fuels. Therefore, two other possibilities were analyzed: one in which hard coal was substituted by brown coal, processed into "brown coal dust" with a low content of water, and one in which hard coal was substituted by natural gas and the extensive use of biomass resources.

When compared to the reference, the analysis of the three variants of the alternative gave the following results:

- *Electricity and heat consumption*: The reference and the three variants of the alternative produced or saved exactly the same amounts of electricity and heat.
- *Primary energy supply*: The three variants of the alternative cut down fuel consumption by 50 percent compared with the reference. This was due to the demand-side efficiency, the use of CHP, and the substitution of electric heating.
- *Environment*: The alternative reduced CO_2 emissions to between 20 and 40 percent of the reference emissions. This was partly due to the low fuel consumption and partly due to the use of renewable energy.
- *Economic feasibility*: Based on the fuel prices of the early 1990s, the alternative had more or less the same total costs as the reference. The variant based on hard coal was 8 percent cheaper, while the variant based on natural gas and biomass was 12 percent more expensive. It may be surprising that the very large investment in district heating and CHP in all variants of the alternative did not make them less cost-effective than the reference. This is because the reference also implied huge investments in changing the house heating system: natural gas boilers, central heating, and natural gas pipe systems.
- *Employment*: The hard coal variant created fewer jobs than the reference. On average, over the 20-year period, the employment effects of the reference corresponded to 55,000 jobs. The brown coal and the biomass variants of the alternative had more or less the same employment effects, while the hard coal variant was expected to provide 46,000 jobs.
- *Energy security*: The reference strategy would need to import 72 PJ/year of oil, while the alternative strategy required only 14 PJ/year. Moreover, the reference emptied the storage of domestic brown coal resources in a rate of 485 PJ/year, contrary to the alternative that used only between 0 and 209 PJ/year. The hard coal variant of the alternative was dependent on an import of hard coal of 190 PJ/year. In case of supply failures in the future, the hard coal could be substituted by brown coal processed to "brown coal dust" with low water content. Therefore, the brown coal variant as well as the hard coal variant provided a better solution in terms of energy security than the reference. The natural gas and biomass variants resulted in a dependence on imported natural gas of 209 PJ/year versus the other variants and the reference using only 99 PJ/year.

Conclusions and reflections

The East German energy system ended up being restructured but, unfortunately, not modernized. The system was changed from a system based completely on brown coal to a system in which electricity production was still based on brown coal and house heating was based on oil and natural gas. Only a very small percentage of house heating was based on CHP. The restructuring would bring the CO_2 emission down from 20 ton/inhabitant in 1989 to 13 ton/inhabitant in year 2010, which was still much higher than the EU average of 8 ton/inhabitant.

By the design and promotion of a concrete technical alternative, it was shown that alternatives did exist, which would not only restructure but also modernize the energy system. The implementation of such a radically different technological alternative could reduce East German CO_2 emissions to below 7 tons per inhabitant. The alternative strategy was based on CHP, conservation, and renewable energy. Total costs and employment effects were approximately the same in the two strategies.

However, the alternative was not implemented. Instead, it was heavily criticized by the coal mining industry (Hvelplund et al., 1994). My colleagues and I used the discussion of the concrete technical alternative to examine and analyze the institutional reasons for not modernizing the East German energy system while rebuilding it. This analysis is described in the references listed in the beginning of this section.

The analysis showed that the economy of the East German power companies depended on a continuation of brown coal-based centralized power production and separated heat production based on oil and gas. Furthermore, these companies had a cost structure with a high element of "fixed costs," and therefore low short-term marginal costs. In periods with excess capacity, this cost structure motivated a rather aggressive behavior on the market. This tendency was enforced by the strong obligation imposed on the power companies to be "stock price locomotives" for their parent companies.

As Hvelplund (2005) described in detail, the German VEBA concern owned both electricity and oil companies, and the economically optimal solution to the concern as a whole was both to operate power stations on brown coal and, at the same time, sell oil and natural gas to private households. Consequently, even though the expansion of CHP had the potential of increasing sales from the power production units, it would still mean a loss compared to the sale of oil and gas. The optimal solution was to avoid CHP.

Again, the main lesson from this case is that, given the institutional set-up, society could not identify and implement the best alternatives. The discourse of the brown coal power and mining companies aimed to maintain and renew the existing technologies. The alternatives, including decentralized CHP stations, energy conservation, and renewable energy, did not form any part of their interest or perception. Such alternatives representing radical technological change could not originate from the organizations linked to the brown coal technologies.

The discourse of the VEBA concern focused on the optimization of the earnings of the concern as a whole, and since that involved power and mining companies as well as oil companies, the alternatives based on CHP, energy conservation, and renewable energy were not in their interest.

The description of alternatives based on radical technological change had to come from outside the existing organizations, and when they were promoted by local citizens and NGOs, they were heavily criticized by the brown coal industry. Based on this case, we identified a number of institutional barriers to the implementation of new technologies, and we advocated that clear and non-bureaucratic rules had to be defined regarding the conditions for technical connection to the public grid. Price conditions based on the long-term "avoided marginal cost" principle (including external costs) had to be established. Likewise, financial possibilities had to be arranged for organizations that wanted to invest in new technologies. Institutional and organizational preconditions of this kind do not automatically evolve on the market.

8   Case VIII: The Green Energy Plan (1996)[ad]

The case of the Green Energy Plan is the story of how the description of concrete technical alternatives can be used to raise Choice Awareness in the public debate on national energy policy. The Green Energy Plan was published by the Danish General Workers' Union and introduced in the spring of 1996 as an input to the public debate on the future energy system of Denmark. The official Danish energy plan, Energy 21 (Danish Ministry of Environment and Energy, 1996), was adopted soon after the public debate. The case description is based on the books *Elements of a Green Energy Plan Which Can Create Job Opportunities* (Lund, 1996), *Feasibility Studies and Public Regulation in a Market Economy* (Hvelplund and Lund, 1998a), and *A Green Energy Plan for Denmark. Job Creation as a Strategy to Implement Both Economic Growth and CO_2 Reduction* (Lund, 1999b).

The Green Energy Plan is based on the observation that normally the cost of implementing CO_2 reduction policies is considered a threat to both economic growth and employment. The plan shows that, to some extent, CO_2 reduction goals can be implemented that create jobs, and such strategies can help economic growth.

The Green Energy Plan involved a number of initiatives, which Denmark could choose to implement as a supplement to the official energy policies. The plan was made on the basis of a total evaluation of environment, economics, balance of payment, consumption, and employment. Primarily, the aim was to enable Denmark to continue the reduction of CO_2 emissions. This should be achieved by means of long-term, investment-dependent alterations in the infrastructure, which would thus develop toward a more decentralized renewable energy system. The changes in question would require investments, but these would provide an improved employment situation and a better environment.

The design of the concrete technical alternative

The public discussion of a new Danish energy policy in 1995 was based on the fact that the former official energy plan from 1993 had proven insufficient in terms of meeting the CO_2 emission targets defined by the Danish Parliament. According to these targets, CO_2 emissions should be cut by 20 percent in 2005 compared to 1988. Consequently, the environmental consideration of the Green Energy Plan focused on the establishment of an energy system that would reduce CO_2 emissions and meet the

---

[ad] Excerpts reprinted from *Environmental & Resource Economics*, 14/3, Henrik Lund, "A Green Energy Plan for Denmark", pp. 431–440 (1999), with permission from Springer Science+Business Media.

targets. Thus, the Green Energy Plan defined the 1993 official strategy as a reference and presented an alternative comprising the following elements:

- *Switch from electric heating to central heating*: In the Danish energy system, the fuel consumption of electric heating was twice as high as the fuel consumption of central heating from individual gas or oil boilers and four times as high as the marginal fuel consumption of heat from CHP. In the reference, it was intended to switch from electric heating to central heating in areas with natural gas or district heating. The Green Plan proposed that the switch from electric heating should also be implemented outside these areas.
- *Improved insulation and low-temperature district heating*: In 1996, the large Danish urban centers were supplied with district heating from large coal-fired CHP stations with steam turbines. The fuel consumption for district heating in these stations depended on the temperature level of the district heating system. Therefore, a double gain could be achieved by improving the insulation of houses in these cities and towns, and thereby reduce the demand for district heating, and by making the heat production of steam turbine stations more effective.
- *Utilization of natural gas in district heating*: In 1996, natural gas used for individual house heating without CHP was widespread in Denmark. Fuel savings could be achieved by installing individual micro-CHP units in the houses or by introducing district heating in combination with CHP. The latter solution was suggested in the Green Energy Plan.
- *Dispersed use of biomass*: The Green Energy Plan proposed the construction of biogas stations corresponding to the use of 100 percent of the organic industrial waste produced and 50 percent of the manure from livestock farming.
- *Wind turbines*: The Green Energy Plan chose a development in which a wind power capacity of 3000 MW would be reached by 2015. The economic calculations assumed that a number of the turbines were established at offshore wind farms, where construction prices were higher.
- *Further training and energy conservation*: The Danish industrial sector had a great potential for reducing its electricity consumption. The problem was to identify and implement these savings. Therefore, the Green Energy Plan suggested that electricity savings were implemented by means of an environmental training program for all industrial workers. On the basis of the Danish experience, the Green Energy Plan expected that such a training program would result in a 20 percent reduction of the electricity consumption of industry.

Evaluation and comparisons

A computer model was constructed to calculate the Green Energy Plan and the reference. Later, the model served as the basis for the development of both the EnergyBALANCE and the EnergyPLAN models. The model was calibrated by applying the figures of the years 1988 and 1993; thus, the energy balances used corresponded to the Danish Energy Agency's energy statistics.

The cost estimation was based on the investment and operation costs of 1995. A certain technological development was considered in the case of heat and power production, and an improvement in efficiency was assumed throughout the period. In the cases of wind turbines and solar thermal technology as well as biogas and gasification, technological development was expressed in the form of falling construction and operation prices. Furthermore, calculations were made for two different developments in fuel prices. According to the first development, all fuel prices were constant, corresponding to the 1995 world market level. In the other scenario, an actual price increase for fossil fuels was included (in accordance with the projections of the Danish Energy Agency).

The Green Energy Plan was compared with the reference on the assumption that the Green Energy Plan was implemented from 1996 to 2015, and both years were included in the model. Detailed technical calculations were made for the years 2005 and 2015, assuming that half of the above-mentioned measures would be implemented by 2005, and all of them by 2015. In comparison with the reference, the implementation of the Green Energy Plan had the following consequences:

- *Environment and fuel consumption*: The total fuel consumption in the Danish energy sector (excluding transportation) would be reduced by 5 percent, and fossil fuels would be reduced by 9 percent by 2005. When including fuel consumption for transportation, CO_2 emissions would be reduced by 20 percent by 2005 compared with 1988. For 2015, the Green Energy Plan proposed an additional reduction of CO_2 emissions by 34 percent in comparison with 1988.
- *New power generation capacity*: In addition to the two power stations, Nordjyllandsværket and Skærbæk, which were already under construction, the reference involved the establishment of two more power stations each of 400 MW before 2005 and six or seven additional stations from 2005 to 2015. However, the Green Energy Plan proved that new power production capacity (except for the stations under construction) would not be needed until 2015. The increased development of decentralized CHP, the implementation of low-temperature district heating in larger urban areas, energy conservation in the industrial sector, and the replacement of electric heating made the construction of new power stations before 2015 unnecessary.
- *Costs, foreign exchange, and employment*: With constant fuel prices, the Green Energy Plan would increase the total annual costs (investments, fuel, and operation) from 27 to 32 billion DKK (not including taxes or subsidies). This was largely due to the increased investments. Foreign exchange consumption would experience a minor rise in the beginning (0.7 billion DKK p.a. in 1996) but would fall thereafter; thus, in 2015, foreign exchange savings would be achieved (0.6 billion DKK p.a. in 2015). With minor increases in fuel prices (according to the latest forecasts from the Danish Energy Agency, where prices were expected to rise from 25 to 50 percent), the additional expenditures would fall to approximately 4 billion DKK p.a., and the savings in foreign exchange would be 1.4 billion DKK p.a. in 2015.

Thus, the Green Energy Plan involved a number of additional Danish investments. In the reorientation period, the plan's effect on the balance of payments would, to a large extent, be neutral.

Regardless of the fuel prices, the reference used an employment parameter of slightly less than 40,000 people. The implementation of the Green Energy Plan would increase employment by ~12,000 in 1996 and rise to the additional employment of ~17,000 by 2015.

- Public finance accounts: In the model, a calculation was made of the net effect that the implementation of the Green Energy Plan would have on public finances. The calculation included increased taxation revenues and reduced social benefit payments, reductions in energy levies, and the resulting profits of the natural gas companies (which in Denmark were owned by the government and the municipalities). Furthermore, the secondary effects from the increased turnovers in the national community were included. A Danish subsidy of between 0.10 and 0.27 DKK/kWh to electricity produced from renewable energy and decentralized

CHP was calculated, assuming that the existing level would remain constant until 2005, thereafter being gradually reduced to 50 percent by 2015.

An estimate of the consequences on public finances was made presupposing that the employments of 12,000–17,000 additional people would cause neither bottleneck problems nor rising wages. At that time, there were 250,000–300,000 unemployed in Denmark. The result of the estimate is shown in Table 9.9. As seen, the implementation of the Green Energy Plan would mean a net additional income to public finances of 1 billion DKK/year in 2005 rising to 1.2 billion DKK in 2015.

The additional income shown in Table 9.9 was calculated on the assumption that consumer prices (total energy costs for households and industry) would remain the same in the Green Energy Plan as well as in the reference. It was calculated to which extent public finances would have to contribute to maintain the average purchasing power and competitiveness at the same level in the two alternatives. The public regulation means were as follows:

- Changes in the regulation of electricity tariffs
- Prohibition of the use of electrical residential space heating after the year 2000
- Specifically aimed subsidies
- Improved insulation standards for houses offered for sale
- Continuation of the CO_2 levy and the subsidies for renewable energy (but with a gradual reduction so that the subsidies would be reduced by 50 percent through a 20-year period)
- Conversion from natural gas in individual boilers to district heating based on CHP
- Training programs in electricity-saving behavior for industrial employees
- More research, such as in straw gasification
- Subsidies for biogas and gasification stations

According to the Green Energy Plan, an implementation of these measures would cost approximately 1.3 billion DKK p.a. in public finances. Thus, these costs were similar to the expected increase in revenue to public finances, as shown in Table 9.9. Therefore, the Green Energy Plan could possibly be implemented in such a way that the effect on public finances would be negligible.

Table 9.9 Green energy plan net effect on public finances, not including public regulation costs.

Billion DKK	2005	2015
Additional costs	4.7	5.2
Foreign exchange costs	0.2	−0.6
Employment effect	13,000	18,000
Reduced benefits	+1.5	+2.0
Increased tax	+1.3	+1.7
Natural gas company deficit	−0.2	−0.5
Reduced energy levies	−1.1	−1.4
Energy subsidies	−1.1	−1.4
Secondary effects	+0.6	+0.8
Effect on public finances	+1.0	+1.2

Conclusions and reflections

The Green Energy Plan illustrated how a concrete technical alternative could be designed that would improve the environment and energy efficiency of Denmark by replacing imports of fossil fuels by domestic investments in energy conservation, efficiency improvements, and renewable energy. The feasibility study showed how this reorientation would mean a small increase in the total direct costs of energy supply. However, it would also allow Danish society to use the available resources of 250,000–300,000 unemployed individuals to both gain an increase in economic growth and improve the environment. The plan was developed by the General Workers' Union to influence the design of Danish energy policies in the public debate prior to the governmental energy plan Energy 21, which was decided soon thereafter.

This case illustrates that while alternatives based on radical technological changes cannot be expected to originate from the organizations linked to the existing technologies, as shown in the previous cases, such alternative proposals can come from other organizations, such as the General Workers' Union.

In the case of Nordjyllandsværket, the institutional set-up made the local workers' unions support a coal-fired power station because of the related operation and construction work, even though the job creation figures were low compared with the alternative of CHP, energy conservation, and renewable energy. The problem was that the investments required were not supported by the power companies that had the money. Moreover, the jobs would be spread all over, and many would be outside the area of the local unions. Therefore, these alternatives were out of reach and were not part of the perception of the local organizations.

This case shows that, when the government enables a discussion of alternatives at a higher level and when this discussion is not linked to specific jobs in specific local regions, it becomes possible to describe, analyze, and discuss how to implement alternatives that fulfill the objectives of energy security, environment, and job creation.

Also illustrated by this case is that the design and economic evaluations of alternatives cannot be performed with applied neoclassical economics. These methodologies cannot provide the relevant information on whether the different alternatives fulfill relevant political objectives. However, the information can be found by applying the methodology of concrete institutional economics.

9   Case IX: The Thai power station case (1999)[ae]

The case of Prachuap Khiri Khan in Thailand is a story about the importance of including all relevant objectives in a feasibility study. This case represents a feasibility study in which a planned new 1400 MW coal-fired power station and a proposed technical alternative were assessed in relation to a wide range of specific and general official development objectives for Thailand. An interesting aspect of the Thai case is that

[ae] Excerpts reprinted from *Applied Energy*, 76/1–3, Henrik Lund et al., "Feasibility of a 1400 MW Coal-Fired Power Plant in Thailand", pp. 55–64 (2003), with permission from Elsevier.

the government had defined clear political objectives for Thai society. Consequently, it was possible to make a strict evaluation of the project and compare it to a concrete alternative. The case shows that the plans to implement a coal-fired power station at Prachuap Khiri Khan were indeed not very rational regarding the fulfillment of the goals of Thai society and that alternatives could be identified that were more suitable. The case description is based on the articles "Sustainable Energy Alternatives to the 1400 MW Coal-Fired Power Plant Under Construction in Prachuap Khiri Khan. A Comparative Energy, Environmental and Economic Cost-Benefit Analysis" (Lund et al., 1999b) and "Feasibility of a 1400 MW Coal-Fired Power-Plant in Thailand" (Lund et al., 2003), and the book *Feasibility Study Cases* (Lund et al., 2007c).

This case is based on the result of a workshop arranged by the Thai-Danish Cooperation on Sustainable Energy and Sustainable Energy Network for Thailand, which took place in Bangkok in 1999. The workshop resulted in the report "Sustainable Energy Alternatives to the 1400 MW Coal-Fired Power Plant Under Construction in Prachuap Khiri Khan" (Lund et al., 1999b). Power sector development projects are well known to be contentious issues in many countries around the world. This was also the case of the planned 1400 MW coal-fired power station in Prachuap Khiri Khan. On the one hand, an institutionalized community of public power companies, private power producers, and industrial companies was setting the pace for establishing a power sector based on fossil fuels. On the other hand, Thai society was badly affected by an economic crisis, unemployment, and the degradation of natural habitats. The political system was facing the difficult task of regulating the energy sector to meet the social and economic objectives of Thailand in the most efficient way. Politicians and stakeholders who wanted to establish a rational basis for decision making needed to continuously examine whether the situation and development were rational and, if not, whether alternatives existed that could accommodate the societal and economic targets more effectively.

The Hin Krut power station in Prachuap Khiri Khan

The Hin Krut power station project in Prachuap Khiri Khan was located at the Kok Ta Horm village, in the Thongchai subdistrict of the Prachuap Khiri Khan province. The owner of the project was Union Power Development Co., Ltd., a joint venture based on 85 percent foreign capital. The joint venture proposed its project in June 1995, received the initial power purchase agreement in December 1996, and signed a 25-year independent power production contract with the Electricity Generating Authority of Thailand (EGAT) in June 1997.

In the meantime, locals and environmentalists were increasingly concerned about the project's negative impact on the surrounding environment. The coal-based power station would not contain scrubbers to reduce sulfur dioxide emissions and was therefore expected to become a hazard to the local people's health. Because of disputes surrounding the project, it was still awaiting the government's permission when the feasibility study was made in 1999. Though already contracted, the electricity authority EGAT was able to cancel the project by paying a compensation of approximately THB 5–6 billion (110–140 million USD) to Union Power.

The coal-fired station was expected to have an installed capacity of 1400 MW from two 700 MW units. Union Power expected the station to consume 3.85 million tons of hard coal per year, producing 7125 GWh/year at an efficiency of 44.77 percent. The construction costs of the turnkey station were 900 million USD plus an additional 300 million USD to cover the costs of, for instance, financing and consulting. Thus, the expected construction price was THB 45 billion (1.2 billion USD). The expected operation and maintenance costs were THB 450 million per year (12 million USD per year).

Union Power projected a coal price of 40 USD/ton, which, according to the electricity authority EGAT, was similar to the prevailing cost of coal. Furthermore, EGAT expected the real price for coal to increase by 2.08 percent per year. According to the contract, Union Power would receive both a fixed payment, covering the available power supply of the station (capacity payment), and a reimbursement of variable costs associated with fuel consumption and operation and maintenance.

The actual payment terms were kept confidential, but the agreement between the electricity authority EGAT and another producer was known to include a capacity payment of 422 THB/kW per month (11.3 USD). According to the agreement, EGAT was obliged to pay this amount whether or not the power station produced electricity. This favorable contract was intended to secure an internal rate of return of 15 percent to the investor.

Official economic objectives for Thailand

The official social and economic objectives of developing Thai society were formulated by the Thai government in the 8th National Development Plan 1997–2001 (NESDB, 1996). The National Energy Policy Office of Thailand (NEPO) was obliged to develop policies, management plans, and measures related to Thailand's energy sector in accordance with the National Economic and Social Development Plan and government policies to be presented to the National Energy Policy Council. In other words, NEPO was required to relate all plans and base all decisions on the wide range of official objectives defined in the National Economic and Social Development Plan and other government policies.

The 8th National Development Plan was influenced by an economic crisis in Thailand in 1997 and was considered to be a turning point in the way in which it recognized the globalization process and the need for a continuous and long-range process of planning, decision making, implementation, monitoring, and evaluation. The plan was said to be moving toward a holistic, people-centered development. The social and economic objectives formulated in the plan were intended to form the basis for planning and decision making in all economic sectors, and thus issues including economic policy, security of supply, employment, rural development, technological innovation, the environment, and others. Energy plans and projects were also to be assessed on this basis.

In the feasibility study (Lund et al., 1999a,b), a careful assessment was made of the different official social and economic objectives of Thai society. All private and

public energy sector projects should be assessed in terms of their ability to provide the
following:

- A sufficient energy supply
- Reasonable energy prices
- High energy efficiency
- High-cost efficiency
- Low import content
- New products for export
- More and better employment
- Positive effect on public budgets
- Rural development
- Decentralization of the planning and decision-making process
- Technological innovation
- A healthy environment

The design of a concrete technical alternative

To show the influence that the preceding objectives could have on the results of the
feasibility study, an alternative to the coal-fired reference was designed for comparison. The alternative consisted of three components: industrial CHP based on biomass,
demand-side management (DSM), and micro-hydro power. The alternative combined
the use of indigenous fuels with Thai industrial and technological development and
intended to create employment, technological innovation, export opportunities, and
other benefits. The alternative had been carefully composed to consist of 1000 MW
industrial CHP based on biomass, 350 MW DSM, and 40 MW micro-hydro power
(see Table 9.10). Consequently, the alternative provided exactly the same capacity
as the 1400 MW reference coal-fired power station.

- The alternative was generally composed to meet the official social and economic objectives
  for Thailand—that is, it should be cost-effective, reliable, stimulate employment, and rely on
  indigenous resources. The micro-hydro power component of the alternative was not likely to
  prove cost-effective in the short term. It was, however, included due to its potential positive
  impact on technology development, employment, and others, in the long term.
- Each element of the alternative exploited only a minor part of the estimated technical potential for that element, which meant that this type of alternative was generally applicable to
  future coal-fired projects in Thailand (see Table 9.10).
- From 2001 to 2010, the coal-fired power station was expected to replace electricity produced
  by oil-fired steam turbines with an efficiency of 33 percent. Over the 10-year construction

Table 9.10 Alternative compared to the estimated technical potential of Thailand.

	Alternative	Estimated Thai potential
Industrial CHP	1000 MW	10,000 MW
Biomass resources	2200 ktoe/year	20,000 ktoe/year
DSM	350 MW	2200 MW
Micro-hydro power	40 MW	8000 MW

period, the alternative was consequently supplemented by such units. All costs of the supplemental electricity production were included in the feasibility study.
- The alternative and the coal-fired power station produced (or saved) identical amounts of electricity, both in total and year by year.
- While the coal-fired power station was to be constructed over a maximum of 2 years, the alternative exploited the surplus capacity in power production, which was expected to be available until 2010, spreading the construction efforts over a 10-year period from 2000 to 2010. This would also give industry time to develop and market even better biomass, CHP and micro-hydro power technologies.

Comparative feasibility study

The alternative and the coal-fired reference were modeled in a period of 25 years and compared in terms of their energy and environmental and economic characteristics. Economic costs were calculated using factor prices, that is, investment costs and operation and maintenance (O&M) costs excluding VAT and other taxes. The economic investment costs of the coal-fired power station were represented by the capacity payment that the electricity authority EGAT had agreed to pay Union Power as an independent power producer instead of implementing the power station project themselves. However, for the employment calculation, the actual investment costs according to Union Power were used. Please refer to the report "Sustainable Energy Alternatives to the 1400 MW Coal-Fired Power Plant Under Construction in Prachuap Khiri Khan" for further details on conditions and assumptions, as well as specific technoeconomic data used in the analyses (Lund et al., 1999a,b).

Regarding the primary energy supply, the final energy consumption of the coal-fired station over a 25-year period was found to be 68 PJ/year, adding up to 1689 PJ. Meanwhile, the alternative was to be introduced over a period of 10 years, and the final energy consumption would be at the highest level in the beginning, when the alternative depended on the electricity production from older, low-efficiency oil-fired power stations. Consequently, the fuel consumption would decrease from an initial 86 PJ/year in 2001 to approximately 2 PJ/year in 2010 and onward. The very low net fuel consumption of approximately 2 PJ/year was partly due to the high savings achieved by replacing old boilers in the industry by new CHP stations and partly from introducing DSM and hydro power with no fuel consumption. The total energy consumption of the alternative reached 468 PJ.

Regarding environmental consequences, the implementation of the alternative would lead to the following reductions over a period of 25 years:

- Reduce fuel consumption by 72 percent
- CO_2 emissions by 670 million tons
- SO_2 emissions by 1.5 million tons
- NO_x emissions by 1.5 million tons

Regarding production costs and consequences for the balance of payment, the net present value of economic costs was almost the same for the coal-fired power station at Prachuap Khiri Khan and the alternative, totaling approximately THB 150 billion when using an annual discount rate of 7 percent on 1.60 THB/kWh of electricity produced. The division between capital costs, O&M costs, and fuel costs influenced the effect on, for example, employment and rural economy. While the production costs of the coal-fired power station at Prachuap Khiri Khan included 53 percent capital costs, 44 percent fuel costs, and only 3 percent O&M costs, the alternative involved 39 percent capital costs, 40 percent fuel costs, and 22 percent O&M costs. The total cost of the coal-fired power station at Prachuap Khiri Khan was THB 117 billion in foreign currency, while the import cost of the alternative was only THB 39 billion. Furthermore, the feasibility study showed that the coal-fired power station at Prachuap Khiri Khan only contributed THB 35 billion to Thailand's economic wealth (GDP), while the alternative contributed THB 115 billion. The implementation of the alternative would thus do the following:

- Imply practically identical economic costs of 1.60–1.62 THB per kWh of electricity produced over a period of 25 years
- Incur lower capital and fuel costs but higher O&M costs, which indicated an advantageous contribution to employment and rural economy
- Save foreign currency worth THB 78 billion (2.1 billion USD), thereby reducing the negative impact on Thailand's balance of payment by 67 percent
- Contribute an additional THB 80 billion to Thailand's GDP

The coal-fired power station only created 0.2 million man-years of employment over a period of 25 years, an average of approximately 7000 man-years per year. The alternative created 1.8 million man-years over a period of 25 years, an average of approximately 71,000 man-years for every year of the period. The implementation of the alternative would consequently do the following:

- Create an additional 1.6 million man-years over a period of 25 years; that is, when the coal-fired power station at Prachuap Khiri Khan created 1 man-year, the alternative would create 10 man-years
- Create an additional approximately 64,000 man-years for every year over the period

Regarding the consequences for the rural economy (i.e., economic activity that stems from biomass production and O&M activities), the net present value of the economic activities contributing to the rural economy was THB 5 billion for the coal-fired power station, while the alternative contributed THB 93 billion. The implementation of the alternative would thus do the following:

- Contribute to the rural economy an additional THB 88 billion; that is, every time the coal-fired power station contributed THB 1 billion to the rural economy, the alternative would contribute THB 19 billion

Regarding the consequences of public finances (primarily from personal income taxes), the net present value of the public revenues created by the coal-fired power station was THB 7 billion, while the alternative would create public revenues of THB 22 billion. The implementation of the alternative would consequently do the following:

- Contribute an additional THB 15 billion to the public revenues; that is, every time the coal-fired power station at Prachuap Khiri Khan contributed THB 1 billion to public revenues, the alternative would contribute THB 3.3 billion

Conclusions and reflections

Two energy projects were analyzed and compared in relation to national Thai objectives: the planned coal-fired power station at Prachuap Khiri Khan (reference) and the biomass, energy efficiency, and micro-hydro power alternative (alternative). The feasibility study showed that the proposed alternative in all aspects was equal to or better than the reference. The alternative was noticeably better in terms of imports; creation of employment; and contribution to Thailand's GDP, public revenues, rural economy, technology development, and the environment, whereas the alternative was almost equal to the reference in terms of economic production costs. The main economic results are summarized in Table 9.11.

The feasibility study of the Prachuap Khiri Khan power station had three main implications for technological innovation in the Thai energy sector. First, it enabled energy planning to link with broader aspects of sustainable development and national development goals. Therefore, energy planning was no longer recognized as a stand-alone exercise that aimed only to meet system demand and reliability with the old technological system. The study stressed very clearly the importance of pursuing an intersectoral policy. Second, it raised public awareness of the potential benefits of renewable energy development in the longer term and broader perspective, especially the social benefits that had previously been little discussed in Thai society. The increasing public awareness urged the Thai government to introduce more programs that supported renewable energy development. In 2001, the Thai government developed a strategic plan for renewable energy. Two years later, it decided to set-up targets for overall renewable energy development and for each specific renewable technology. Third, the feasibility study provided the opportunity and tools for meaningful public participation in validating governmental decision making, which could prevent excessive and inappropriate investments. In this case, in 2001, the Thai government opened its doors to public discussion, partly based on this feasibility study. In 2002, the Thai government decided to postpone the construction of the power station and change its location and fuel. In short, the feasibility study represented one of the essential tools for Thai society to collectively find its most appropriate development path and reach its own development goals and objectives.

The main lesson to be learned from this case is that it supports several of the previous cases in the observation that organizations linked to existing technologies cannot be expected to invent or to implement alternatives representing radical technological change. These alternatives do not form part of their discourses, interests, and perceptions. Moreover, such organizations will automatically use economic evaluation methods (based on applied neoclassical economics) focusing on the objectives that are included in the interests and perception of the existing technologies. This perception does not include political objectives if these imply radical technological change. The latter aspect becomes especially visible in the Thai case in which the political objectives were well defined.

The design and promotion of alternatives based on radical technological changes had to come from NGOs and local citizens. To show how these alternatives would meet official political objectives better than the coal-fired power station, feasibility

Table 9.11 Summary of main economic and employment results.

Main economic results (billion baht)	Reference	Alternative	Difference
Economic costs	152	153	1
Import costs	117	39	−78
GDP contribution	35	115	80
Income revenue contribution	7	22	15
Rural economy contribution	5	93	88
Main employment results (discounted at 7 percent per year)			
Employment (man-years/year)	7009	34,976	27,968
Rural (man-years/year)	1527	28,311	26,784

studies had to be made on the basis of concrete institutional economics including the identification of relevant political aims and objectives. The national energy policy office did not conduct such a systematic analysis, even though the legal procedures required that an evaluation of the project should be made regarding the national Thai development plan. Instead, the analysis had to come from outside the energy authorities.

10   Case X: The economic council case (2002–2003)

In 2002, the Danish Economic Council made an evaluation of the Danish energy policies of the 1990s. This evaluation represents a story of how applied neoclassical economics may be blind to relevant real-life technical and economic circumstances as well as relevant political goals. Instead of analyzing and recognizing the real-life economic situation of the 1990s, in which unemployment, among other issues, was a major problem, the methodology disregarded such issues and implicitly presupposed that the economic situation was optimal. Instead of measuring the success of energy policies against politically well-defined economic objectives, such as decreasing unemployment, improving the balance of payment, and creating technological and industrial development, the methodology excluded these issues from the evaluation. The case description is based on the book chapter "Feasibility Study Cases" (Lund et al., 2007c) and the articles "Economic Council Assessment of Wind Power and Small CHP Is Based on Incorrect Assumptions" (Lund, 2002a),[af] "Saved Operation Costs and Saved Capacity Are Missing" (Lund, 2002b),[ag] and "A Better Society without Wind Power and CHP?" (Lund, 2002c).[ah]

[af] Translated from Danish: *Vismands vurdering af vindkraft og decentral kraft/varme bygger på fejl i forudsætningen.*

[ag] Translated from Danish: *Sparet drift og anlæg medregnes ikke.*

[ah] Translated from Danish: *Et bedre samfund uden vindkraft og kraft/varme?*

This case study is supplemented with alternative calculations in accordance with the Choice Awareness strategies. Thus, the case illustrates both the difference between applied neoclassical ideology and feasibility studies based on a concrete institutional economy and the consequences of designing feasibility studies in one way or the other.

As already explained in the previous sections and chapters, Denmark has conducted an active and innovative energy policy for many years. From 1972 to 1990, the major objective of the policy was to decrease dependency on oil imports, while during the 1990s, Danish energy policy mainly aimed at reducing CO_2 emissions, as expressed by several national energy plans and further supported by the Kyoto Protocol. In the 1970s and 1980s, the strategic objective of protecting energy security was met by energy savings in combination with an increased domestic production of oil and natural gas and the replacement of oil with other fuels, mainly coal and natural gas. Houses were insulated and power stations replaced oil with imported coal. In the 1990s, a number of environmental energy policy measures were adopted, including the expansion of wind power and the replacement of more than a hundred district heating boilers with small-scale CHP stations distributed throughout the country.

In the spring of 2002, the Danish Economic Council (DEC) published an evaluation of the environmental impact of Danish energy policy measures implemented during the 1990s. The analysis was carried out as a cost-benefit analysis based on applied neoclassical economics. Costs were calculated as investment and maintenance costs, while benefits were calculated as saved fuel costs and environmental benefits. Additionally, the analysis calculated tax and consumer misallocations, which were included as costs. The study included a number of different policy measures and reached the general conclusion that resources had not been used in the most cost-effective way, since many of the elements revealed negative net results. Consequently, the main recommendation of the DEC was that cost-benefit analyses should be used much more in the future to avoid making the same mistake again.

The expansion of wind power and small-scale CHP stations was included in the analysis, and as a side effect of the study, the results showed that these investments had a negative net result, meaning that Denmark should never have made such investments. However, this statement was the result of an analysis that included a premise of overcapacity that would lead to negative results for almost any new investment, including traditional technologies such as coal-fired steam turbines. Benefits relating to employment, balance of payment, and technological innovation were either not included or were underestimated due to the premises of the applied neoclassical cost-benefit methodology.

Consequently, the study provides an excellent example of the consequences that applied neoclassical premises may have on the implementation of elements of radical technological change. The following description is based on a critique of the Economic Council's evaluation that my colleagues and I presented (Lund, 2002a,b,c; Serup and Hvelplund, 2002). The critique was discussed in the Danish public debate after the publication of the study, together with critiques from others as well. Two important elements were emphasized in the discussion. The first was the exclusion of the capacity benefits of wind power and small CHP stations, illustrating how the new technologies were evaluated on the premises of the old technologies to such an extent that the old technologies became even less cost-effective when evaluated on the same premises. The second was the omission of benefits of employment, import savings, and technological innovation.

Missing capacity benefits (unfair premises)

The results of the cost-benefit analyses were divided into a number of Danish energy policy measures in the 1990s. Three of the elements are discussed here: namely, the investments in small-scale CHP stations totaling 826 MW, private wind turbines totaling 1769 MW, and wind turbines owned by the utility companies with a total capacity of 1098 MW. The results of the analysis made by the DEC are shown in Table 9.12.

As Table 9.12 shows, the DEC concluded that neither small CHP stations nor wind power were economically feasible to Danish society. They all resulted in a deficit in the cost-benefit analysis. In the cost-benefit study, all prices are given in terms of *consumer prices*. In this specific case, the consumer price derived from factor prices (market prices of construction, maintenance, and fuel costs) multiplied by a factor of 1.25, equal to the Danish VAT percentage. Moreover, the prices were converted into 2002 prices by applying an annual inflation of 2 percent, while the net present value was calculated on the basis of an interest rate of 6 percent.

Small-scale CHP stations totaling 826 MW were built between 1992 and 1998. The cost of this investment was calculated on the basis of a survey made by the Danish

Table 9.12 Results of the Danish Economic Council's cost-benefit analysis.

Billion DKK consumer prices year 2002	826 MW small-scale chp stations built 1992–1998	1769 MW private wind turbines built 1992–2011	1098 MW utility wind turbines built 1992–2008
Cost			
Investment	12.2	20.6	8.0
Maintenance	8.9	6.7	2.6
Tax dislocation	1.7	0.6	0.1
Consumer misallocation	0.5	0.8	0.4
Sum	23.3	28.7	11.1
Benefit			
Saved fuel	2.1	9.9	4.4
Saved capacity	–	–	–
Saved maintenance	–	–	–
Environmental benefit	16.5	14.8	6.0
Sum	18.6	24.7	10.4
Net benefit	−4.7	−4.0	−0.7

Source: Danish Economic Council, 2002. Dansk Økonomi, Forår 2002. Danish Economic Council, Copenhagen.

Energy Agency, leading to a total cost of 5.3 billion DKK. These investments corresponded to approximately 12 billion DKK, when calculated as described previously.

The crucial premise of the cost-benefit calculation was this sentence in the Economic Council's main report:

> *The reason for the negative net result is first of all the fact that Denmark had plenty of electricity production capacity.*[ai]

Furthermore, in the attachment, it states:

> *Consequently, only variable fuel costs at the power stations are saved.*[aj]

The analysis did not include saved variable maintenance costs, and no explanation was given for this omission. To illustrate the importance of this crucial premise, the critics made a calculation including the benefits of both saved investments and saved O&M costs (Lund, 2002a). Table 9.13 shows the result. As can be seen, the deficit of Table 9.12 is now turned into a surplus for all three technologies.

Table 9.13 Same calculations as in Table 9.12 but with saved capacity and maintenance included.

Billion DKK consumer prices year 2002	826 MW small-scale chp stations built 1992–1998	1769 MW private wind turbines built 1992–2011	1098 MW utility wind turbines built 1992–2008
Cost			
Investment	12.2	20.6	8.0
Maintenance	8.9	6.7	2.6
Tax dislocation	1.7	0.6	0.1
Consumer misallocation	0.5	0.8	0.4
Sum	23.3	28.7	11.1
Benefit			
Saved fuel	2.1	9.9	4.4
Saved capacity	14.8	4.8	2.4
Saved maintenance	7.2	4.3	2.9
Environmental benefit	16.5	14.8	6.0
Sum	40.6	33.8	15.7
Net benefit	+17.3	+5.1	+4.6

[ai] Translated from Danish: *Årsagen til tabet er først og fremmest, at der i udgangspunktet var rigelig elproduktionskapacitet i Dansmark.* Danish Economic Council (2002), p. 16.

[aj] Translated from Danish: *… og dermed kun har sparet brændselsudgifter på kraftværkerne.* Danish Economic Council (2002), p. 210.

The calculation was based on the alternative costs related to the production of electricity at coal-fired power stations, with capital costs of 8 million DKK/MW and maintenance costs of 60 DKK/MWh, as per the official expectations of the Danish Energy Agency. In these calculations, small-scale CHP stations were assigned a capacity factor of 100 percent, while wind power was assigned a capacity factor of 20 percent. The 20 percent represents the capacity likely to be available with the same probability as the capacity of big power stations.

For small CHP stations, the saved capacity costs amounted to 6.6 billion DKK from 1992 to 1998. Meanwhile, after including inflation and projected consumer prices according to a 2002 value, as previously described, the costs corresponded to 14.8 billion DKK in the analysis. Following the same procedure, the saved maintenance costs totaled 7.2 billion DKK. Table 9.13 shows how much this influences the results. A negative result totaling 9.4 billion DKK for all three elements was converted into a surplus of 27 billion DKK. Consequently, the premise of not including capacity payment due to overcapacity totally dominated the analysis.

However, this premise is not related to the question of whether small CHP stations are better or worse than big power stations; it only expresses that if no new power stations are needed, it cannot pay to build them. This premise would probably result in negative figures of almost any power station investment. To illustrate this point, the same cost-benefit analysis was conducted for two large power stations built in the same period, 1996 and 1998, respectively. Each power station had a capacity of 400 MW, which made them comparable to the 826 MW capacity of small-scale CHP stations. The two power stations were evaluated on exactly the same premises and according to the same methodology as shown before in the DEC study, with the results as shown in Table 9.14.

Table 9.14 Cost-benefit analysis of 800 MW large power stations conducted by applying the same method as in Table 9.12.

Billion DKK consumer prices year 2002	400 MW coal-fired power station year 1998	400 MW natural gas-fired power station year 1996	Sum
Cost			
Investments	5.3	4.2	9.5
Maintenance	2.1	0.8	2.9
Tax dislocation	–	–	–
Consumer misallocation			
Sum	7.4	5.0	12.4
Benefit			
Saved fuel	1.0	−4.7	−3.7
Saved capacity	–	–	–
Saved maintenance	–	–	–
Environmental benefit	0.1	3.4	3.5
Sum	1.1	−1.3	−0.2
Net benefit	−6.3	−6.3	−12.6

As Table 9.14 shows, the results of building large power stations totaling 800 MW were even more negative than the results of building small-scale CHP stations with a total capacity of 826 MW. The large power stations with a total capacity of 800 MW had a deficit of 12.6 billion DKK, while the small-scale CHP stations with a total capacity of 826 MW had a deficit of only 4.7 billion DKK. It should again be emphasized that both calculations assumed that no new capacity was needed and that no maintenance costs could be saved.

The cost-benefit analysis concluded that, on the basis of the premise of overcapacity (because both wind turbines and large and small power stations were built), no power production units should have been built; they all came out with a negative result. The problem was that *if* society had followed the advice and no power capacity had been built, then the premise of overcapacity would not be valid, and, consequently, the results made no sense.

Meanwhile, one can ask the question: "How should Denmark have provided new capacity during the 1990s to achieve the lowest possible cost?" The answer can be found in Tables 9.12 and 9.14. Small-scale CHP stations should be preferred to big power stations. Consequently, one can conclude that the premise of overcapacity is not valid, as it leads to the erroneous conclusion that small-scale CHP is not a cost-effective option.

Balance of payment, employment, and technological innovation

The DEC made the assumption that the effects that innovation, employment, and the balance of payment had on the green energy policy in the 1990s had no societal value. The Danish energy policy of the 1990s resulted in an increase in the export of green energy technologies from 4 billion DKK in 1992 to 30 billion DKK in 2001. Using an import share of 50 percent, the net effect of this export on the balance of payment was around 15 billion DKK in 2001. Thus, the export of these technologies became as large as, for instance, the very important export of Danish bacon or as important as the effects of the Danish North Sea oil adventure in relation to the balance of payment.

The DEC was criticized (Serup and Hvelplund, 2002) for not including such positive effects in their cost-benefit analysis, but they refused to attribute these effects to the energy policy in the 1990s, and they would not accept the suggestion that export should be accredited a benefit value in their analyses. They argued that the Danish surplus on the balance of payment and the international financial situation at the time meant that the positive balance of payment effects of a given technology were of no importance.

The counterarguments to this position were (1) Denmark still had a rather considerable foreign debt of ~200 billion DKK, and (2) the positive Danish balance of payment was caused by, among other factors, self-sufficiency in oil and gas, resulting in a balance of payment net income of ~15 billion DKK per year. This positive effect was expected to disappear, as Danish oil wells would gradually run dry from 2005 to 2020. Finally, the positive Danish balance of payment was also the result of the preceding net effect of the export of green energy technologies, corresponding to 15 billion DKK.

A former member of the DEC presented similar arguments, and the Danish Minister of Foreign Affairs even stated that successful export was the backbone of Danish economic development. The export subsidies from the Danish Export Council showed exports of 6 DKK for each DKK of subsidy. This number was an indication of the politicians' *willingness to pay*, leading to the conclusion that an export of 6 DKK had a societal value of 1 DKK. This meant that an export of 30 billion DKK in 2001 could be accredited an extra social value of 5 billion DKK and, consequently, the export values could be included in the analysis based on a political "willingness to pay" principle.

If the annual export of green energy technologies from 1992 to 2001 and the expected effects until 2011 were included in the calculations, the accumulated 2001 value of these benefits amounted to between 40 and 60 billion DKK, depending on the interest rate and prognoses for export after 2001. These 40–60 billion DKK should be added to the benefits shown in Tables 9.12 and 9.13. As can be seen, the inclusion of the balance of payment effect of the energy policy of the 1990s totally changed the results of the calculations.

Moreover, the employment effects were also important to consider. The development of the green energy sector in Denmark created around 30,000 new Danish jobs. Thus, among others, these technologies contributed to the relatively low unemployment rate in Denmark in 2002 compared with other European countries. The employment linked to these technologies was, to a large extent, located in rural areas, where unemployment was relatively high compared with the Danish average.

The DEC argued that unemployment in general was not a problem in Denmark and that the people employed in the green energy sector, especially the wind power sector, would have been employed in other sectors if the green energy sector had not existed. They also argued that the energy policy of the 1990s had no specific innovation effects and therefore excluded this type of consideration from their analysis. Thus, no systematic analysis was made of the innovation effects of the energy policy in the 1990s. It is rather obvious that the analytical tools linked to neoclassical economic theory cannot be used for such analyses. Nevertheless, the hypothesis assuming that innovation effects may be of great significance is still valid, and an analysis based on concrete institutional economics involving innovation theory may be able to qualify and quantify these effects.

Conclusions and reflections

The discussion on the results of the Danish Economic Council's analysis in 2002 illustrates how traditional cost-benefit thinking may experience severe problems when applied to the evaluation of political strategies and technological change. Thus, this method was unable to integrate objectives of decreasing unemployment and improving the balance of payment by increasing exports in the Danish case.

Based on the premise of overcapacity, the Economic Council concluded that wind power and small-scale CHP were not cost-effective options. However, this premise would characterize other production units, including traditional power stations, as even less cost-effective. Consequently, the premise of overcapacity was not valid for the analysis of wind power and small-scale CHP. After correcting these misleading assumptions, the result changed its characterization of wind power and CHP from "not cost-effective" to "cost-effective." If the value of exports was included on the basis of a "willingness to pay" principle, the feasibility would even be improved. This case illustrates the importance of relating the feasibility studies of a certain case to the context and the objectives of the decision makers for whom the analyses are intended.

The main point of this case is that applied neoclassical cost-benefit analyses are blind to the relevant benefits of new technologies in terms of fulfilling political objectives, including balance of payment, job creation, and industrial innovation. Moreover, the interests and perceptions of the organizations linked to the old technologies seem to influence the concrete design of applied cost-benefit analyses to such an extent that new technologies, such as wind power and small CHP stations (but not old technologies such as coal-fired power stations), are evaluated on premises that would characterize any technology as "not cost-effective."

The institutional set-up of the Danish Economic Council was not capable of applying the same premises to a similar analysis of the old technologies. This analysis had to come from outside. When promoted by others, this analysis faced heavy resistance from the representatives of the DEC. Even the obviously incorrect premise that *variable* operation costs were not saved in the DEC analysis, when CHP and wind power replaced the electricity production of larger coal-fired power stations, was defended by the Economic Council.

Instead, the main purpose and main conclusion of the DEC study seemed to be to advocate for the use of the same type of applied neoclassical cost-benefit analyses, presumably to avoid that the politicians "made the same mistakes again." In this case, this meant that the politicians should be prevented from promoting new technologies to fulfill the political objectives of, among others, job creation and industrial development. Instead, they should accept the premises of applied neoclassical economics, meaning that the benefit of such objectives was not included in the calculation.

The DEC study methodology leaves the politicians with a Catch-22 choice (see Chapter 2). Based on the assumption of overcapacity, the DEC advocates that the politicians should never have allowed the implementation of wind power and small CHP plants nor should they have allowed big power stations. However, if they had followed such advice and not allowed any of the three investments, no overcapacity would have occurred in the system, and then all three investments should have been made. Nevertheless, the least-cost solution would have been to invest in wind power and small CHP and not in large-scale power stations. Unfortunately, the DEC study failed to identify that fact and instead advocated the opposite solution.

11   Case XI: The Ida Energy Plan 2030 (2006–2007)

The case of the IDA Energy Plan 2030 illustrates how concrete descriptions of a technical alternative plan can provide information that macroeconomic models, based on applied neoclassical theory, are unable to identify. The IDA Energy Plan 2030 was made by the Danish Society of Engineers, as already described in Chapter 7. The plan proposed concrete technical measures to decrease CO_2 emissions, maintain the security of supply, and exploit Danish business potentials. If implemented, the plan would imply more or less the same direct costs as the business-as-usual reference presented by the Danish Energy Agency. Consequently, the benefits achieved regarding the environment, energy security, and business potentials were additional and did not involve costs from a pure economic point of view, when compared to the reference. However, the plan emphasized that the measures proposed would not be implemented per se under the existing institutional conditions. The parliament and the government would have to pursue an active energy policy. If implemented, the IDA Energy Plan 2030 would increase the share of renewable energy from 15 to 45 percent. The case description is based on the books *The Energy Plan 2030 of the Danish Society of Engineers, Background Report. Technical Energy Systems Analysis, Socio-Economic Feasibility Studies and Estimation of Business Potentials* (Lund and Mathiesen, 2006)[ak] and *Socio-Economic Costs of Increasing Renewable Energy and Energy Conservation* (Danish Ministries of Transport and Energy, Taxation and Finance, 2007).[al]

As already described in Chapter 7, the plan was the result of the Organization's Energy Year 2006 with more than 40 seminars involving more than 1600 participants. The plan was published in December 2006, soon after the prime minister, in his opening speech to the Danish Parliament in October 2006, pronounced that a 100 percent renewable energy share was the long-term target for Danish society.

Later, in January 2007, the Danish government announced a proposal of the first steps to be taken. This was published as a small paper describing some overall targets to be reached by the year 2025. The paper was to be followed by concrete proposals for public regulation measures to be discussed by the government and the opposition parties to develop a common Danish energy plan and define a policy based on a huge majority.

One of the proposed targets announced by the Danish government in January 2007 was to increase the share of renewable energy in Denmark from 15 to 30 percent by 2025. When asked by the press why the increase should not be larger if the overall goal was to reach 100 percent, the representatives of the government answered that this would be too expensive for Danish society, referring to a socioeconomic loss of 5 billion DKK.

This answer provided an interesting contradiction to the IDA Energy Plan in the public discussions. On one hand, the government claimed that it would cost 5 billion DKK to reach a 30 percent share of renewable energy; on the other hand, the IDA Energy Plan claimed that a share of 45 percent could be reached without any extra costs. In fact, Danish society would save approximately 15 billion DKK/year, if the IDA Energy Plan was implemented.

The debate led to a number of interesting discussions. One focused on the interest rate used in the calculations. The IDA Energy Plan applied a 3 percent rate, while the

---

[ak] Translated from Danish: *Ingeniørforeningens Energiplan 2030, baggrundsrapport. Tekniske energisystemanalyser, samfundsøkonomiske konsekvensvurdering og kvantificering af erhvervspotentialer.*

[al] Translated from Danish: *Samfundsøkonomiske omkostninger forbundet med udbygning med vedvarende energi samt en øget energispareindsats.*

government insisted on using a rate of 6 percent. This discussion raised the critique from several Danish economists, who claimed that the 6 percent rate was too high and would mean that the benefit of long-term investments was underestimated. Moreover, the rate of 6 percent was high compared to the rate used in similar countries such as Sweden, Germany, the United Kingdom, and many other European countries (Andersen, 2007).

Another issue was the fuel price. The IDA Energy Plan used the average oil prices in 2006 equal to an oil price level of 68 USD/barrel, while the government used only 50 USD/barrel. However, in the beginning, it was impossible to provide a specific explanation for the difference between the IDA Energy Plan and the government's plan for the simple reason that the governmental calculations were not published. However, after being criticized for referring to secret calculations, three ministries published a small paper of seven pages on February 8, 2007 (referred to in the beginning of this section). This paper is indeed an interesting subject of investigation. It represents a classic example of applied neoclassical economics and is based on the existing institutional set-up to such a degree that the outcome becomes irrelevant to the decision-making process.

The main difference between the two calculations is that the ministries assumed that no change would take place in existing institutions, while the IDA Energy Plan calculated the costs independently from present taxation, and so forth. Thus, as described in Chapter 7, the IDA Energy Plan was based on a specific identification of a number of investments and was therefore able to account for the exact number of wind turbines and CHP stations included in the analysis. It was thus able to identify the resulting costs in terms of investment as well as fuel, operation, and maintenance. However, the ministries' calculations were, according to their own statement, the result of

a macroeconomic model based on simplified and general assumptions.[am]

Consequently, the ministries were not able to account for the renewable energy sources used in their calculations or the investment costs assumed. When calculating on the basis of the macroeconomic model, the ministries, among other things, assumed that

the existing regulation of the fuels chosen is maintained "and renewable energy is promoted" by applying a consistent subsidy rate.[an]

[am] Translated from Danish: *Omkostningsberegningerne er fremkommet i en generel økonomisk model, der bygger på forenklede og generelle antagelser.* Danish Departments of Transport and Energy, Taxation and Finance (2007), p. 3.

[an] Translated from Danish: ... *de nuværende reguleringer af brændselsvalg fastholdes* ... and ... *VE fremmes ved en ensartet støttesats...* . Danish Departments of Transport and Energy, Taxation and Finance (2007), p. 2.

It seems as if the whole evaluation was made on the basis of existing institutions and taxation rates. Regarding the calculation of energy conservation, such assumptions became particularly obvious in the following statement.

> *Fundamentally, the calculations are based on the assumption that a number of socio-economically cost-effective energy conservation investments are not implemented due to different barriers, inexpedient market incentives, or the lack of knowledge among the actors.*[ao]

Thus, the ministries acknowledged both the presence of institutional barriers to the implementation of socioeconomically cost-effective investments and the existence of such investments. However, they made the calculations on the assumption that these barriers would remain and came to the conclusion that the implementation of energy savings and renewable energy was costly to society because it would require a lot of subsidies. The subsidy itself was not the main problem, but in the macroeconomic models, the idea that such subsidies may cause a distortion in relation to market equilibrium generally created a problem. However, the ministries did not comment on the fact that the above-mentioned barriers to the implementation of socioeconomically cost-effective investments did actually constitute such a distortion. Nor did they comment on the possibility that the removal of the barriers would likely remove a distortion rather than create one.

Conclusions and reflections

The IDA Energy Plan and the ministries agreed on the fundamental fact that potential socioeconomically cost-effective investments did exist which were not implemented due to different institutional market barriers.

The IDA Energy Plan identified a number of cost-effective investments in renewable energy, energy conservation, and efficiency measures simply by comparing the needed investments to the saved fuel and operation costs. Based on these observations, the IDA Energy Plan recommended that both the Danish Parliament and the government led an active energy policy, thus removing barriers to the implementation of economically feasible investments.

The ministries used applied neoclassical theory in the form of macroeconomic models when they calculated the costs of implementing renewable energy and energy conservation. In such models, no changes in existing market institutions are assumed. Moreover, a present situation of equilibrium is assumed: "We are living in the best of all worlds." Thereby, the models assume that the present market institutions provide the optimal use of resources in society. Based on these assumptions, the introduction of subsidies is expected to distort the market mechanism, which will then increase the

---

[ao] Translated from Danish: *Beregningerne er grundlæggende baseret på en antagelse om, at en række samfundsøkonomisk fordelagtige energibesparelser ikke gennemføres som følge af forskellige blokeringer, uhensigtsmæssige incitamenter på markedet eller manglende viden om mulighederne hos aktørerne.* Danish Departments of Transport and Energy, Taxation and Finance (2007), p. 6.

costs of society. In the Danish case, the government decided to go for a moderate objective on the basis of this calculation.

In short, the ministries' calculations provided the information that if Denmark had no institutional market barriers (and consequently no socioeconomically feasible investments to promote), the cost of increasing the share of renewable energy and other investments would be high. However, since Denmark actually had institutional market barriers (which both parties agreed to be the case), the IDA Energy Plan was able to identify socioeconomically feasible investments, which could be implemented if these barriers were removed in the proper way.

The mistake arose when the assumptions behind the ministries' calculations were not communicated, and the politicians drew the conclusion that renewable energy could not be increased without substantial socioeconomic costs. This conclusion was wrong. The IDA Energy Plan 2030 showed that the politicians indeed did have a choice. Socioeconomically feasible investments could be made if the institutional barriers were removed—in other words, if the politicians decided for an active energy policy.

Again, the main lesson to be learned here is that applied neoclassical economics—in this case in the form of macroeconomic models assuming that we are already living in the best of all worlds—are blind to the benefits that could be achieved by introducing alternatives based on radical technological change. Moreover, these models are blind to the identification of institutional barriers and therefore cannot contribute with relevant information to the political decision-making process. The models cannot define the best alternative, nor can they provide relevant information on how to implement the alternatives.

12   Summary

In this chapter, a number of cases were examined that all represent empirical examples of applying Choice Awareness strategies to specific decision-making processes. All of the cases concern energy investments since 1982. They all focus on collective decision making in a process involving many people and organizations that represent different interests and discourses, as well as different levels of power and influence. Moreover, they all involve political objectives of implementing often radical technological change in society, that is, measures that imply significant institutional changes. Regarding the Choice Awareness theses and strategies formulated in Chapters 2 and 3, the following can be learned from these cases.

Existing organizations initiate old technology proposals

In most cases, proposals were made by organizations that were linked to existing technologies, and projects were proposed that fit well into these organizations. Typically, only one proposal was put forward. No alternatives representing radical technological change were proposed by the organizations.

- *Nordkraft*: The power companies proposed a coal-based and centralized solution. No alternatives involving energy conservation, the expansion of CHP, or renewable energy were put forward by the power companies.
- *Aalborg Heat Planning*: Three alternatives were put forward; none of these representing the combination of CHP and other fuels than coal.
- *Nordjyllandsværket*: The power companies proposed a coal-fired power station. A natural gas alternative suggested by members of the board of representatives was left out of the decision-making process. No alternatives based on renewable energy were put forward by the power companies.
- *Transmission Line*: The power companies proposed a 400 kV airborne connection, which was originally part of a centralized power supply plan from the 1960s. Technical alternatives that fit better into decentralized energy supply systems were never proposed by the power companies.
- *Lausitz*: The power and mining companies proposed a renovation and expansion of existing coal-fired power stations and thereby implicitly suggested to heat the houses without using CHP. No proposals to combine the introduction and expansion of CHP with savings and renewable energy were made by the power and mining companies.
- *Prachuap Khiri Khan*: The power companies proposed to build a large coal-fired power station based on imported hard coal. No technical alternative involving DSM, CHP, and the use of domestic biomass resources was presented by the power companies.

From these cases, it can be observed that organizations linked to existing technologies will initiate project proposals within their organizational framework. One cannot expect alternatives representing radical technological change to originate from these organizations. It is outside their discourse, their interest and their perception.

Objectives of radical technological change are disregarded

In several cases, the proposals have been contradictory to the parliamentary energy objectives expressed or to other politically defined objectives or wishes.

- *Nordkraft*: Local city council members expressed preference for an alternative based on natural gas but did not have the power or the resources to even ensure a proper description of this alternative.
- *Aalborg Heat Planning*: The law told the municipality to choose the socioeconomically best solution. However, this solution did not fit well into the preference of the city council and was disregarded in the definition of potential alternatives discussed in the public participation phase.
- *Nordjyllandsværket*: The proposal was in conflict with the official parliamentary energy plan, Energy 2000, which did not suggest that Denmark should build new big coal-fired power stations. Instead, Denmark was to expand CHP.
- *Transmission Line*: The official parliamentary energy plan, Energy 2000, advocated changes from a centralized to a decentralized supply system.
- *Prachuap Khiri Khan*: The power station proposal ignored the fulfillment of a number of relevant and well-defined political objectives of the Thai society regarding economics, environment, rural development, job creation, and industrial innovation.

From these cases, it can be observed that, even when political decisions have been made implying the wish for radical technological change, the organizations linked to existing technologies will continue to initiate project proposals within their organizational framework.

Alternatives must come from someone else

Alternatives representing radical technological change had to come from others. When these alternatives were introduced, the existing organizations sought to eliminate them from the decision-making processes:

- *Nordkraft*: Local citizens were the ones who described and promoted a concrete technical alternative. The alternative was analyzed by the citizens themselves, and the analysis showed that it would have clear advantages compared to the projected conversion to coal.
- *Aalborg Heat Planning*: Alternatives representing radical technological change came from the university and local citizens. When these alternatives were introduced, the municipality responded by disregarding them in the comparative analyses. Then (when comparative analyses showed an inconvenient result), new analyses were made, and the inconvenient analysis was withheld from the city council, and, finally, a discussion of the letterhead used was initiated instead of a discussion of the contents of the proposal.
- *Nordjyllandsværket*: Alternatives involving energy conservation and the expansion of CHP and renewable energy came from the university and local citizens. When the alternatives were introduced, the power companies tried to eliminate the choices from the debate. The choice-eliminating strategies included hiding differences in the prognoses of the electricity demand.
- *Transmission Line*: The proposal of technically different alternatives came from local citizens. When these alternatives were introduced, the power companies tried to eliminate the choice from the debate. The choice-eliminating strategies included (1) claiming that the alternatives were not technically possible; (2) when technical calculations made by the citizens proved the opposite, claiming that the analyses were based on incorrect data; and (3) withholding the correct data by referring to the risk of endangering "national security."
- *Lausitz*: Local civic organizations involved foreign researchers in the design of alternatives based on energy conservation, CHP, and renewable energy, and the local civic organizations had to promote these alternatives themselves. When the alternatives were introduced into the debate, they were heavily opposed by the power and mining companies.
- *Prachuap Khiri Khan*: Local energy organizations designed and promoted an alternative based on energy demand management, CHP, and domestic biomass resources. When the alternative was introduced into the debate, it was heavily opposed by the power companies and potential investors.

Institutional change is essential

For society, it is not an efficient public regulation measure to force existing organizations to act at variance with their interests. This measure has to be supplemented by institutional changes.

- *Aalborg Heat Planning*: The authorities failed to force the municipality to choose the socio-economically best solution, when this solution resulted in increasing consumer heat prices compared to less favorable alternatives.

- *Nordjyllandsværket*: The authorities failed to stop new coal-fired power stations, as prescribed by Energy 2000.
- *European EIA Procedures*: The legal procedures failed to ensure the description of cleaner alternatives, when these alternatives represented radical technological change.

Applied neoclassical economics provide irrelevant information

When the evaluation of technological change, such as the introduction of renewable energy systems, is based on applied neoclassical economics, the analyses often ignore relevant political objectives and instead provide information that is irrelevant to the political decision-making process.

- *Biogas*: The cost-benefit analysis based on applied neoclassical economics was not capable of relating the case to the most important economic objectives of the government and the Danish Parliament of the time, that is, job creation and the improvement of the balance of payment.
- *Economic Council*: The cost-benefit analysis was not able to relate the case to the government's political objectives of job creation, improvement of the balance of payment, and industrial innovation. Moreover, the evaluation of new technologies was made solely on the premises of coal-fired power stations. For example, not even the saved variable operation costs of coal-fired power stations were included as a benefit when electricity production was replaced by wind power and small CHP. Wind power and small CHP were analyzed on an assumption of overcapacity that would characterize any investment as not cost-effective. The same assumption was not used in relation to coal-fired power stations.
- *IDA Energy Plan 2030*: The analyses made by the ministries were based on the assumption that barriers to socioeconomically feasible investments would not change. Moreover, the macroeconomic models generally assumed that society found itself in an optimal situation in which the market would provide the best allocation of resources, even though institutional barriers hindering the implementation of socioeconomically feasible investments had been defined.

Concrete institutional economics provide relevant information

When the evaluation of technological change is based on concrete institutional economics, it is possible to relate the change in question to relevant political objectives and thus provide information that is relevant to the political decision-making process.

- *Biogas*: The analysis showed how the implementation of biogas could help fulfill all relevant political objectives of Danish society, including job creation and economic growth, better than the reference.
- *Prachuap Khiri Khan*: The analyses revealed that an alternative consisting of energy demand management, CHP, and renewable energy could meet the relevant and well-defined political objectives of Thai society better than a coal-fired power station, when these political objectives were included in the design of the analysis.
- *Economic Council*: The method based on concrete institutional economics showed that when relating the analysis to relevant political objectives, the overall characterization of the technologies in question changed from "not cost-effective" to "cost-effective."

- *IDA Energy Plan 2030*: The economic calculation based on well-described specific invest-ments identified an alternative that would be able to improve economic growth and create industrial development. However, institutional change was required for society to implement the strategy and achieve these benefits.

Concrete alternatives raise Choice Awareness

The design and promotion of concrete technical alternatives raise the awareness that society has a choice and make it possible to discuss choices in public.

- *Nordkraft*: An alternative representing energy conservation, the expansion of CHP, and renewable energy was put forward and was subject to public discussion.
- *Aalborg Heat Planning*: An "alternative 4" representing the combination of CHP and natural gas was put forward and was subject to public discussion.
- *Nordjyllandsværket*: An alternative based on small CHP and renewable energy was put for-ward and was subject to public discussion.
- *Transmission Line*: Several technical alternatives, including proposals that fit better into decentralized energy supply systems, were put forward and were subject to public discussion.
- *Lausitz*: An alternative combining the introduction and expansion of CHP in combination with savings and renewable energy was put forward and was subject to public discussion.
- *Green Energy Plan*: A plan was designed that would improve the environment and, at the same time, create the basis for economic growth. The plan was put forward and was subject to public debate prior to the introduction of the governmental energy plan, Energy 21.
- *Prachuap Khiri Khan*: A technical alternative involving DSM, CHP, and the use of domestic biomass resources was put forward and was subject to public discussion.
- *IDA 2030 Energy Plan*: A concrete alternative aimed at decreasing CO_2 emissions, improv-ing energy security, and creating industrial development was put forward and was subject to public discussion in the debate of new Danish energy policies.

Concrete alternatives help identify institutional barriers

The existence of concrete technical alternatives makes it possible to identify institu-tional barriers to the implementation of radical technological change and design pub-lic regulation measures to overcome these barriers.

- *Aalborg Heat Planning*: It was possible to identify fuel taxation rules that constituted a bar-rier to the implementation of the socioeconomically least-cost solutions prescribed by the law on heat supply and to design concrete public regulation measures to overcome this barrier.
- *Nordjyllandsværket*: It was possible to identify how the existing organization of the power companies involved mechanisms that would automatically lead to the wish for a new power station. Moreover, a number of mechanisms were revealed that led to the proposals of "new-corporative" as opposed to "old-corporative" regulation (see Chapter 3).
- *Transmission Line*: The discussion on "national security" led to an amendment of the law, thus including utility companies in the Danish legal act of public access to information.

13   Conclusions

True choices between relevant alternatives are essential when society is to implement political objectives implying radical technological change. However, the cases presented clearly indicate that the true choice will not appear by itself.

The organizations linked to existing technologies are typically those that take on the responsibility for proposing new projects. However, the institutional set-up of such organizations implicitly entails that they cannot generate proposals that imply radical technological changes; it is outside their discourse, interests, and perception. Even if they made such proposals, it would be out of their reach to implement these. Consequently, the set-up by default does not involve the true choice. On the contrary, it will repeatedly result in Hobson's Choice: choose a project that fits well into the existing organizations or no project at all. Also, when political decisions have been made implying the wish for radical technological change, the organizations linked to existing technologies will still continue to initiate project proposals within their organizational framework.

In all cases presented, alternatives representing radical technological change had to come from outside the organizations representing the existing technologies. When these alternatives were introduced, the existing organizations sought to eliminate them from the decision-making processes by use of various means.

Consequently, the elimination of true choice is twofold. First, those organizations that typically make the proposals cannot generate proposals of radical technological change. It is outside their discourse. Second, if such proposals are promoted by others, the same organizations will seek to eliminate these proposals from the decision-making process. It is not in their interest to do otherwise.

The proper evaluation of alternatives representing radical technological change cannot be done by using applied neoclassical economics in the form of applied cost-benefit analyses and macroeconomic equilibrium models. As illustrated by the preceding cases, the existing institutional set-up seems to be tailored to the practical application of these models to such an extent that they cannot analyze other alternatives and identify those that will fulfill relevant political economic objectives in the best way. Moreover, the models are not designed to assist the identification of relevant market barriers and failures to implement the socioeconomically best solutions.

The introduction of concrete technical alternatives contributes to raising the public awareness of true choice. Discussions arising from this awareness and public debates will contribute to the identification of institutional barriers at all levels: from market barriers, such as taxation rules, to barriers within the democratic infrastructure, such as the design of the committees that provide information to the political decision-making processes.

Conclusions and recommendations 10

This book has unified and deduced the learning and results of a number of separate studies related to the coherent understanding of how society can perceive and implement renewable energy systems in the context of achieving a carbon neutral society. The subject has been dealt with by combining two aspects: the formulation of a *Choice Awareness* theory used as a theoretical framework approach to understanding how major technological changes, such as renewable energy, can be implemented both at the national and international levels, and the development of an energy system analysis method and tool for the design of *renewable energy systems*, including the results of various analyses. In this chapter, the results are presented as conclusions related to principles and methodologies as well as recommendations for the practical implementation of renewable energy systems in Denmark and in similar countries. Conclusions and recommendations are made first regarding Choice Awareness and then with regard to renewable energy systems.

1 Choice Awareness

Two Choice Awareness theses were formulated in Chapter 2. Based on discourse and power theory, the theses assume that the perception of reality and the interests of existing organizations will seek to hinder radical institutional changes by which these organizations will lose power and influence. Choice Awareness theory states that one key factor in this manifestation is the societal feeling of having either *a choice* or *no choice*.

The first thesis states that when society defines and seeks to implement objectives that will imply radical technological change, the influence of existing organizations will often seek to create a perception indicating that society has no choice but to implement technologies that constitute existing positions. The results of such influence will take various forms and will typically be based on the applied neoclassical assumption that the existing institutional and technological set-up is defined by the market and that the market works in such a way that it will, by definition, identify and implement the best solutions.

The second thesis of the Choice Awareness theory argues that in this situation, society will benefit from focusing on Choice Awareness; that is, raising the awareness that alternatives *do* exist and that it is possible to make a choice. Four key strategies for raising Choice Awareness were presented.

The *description and promotion of concrete alternatives* are core strategies in Choice Awareness. They are the essential first steps that must be taken to change the focus of a public discussion. Typically, the promotion of a concrete alternative will lead to two changes. First, it becomes obvious that society indeed does have a

Renewable Energy Systems. https://doi.org/10.1016/B978-0-443-14137-9.00010-3

choice, and second, the focus of the discussion changes from "Yes, it is bad, and so what?" to "Which of the alternatives is the best solution?"

The next strategy is to *consider the relevant economic objectives of the society in question* and include these in economic feasibility studies. Choice Awareness theory is based on the observation that in societal decision-making processes involving radical technological change, existing institutional interests will try to influence the process in the direction of no choice. This influence includes the use of feasibility studies based on methodologies and assumptions supporting existing organizational interests. Consequently, Choice Awareness involves the awareness of how feasibility studies are and should be carried out. The combination of business and socioeconomic analyses can reveal institutional barriers to the implementation of suitable new technologies.

The third strategy is to *propose concrete public regulation measures*. These measures to implement radical technological change cannot be designed on the basis of the aforementioned preconditions of applied neoclassical economic theory. The main problem is that the necessary technical solutions often require new organizations and new institutions. In general, the applied neoclassical model considers the institutional conditions as given and does not consider them to be modifiable via public regulation.

The last strategy emphasizes that *public decision making does not occur in a political vacuum*. The decision-making process is shaped by various political and economic interest groups in society that strive to protect their profits or pursue their values. It is therefore important to be aware of the fact that, typically, existing technologies are well represented in the democratic decision-making infrastructure, whereas potential future technologies are weakly represented, if they are represented at all. Therefore, the strategy advocates leading a "new-corporative" regulation in which the representatives of new technologies are given high priority in various committees.

Chapter 9 presented 11 empirical cases that applied Choice Awareness strategies to specific decision-making processes related to energy investments since 1982. They all concerned collective decision making in a process involving many persons and organizations representing different interests and discourses as well as different levels of power and influence. In addition, they all involved political objectives of implementing radical technological change in society, that is, changes that would imply significant institutional changes.

The cases represent situations in which politically decided objectives call for the proposal of alternatives representing radical technological change. Regarding Choice Awareness theses and strategies, the following can be learned from the cases.

In general, the cases started out with one, and only one, alternative introduced by existing organizations and fitting well into such organizations. No alternatives representing radical technological change were ever proposed by the same organizations. The cases clearly illustrate how alternatives representing radical technological change cannot be expected to come from the organizations linked to existing technologies. Such alternatives simply do not form part of their perception of the problem and its solution. Furthermore, even if these alternatives were proposed, they could hardly be implemented by the same organizations, since such activities would be out of their reach.

In several cases, the proposals presented have been contradictory to parliamentary energy objectives or other politically defined objectives or wishes. Existing organizations disregarded the political objectives if they implicated radical technological change. Instead they presented proposals that fit well into their own institutional and organizational set-up.

In all cases, it was possible to identify alternatives based on radical technological changes that were likely to fulfill the political objectives better than the proposals of existing organizations. However, these alternatives were presented by others, such as universities, nongovernmental organizations, or local citizens.

When radical alternatives were introduced, the existing organizations sought to eliminate the choice of these alternatives from the decision-making process. The cases show a wide range of such eliminating mechanisms including the following:

- Disregarding inconvenient alternatives
- Downplaying contradictions to official forecasts of energy demands
- Withholding technical data with reference to national security
- Withholding inconvenient results from political decision makers
- Discussing the letterhead used instead of the content of the letter to incriminate the senders

The socioeconomic feasibility of a radical technological change is often evaluated on the basis of applied neoclassical economics. The cases show how such analyses ignore relevant political objectives and provide information that is irrelevant to the political decision-making process. However, when conducting feasibility studies based on concrete institutional economics, it is possible to relate the case in question to relevant political objectives and provide information that is relevant to the actual political decision-making process.

Consequently, the cases confirm the first Choice Awareness thesis that states that in a situation of radical technological change, the influence of existing organizations will often seek to create a perception indicating that society has no choice but to implement technologies that will save and constitute existing positions.

Regarding the second Choice Awareness thesis, the cases show that the design and promotion of concrete technical alternatives raise the awareness that society has a choice and makes it possible to discuss this choice in public. In almost all cases, the existence of a concrete alternative has given rise to public debates. Moreover, the existence of concrete technical alternatives makes it possible to identify institutional barriers to the implementation of radical technological change and design public regulation measures to overcome these barriers.

The cases show that it is not an efficient public regulation measure for society to force existing organizations to act against their interests. Thus, in the Aalborg Heat Planning case, the authorities failed to force the municipality to choose the socioeconomically best solution, when this solution would result in higher consumer heat prices compared to less socioeconomic alternatives. In the case of Nordjyllandsværket, the authorities failed to implement the Energy 2000 to stop new coal-fired power stations, and in the case of European Environmental Impact Assessment procedures, Danish practice failed to ensure that proper analyses of relevant cleaner alternatives were made. Such a measure has to be supplemented by institutional changes in the form of changing the market conditions and making it possible for the organizations of new business areas to enter. Consequently, these cases confirm the second thesis of the Choice Awareness theory by showing that society will benefit from raising the awareness that alternatives **do** exist and that it is possible to make a choice.

The article "Choice Awareness: The Development of Technological Choice in the Public Debate of Danish Energy Planning" (Lund, 2000) described how Danish energy policy, for a period of 25 years, was formed by a process of conflicts. This process led to the implementation of radical technological changes, and Denmark shows remarkable results on the international stage. This ability to act as a society has been possible despite the conflicts between representatives of the old and the new technologies. Official energy objectives and plans have been developed in a constant interaction between Parliament and public participation, in which the description of new technologies and alternative energy plans has played an important role. Public participation, and thus the awareness of choices, has been an important factor in the ultimate decision-making process. The conflict-ridden debates should therefore be seen as necessary conditions for further improvements of energy initiatives and programs.

2   Renewable energy systems

The method of designing concrete technical alternatives based on renewable energy technologies has been described and applied to the preceding cases. A distinction has been made between three implementation phases: introduction, large-scale integration, and 100 percent renewable energy systems. In the development of tools and methods for the design and evaluation of renewable energy system alternatives, the two latter phases have been emphasized.

The energy system analysis tool, EnergyPLAN, has been described, and the method and tools have been discussed in relation to the theoretical framework of Choice Awareness. In accordance with the idea of Choice Awareness, the overall aim of the EnergyPLAN model is to analyze energy systems with the purpose of assisting the design of alternatives based on renewable energy system technologies. Regarding Choice Awareness, the following key considerations can be highlighted.

The EnergyPLAN model can make a consistent and comparative analysis of different energy systems based on fossil fuels and nuclear as well as renewable energy systems. When the reference energy system is described, EnergyPLAN makes it possible to conduct a fast and easy analysis of radically different alternatives without losing coherence and consistence in the technical assessment of even complex renewable energy systems.

The EnergyPLAN model seeks to enable the analysis of radical technological changes. The model describes existing fossil fuel systems in aggregated technical terms, which can thereby be changed fairly easily into radically different systems, for example, systems based on 100 percent renewable energy sources (RES). The model divides the input in market-economic analyses into taxes and fuel costs, thereby making it possible to analyze different institutional frameworks in the form of different taxes. Moreover, if more radical institutional structures are to be analyzed, the model can provide purely technical optimizations. This makes it possible to separate the discussion of institutional frameworks, such as specific electricity market designs, from the analysis of fuel and/or CO_2 emissions alternatives. EnergyPLAN has not incorporated the institutional set-up of the electricity market of today as the only optional institutional framework.

The model can calculate the costs of the total system divided into investment costs, operational costs, and taxes, such as CO_2 emission trading costs. Thus, the model is able to provide data for other socioeconomic feasibility studies, such as data including balance of payment, job creation, and industrial innovation. Examples of such studies were given in Chapters 7 and 9.

Regarding the three different implementation phases, the model includes a high number of different technologies relevant for renewable energy systems. Consequently, it serves as a useful tool for making detailed and comprehensive analyses of a wide spectrum of large-scale integration possibilities as well as 100 percent renewable energy systems. Chapter 5 discussed the essence of a wide range of studies of the Danish energy system in which the EnergyPLAN model was applied to the analysis of large-scale integration of renewable energy. At present, the Danish energy system is already characterized by a relatively high share of renewable energy and is therefore suitable for the analysis of further large-scale integration.

Following the presentation of these studies, Chapter 5 presented a method of comparing different energy systems in terms of their ability to integrate RES on a large scale. The question in focus was how to design energy systems with a high ability to use intermittent RES. These systems must be designed in such a way that they can cope with the fluctuating and intermittent nature of RES, especially with regard to the electricity supply.

From a methodological point of view, this raises the problem of how to deal with fluctuations of the same installed capacity of RES, such as photovoltaic (PV), wind, wave, and tidal power, since they differ from one year to the next. To deal with these problems, excess electricity production diagrams have proven to be a suitable and workable methodology. In these diagrams, each energy system is represented by only one curve, which shows the ability of the system to integrate fluctuating RES representing several years of different hour-by-hour fluctuations. The diagrams have proven usable for PV, wind, and wave power as well as combinations of these.

Moreover, the analyses show that in the design of systems that are suitable for large-scale integration of RES, it is important to distinguish between two different issues. The first issue is concerned with *annual amounts* of energy; on an annual basis, the supply of electricity, heat, and fuel must meet the corresponding demands. The other issue is concerned with *time*; the supply of energy must meet the temporal duration of the same demands. The latter is of special importance to the electricity supply.

Regarding such recommendations, there is much to learn from the analyses. Conversion technologies may improve the efficiency of the system, such as heat pumps, and at the same time, feature the possibilities of cost-effective and efficient storage options. On the other hand, "pure" electricity storage technologies, such as compressed air energy storage and hydrogen/fuel cell systems, only contribute marginally to the integration of fluctuating RES, and they also have a low economic feasibility.

Chapter 6 discussed the essence of a number of studies related to the challenge for future energy infrastructures, and defined concepts of *smart electricity grid, smart thermal grids*, and *smart gas grid*. These three concepts have similarities in terms of grid structures allowing for distributed activities involving interaction with consumers and bidirectional flows. To meet this challenge, all grids will benefit from the use of modern information and communication technology as an integrated part of the grids at all levels. However, they also differ in their major challenges. Smart thermal grids face major challenges in the temperature level and the interaction with low-energy buildings. Smart electricity grids face major challenges in the reliability and in the integration of variable renewable electricity production. Smart gas grids face major challenges in mixing gases with different heating value and in the efficient use of limited biomass resources. This includes that future energy system may benefit from new gas grids such as a hydrogen grid or hydrogen may be integrated into the use of existing gas grids.

From a methodological point of view, the main point in Chapter 6 is that each of the three types of smart grids is an important contribution to future renewable energy systems, but each individual smart grid should not be seen as separate from the others or separate from the other parts of the overall energy system. Consequently, Chapter 6 promoted the concept of *smart energy systems* defined as an approach in which smart electricity grids and smart thermal grids as well as smart gas grids are combined and coordinated to identify synergies between them in order to achieve an optimal solution for each individual sector as well as for the overall energy system.

One important example of such a synergy was already highlighted in Chapter 5, that is, the option of using heat storage instead of electricity storage when realizing that some electricity has to be converted to heat preferably by the use of heat pumps to implement an efficient and least-cost overall renewable energy systems solution.

Chapter 6 highlighted another important synergy when coordinating smart electricity and gas grids. A survey of potential pathways for providing gas or liquid fuel to the transportation sector to supplement the direct use of electricity points to the need for power-to-X from electrolysis to boost the conversion of biomass into gas. This need for power to gas for transportation makes it possible to replace the potential long-term need for electricity storage with gas storage, which is both cheaper and more efficient. Consequently, the identification of integration of renewable energy into the electricity supply should not only include the measure of direct use of electricity for transportation but the power-to-X need as well.

The concept of Smart Energy Systems is not only a set of definitions as expressed above; it is also an approach to address the design of affordable and liable transitions of the energy system into a carbon neutral society. Thus in Chapter 6, the hypothesis was formulated that given the complexity of the situation, one cannot find the best solutions of affordable and liable transitions of the energy system into a carbon neutral society within each subsector of the energy system. One must take a holistic and cross-sectoral smart energy systems approach in order to be able to identify the best solution. Therefore, the purpose of subsector studies (no matter if they concern the role of a specific technology or the role of a region or country) should aim at identifying the role to play in context of the overall transition of the whole system rather than aim at decarbonizing the sub-sector on its own.

Chapter 7 presented recent results achieved by applying the EnergyPLAN tool to the design of 100 percent renewable energy systems. The question was how to compose and evaluate these systems. The chapter treated the principal changes in the methods of analysis and evaluation of such systems compared to systems based on fossil fuels with or without large-scale integration of renewable energy. The implementation of 100 percent renewable energy systems adds to the challenges of integrating RES into existing energy systems on a large scale as well as designing future smart energy infrastructures. Not only must fluctuating and intermittent renewable energy production be coordinated with the rest of the energy system, but the energy demands must also be adapted to an economically realistic integration of potential renewable sources. Furthermore, this adjustment must address the differences in the characteristics of different sources, such as biomass fuels and electricity production from wind power.

Based on the case of Denmark, Chapter 7 presented three studies that analyzed the problems and perspectives of converting the traditional energy system into a 100 percent renewable energy system. The first study was a one-person university study that applied the analyses in Chapter 5 to the analysis of a coherent renewable energy system. The second study was based on the technical input from members of the Danish Society of Engineers (IDA). In the third study CEESA, a group of interdisciplinary researchers, made further investigations into, among other things, pathways to produce gas or liquid fuel to supplement the direct use of electricity in the transportation sector. All three studies analyzed the design of coherent and complex renewable energy systems, including the suitable integration of energy conversion and storage technologies. Furthermore, the studies were based on detailed hour-by-hour simulations carried out by use of the EnergyPLAN model.

From a methodological point of view, it can be concluded that the design of future 100 percent renewable energy systems is a very complex process. On the one hand, a broad variety of measures must be combined to reach the target, and on the other hand, each individual measure has to be evaluated and coordinated with the new overall system. In the case of the IDA Energy Plan, the process has been achieved by combining a creative phase, involving the inputs of a number of experts, and a detailed analysis phase with technical and economic analyses of the overall system, providing feedback on the individual proposals. In a back-and-forth process, each proposal was formed in such a way that it combined the best of the detailed expert knowledge with the ability of the proposal to integrate into the overall system, in terms of technical innovation, efficient energy supply, and socioeconomic feasibility.

Chapter 8 added to Chapter 7 by presenting guidelines for the design of 100 percent renewable energy systems in the context of achieving carbon neutral societies. As a core element of the guidelines, a country would have to fulfill its objectives of increased renewable energy and CO_2 reductions in a way that would fit well into a context in which the rest of the world realistically would be able to do the same. Such overall criteria imply several choices and issues. Some of the most important are to include the country share of *international aviation and shipping* and not exceed the country share of a global *sustainable use of biomass*, as well as enabling the country's contribution in terms of *flexibility and reserve capacity* to increase the integration of non-continuous RES such as wind and solar into the electricity supply.

In Chapter 8, these guidelines were applied to the case of a Smart Energy Denmark 2045 scenario. The key findings of the case are that a fully decarbonized energy systems in the context of a carbon neutral society by 2045 is technically doable within the limitations of Denmark's share of global sustainable biomass resources. Compared to a business-as-usual reference, the societal costs do not have to increase. Hence, the case illustrates that a green transition of the Danish economy and society is doable, economically responsible, but also realistic within a sufficiently ambitious deadline.

In order to achieve the best role for bioenergy, it is recommended to conduct a multiple technology solution with a focus on the use of different biomass technologies by enabling synergies between them and the energy system. With regard to the transportation sector, a series of measures are necessary to implement. The main focus should be on lowering the growth of the transportation demand in inefficient modes of transportation, while not compromising the development and growth of mobility. The issues of balancing and stabilizing the power grid become essential in a 100 percent renewable energy system. On one hand, this is a challenge that should not be neglected; on the other hand, this is also fully doable and many of the technologies are already implemented. In Chapter 8, it was illustrated how to deal with the matter using the Smart Energy Denmark 2045 scenario as a case.

References

Amin, S.M., Wollenberg, B.F., 2005. Toward a smart grid. IEEE Power Energ. Mag. 2005, 34–41.

Andersen, M.S., 2007. Responsum angående samfundsøkonomiske analyser af vedvarende energi. Danish Society of Engineers, Copenhagen.

Andersen, A.N., Lund, H., 2007. New CHP partnerships offering balancing of fluctuating renewable electricity productions. J. Clean. Prod. 15 (3), 288–293.

Andersen, A.N., Lund, H., Pedersen, N., 1995. Vurdering af behovet for en højspændingsledning mellem aalborg og Århus. Himmerlands Energi og Miljøkontor, Skørping.

Bachrach, P., Baratz, M.S., 1962. Two faces of power. Am. Polit. Sci. Rev. 56 (4), 947–952.

Blarke, M.B., Lund, H., 2008. The effectiveness of storage and relocation options in renewable energy systems. Renew. Energy 33 (7), 1499–1507.

Bromberg, L., Cheng, W.K., 2010. Methanol as an Alternative Transportation Fuel in the US: Options for Sustainable and/or Energy-Secure Transportation. Sloan Automotive Laboratory, Massachusetts Institute of Technology. http://www.afdc.energy.gov.

Brynolf, S., Taljegard, M., Grahn, M., Hansson, J., 2017. Electrofuels for the transport sector: a review of production costs. Renew. Sust. Energ. Rev. 81, 1887–1905.

Chang, M., Thellufsen, J.Z., Zakeri, B., Pickering, B., Pfenninger, S., Lund, H., Østergaard, P. A., 2021. Trends in tools and approaches for modelling the energy transition. Appl. Energy 290.

Christensen, J., 1998. Alternativer—Natur—Landbrug. Akademisk Forlag, Copenhagen.

Christensen, S., Jensen, P.-E.D., 1986. Kontrol i det stille—om magt og deltagelse. Samfundslitteratur, Copenhagen.

Christensen, P., Lund, H., 1998. Conflicting views of sustainability: the case of wind power and nature conservation in Denmark. Eur. Environ. 8, 1–6.

Clark II, W.W., Bradshaw, T., 2004. Agile Energy Systems: Global Lessons From the California Energy Crisis. Elsevier Press.

Clark II, W.W., Eisenberg, L., 2008. Agile sustainable communities: on-site renewable energy generation. Util. Pol. 16 (4), 262–274.

Clark II, W.W., Lund, H., 2001. Civic markets: the case of the California energy crisis. Int. J. Global Energ. Issues 16 (4), 328–344.

Clark, I.I., Woodrow, W., Fast, M., 2008. Qualitative Economics: Toward a Science of Economics. Coxmoor Press, Oxford, UK.

Connolly, D., Lund, H., Mathiesen, B.V., Leahy, M., 2010. A review of computer tools for analyzing the integration of renewable energy into various energy systems. Appl. Energ. 87, 1059–1082.

Connolly, D., Mathiesen, B.V., Ridjan, I., 2014. A comparison between renewable transport fuels that can supplement or replace biofuels in a 100% renewable energy system. Energy 73, 110–125.

Connolly, D., Lund, H., Mathiesen, B.V., 2016. Smart energy Europe: the technical and economic impact of one potential 100% renewable energy scenario for the European Union. Renew. Sust. Energ. Rev. 60, 1634–1653.

Copenhagen County, 1996. Avedøreværket, blok 2, VVM-redegørelse, vurdering af miljømæssige konsekvenser Copenhagen County.

Creutzig, F., Ravindranath, N.H., Berndes, G., Bolwig, S., Bright, R., Cherubini, F., et al., 2015. Bioenergy and climate change mitigation: an assessment. GCB Bioenergy 7, 916–944.

Crossley, P., Beviz, A., 2009. Smart energy systems: transitioning renewables onto the grid. Renew. Energy Focus 11 (5), 54–59.

Dahl, R.A., 1961. Who Governs? Yale University Press, New Haven.

Danish Economic Council, 2002. Dansk Økonomi, Forår 2002. Danish Economic Council, Copenhagen.

Danish Energy Agency, 1991. Biogasfællesanlæg, hovedrapport fra koordineringsudvalget for biogasfællesanlæg. Danish Energy Agency, Copenhagen.

Danish Energy Agency, 1996. Danmarks vedvarende energiressourcer. Danish Energy Agency, Copenhagen.

Danish Energy Agency, 2001. Rapport fra arbejdsgruppen om kraftvarme- of VE-electricitet. Danish Energy Agency, Copenhagen.

Danish Energy Agency, 2008. Alternative drivmidler i transportsektoren (Alternative Fuels in the Transport Sector). Danish Energy Agency, Copenhagen. http://www.ens.dk.

Danish Energy Agency, 2010. Technology Data for Energy Plants. Danish Energy Agency, Copenhagen.

Danish Energy Agency, 2011. Forudsætninger for samfundsøkonomiske analyser på energiområdet, April 2011. Danish Energy Agency, Copenhagen. http://www.ens.dk, 978-87-7844-895-8.

Danish Energy Agency, 2012. Energy Policy in Denmark. Danish Energy Agency, Copenhagen.

Danish Energy Agency, 2014. Energiscenarier frem mod 2020, 2035 og 2050 (Energy Scenario Towards 2020, 2035 and 2050). Danish Energy Agency, Copenhagen.

Danish Energy Agency, 2020a. Data, tabeller, statistikker og kort. In: Energistatistik 2019. Danish Energy Agency, Copenhagen.

Danish Energy Agency, 2020b. Biomass Analysis. Danish Energy Agency, Copenhagen.

Danish Energy Agency, 2022. Technology Data Catalogue. Danish Energy Agency, Copenhagen.

Danish Forest Industry (Dansk Skovforening), 2020. Træ Til Energi. Fremtiden for dansk skovflis med 2050—Analyse og anbefalinger (The future of Danish wood chips by 2050—Analysis and recommendation).

Danish Government, 2008. Aftale mellem regeringen (Venstre og Det Konservative Folkeparti), Socialdemokraterne, Dansk Folkeparti, Socialistisk Folkeparti, Det Radikale Venstre og Ny Alliance om den danske energipolitik i årene 2008–2011. 21 February 2008. Danish Government, Copenhagen.

Danish Government, 2011. Et Danmark der står sammen, regeringsgrundlag. (October 2011).

Danish Ministries of Transport and Energy, Taxation and Finance, 2007. Samfundsøkonomiske omkostninger forbundet med udbygning med vedvarende energi samt en øget energispareindsats. Danish Ministries of Transport and Energy, Copenhagen.

Danish Ministry of Climate, Energy and Building, 2012. The Danish Energy Agreement of March 2012, Copenhagen.

Danish Ministry of Climate, Energy and Utilities, 2020. The Danish Climate Act. No 965 of 26 June 2020, Copenhagen.

Danish Ministry of Energy, 1981. Energiplan 81. Danish Ministry of Energy, Copenhagen.

Danish Ministry of Energy, 1990. Energy 2000: A Plan of Action for Sustainable Development. Danish Ministry of Energy, Copenhagen.

Danish Ministry of Environment and Energy, 1996. Energy 21, The Danish Government's Action Plan for Energy. Danish Ministry of Environment and Energy, Copenhagen.

Danish Ministry of Finance, 1991. Finansredegørelse 91. Budgetdepartementet, Danish Ministry of Finance, Copenhagen.

De Barbosa, L.S.N.S., Bogdanov, D., Vainikka, P., Breyer, C., 2017. Hydro, wind and solar power as a base for a 100% renewable energy supply for South and Central America. PLoS One 12.

Duke Energy Carolinas, 2007. Notice Announcing the Availability of an Analysis of Renewable Portfolio Standard for the State of North Carolina and Request for Public Comment. North Carolina Utilities Commission, Raleigh.

ELSAM, 1991. Netudvidelsesplan 91. ELSAM, Skærbæk.

ELSAM, 2005. VEnzin Vision 2005. ELSAM, Skærbæk. www.elsam.com.

European Commission, 2011. White Paper on Transport Roadmap to a Single European Transport Area—Towards a Competitive and Resource-Efficient Transport System, 2011. Publications Office of the European Union, Luxembourg.

European Commission, 2011a. Smart Grid Mandate. Standardization Mandate to European Standardisation Organisations (ESOs) to Support European Smart Grid Deployment. European Commission, Directorate General for Energy. Ares (2011)233514, 02/03/2011.

European Commission, 2011b. Energy 2020—A Strategy for Competitive, Sustainable and Secure Energy. Publications Office of the European Union, Luxembourg.

European Commission, 2018. A Clean Planet for all. A European long-term strategic vision for a prosperous, modern, competitive and climate neutral economy. In: In-Depth Analysis in Support of The Commission Communication COM(2018) 773, Brussels, 28 November 2018.

European Council, 1985. Council directive of June 27, 1985 on the assessment of certain public and private projects on the environment. European Council, 85/337/EEC.

Flyvbjerg, B., 1991. Rationalitet og magt. Det konkretes videnskab. Akademisk Forlag, Copenhagen.

GDS Associates, 2006. A Study of the Feasibility of Energy Efficiency as an Eligible Resource as Part of a Renewable Portfolio Standard for the State of North Carolina. North Carolina Utilities Commission, Raleigh.

Gollakota, A.R.K., Kishore, N., Gu, S., 2018. A review on hydrothermal liquefaction of biomass. Renew. Sust. Energ. Rev. 81, 1378–1392.

Grahn, M., 2004. Why Is Ethanol Given Emphasis Over Methanol in Sweden? Department of Physical Resource Theory, Chalmers University of Technology, Göteborg.

Hannula, I., 2016. Hydrogen enhancement potential of synthetic biofuels manufacture in the European context: a techno-economic assessment. Energy 104, 199–212.

Hansen, K., Mathiesen, B.V., Connolly, D., 2011. Technology and Implementation of Electric Vehicles and Plug-In Hybrid Electric Vehicles. Aalborg University, Copenhagen.

Heller, J., 1961. Catch-22. Simon & Schuster, New York.

Hrbek, J., 2019. Status Report on Thermal Gasification of Biomass and Waste 2019. International Energy Agency, Paris.

Hvelplund, F., 2005. Erkendelse og forandring. Teorier om adækvat erkendelse og teknologisk forandring med energieksempler fra 1974–2001. Aalborg University, Department of Development and Planning.

Hvelplund, F., Lund, H., 1988. De lave kulafgifter ødelægger varmeplanlægningen—en kommenteret aktsamling fra varmeplanlægningen i Aalborg. Aalborg University Press.

Hvelplund, F., Lund, H., 1998a. Feasibility Studies and Public Regulation in a Market Economy. Aalborg University, Department of Development and Planning.

Hvelplund, F., Lund, H., 1998b. Rebuilding without restructuring the energy system in East Germany. Energy Policy 26 (7), 535–546.

Hvelplund, F., Lund, H., 1999. Energy planning and the ability to change. The East German example. In: Lorentzen, A., Widmaier, B., Laki, M. (Eds.), Institutional Change and Industrial Development in Central and Eastern Europe. Ashgate, pp. 117–141.

Hvelplund, F., Illum, K., Lund, H., Mæng, H., 1991. Dansk energipolitik og ELSAMs udvidelsesplaner—et oplæg til en offentlig debat. Aalborg University, Department of Development and Planning.

Hvelplund, F., Knudsen, N.W., Lund, H., 1993. Erneurung der Energiesysteme in den neuen Bundesländern—aber wie? Netswerk Dezentrale EnergieNutzung e.V, Potsdam.

Hvelplund, F., Knudsen, N.W., Lund, H., 1994. Kommentar zur Kritik der Lausitzer Braunkohlen AG an der AalborgUniversität Studie. Department of Development and Planning, Aalborg University.

Hvelplund, F., Lund, H., Serup, K.E., Mæng, H., 1995. Demokrati og forandring. Aalborg University Press, Aalborg.

Hvelplund, F., Lund, H., Sukkumnoed, D., 2007. Feasibility studies and technological innovation. In: Kørnøv, L., Thrane, M., Remmen, A., Lund, H. (Eds.), Tools for Sustainable Development. Aalborg University Press, Copenhagen, pp. 593–618.

IEA (International Energy Agency), 2013. Smart Grid. Paris. http://www.iea.org/topics/smartgrids/. (Accessed January 2013).

IEA (International Energy Agency), 2016. World Energy Outlook 2016. Paris.

IEA (International Energy Agency), 2021a. Net Zero by 2050. A Roadmap for the Global Energy Sector. Paris.

IEA (International Energy Agency), 2021b. Energy Technology Perspectives 2012: Pathways to a Clean Energy System. Paris.

Ingeniøren, 2008a. Klimakrav anno 2050: Elbiler og underjordiske CO2-lagre. Ingeniøren, Copenhagen. 11 January, 1st section, p. 14.

Ingeniøren, 2008b. Undergrunden bliver den nye CO2-losseplads. Ingeniøren, Copenhagen. 18 January, front page.

Janis, I.L., 1972. Victims of Groupthink: A Psychological Study of Foreign-Policy Decisions and Fiascoes. Houghton Mifflin Company, Boston.

Janis, I.L., 1982. Groupthink. Houghton Mifflin Company, Boston.

Johannsen, R.M., Mathiesen, B.V., Kermeli, K., Crijns-Graus, W., Østergaard, P.A., 2023. Exploring pathways to 100% renewable energy in European industry. Energy 268, 126687.

Kany, M.S., Mathiesen, B.V., Skov, I.R., Korberg, A.D., Thellufsen, J.Z., Lund, H., Sorknæs, P., Chang, M., 2022. Energy efficient decarbonisation strategy for the Danish transport sector by 2045. Smart Energy 5.

Kaspersen, L.B. (Ed.), 2020. Klassisk og Moderne Samfundsteori. Hans Reitzels Forlag, Copenhagen.

Kleindorfer, G.B., O'Neill, L., Ram, G., 1998. Validation of simulation: various positions in the philosophy of science. Manag. Sci. 44 (8), 1087–1099.

Korberg, A.D., Mathiesen, B.V., Clausen, L.R., Skov, I.R., 2021. The role of biomass gasification in low-carbon energy and transport systems. Smart Energy 1.

La Capra Associates, 2006. Analysis of a Renewable Portfolio Standard for the State of North Carolina. North Carolina Utilities Commission, Raleigh.

Lackner, K.S., 2009. Capture of carbon dioxide from ambient air. Eur. Phys. J Spec. Top. 176, 93–106.

Laclau, E., Mouffe, C., 1985. Hegemony and Socialist Strategy. Verso, London.

Liu, W., Lund, H., Mathiesen, B.V., Zhang, X., 2011a. Potential of renewable energy systems in China. Appl. Energ. 88 (2), 518–524.

Liu, W., Lund, H., Mathiesen, B.V., 2011b. Large-scale integration of wind power into the existing Chinese energy system. Energy 36 (8), 4753–4760.

Lukes, S., 1974. Power. A Radical View, Macmillan, London.

Lund, H., 1984. Da ELSAM lærte Aalborg om planlægning. Byplan, January 1984.

Lund, H., 1988. Energiafgifterne og de decentrale kraft/varme-værker. Aalborg University Press, Aalborg.

Lund, H., 1990. Implementaring af bæredygtige energisystemer (Implementation of Sustainable Energy Systems) (PhD dissertation). Department of Development and Planning, Aalborg University. Skriftserie nr. 50, 1990.

Lund, H., 1992a. Samfundsøkonomisk Projektvurdering og virkemidler med biogasfællesanlæg som eksempel. Aalborg University Press, Aalborg.

Lund, H., 1992b. Et Miljø- og Beskæftigelses Alternativ til ELSAMs planer om 2 kraftværker— en viderebearbejdelse af det tidligere fremsatte "Alternativ til et kraftværk i Nordjylland". Department of Development and Planning, Aalborg University.

Lund, H., 1996. Elements of a Green Energy Plan Which Can Create Job Opportunities. General Workers Union (SID), Copenhagen.

Lund, H., 1999a. Implementation of energy-conservation policies: the case of electric heating conversion in Denmark. Appl. Energ. 64, 117–127.

Lund, H., 1999b. A green energy plan for Denmark, job creation as a strategy to implement both economic growth and a CO_2 reduction. Environ. Resour. Econ. 14, 431–439.

Lund, H., 2000. Choice awareness: the development of technological and institutional choice in the public debate of Danish energy planning. J. Environ. Pol. Plann. 2, 249–259.

Lund, H., 2002a. Vismands vurdering af vindkraft og decentral kraft/varme bygger på fejl i forudsætningen. Department of Development and Planning, Aalborg University.

Lund, H., 2002b. Sparet drift og anlæg medregnes ikke. Naturlig. Energi. 24 (11), 16.

Lund, H., 2002c. Et bedre samfund uden vindkraft og kraft/varme? Naturlig Energi 25 (2), 19–20.

Lund, H., 2003a. Excess electricity diagrams and the integration of renewable energy. Int. J. Sustain. Energ. 23 (4), 149–156.

Lund, H., 2003b. Flexible energy systems: integration of electricity production from CHP and fluctuating renewable energy. Int. J. Energ. Tech. Pol. 1 (3), 250–261.

Lund, H., 2003c. Distributed generation. Int. J. Sustain. Energ. 23 (4), 145–147.

Lund, H., 2004. Electric grid stability and the design of sustainable energy systems. Int. J. Sustain. Energ. 24 (1), 45–54.

Lund, H., 2005. Large-scale integration of wind power into different energy systems. Energy 30 (13), 2402–2412.

Lund, H., 2006a. The Kyoto mechanisms and technological innovations. Energy 31 (13), 1989–1996.

Lund, H., 2006b. Large-scale integration of optimal combinations of PV, wind and wave power into the electricity supply. Renew. Energy 31 (4), 503–515.

Lund, H., 2007a. Renewable energy strategies for sustainable development. Energy 32 (6), 912–919.

Lund, H., 2007b. Introduction to sustainable energy planning and policy. In: Kørnøv, L., Thrane, M., Remmen, A., Lund, H. (Eds.), Tools for Sustainable Development. Aalborg University Press, Aalborg, pp. 439–462.

Lund, H., Andersen, A.N., 2005. Optimal designs of small CHP plants in a market with fluctuating electricity prices. Energ. Convers. Manag. 46 (6), 893–904.

Lund, H., Bundgaard, P., 1983. Når ELSAM Planlægger. Ålborg, Brønderslev … brikker i spillet. OOA Aalborg.

Lund, H., Hvelplund, F., 1993. Aktsamling om Nordjyllandsværket del I og II. Department of Development and Planning, Aalborg University.

Lund, H., Hvelplund, F., 1994. Offentlig Regulering og Teknologisk Kursændring. Aalborg University Press.

Lund, H., Hvelplund, F., 1997. Does environmental impact assessment really support technological change? Analyzing alternatives to coal-fired power stations in Denmark. Environ. Impact Assess. Rev. 17 (5), 357–370.

Lund, H., Hvelplund, F., 1998. Energy, employment and the environment: towards an integrated approach. Eur. Environ. 8 (2), 33–40.

Lund, H., Hvelplund, F., 2012. The economic crisis and sustainable development: the design of job creation strategies by use of concrete institutional economics. Energy 43, 192–200.

Lund, H., Kempton, W., 2008. Integration of renewable energy into the transportation and electricity sectors through V2G. Energy Policy 36 (9), 3578–3587.

Lund, H., Mathiesen, B.V., 2006. Ingeniørforeningens Energiplan 2030, baggrundsrapport. Tekniske energisystemanalyser, samfundsøkonomiske konsekvensvurdering og kvantificering af erhvervspotentialer. Danish Society of Engineers, Copenhagen.

Lund, H., Mathiesen, B.V., 2009. Energy system analysis of 100 percent renewable energy systems. Energy 34 (5), 524–531.

Lund, H., Mathiesen, B.V., 2012. The role of carbon capture and storage in a future sustainable energy system. Energy 44, 469–476.

Lund, H., Münster, E., 2001. AAU's Analyser. Rapport fra arbejdsgruppen om kraftvarme- og VE-elektricitet, Bilagsrapport. Danish Energy Agency, Copenhagen.

Lund, H., Münster, E., 2003a. Modelling of energy systems with a high percentage of CHP and wind power. Renew. Energy 28, 2179–2193.

Lund, H., Münster, E., 2003b. Management of surplus electricity-production from a fluctuating renewable-energy source. Appl. Energ. 76 (1–3), 65–74.

Lund, H., Münster, E., 2006a. Integrated energy systems and local energy markets. Energy Policy 34 (10), 1152–1160.

Lund, H., Münster, E., 2006b. Integrated transportation and energy sector CO_2 emission control strategies. Transpor. Pol. 13 (5), 426–433.

Lund, H., Salgi, G., 2009. The role of compressed air energy storage (CAES) in future sustainable energy systems. Energ. Convers. Manag. 50 (5), 1172–1179.

Lund, H., Hvelplund, F., Kass, I., Dukalskis, E., Blumberga, D., 1999a. District heating and market economy in Latvia. Energy 24, 549–559.

Lund, H., Hvelplund, F., Sukkumnoed, D., Lawanprasert, A., Natakuatoong, S., Nunthavorakarn, S., Blarke, M.B., 1999b. Sustainable energy alternatives to the 1,400 MW coal-fired power plant under construction in Prachuap Khiri Khan. In: A Comparative Energy, Environmental and Economic Cost-Benefit Analysis. Thai-Danish Cooperation on Sustainable Energy With Sustainable Energy Network for Thailand (SENT), Nakhon Ratchasima.

Lund, H., Hvelplund, F., Ingermann, K., Kask, Ü., 2000. Estonian energy system. Proposals for the implementation of a cogeneration strategy. Energy Policy 28, 729–736.

Lund, H., Hvelplund, F., Nunthavorakarn, S., 2003. Feasibility of a 1400 MW coal-fired power-plant in Thailand. Appl. Energ. 76 (1–3), 55–64.

Lund, H., Østergaard, P.A., Andersen, A.N., Hvelplund, F., Mæng, H., Münster, E., Meyer, N.I., 2004. Lokale energimarkeder. Department of Development and Planning, Aalborg University.

Lund, H., Šiupšinskas, G., Martinaitis, V., 2005. Implementation strategy for small CHP-plants in a competitive market: the case of Lithuania. Appl. Energ. 82 (3), 214–227.

Lund, H., Andersen, A.N., Antonoff, J., 2007a. Two energy system analysis tools. In: Kørnøv, L., Thrane, M., Remmen, A., Lund, H. (Eds.), Tools for Sustainable Development. Aalborg University Press, Aalborg, pp. 519–539.

Lund, H., Duić, N., Krajačić, G., Carvalho, M., 2007b. Two energy system analysis models: a comparison of methodologies and results. Energy 32 (6), 948–954.

Lund, H., Hvelplund, F., Sukkumnoed, D., 2007c. Feasibility study cases. In: Kørnøv, L., Thrane, M., Remmen, A., Lund, H.L. (Eds.), Tools for Sustainable Development. Aalborg University Press, Aalborg, pp. 619–638.

Lund, H., Möller, B., Mathiesen, B.V., Dyrelund, A., 2010. The role of district heating in future renewable energy systems. Energy 35 (3), 1381–1390.

Lund, H., Hvelplund, F., Mathiesen, B.V., Østergaard, P.A., Christensen, P., Connolly, D., Schaltz, E., Pillay, J.R., Nielsen, M.P., Felby, C., Bentsen, N.S., Meyer, N.I., Tonini, D., Astrup, T., Heussen, K., Morthorst, P.E., Andersen, F.M., Münster, M., Hansen, L.-L.P., Wenzel, H., Hamelin, L., Munksgaard, J., Karnøe, P., Lind, M., 2011a. Coherent Energy and Environmental System Analysis. Aalborg University.

Lund, H., Marszal, A., Heiselberg, P., 2011b. Zero energy buildings and mismatch compensation factors. Energ. Build. 43 (7), 1646–1654.

Lund, H., Andersen, A.N., Østergaard, P.A., Mathiesen, B.V., Connolly, D., 2012. From electricity smart grids to smart energy systems—a market operation based approach and understanding. Energy 42, 96–102.

Lund, H., Werner, S., Wiltshire, R., Svendsen, S., Thorsen, J.E., Hvelplund, F., Mathiesen, B.V., 2014. 4th Generation District Heating (4GDH). Integrating smart thermal grids into future sustainable energy systems. Energy 68, 1–11.

Lund, H., Østergaard, P.A., Connolly, D., Ridjan, I., Mathiesen, B.V., Hvelplund, F., et al., 2016. Energy storage and smart energy systems. Int. J. Sustain. Energy Plan. Manag. 11.

Lund, H., Østergaard, P.A., Chang, M., Werner, S., Svendsen, S., Sorknæs, P., Thorsen, J.E., Hvelplund, F., Mortensen, B.G., Mathiesen, B.V., Bojesen, C., Duic, N., Zhang, X., Möller, B., 2018. The status of 4th generation district heating: research and results. Energy 164, 147–159.

Lund, H., Mathiesen, B.V., Thellufsen, J.Z., Sorknæs, P., Chang, M., Kany, M.S., et al., 2021a. IDAs Klimasvar 2045—Sådan bliver vi klimaneutrale. Ingeniørforeningen IDA, Copenhagen.

Lund, H., Østergaard, P.A., Nielsen, T.B., Werner, S., Thorsen, J.E., Gudmundsson, O., Arabkhoohsar, A., Mathiesen, B.V., 2021b. Perspectives on fourth and fifth generation district heating. Energy 227.

Lund, H., Thellufsen, J.Z., Østergaard, P.A., Sorknæs, P., Skov, I.R., Mathiesen, B.V., 2021c. EnergyPLAN—advanced analysis of smart energy systems. Smart Energy 1.

Lund, H., Skov, I.R., Thellufsen, J.Z., Sorknæs, P., Korberg, A.D., Chang, M., Mathiesen, B.V., Kany, M.S., 2022a. The role of sustainable bioenergy in a fully decarbonised society. Renew. Energy 196, 195–203.

Lund, H., Thellufsen, J.Z., Sorknæs, P., Mathiesen, B.V., Chang, M., Madsen, P.T., Kany, M.S., Skov, I.R., 2022b. Smart energy Denmark. A consistent and detailed strategy for a fully decarbonized society. Renew. Sust. Energ. Rev. 168.

Mai-Moulin, T., Hoefnagels, R., Grundmann, P., Junginger, M., 2021. Effective sustainability criteria for bioenergy: towards the implementation of the european renewable directive II. Renew. Sust. Energ. Rev. 138, 110645.

March, J.G., 1966. The power of power. In: Easton, D. (Ed.), Variations of Political Theory. Prentice-Hall, Englewood Cliffs, NJ.

Mathiesen, B.V., Lund, H., 2009. Comparative analyses of seven technologies to facilitate the integration of fluctuating renewable energy source. IET Renew. Power Gener. 3 (2), 190–204.

Mathiesen, B.V., Lund, H., Nørgaard, P., 2008. Integrated transport and renewable energy systems. Util. Pol. 16 (2), 107–116.

Mathiesen, B.V., Lund, H., Karlsson, K., 2011. 100% renewable energy systems, climate mitigation and economic growth. Appl. Energ. 88 (2), 488–501.

Mathiesen, B.V., Connolly, D., Lund, H., Nielsen, M.P., Schaltz, E., Wenzel, H., Bentsen, N.S., Felby, C., Kaspersen, P., Ridjan, I., Hansen, K., 2014. CEESA 100% Renewable Energy Transport Scenarios Towards 2050. Aalborg University.

Mathiesen, B.V., Lund, H., Hansen, K., Ridjan, I., Djørup, S., Nielsen, S., et al., 2015. IDA's energy vision 2050. In: A Smart Energy System Strategy for 100% Renewable Denmark. Aalborg University.

Methanol Institute China, 2011. The Leader in Methanol Transportation. Methanol Institute. http://www.methanol.org.

Möller, B., 2008. A heat atlas for demand and supply management in Denmark. Manag. Environ. Qual. 19 (4), 467–479.

Möller, B., Lund, H., 2010. Conversion of individual natural gas to district heating: geographical studies of supply costs and consequences for the Danish energy system. Appl. Energ. 87 (6), 1846–1857.

Mouffe, C., 1993. The Return of the Political. Verso, London.

Müller, J., 1973. Choice of Technology in Underdeveloped Countries. Technical University of Denmark, Copenhagen (Ph.D. dissertation).

Müller, J., 2003. A conceptual framework for technology analysis. In: Kuada, J. (Ed.), Culture and Technological Transformation in the South: Transfer or Local Innovation. Samfundslitteratur, Copenhagen.

Müller, J., Remmen, A., Christensen, P., 1984. Samfundets Teknologi—Teknologiens Samfund. Systime, Herning.

Murawski, J., 2006. Report: state could go more green. In: The News and Observer. 14 December 2006.

Nature Protection Appeal Board, 1993. Decision in the Case of a Supplement to the Regional Plan to Enable the Construction of a Combined Power and Heat Generating Station on the Limfjord (Nordjyllandsværket). Nature Protection Appeal Board, Copenhagen.

Nature Protection Appeal Board, 1994. Decision in the Case of a Supplement to the Regional Plan to Enable the Construction of a Combined Power and Heat Generating Station on the Limfiord (Nordjyllandsværket). Nature Protection Appeal Board, Copenhagen.

NESDB, National Economic and Social Development Board, 1996. The 8th Economic and Social Development Plan. Office of the National Economic and Social Development Board, Bangkok.

Nielsen, L.H., Jørgensen, K., 2000. Electric Vehicles and Renewable Energy in the Transport Sector—Energy System Consequences. Risø National Laboratory, Roskilde.

Nord-Larsen, T., Johannsen, V.K., Riis-Nielsen, T., Thomsen, I.M., Jørgensen, B.B., 2020. Skovstatistik 2018 (Forest Statistics 2018). University of Copenhagen, Copenhagen.

North Carolina Utilities Commission, 2008. Annual report of the North Carolina utilities commission: regarding long range needs for expansion of electric generation facilities for service in North Carolina. In: Report to the Joint Legislative Utility Review Committee of the North Carolina General Assembly, North Carolina Utilities Commission, Raleigh.

North Jutland Regional Authority, 1993. Regional Plan for North Jutland, Supplement No. 26 and EIA Analysis for Nordjyllandsværket. North Jutland Regional Authority, Aalborg.

O'Brian, M., 2000. Making Better Environmental Decisions. An Alternative to Risk Assessment. MIT Press, Cambridge, MA; London.

OOA, 1980. Fra Atomkraft til Solenergi. OOAs forlag, Århus.

OOA, 1996. Vedrørende forslag til regionsplantillæg. In: Avedøreværket, blok 2. vol. 6. OOA, Copenhagen.

Orecchini, F., Santiangeli, A., 2011. Beyond smart grids—the need of intelligent energy networks for a higher global efficiency through energy vectors integration. Int. J. Hydrogen Energ. 36 (13), 8126–8133.

Østergaard, P.A., Lund, H., Blaabjerg, F., Mæng, H., Andersen, A.N., 2004. MOSAIK Model af samspillet mellem integrerede kraftproducenter. Department of Development and Planning, Aalborg University.

Østergaard, P.A., Lund, H., Thellufsen, J.Z., Sorknæs, P., Mathiesen, B.V., 2022a. Review and validation of EnergyPLAN. Renew. Sust. Energ. Rev. 168.

Østergaard, P.A., Werner, S., Dyrelund, A., Lund, H., Arabkoohsar, A., Sorknæs, P., Gudmundsson, O., Thorsen, J.E., Mathiesen, B.V., 2022b. The four generations of district cooling—a categorization of the development in district cooling from origin to future prospect. Energy 253.

OVE, 2000. Vedvarende energi i Danmark. In: En krønike om 25 opvækstår. OVEs Forlag, Århus, pp. 1975–2000.

Oxford English Dictionary, 2008. www.aub.aau.dk. (Accessed January–March 2008).

Persson, U., Werner, S., 2011. Heat distribution and the future competitiveness of district heating. Appl. Energ. 88 (3), 568–576.

Pidd, M., 2010. Why modelling and model use matter. J. Oper. Res. Soc. 61, 14–24.

Pillai, J.R., Bak-Jensen, B., 2011. Integration of vehicle-to-grid in western Danish power system. IEEE Trans. Sustain. Energy 2 (1), 12–19.

Pontzen, F., Liebner, W., Gronemann, V., Rothaemel, M., Ahlers, B., 2011. CO2-based methanol and DME—efficient technologies for industrial scale production. Catal. Today 171 (1), 242–250.

Qudrat-Ullah, H., Seong, B.S., 2010. How to do structural validity of a system dynamics type simulation model: the case of an energy policy model. Energy Policy 38 (5), 2216–2224.

Rasburskis, N., Lund, H., Prieskienis, Š., 2007. Optimization methodologies for national small-scale CHP strategies (the case of Lithuania). Energetika 53 (3), 16–23.

Rasmussen, M.G., Andresen, G.B., Greiner, M., 2012. Storage and balancing synergies in a fully or highly renewable pan-European power system. Energy Policy 51, 642–651.

Ren, J., Liu, Y., Zhao, X., Cao, J., 2020. Methanation of syngas from biomass gasification: an overview. Int. J. Hydrog. Energy 45, 4223–4243.

Risø, 1991. Høringsudkast til Samfundsøkonomiske analyser af Biogasfællesanlæg, bilag til hovedrapport fra Koordineringsudvalget for Biogasfællesanlæg. Risø og Statens Jordbrugsøkonomiske Institut.

Rosager, F., Lund, H., 1986. Analyse af eloverløbs- og elkvalitetsproblemer. Bornholms Amt, 1986.

Serup, K.E., Hvelplund, F., 2002. Vindkraftens eksport og beskæftigelses-effekt uden værdi? Naturlig Energi 24 (11), 12–13.

Smart Grids European Technology Platform, 2006. www.smartgrid.eu. Retrieved January 2013.

Smith, G., 1882. The Dictionary of National Biography, Founded in 1882 and Published by Oxford University Press since 1917.

Sorknæs, P., Nielsen, S., Lund, H., Mathiesen, B.V., Moreno, D., Thellufsen, J.Z., 2022. The benefits of 4th generation district heating and energy efficient datacentres. Energy 260.

Stokes, G., Whiteside, D., 1984. One Brain. Dyslexic Learning Correction and Brain Integration. Three In One Concepts. Burbank, CA.

Stokes, G., Whiteside, D., 1986. Advanced one brain. Dyslexia—the emotional cause. In: Three in One Concepts. Burbank, CA.

The Danish Energy Agency, 2019. Basisfremskrivning 2019. Danish Energy Agency, Copenhagen.

Thellufsen, J.Z., Lund, H., Sorknæs, P., Østergaard, P.A., Chang, M., Drysdale, D., Nielsen, S., Djørup, S.R., Sperling, K., 2020. Smart energy cities in a 100% renewable energy context. Renew. Sust. Energ. Rev. 129.

Thellufsen, J.Z., Lund, H., Mathiesen, B.V., Madsen, P.T., Østergaard, P.A., Nielsen, S., Sorknæs, P., Wenzel, H., Münster, M., Rosendal, M.B., Madsen, H., Østergaard, J., Morthorst, P.E., Sørensen, P.B., Andresen, G.B., Victoria, M., 2022. Fakta om Atomkraft: Input til en faktabaseret diskussion af fordele og ulemper ved atomkraft som en del af den grønne omstilling i Danmark. Aalborg Universitet, Aalborg.

Thellufsen, J.Z., Lund, H., Korberg, A.D., Sorknæs, P., Nielsen, S., Chang, M., Mathiesen, B.V., 2023. Beyond sector coupling: utilizing energy grids in sector coupling to improve the European energy transition. Smart Energy 5.

Thomsen, J., Frølund, P., Andersen, H., 1996. In: Andersen, H. (Ed.), Nyere Marxistisk Teori.

United Nations, 2015. Paris Agreement. In: United Nations Climate Change Conference, COP 21 or CMP 11, Paris, France.

United Nations, World Commission on Environment and Development, 1987. Our Common Future. Oxford University Press, Oxford.

US Department of Energy, 2012. Smart Grid/Department of Energy. http://energy.gov/oe/tech nology-development/smart-grid. Retrieved 2012-06-18.

Wenzel, H., Østergaard, N., Triolo, J.M., Toft, L.V., 2020. Energiafgrødeanalysen (The Energy Crop Analysis). University of Southern Denmark, Odense.

Werner, S., 2004. District heating and cooling. In: Encyclopedia of Energy, pp. 841–848. Elsevier Science.

Werner, S., 2005. The European heat-market. In: Ecoheatcool Project. December, 2005. www. ecoheatcool.org.

Wikipedia, 2008. www.wikipedia.dk. (Accessed January–March 2008).

Wiltshire, R., 2011. Towards Fourth Generation District Heating. (Discussion paper).

Wiltshire, R., Williams, J., 2008. European DHC research issues. In: The 11th International Symposium on District Heating and Cooling, 2008, Reykjavik, Iceland.

Index

Note: Page numbers followed by *f* indicate figures and *t* indicate tables.

Printed in the United States
by Baker & Taylor Publisher Services

Electrochemical batteries

The Daniel Cell. The various types of electrochemical devices that have been developed since the early work of Volta, Faraday and their contemporaries will now be discussed, starting with a description of a variety of types of battery. The first example of a cell construction in which hydrogen evolution does not occur and which demonstrates the use of the data of Table 7.1 is the so-called Daniell cell. This cell, invented in 1836 by the British scientist, John Frederic Daniell (see Fig. 7.7), uses the same Cu and Zn electrodes as in the Volta pile but these are positioned in two separate vessels connected by a membrane barrier to allow the passage of ions between the two compartments; a salt bridge can also be used (See Box 7.2 and Fig. 7.3).

The anode is Zn metal and this is immersed in a solution of zinc sulphate; the cathode is Cu and is immersed in a solution of copper sulphate. When the cell is operated as a battery, Zn^{2+} ions are liberated at the anode surface

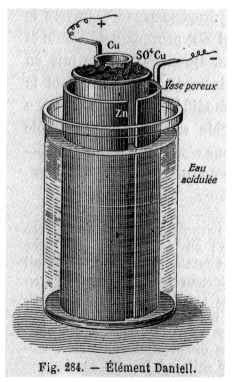

Fig. 284. — Élément Daniell.

Fig. 7.7 An historical diagram of the Daniell cell from 1904. *(https://en.wikipedia.org/wiki/Daniell_cell).*

while Cu^{2+} ions in the cathode compartment are simultaneously reduced to Cu metal that deposits on the cathode. The porous membrane allows the transport of Zn^{2+} and $SO_4^=$ ions between the two cells but without significant mixing of the sulphate solutions, thus maintaining charge neutrality in the two vessels. The total voltage created by the combined reactions, the so-called 'open-cell voltage', is the sum of the two half-cell potentials, $0.7618 + 0.340 = 1.1018\,V$, and a current is generated which passes through the external circuit. The allover reaction is:

$$Zn_{(s)} + Cu^{2+}{}_{(aq)} \rightarrow Zn^{2+}{}_{(aq)} + Cu_{(s)}$$

The zinc electrode is thus corroded during operation while the copper electrode accumulates extra copper. The reaction can be reversed if a suitable potential source is applied across the cell so that the voltage is greater than $1.1\,V$. As a result, the Daniel cell can also be used as a storage device. The EMF of the Daniell cell gradually drops off with time during use due to the deposition of copper in the pores of the membrane barrier and also because of the gradual mixing of Cu and Zn ions between the two vessels.

When the Daniel cell is operated to provide power, acting as a battery, the reaction will continue until all the Cu^{2+} ions have been reduced. The actual voltage created will be less than the open-cell voltage due to the over-voltages required to give the necessary rate of reaction and due to factors such as the internal resistance of the assembly and of the external circuit; the voltage will also gradually decrease as the concentration of the Cu^{2+} solution is depleted. As the allover reaction between metallic zinc and Cu^{2+} ions is exothermic, the temperature of the cell increases, this also causing a decrease in cell voltage.

The Lead Acid Battery. The lead–acid battery will be familiar to the reader as being an important component of the majority of all motor vehicles currently in use, supplying the power for startup and for the operation of on-board instrumentation(Chapter 6). It was first invented by Gaston Planté (a French physicist) in 1859 and it was the first practical example of a truly rechargeable battery. Both the electrodes are composed of lead in the form of a lead–alloy grid. The cathode is coated with sponge lead and the anode is coated with lead dioxide (this having metallic conductivity), both being immersed in a solution of sulphuric acid. The reaction occurring at the negative plate (cathode) is:

$$Pb(s) + HSO_4{}^-{}_{(aq)} \rightarrow PbSO_{4(s)} + H^+{}_{(aq)} + 2e^-$$

and the reaction at the positive plate (anode) is:

$$PbO_{2(s)} + HSO_4{}^-{}_{(aq)} + 3H^+ + 2e^- \rightarrow 2PbSO_{4(s)} + 2H_2O_{(l)}$$

The total reaction is thus:

$$Pb_{(s)} + PbO_{2(s)} + 2H_2SO_{4(aq)} \rightarrow 2PbSO_{4(s)} + 2H_2O_{(l)}$$

The total standard cell voltage, $E_{cell} = 2.05\,V$.

When the battery is fully discharged, both electrodes have become coated with lead sulphate and the sulphuric acid contained in the volume between them is significantly more dilute than with the fully charged battery as a result of the formation of water. When the battery is recharged, this water is used up and the strength of the acid is again increased; at the same time, the positive plate regains its coating of lead dioxide and the negative plate becomes again pure lead. If the recharging process is carried out too quickly, the overvoltage increases to a level at which hydrogen can be formed by electrolysis and the total water content of the battery is thus depleted. Because of this possibility of hydrogen evolution, thus creating an explosive hazard, the recharging process has to be carefully controlled in such a way that the rate is not too high.

Most lead-acid batteries have either six or twelve cells in series, giving output potentials of 12 or 24 V. Apart from their use in automobiles, where their ability to supply the large currents required for the start-up ignition, they are used for backup power supplies. They are also used as power supplies for the electrical motors of traditional submarines for which the use of internal combustion engines would be undesirable. For such uses, it is important to be able to measure the level of charge of the bank of batteries. This is often done for the batteries by measuring the specific gravity of the sulphuric acid, this being directly related to its concentration. Lead-acid batteries have a relatively low electrical capacity as a function of their weight and so it is much more preferable for many applications to use one of the more modern types of battery such as the Li-ion battery to be discussed below. Lead-acid batteries generally have reasonable lifetimes during which they can be recharged many times. However, these batteries do age, particularly if left unused at a low level of charge, when a process known as sulphation occurs; although lead sulphate in a finely divided form is a participant in the reversible process that produces an electric current (see equations above), crystallisation of the sulphate occurs when the battery is left uncharged for a long period. This crystalline fraction is then no longer able to participate in the charging/discharging cycle, the result being a decrease in the power output of the battery to such an extent that it cannot any longer be used. Fortunately, however, it is relatively easy to recycle used lead–acid batteries and the process used for such recycling is generally very efficient; for example, it has been reported that 99% of the lead from used batteries in the USA is recycled. The size of the industry involved in lead–acid battery manufacture and recycling can be seen from the total worldwide use of lead in batteries. This has been estimated as being more than 1,00,000 metric tons per year.

Dry Cell Batteries. An important battery type is the so–called dry-cell battery that was first developed by Carl Gassner in 1886 on the basis of a wet zinc-carbon battery that had been developed twenty years earlier by Georges Leclanché. Fig. 7.8 shows a schematic representation of the structure of such a battery, now also known as an 'alkaline battery'. The outer zinc casing contains a slurry of electrolyte contained in a volume bounded by a porous cardboard layer. This electrolyte is composed of two layers: the first is of a paste of ammonium chloride that is positioned next to the zinc anode casing; and the second is a paste of ammonium chloride and manganese dioxide, the latter acting as a cathode. The second layer surrounds a carbon rod set in the centre of the structure.

The anode reaction is:

$$Zn + 2\,Cl^- \rightarrow ZnCl_2 + 2\,e^-$$

and the cathode reaction is:

$$2\,MnO_2 + 2\,NH_4Cl + H_2O + 2\,e^- \rightarrow Mn_2O_3 + 2\,NH_4OH + 2\,Cl^-$$

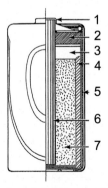

Fig. 7.8 Schematic diagram of a dry cell. 1, brass cap; 2, plastic seal; 3, expansion space; 4, porous cardboard; 5, zinc can; 6, carbon rod; 7, chemical mixture. *(https://en.wikipedia.org/wiki/Dry_cell.)*

so that the overall reaction is:

$$Zn + 2\,MnO_2 + 2\,NH_4Cl + H_2O \rightarrow ZnCl_2 + Mn_2O_3 + 2\,NH_4OH.$$

It is also possible to substitute zinc chloride for the ammonium chloride as the electrolyte when the overall reaction becomes:

$$Zn + 2\,MnO_2 + H_2O \rightarrow Mn_2O_3 + Zn(OH)_2$$

The cell voltage is obtained is 1.5 V but this drops under load and also as a result of the depletion of the reactants. The great advantage of this structure is that it can operate in any orientation without spilling and this allows it to be used in mobile devices such as torches. It requires no maintenance but cannot be recharged.

Rechargeable Ni-Cd and Ni-MH Batteries. The nickel-cadmium (Ni-Cd) and nickel-metal hydride (Ni-MH) batteries are both rechargeable and have voltages that are similar to those provided by the alkaline battery discussed in the previous section: 1.2 V for Ni-Cd and 1.25 V for Ni-MH. As a consequence, both have been commonly used as rechargeable substitutes for the dry-cell batteries described in the previous section. The Ni-Cd battery has been gradually developed over a long period, starting from its original creation in 1899 by the Swede, Waldemar Jungner. By 2000, some 1.5 billion batteries were sold globally per year. The currently available type for portable use is in the form of a Swiss roll with several layers of positive and negative material held in a cylindrical container. Larger variants are also constructed in the form of ventilated cells, these being used for standby power or in electric vehicles (Chapter 6). As cadmium is considered to be a toxic material, its use is now banned in Europe, except for use in a limited number of applications such as medical devices. Furthermore, European regulations require that any producer of these devices is responsible for recycling the cadmium.

The Ni-Cd cell, when producing a current, operates with the following electrode reactions:

$$Cd + 2OH^- \rightarrow Cd(OH)_2 + 2e^-$$

and

$$2NiO(OH) + 2H_2O + 2e^- \rightarrow 2Ni(OH)_2 + 2OH^-$$

The allover reaction is thus:

$$2NiO(OH) + Cd + 2H_2O \rightarrow 2Ni(OH)_2 + Cd(OH)_2$$

During charging, the reaction occurs in the opposite direction. Compared with the dry-cell carbon-zinc batteries, the voltage of the Ni-Cd battery declines very little with the extent of discharge and it can also tolerate very high discharge rates, making it preferable to lead-acid batteries. It is also much lighter than the lead-acid batteries and so is now used in situations where weight is an important factor, for example in aircraft. Because both Ni and Cd are expensive, however, the higher cost of these batteries is also an important factor. As a result, the Ni-Cd battery is currently losing out in most applications to two other types of batteries: either the nickel-metal-hydride (NiMH) that will now be discussed, or the lithium-ion batteries, discussed in the next section.

The Ni-MH battery is a much more recent arrival than the Ni-Cd battery, having been gradually developed over the period since its discovery in 1967. The electrochemical half-cell reactions occurring in the discharging process are:

$$OH^- + MH \rightarrow H_2O + M + e^-$$

and

$$NiO(OH) + H_2O + e^- \rightarrow Ni(OH)_2 + OH^-$$

(Both these reactions occur in the reverse direction during the charging process.) The allover reaction is thus:

$$NiO(OH) + MH \rightarrow Ni(OH)_2 + M$$

Hence, the inter-conversion of $NiO(OH)$ and $Ni(OH)_2$ by reaction with water occurs in both the Ni-Cd and the Ni-MH batteries. However, with the Ni-MH combination, hydrogen is produced and this is absorbed by the Ni-M component of the positive electrode, where M is a rare-earth metal, most commonly Lanthanum (La). It is well established that the alloy $LaNi_5$ readily absorbs hydrogen reversibly to form a stable hydride.[c] Hence, hydrogen gas is not formed within the cell as long as the hydride can be formed. As pure lanthanum is expensive due to the high cost of isolating its oxide from mixtures of the rare earth oxides in which it occurs, so-called mischmetal is often used, this being directly derived from the rare-earth ores without separation and typically having the composition of 55% cerium (Ce), 25% La and 15%–18% neodymium (Nd) in addition to about 5% iron.

---

[c] For a detailed description of the $LaNi_5$ system, see a paper by J. Reilly (Brookhaven National Laboratory) available for download at https://www.osti.gov/servlets/purl/6084207.

(The variability of the composition of the mischmetal explains the rather variable ratios that are listed on various related web sites for the components of commercial Ni-M-H batteries.)

It is important not to overcharge Ni-M-H batteries and hence smart chargers have been developed to prevent the occurrence of such damage. As one of the consequences of overcharging is the evolution of hydrogen, with the consequent possible rupture of the casing, modern cells often contain a catalyst to cause the recombination of this hydrogen with oxygen to form water. Another problem that can occur in a battery pack containing a number of separate cells in series is that if the complete discharge occurs, one or more of these cells may suffer from polarity reversal, this resulting in irreversible damage of the combined arrangement.

One of the great advantages of Ni-MH batteries is that they are much lighter than the Ni-Cd batteries discussed above and another is that they have none of the latter's environmental problems. They are very useful for applications in which high currents are drawn as they have low internal resistance and therefore are less prone to overheat. The Ni-MH battery was the predominant choice for use in early electric vehicles but they have now been largely superseded by lithium batteries. However, patent problems have rather limited their use. It appears that their sale is controlled by Cobasys, a subsidiary of Chevron, which only provides large orders; one consequence of this has been that General Motors has shut down production of its EV1, citing problems with battery availability.

Li-Ion Batteries. The idea of a lithium-based battery system (see Box 7.4) was first investigated in the early 1970s by Stanley Whittingham, then at Stanford University, who showed that it was possible to store Li^+ ions in the layers of a disulphide. He was then hired by Exxon where he showed that it was possible to make a battery consisting of titanium (IV) sulphide and lithium metal as the electrodes. However, this battery was found not to be practicable due to the insurmountable problems such as the need to ensure that the Li did not encounter water and thus release hydrogen gas. Further work discussed in Box 7.4 led to the Li-ion battery as we now know it and, finally, to the award of the 2019 Nobel Prize in Chemistry to Stanley Whittingham (now at Binghampton University), John Goodenough (now at the University of Texas) and Akira Yoshino (now at Meija University, Nagoya and Asahi Kasei Corporation). The citation for the award said of the lithium-ion battery: 'This lightweight, rechargeable and powerful battery is now used in everything from mobile phones to laptops and electric vehicles. It can also store significant amounts of energy from solar and wind power, making possible a fossil-free society.'

BOX 7.4 The 2019 Nobel Prize for Chemistry

The Nobel Prize in Chemistry for 2019 was awarded to John B. Goodenough, M. Stanley Whittingham and Akira Yoshino 'for the development of lithium-ion batteries.' (https://www.nobelprize.org/prizes/chemistry/2019/press-release/).

The work leading to the award to this trio of scientists commenced when Whittingham, who had just completed his postdoc at Stanford University on aspects of intercalation started to work at Exxon where his work involved a search for materials that would intercalate lithium ions for possible battery applications. He found that titanium disulphide, a material with a layered structure, could reversibly accommodate the Li^+ ions. This material is electrically conductive and was found not to interact with other materials such as electrolyte molecules, thus making it ideal as a cathode (in the discharge mode) that would take up Li ions during the discharge process. Exxon funded a major project on the work that led to the development of a battery consisting of a lithium-metal anode, lithium perchlorate in dioxolane as electrolyte, and a titanium disulphide cathode. This battery provided a voltage of 2.5 V and stored about ten times as much energy as a lead-acid battery or five times as much as a Ni-Cd battery. However, it was found that this battery design had serious problems as the lithium deposited spikey formations ('dendrites') that intruded into the electrolyte gap, ultimately causing short circuits that resulted in the combustion of the electrolyte material and even the Li itself.*

Goodenough, a physicist who had previously worked in the Lincoln Laboratory at MIT and is now still working at the University of Texas in Austin, was appointed in 1976 as head of the Inorganic Chemistry Laboratory at Oxford University. He started to work on cathode materials in an effort to replace Whittingham's titanium disulphide. In 1980, he recognised that it might be possible to replace the sulphur species with oxygen. Koichi Mizushima (now of Toshiba Corporation, who was working with him at the time) started to screen a range of metal oxides and found that there existed a series of ternary layer compounds ($LiMO_2$), containing metals such as V, Cr, Fe, Co and Ni, which incorporated Li ions reversibly. He found that $LiCoO_2$ was the most effective of these compounds and that the structure was maintained as the Li ions moved in and out of the structure during the charging and discharging steps. The potential of a battery constructed with a $LiCoO_2$ electrode was significantly increased over that of the Whittingham battery, reaching 4 V. This battery still used a Li metal electrode.

The final step in achieving a usable design was achieved by Yoshino (now at Meijo University, Nagoya) who at that stage worked with Asahi Kasei in Japan. The problem with Goodenough's battery was still that the Li electrode was susceptible to combustion. Yoshino focused on a so-called 'rocking-chair principle' in which the lithium is not converted to metallic species at either electrode but is incorporated into the electrode structures, one of these being

BOX 7.4 The 2019 Nobel Prize for Chemistry—cont'd

the $LiCoO_2$ discussed above. He found that a petroleum-derived coke was an ideal component for the second electrode. Following a number of years of development work by Asahi Kesei, in collaboration with Sony, that was aimed at getting the electrodes into the correct form with adequate strength, the first commercial lithium-ion battery was launched by Sony in 1991.**

  * See the section on Lithium Metal Batteries below.

  ** The fascinating history of the development of the Li-ion battery is covered by a number of useful references, including the Nobel Prize press release referred to above and an article by Katrina Krämer in Chemistry World entitled 'The Lithium Pioneers' (Chemistry World, November 2019, 24–30.)

The Li-ion battery as we now know it has a number of different variants. However, the essential feature of all such batteries that derive from the work of Whittingham, Goodenough and Yoshino is that Li^+ ions are transported to and from the electrodes by reducing the cobalt species in a $Li_{1-x}CoO_2$ material from Co^{4+} to Co^{3+} during the discharge process and re-oxidising the Co^{3+} to CO^{4+} during the charging process. The reaction occurring at the positive electrode (cathode) during the discharging reaction can be depicted as follows:

$$CoO_2 + Li^+ + e^- \rightarrow LiCoO_2$$

At the negative electrode, Li metal, intercalated in a graphite matrix, is oxidised in the reaction:

$$LiC_6 \rightarrow C_6 + Li^+ + e^-$$

The allover discharging reaction is thus:

$$LiC_6 + CoO_2 \rightarrow C_6 + LiCoO_2$$

These reactions are reversible and occur in the opposite direction in the charging process. Water cannot be used as a solvent for the electrolyte by which the Li + ions are transported since hydrolysis of the water to hydrogen and oxygen would then be preferred to the liberation of lithium metal at the CoO_2 electrode. Hence, a variety of different organic solvents have been used, these including ethylene carbonate, dimethyl carbonate and diethyl carbonate. Recently developed variants use solid ceramic ion conductors such as perovskites for the transport of the Li^+ ions. The Li^+ ions are generally included in lithium salts such as $LiPF_6$, $LiBF_4$ or $LiClO_4$, all of which are soluble in the organic solvents mentioned above.

Charging of the Li-ion battery must be carefully controlled by limiting the peak voltage in order to avoid irreversible damage. For example, overcharging at relatively low voltages causes supersaturation of the lithium cobalt oxide electrode material, thus resulting in the formation of lithium oxide by the irreversible reaction:

$$Li^+ + e^- + LiCoO_2 \rightarrow Li_2O + CoO$$

Overcharging at higher voltages of up to 5.2 V results in the production of Co(IV) oxide by the following reaction:

$$LiCoO_2 \rightarrow Li^+ + CoO_2 + e^-$$

and this again occurs irreversibly.

Another limitation of the Li-ion system is that the Co electrode reaction described above is only reversible for $x < 0.5$ and this limits the depth of discharge allowable. It is therefore critical in the use of a Li-ion battery that charging and discharging are very carefully controlled and it is therefore necessary to use specially designed charging systems and also to monitor and control the discharge process. Nevertheless, even though Li batteries are more expensive than the Ni-Cd batteries discussed in the previous section, they have the advantage that they operate over a wider temperature range and give higher energy densities. They are also significantly lighter, an important advantage in applications such as their use to power electric vehicles as discussed in Chapter 6.

Lithium Metal Batteries. Also depending on the Li chemistry discussed above and also based on the work of Whittingham in Exxon is the class of cell referred to as a lithium metal battery. These batteries are irreversible and therefore disposable, consuming the lithium metal to form Li ions. They are able to provide high power and are commonly preferred for use in small portable electronic devices such as watches and digital cameras. They generally have long lives and, although more expensive, are therefore often used in place of the more common alkaline cells described in an earlier section. Because of the danger that the lithium metal in the cell can react with water vapour to produce explosive hydrogen mixtures if the cell is damaged, restrictions are in place for the transport of these lithium metal batteries. They are also susceptible to the growth of dendrites in the non-aqueous electrolyte between the Li anode and the cathode and this can cause short-circuiting and failure of the cells, a problem that can be avoided with the correct choice of electrolyte. The interested reader should search the web for further information on lithium metal batteries and how they are distinguished from lithium-ion batteries.

Flow batteries

In the battery types described above, the reactants providing the power are components of the constituent cells. In contrast, flow batteries depend on the continuous supply of reactants from outside the assembly and the simultaneous continuous collection of the products as shown schematically in Fig. 7.9.

The cell works in exactly the same way as an electrochemical cell as described in a previous section apart from the continuous replacement of reactant. For a redox cell, the reactants are dissolved in the electrolyte on one side of the cell and the products of the electrochemical reaction are collected in the electrolyte on the other side of a separating proton exchange membrane. Many different combinations of reactants have been examined for use and these include the Mn^{6+}/Mn^{7+} and V^{3+}/V^{5+} redox couples. For the vanadium cell, four different oxidation states occur, the electrode reactions occurring during discharge at the positive electrode being:

$$VO_2{}^+ + 2\,H^+ + e^- \rightarrow VO^{2+} + H_2O$$

and the simultaneous reaction at the negative electrode being:

$$V^{2+} \rightarrow V^{3+} + e^- \;(\text{negative electrode})$$

The allover discharge reaction is therefore:

$$VO^{2+} + 2\,H^+ + V^{2+} \rightarrow VO^{2+} + H_2O + V^{3+}$$

All the reactions shown are reversible and the recharging reaction, therefore, goes in the opposite direction. During both discharging and

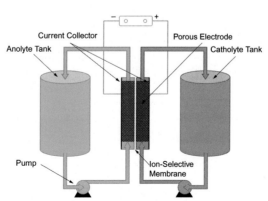

Fig. 7.9 Schematic representation of a flow battery. *(https://upload.wikimedia.org/ wikipedia/commons/5/5b/Redox_Flow_Battery.jpg.)*

charging, a proton is transferred across the membrane and an electron is transferred between the electrodes. A typical open-circuit voltage at room temperature is 1.41 V. Typical electrodes consist of a felt or cloth made of either carbon or graphite.

The advantage of such a set-up is that the extents of the reactions occurring are determined not by the contents of the cell but by the amounts of reactants and products that can be transmitted through the cell assembly and so they depend only on the volume of the storage tanks. Such systems are used for stationary applications such as load balancing for an electrical grid, when excess production of power can be transferred by the electro-chemical process to chemicals, these being stored until power is required when it is generated by operating the reverse process. The advantage of such a system is that there is none of the gradual loss of electrical energy due to leakage that there would be with a battery storage system. However, all-over energy losses do still occur due to the need to supply energy for the pumps in the so-called 'analyte' and 'catholyte' storage systems.

Fuel cells

A fuel cell has a very similar function to a flow battery but it is generally used to convert fuel directly and irreversibly to electrical energy. Fuel cells are most commonly used for the oxidation of hydrogen (forming water) although they can also be used for the conversion of a number of other fuels. The structure of a fuel cell for use with hydrogen is shown schematically in Fig. 7.10.

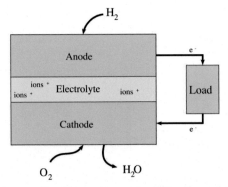

Fig. 7.10 A block diagram for the structure of a fuel cell using hydrogen as a fuel. (https://upload.wikimedia.org/wikipedia/en/1/1b/Fuel_Cell_Block_Diagram.svg.)

Hydrogen and oxygen (or air) are fed separately to the anode and cathode compartments respectively. Hydrogen dissociates at the surface of the anode and loses two electrons, forming H^+ ions:

$$H_2 \rightarrow 2H^+ + 2e^-$$

The hydrogen ions then transfer through the electrolyte material, the nature of which depends on the type of fuel cell in use (see below), and they then combine with oxygen at the cathode, producing water:

$$2H^+ + 0.5O_2 \rightarrow H_2O$$

The allover reaction is thus:

$$2H_2 + O_2 \rightarrow 2H_2O$$

As the half-cell potential for the reaction $O_2 + 4H^+ \rightarrow 2H_2O$ is $1.229\,V$ and that for the hydrogen ionisation reaction is $0.0\,V$, a typical single hydrogen fuel cell produces a voltage of between 0.6 and $0.7\,V$, this dropping with increasing load. A series of such cells can be built into a stack with the result that significantly higher voltages can be achieved, delivering powers up to several MW. Heat is also produced and so fuel cells can be used very effectively for combined heat and power applications.

The electrolyte in the fuel cell allows the transport of the cations, in most cases H^+, from the anode to the cathode. For the simplest type of fuel cell, first described by Sir William Grove in 1838, the electrolyte was simply a solution of copper sulphate and dilute sulphuric acid while the electrodes were sheets of iron and copper. This design was similar to what is now known as the phosphoric acid fuel cell in which the phosphoric acid replaces the sulphuric acid used by Grove. The various types of fuel cells generally available, as well as their uses, are summarised in Table 7.3. As well as in military and space applications, for which much of the background development work was carried out by NASA and other space agencies, the principal uses of all these fuel cell types are in backup power and aspects of electricity distribution. The limited use of transportation was discussed in Chapter 6.

Polymer electrolyte membrane fuel cell (PEM). The construction of all the different types of fuel cell assemblies are similar and they differ mostly in the nature of the electrolyte that transfers ions between the electrodes. The structure of the polymer electrolyte membrane (PEM) fuel cell is shown in Fig. 7.11.

Hydrogen ions are formed at the anode, which is composed of platinum, and these are then transported through the polymer electrolyte membrane to

Table 7.3 Fuel cell types.

Fuel cell type/ abbreviation	Electrolyte	Operating temperature/°C	Electrical efficiency/%	Applications
Alkaline fuel cell (AFC)	Aqueous KOH or alkaline polymer membrane	<100	60%	Military Space Backup power Transportation
Polymer electrolyte membrane fuel cell (PEM)	Perfluorosulphonic acid	<120	60%	Backup power Portable power Distributed generation Transportation Speciality vehicles
Phosphoric acid fuel cell (PAFC)	Phosphoric acid in a porous matrix or in a polymer membrane	150–200	40%	Distributed generation
Molten carbonate fuel cell (MCFC)	Molten lithium, sodium and/ or potassium carbonates in a porous matrix	600–700	50%	Electric utility Distributed generation
Solid oxide fuel cell (SOFC)	Yttria stabilised zirconia	500–1000	60%	Auxiliary power Electric utility Distributed generation

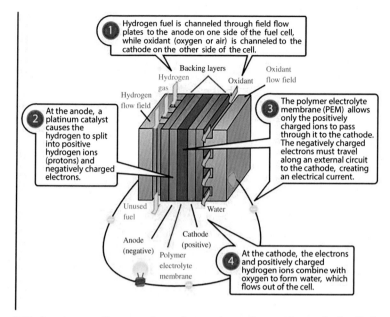

① Hydrogen fuel is channeled through field flow plates to the anode on one side of the fuel cell, while oxidant (oxygen or air) is channeled to the cathode on the other side of the cell.

Backing layers

Hydrogen
gas

Oxidant

Oxidant
flow field

Hydrogen
flow field

② At the anode, a platinum catalyst causes the hydrogen to split into positive hydrogen ions (protons) and negatively charged electrons.

③ The polymer electrolyte membrane (PEM) allows only the positively charged ions to pass through it to the cathode. The negatively charged electrons must travel along an external circuit to the cathode, creating an electrical current.

Unused
fuel

Water

Anode
(negative)

Cathode
(positive)

Polymer
electrolyte
membrane

④ At the cathode, the electrons and positively charged hydrogen ions combine with oxygen to form water, which flows out of the cell.

Fig. 7.11 A cut-away diagram of a polymer electrode membrane fuel cell. *(https://upload.wikimedia.org/wikipedia/commons/0/0d/PEM_fuelcell.svg)*

the cathode where they combine with $O^=$ ions to form water, this then flowing out of the cell. As the platinum electrode material can be irreversibly poisoned by the adsorption of impurities in the feed gas, particularly of any CO remaining from the steam reforming process used to create it, the feed must contain very low concentrations of such impurities. However, this restriction becomes significantly less important as the temperature of operation of the fuel cell is increased. For example, the molten carbonate fuel cell, operating at about 650°C, is much less susceptible to CO poisoning.

A relatively recent application of the PEM is the Direct Methanol Fuel Cell (DMFC) in which an aqueous solution of the methanol fuel is fed to a cathode containing platinum and ruthenium which then reforms the methanol-water mixture directly to give hydrogen and carbon dioxide. The allover anode reaction can therefore be expressed as follows:

$$CH_3OH + H_2O \rightarrow 6\,H^+ + 6\,e^- + CO_2.$$

The DMFC is most commonly used currently in place of more conventional batteries in applications such as fork-lift trucks used in warehouses but it has potential for use in automobiles if the structure exists for the distribution of the methanol fuel. Another possible fuel for such purposes is formic acid which can be decomposed directly in situ to give hydrogen and CO_2. As both methanol and formic acid can be obtained from non-fossil fuels, their use would not contribute to the emission of greenhouse gases. Such an approach is discussed further later in this chapter.

Molten carbonate fuel cell (MCFC). In the molten carbonate fuel cell (see Fig. 7.12), the electrolyte is a mixture of molten carbonates, Li_2CO_3, K_2CO_3 and/or Na_2CO_3, suspended in a ceramic matrix. Carbonate ions are formed at the cathode and pass through the carbonate mixture of the matrix to the anode where they react with syngas (carbon monoxide and hydrogen) to give carbon dioxide and water.

The reactions occurring are as follows:

Anode reaction:

$$H_2 + CO_3^{=} \rightarrow H_2O + CO_2 + 2e^{-}$$

Cathode reaction:

$$1/2O_2 + CO_2 + 2e^{-} \rightarrow CO_3^{=}$$

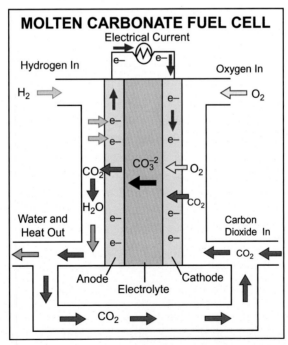

Fig. 7.12 A molten carbonate fuel cell. *(United States Department of Energy.)*

The cell reaction is thus:

$$H_2 + 1/2O_2 \rightarrow H_2O$$

The high temperature of operation, between 600°C and 700°C, increases the rates of all the reactions involved and this means that it is not necessary to use noble metals such as platinum for the electrodes. The anode thus generally comprises of an alloy of nickel-containing chromium or aluminium and the cathode is either lithium metatitanate or NiO intercalated with lithium species. Because these electrodes are much less prone to poisoning by CO, a variety of different fuels can be used, including fuel gas (obtained from coal), methane or natural gas. The methane can be reformed to syngas just prior to admission of the fuel to the system (the reaction being termed 'Internal Reforming') in which case the fuel reacting the cathode is approximately a 3:1 mixture of H_2 and CO:

$$CH_4 + H_2O \rightarrow CO + 3H_2$$

The oxidant fed to the cathode compartment is either O_2 or CO_2; in the latter case, the CO_2 formed at the anode in the anode half-cell reaction described above is circulated back to the cathode. Some problems have been encountered due to poisoning and corrosion as a result of passage of the alkali metal ions present in the cell to other parts of the system but these difficulties have largely been overcome. The use of MCFC's is predominantly in large systems such as fuel-cell powered electricity generation stations and combined heat and power plants.

Solid oxide fuel cell (SOFC). The construction of the solid oxide fuel cell is shown schematically in Fig. 7.13. In this, the oxygen fed to the

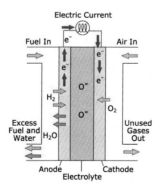

Fig. 7.13 A solid oxide fuel cell. (*https://upload.wikimedia.org/wikipedia/commons/4/42/Solid_oxide_fuel_cell.svg.*)

cathode compartment is transmitted as $O^=$ ions through a ceramic membrane to the anode where these ions oxidise the fuel directly.

The fuel is generally hydrogen but it could also be methane, propane or butane or even gasoline, diesel or jet fuel. The membrane material is a dense ceramic that conducts oxygen ions without having any electronic conduction ability, normally yttrium-stabilised zirconia. The anode is commonly a cermet comprising nickel mixed with the electrolyte material. It performs the function of not only providing the H^+ ions reacting with the $O^=$ ions diffusing through the electrolyte but also reforming the hydrocarbon fuels that are frequently fed instead of pure hydrogen. The cathode material is currently lanthanum strontium manganite, compatible with the yttria-doped zirconia electrolyte, its function being to convert the oxygen feed into $O^=$ ions. The operation temperature is generally high, between 500°C and 1000°C, to facilitate the transport of the $O^=$ ions through the solid electrolyte membrane layer. The main current use of SOFCs is in applications such as stationary power generation, the systems having outputs up to 2 MW. There is currently much research and development work related to the application of SOFCs as well as to the other types of the fuel cell. Some of the most recent work is discussed in Chapter 8.

Electrolysis

The process of electrolysis is most commonly encountered in relation to the production of hydrogen from water. In this case, the reaction is the reverse of that encountered in the operation of the alkaline and polymer electrolyte membrane fuel cells discussed in the previous section. However, electrolysis has many other applications, some of which will also be discussed briefly below.

The electrolysis of water has been known for more than 200 years, much of the early work having resulted from the invention of the Volta pile described above. In common with all other important applications of electrolysis, the electrolysis of water is a reaction that is not allowed thermodynamically to occur without the application of an external voltage. Currently, the cost of hydrogen produced using electrolysis is higher in most developed economies than that of the hydrogen from fossil-fuel-based steam reforming plants of the type discussed in Chapter 2. However, with the steady progress that is currently occurring towards the introduction of renewable energies (Chapter 3) and the consequent reduction of the cost of electricity, it is clear that electrolysis will soon become a competitive and more acceptable route

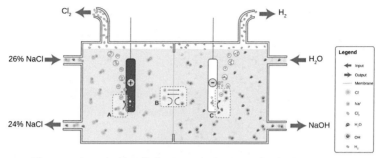

Fig. 7.14 The membrane chloralkali process. *(https://upload.wikimedia.org/wikipedia/commons/7/7a/Chloralkali_membrane.svg.)*

for the production of hydrogen for a wide range of different applications, including for the operation of fuel cell systems. Some of the potential uses of 'green' hydrogen will be discussed in more detail in Chapter 8. For the purposes of the current chapter, it is sufficient to say that the types of electrolyses currently being used are very closely related to the fuel cells described above. In many ways, fuel cells could have been described under the heading 'Reversible Batteries' as they can all occur reversibly when suitable potentials are applied. Electrolysis is a well-established approach to obtaining many of our currently important industrial products. For example, the production of chlorine and sodium hydroxide from brine can be carried out using a number of different technologies, one of the most modern of which is the membrane cell electrolysis process shown schematically in Fig. 7.14.

In this, a solution of sodium chloride is fed to the anode compartment and some of the sodium ions are transmitted through the membrane to the cathode compartment, the sodium ion concentration of the brine solution leaving the anode compartment having been reduced. Chlorine gas is formed simultaneously and is emitted from this compartment. The sodium ions entering the cathode compartment react with hydroxyl ions formed at the cathode and hydrogen is liberated simultaneously. The cathode reaction is:

$$2\,H^{+}_{(aq)} + 2\,e^{-} \rightarrow H_{2\,(g)}$$

and the anode reaction is:

$$2\,Cl^{-}_{(aq)} \rightarrow Cl_{2(g)} + 2\,e^{-}$$

and so the allover reaction is:

$$2\,NaCl + H_2O \rightarrow Cl_2 + H_2 + 2\,NaOH$$

The process can also be carried out with KCl to produce KOH.

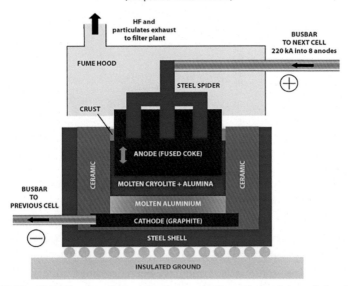

Fig. 7.15 The Hall-Hérault cell used for the production of aluminium. *(https://upload.wikimedia.org/wikipedia/commons/2/24/Hall-heroult-kk-2008-12-31.png)*

Other electrolysis processes of importance include the production of aluminium from alumina by the Hall-Hérault process in which a molten mixture of alumina with cryolite (Na_3AlF_6) and calcium fluoride is electrolysed at temperatures between 950°C and 980°C; see Fig. 7.15.

The system consumes the carbon of the anode. The cathode reaction is:

$$Al^{3+} + 3e^- \rightarrow Al$$

and the anode reaction is:

$$O^{2-} + C \rightarrow CO + 2e^-$$

so that the overall reaction is:

$$Al_2O_3 + 3C \rightarrow 2Al + 3CO$$

In practice, CO_2 is also formed at the anode, the allover reaction then being:

$$2Al_2O_3 + 3C \rightarrow 4Al + 3CO_2.$$

The production of aluminium consumes large amounts of electrical energy and so contributes very significantly to the production of greenhouse gas unless renewable electrical energy is used. Many other metals are also produced from their oxides or other related raw materials by electrolysis processes. These include magnesium, zinc, lead, chromium, manganese and titanium. A useful summary of some of these processes is to be found at https://www.osti.gov/servlets/purl/6063735.

CHAPTER 8

The way forward: Net Zero

Introduction

Chapter 1 outlined the problems associated with greenhouse gas emissions and the resultant global warming which is currently occurring and went on to discuss attempts that are being made to combat this warming. The Paris Accord of 2015, to which the majority of the countries in the world have subscribed, was drawn up under the auspices of the United Nations Convention on Climate Change (UNFCCC). This accord has as its aim the achievement of a limit to the increase in global temperature since pre-industrial levels of not more than 2.0°C; a secondary aim of the accord is that every effort should be made to achieve a lower temperature increase of only 1.5°C. The term 'Net-Zero' has recently been introduced, this corresponding roughly to the aim towards the 1.5°C limit of the Paris Accord. A very recent report[a] from the International Energy Agency, entitled 'Net Zero by 2050—A Roadmap for the Global Energy Sector', supplies pointers as to what must be achieved if the global community is to achieve the Net–Zero target by 2050 and some of its contents are very relevant to the material of this book. Fig. 8.1, taken from that report, shows the various hurdles that must be surmounted in the intervening period. The report stresses that each participating government will need to set its own targets since each country will have differing needs and opportunities and that much will depend on the commitment of all sectors of society in each region of the global community to achieve these targets. It is already quite clear that many countries are already falling far behind the targets that they had set themselves in their action plans drawn up in response to the Paris Accord and that significant changes need to be made if we are even to achieve the higher 2.0°C target.

Some significant pointers from the boxes of Fig. 8.1 are summarised below.

- From today, no new oil or gas fields or coal plants should be approved for development.
- From 2025, no fossil fuel boilers should be sold.

---

[a] https://www.iea.org/reports/net-zero-by-2050.

Sustainable Energy
https://doi.org/10.1016/B978-0-12-823375-7.00005-6

Copyright © 2022 Elsevier B.V.
All rights reserved.

197

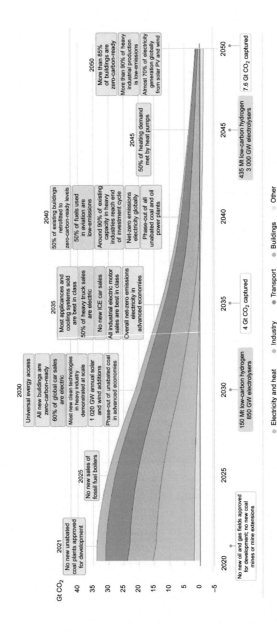

Fig. 8.1 Net zero by 2050. (From *Net Zero by 2050—A Roadmap for the Global Energy Sector. https://www.iea.org/reports/net-zero-by-2050*)

- By 2030, all new buildings should be zero-carbon ready; 60% of car sales should be electric; there should be substantial progress in developing solar and wind electricity generation; and the use of coal without CO_2 capture and storage should be terminated.
- By 2035, 50% of heavy-duty trucks should be electric; there should be no further sales of vehicles with internal combustion engines; and all advanced economies should have only renewable electricity.
- By 2040, 50% of existing buildings should be retrofitted to zero-carbon levels; 50% of fuels for aviation should be low emission; 90% of the heavy industrial equipment now in use should have been replaced; all electricity globally should be renewable; and all unabated coal- and oil-based power plants should have closed down.
- By 2045, 50% of heating should be supplied by heat pumps.
- By 2050, more than 85% of buildings should have reached zero-carbon levels; more than 90% of the heavy industry should use low-emission technology; and almost 70% of electricity globally should be renewable.

Two other important topics featured in the IEA report are shown at the base of Fig. 8.1: low-carbon electrolysers and CO_2 capture. These two topics will now be discussed further in the light of some of the material of Chapters 1–7. This chapter then concludes with a brief mention of several other topics of relevance to the reduction of greenhouse gas emissions.

Hydrogen production using renewable energy

It is now generally accepted that the most desirable approach to enabling the reduction of greenhouse emissions is the generation of 'green' hydrogen by electrolysis of water using renewable electricity. However, it is also possible to produce 'blue' hydrogen by steam reforming of raw materials such as natural gas as described in Chapter 4 if successful methods can be established for the complete capture and successful storage of all the CO_2 formed during the reforming process. Currently, as shown in Box 8.1, the costs of producing hydrogen by conventional electrolysis methods using alkaline cells or polymer electrode cells are currently much higher than those for the production by the steam reforming of methane. However, for some purposes, electrolysis using these methods is already the preferred route when the price of carbon emissions is factored in. As will be discussed further in some detail in a later section, there is also some hope that the use of high-temperature hydrolysis using solids oxide cells of the type currently under development may become commercialised very soon.

BOX 8.1 The cost of producing hydrogen by electrolysis

A recent report (2020) prepared for the European Parliament entitled 'The Potential of Hydrogen for Decarbonising Steel Production' gives some of the costs of producing hydrogen.[b] The cost of 'grey' hydrogen produced by steam reforming varies quite significantly and depends on the cost of the natural gas used. Table 4.1 gives the current prices for hydrogen production using the different methods described in Chapter 4, these ranging from about €1.3–2.0/kg. In the electrolysis process, about 70%–80% of the electricity used is required to produce 'green' carbon-dioxide-free hydrogen and this requires 50–55 kWh (kWh) of electricity per kg. The cost of green hydrogen is thus currently in the range of €3.6–5.3/kg, depending on the exact cost of electricity in the country in question. The price of electricity has fallen by 60% over the last decade and it is predicted that it will continue to fall as the technology for electrolysis is improved and the costs of renewable energy decrease. The report, therefore, suggests that the price of green hydrogen production by electrolysis in Europe could fall as low as €1.8/kg by 2030, although the basic costs will vary somewhat in non-European countries. For example, in the US, where the cost of natural gas produced by fracking is significantly lower[c], electrolysis may be less competitive even when there is a reduction in the cost of renewable energy. In contrast, as will be discussed further below in relation to the costs of steel production using hydrogen, it may soon be commercially attractive for Australia to produce the necessary hydrogen by electrolysis.

[b]https://www.europarl.europa.eu/RegData/etudes/BRIE/2020/641552/EPRS_BRI(2020)641552_EN. pdf.
[c]A recent report from the Fuel Cell and Hydrogen Energy Association (FCHEA). https://www.fchea. org/us-hydrogen-study discusses the importance to the US of hydrogen produced from fracked natural gas by steam reforming, the carbon dioxide produced being captured and stored. The report, compiled by McKinsey, includes details of storage sites as well as showing the infrastructure already existing for the transportation of hydrogen by pipeline. Although a similar infrastructure exists in parts of Europe, this is not the case globally.

Chapter 3 discussed the many different routes to produce renewable energy that have been either fully developed or are currently being investigated. It is clear that in order to reach Net-Zero by 2050, some of these renewable resources must be developed even more rapidly than hitherto and that each jurisdiction must do everything possible to develop these resources in such a way that it will be possible to phase out the use of fossil fuels in the vast majority of current applications. Oil will still most probably have to be used as a source of chemicals for use, for example, in the production of polymers and pharmaceuticals. However, fossil-fuel-based petroleum and diesel will be things of the past, superseded in some instances by fuels derived from biomass. Although nuclear energy is

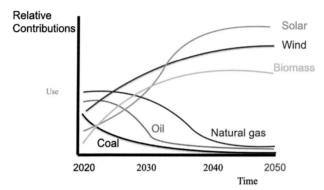

Fig. 8.2 Schematic representation of the changes in the usage of different energy sources that will be required by the year 2050 to achieve Net Zero.

currently considered to be a contributor to the world's renewable energy resources, the problem of the disposal of nuclear waste is insurmountable and it is likely that the number of countries banning the use of nuclear power will increase. It is likely therefore that the majority of the world's economies will rely on the production of their renewable electrical energy using either wind or solar generation. Some countries have significant hydro–electric power and one or two have geothermal resources so that these methods will in these cases contribute to the energy mix. Tidal and ocean current generation systems, although available on some sea-boards, are likely only to have very limited application due to the difficulties of operating in what are often extreme conditions. The use of coal for the production of electricity in countries such as Australia, China and India, where large reserves of coal are available, will only be acceptable if reliable methods are found to trap and store the CO_2 produced and hence much work remains to be done on developing and testing appropriate methods for carbon capture and storage.

Fig. 8.2 shows schematically the decreasing use of coal, oil and natural gas that might be expected over the next 30 years if Net Zero is to be achieved and it also sketches the expected growth in the same period of the use of wind, solar and biomass for the production of energy. As the use of coal will only be acceptable in situations in which its use is unavoidable, it is expected that Australia, China and India will before 2050 have installed sufficient wind and solar power resources to enable them to minimise their use of coal.[d] Oil will no longer be used for traction and natural gas will only be

---

[d] A very useful source of information on the continuous progress of the installation of solar and other green power sources is provided by a newsletter updated daily: daily.newsletter@pv-magazine.com. This newsletter also gives authoritative information on developments in a hydrogen economy.

used when adequate CO_2 capture facilities exist. The net result will be that many products currently made using coal, oil or natural gas will be produced by new routes while biomass–derived fuels will substitute some of the uses of oil–derived products for traction purposes. Box 8.2 lists some of the national reports on possible developments in renewable energy provision and use. Of

---

BOX 8.2 Some national reports on the use of hydrogen

There exist a number of organisations and networks dedicated to the development of the use of hydrogen and of fuel cell technologies. As a result, a large number of reports have been written on the subject. Some of these are described briefly below. Hydrogen Europe (https://www.hydrogeneurope.eu) is an organisation that incorporates a range of some 260+ European industrial companies and 27 national associations. Partnered with the European Commission, it is involved in an innovation programme entitled 'Clean Hydrogen for Europe'. The European Commission has established a fund of €10 billion with the aim of speeding up 'the transition towards a green, climate neutral and digital Europe'. Together with the Hydrogen Alliance (https://www.ech2a.eu), the Commission will devote part of this funding to achieving the aims given in a report entitled 'A Hydrogen Strategy for a Climate-Neutral Europe' (https://ec.europa.eu/energy/sites/ener/files/hydrogen_strategy.pdf). The priority set forward in this report is the achievement of renewable hydrogen formed by electrolysis using largely wind or solar energy but in the short and medium term, also other forms of low-carbon hydrogen.

In the US, a group comprising major oil companies, car makers, hydrogen producers and fuel cell manufacturers is pushing the US government to follow the lead of Europe and has published a report compiled by McKinsey and Company entitled: 'Road Map to a US Hydrogen Economy' (https://www.fchea.org/us-hydrogen-study). This report emphasises that hydrogen has many advantages as an energy carrier and urges that the US government invest funding for research, development, demonstration and deployment of hydrogen technologies. It stresses that 'directing capital to hydrogen is key to enabling growth in the US'.

Japan, despite having no hydrogen-based industry of its own, has also developed a strategy for hydrogen usage. This includes the goals of demonstrating the storing and transportation of hydrogen from abroad by 2022, introducing full-scale hydrogen generation by around 2030 and achieving fully-fledged domestic use of CO_2-free hydrogen by 2050. Japan aims to have 200,000 fuel-cell vehicles by 2025 and 800,000 by 2030, with 320 refueling stations by 2025 and 900 by 2030 (https://www.meti.go.jp/english/press/2017/pdf/1226_003b.pdf). For additional useful information on the Japanese strategy, see a report prepared by the New Zealand Embassy in Tokyo: https://www.mfat.govt.nz/en/trade/mfat-market-reports/market-reports-asia/japan-strategic-hydrogen-roadmap-30-october-2020/.

Continued

> **BOX 8.2 Some national reports on the use of hydroge—cont'd**
>
> An interesting report compiled by the International Energy Association and the Clingendael International Energy Programme entitled 'Hydrogen in North-Western Europe' (https://www.iea.org/reports/hydrogen-in-north-western-europe) considers how the countries of this region, some of which are relatively far advanced in achieving large supplies of renewable energy through hydroelectric schemes and off-shore wind power, can collaborate and benefit from hydrogen developments in their neighbouring countries. Partners in the programme are Belgium, Denmark, France, Germany, The Netherlands, Norway and the United Kingdom. Part of the programme envisaged will develop the use of carbon dioxide capture in off-shore oil wells in the North Sea and elsewhere.

particular interest in the current context are the many potential applications of 'green hydrogen' to be produced by electrolysis methods that are discussed in some of these reports.

Fuel cells to be used for transportation purposes

The topic of fuel cells for use in automobiles and other vehicles was discussed in Chapter 6 for the situation in which the fuel of choice was hydrogen gas, this being transported and stored under pressure. As discussed in that chapter, a severe limitation of the use of hydrogen in such an application is that there is no existing infrastructure for the supply of hydrogen other than in very specific locations such as California. Despite the hope that was prevalent several decades ago that molecular hydrogen would become widely used in transport, the rapid rise in the availability of electric battery vehicles, tied to parallel major advances in battery technology over the same period, has meant that the widespread use of hydrogen as a fuel is now much less likely, at least in the immediate future. As a result of the hope that hydrogen might be used, there has been a very significant research effort devoted to ways in which hydrogen might be transported and stored so that it could be used in transport applications. One possibility under consideration has been that it could be stored in the form of a metallic hydride or another similar material.[e] If a material could be found that had a sufficiently high storage capacity, its use would be much preferable

---

[e] A following is a useful review on hydrogen storage materials: 'Hydrogen storage: the major technological barrier to the development of hydrogen fuel cell cars', D.K. Ross, Vacuum, 80 (2006) 1084–1089. A preprint of this paper is available at http://usir.salford.ac.uk/id/eprint/16768/.

to the use of sturdy and weighty high–pressure vessels used in the currently available fuel-cell vehicles described in Chapter 6. If such storage materials could be developed, the distribution of hydrogen would also become much simpler.

There remains the possibility that fuels other than hydrogen might be used in fuel-cell vehicles. The most highly advanced of such options is the use of methanol as fuel in a methanol fuel cell, of the type mentioned briefly in Chapter 7, which has already been used in a number of demonstration vehicles. Fig. 8.3 shows the all–over construction of a methanol fuel-cell automobile operating with a PEM fuel cell. Some of the constituent parts are also depicted, including a methanol-reformer used to convert the methanol to hydrogen and CO_2.[f]

The reaction occurring in methanol reforming is:

$$CH_3OH + H_2O \rightarrow 3H_2 + CO_2$$

and this takes place in a separate reactor prior to the entrance to the fuel cell. If the methanol is synthesised from renewable hydrogen using, for example, bio-derived CO_2, the use of this approach can be considered to give 'green' or 'Zero Carbon' power.[g] There is currently great interest in using methanol fuel-cells in other applications such as in freight transport, shipping and stationary applications. The work referred to in footnote 'f' was done in an EU–funded project in which the aim was to provide a catalyst for use in a fuel cell for a ship to be operated on Lake Como.

A methanol-fuelled vehicle has some very significant advantages over both electric vehicles and ones powered by hydrogen fuel cells. Some of the advantages to be gained by using fuel cells fed with methanol or with similar fuel cells fuelled by ethanol or formic acid (see Chapters 5 and 7) are given in Box 8.3.

[f] Catalysts for the methanol reforming reaction are generally Cu-containing and there has been significant research devoted to improving their behaviour. A relevant paper from the author's laboratory, 'Methanol reforming for fuel-cell applications: Development of zirconia-containing Cu-Zn-A catalysts' (J.P. Breen and J.R.H. Ross, Catal. Today, 51 (1999) 521–533) describes the preparation and properties of a particularly active and stable catalyst.

[g] As is described below, another route to green methanol uses an SOEC system to produce the syngas used.

BOX 8.3 Advantages of fuel cells powered by methanol

- The range of a car powered by a methanol fuel cell (ca. 1000 km; see Fig. 8.3) is significantly higher than that of most, if not all, battery vehicles and probably also of hydrogen fuel-cell vehicles.
- The refueling time is very short, roughly equivalent to that of a petrol-fuelled vehicle (ca. 3 min), this being much lower than the average recharge time required for battery vehicles or for refueling with high-pressure hydrogen.
- Although there is currently no infrastructure for the provision of methanol, it would require little modification to the current fuel distribution system to allow methanol to be distributed and for suitable pumps to be installed in every refilling station.
- Methanol can be handled more safely than can gaseous hydrogen. It disperses rapidly if spillage occurs, either evaporating or dissolving in water. Although methanol is poisonous if swallowed, it should be recognised that ingestion of petrol is also dangerous.

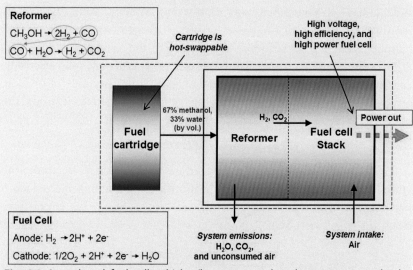

Fig. 8.3 A methanol fuel cell vehicle. *(https:www.methanol.orgwp-contentuploads 202004Methanol-Fuel-Cell-Powering-the-Future-webinar-presentation.pdf.)*

Solid oxide hydrolysis cells (SOEC's) for hydrogen production and their use for the synthesis of green ammonia and methanol

The production of hydrogen by the electrolysis of water using electrical energy avoids the need for an elaborate hydrogen supply network to provide hydrogen for use in more isolated industrial complexes since existing electrical grid connections could be used to transmit electricity to the site at which the hydrogen is required.[h] Such a possibility has focussed attention on the use of 'green' hydrogen in non-transport-related applications that have previously been dominated by the production of 'grey hydrogen' from coal, oil or natural gas (or 'blue hydrogen' produced using carbon capture and storage). Two such applications are the production of 'green ammonia' and 'green methanol'.

Green Ammonia

As discussed in Chapter 4, ammonia is prepared commercially from nitrogen and hydrogen using the Haber-Bosch Process. Chapter 4 also includes a detailed description of the Haldor Topsøe process using hydrogen made from natural gas (Figs. 3.14–3.16). Reports from Haldor Topsøe have recently shown that this process can easily be adapted to use 'green hydrogen' and have also described work on this development. Solid Oxide Electrolysis Cells (SOEC's) for the electrolysis of water to give the needed hydrogen. The SOEC is similar in structure to the Solid Oxide Fuel Cell (SOFC) discussed in Chapter 7. It consists of two permeable solid electrodes separated by a solid oxide membrane that allows transport only of $O^=$ ions formed by dissociation of water from the feed at the negative electrode, this generally being an oxide material containing Ni species:

$$2\,H_2O + 4\,e^- \rightarrow 2H_2 + 2O^=$$

The $O^=$ is then discharged as oxygen gas at the positive electrode, this typically, for higher performance applications, containing mixed conductors such as lanthanum-strontium-ferrite-cobaltite (LSCF) or lanthanum-strontium-cobaltite (LSC):

$$2\,O^= \rightarrow O_2 + 4\,e^-$$

The operation temperature is 600°C to 850°C and, in consequence, the process is much more efficient than the equivalent one occurring in the lower temperature range used for either an alkaline hydrolysis system or a polymer electrolyte membrane hydrolysis system.

---

[h] The grid structures in many countries will still need some modification since the sources of renewable energy such as off-shore wind are often confined to regions well away from the major industries. However, the costs of such modifications are likely to be low in comparison to the provision of additional pipelines. New industries using renewable energy may also become sited closer to the energy sources.

Haldor Topsøe has recently established a demonstration plant for the production of ammonia using their SEOC system. Fig. 8.4 shows a single SOEC assembly of the type used, Fig. 8.5 shows an array of several cells assembled together and Fig. 8.6 shows the complete ammonia synthesis reactor assembly.

Fig. 8.4 A single SEOC cell from Haldor Topsøe. *(Reproduced with the kind permission of Haldor Topsøe.)*

Fig. 8.5 An array of Haldor Topsøe SEOC cells. *(Reproduced with the kind permission of Haldor Topsøe.)*

Fig. 8.6 A complete demonstration plant for ammonia synthesis based on Haldor Topsøe's SEOC units. *(Reproduced with the kind permission of Haldor Topsøe.)*

The possibility of using ammonia as a fuel has recently gained significant attention. In a recent article in Lloyd's Register, Charles Haskell has reported that there are advanced plans for the introduction of ships fuelled by ammonia.[i] The possibility of the use of green ammonia as a fuel would have a great advantage over other green alternatives such as hydrogen that ammonia is a liquid which can be easily handled and stored at an easily used temperature of $-33°C$. Another advantage of using ammonia is that its combustion does not produce any CO_2. However, any N_2O formed during the combustion process is a more damaging greenhouse gas than CO_2 and it therefore has to be collected and treated. As Haskell points out, there may need to be a radical re-design of the vessels that use it; there will also be added storage and handling problems, ammonia being more toxic than the currently used maritime fuels. However, as shipping currently contributes a large proportion of the world's CO_2 emissions (almost 3% of the global quantity), there is a considerable impetus behind the current movement towards the use of ammonia as a fuel. Haskell reports that it is expected that 7% of the fuel used in ships will be ammonia by 2030 and that this proportion will increase to 20% by 2050.

---

[i] Charles Haskell, Lloyd's Register: https://www.lr.org/en/insights/articles/decarbonising-shipping-ammonia/. This article describes all the advantages of ammonia as a fuel and shows a photograph of an ammonia-fuelled ship.

The introduction of ammonia as a fuel would require very significant increases in global production capacity and the majority of this would need to be produced using renewable resources. Haldor Topsøe has investigated the production of ammonia using hydrogen produced by their SEOC systems. This work has been closely associated with the development of the Syncor reactor discussed in Chapter 4. Hansen and Skyøth-Rasmussen have described a joint project currently being funded by the Danish Energy Agency EUDP in cooperation with Aarhus University, the Technical University of Denmark, Energinet.DK, Vestas Wind Systems, Orsted and Equinor using the SEOC technology described above.[j] They point out that the SEOC-based system has the additional advantage of being able to operate as an energy storage system since the ammonia synthesis reaction can be reversed if necessary. Although the work described in the Hansen and Skyøth-Rasmussen paper is related to the production of ammonia for normal purposes, the product can also be used as a fuel as discussed above. The great advantage of using the SEOC system in conjunction with ammonia synthesis is that the nitrogen needed for the reaction is provided by feeding a mixture of air plus water to the electrolysis system. The product gas is then a mixture of hydrogen and nitrogen; the oxygen of the air feed, as well as that formed by the electrolysis of the water, passes through the solid oxide membrane and therefore also provides a stream of pure oxygen gas. The SEOC therefore in effect also acts as an air separation system. The allover efficiency of this SEOC approach to the synthesis of ammonia is claimed to be ca. 70%.

A press release from the Fraunhofer Institute for Microengineering and Microsystems (IMM) in Mainz, Germany, recently described a project entitled ShipFC aimed at developing eco-friendly technologies for the maritime sector.[k] It describes the development of a catalytic reactor for ammonia splitting that is combined with a fuel cell system for power production. The aim of the team collaborating on the project is to complete a small prototype by the end of 2022 and that a ship powered by an ammonia-powered fuel cell ('The Viking Energy', owned by the Norwegian shipping company Eidesvik) will be put to sea in late 2013.

---

[j] J.B. Hansen and M.S. Skjøth-Rasmussen, 'Can ammonia be a future energy storage solution', Hydrocarbon Processing, January 2020, 1–2.

[k] https://www.fraunhofer.de/press/research-news/2021/march-2021/worlds-first-hightemperature-ammonia-powered-fuel-cell-for-shipping.html.

Green Methanol

As discussed in Chapter 4, methanol is currently synthesised industrially from a mixture of carbon monoxide and hydrogen produced in a well-established industrial process by the steam reforming of natural gas or naphtha. A number of companies are now involved in an alternative process for the production of 'e-methanol' using carbon dioxide and electrolytically produced hydrogen as feedstock. (See Box 4.6.) To date, most e-methanol is produced using electricity from the grid and so the product is not green.

An exciting aspect of the SEOC discussed above compared with the other types of electrolysis systems is that it can also be used to produce a mixture of CO and hydrogen if CO_2 is fed to the system in addition to water:

$$CO_2 + 2e^- \rightarrow CO + O^=$$

The syngas ratio obtained can be adjusted as required by adjusting the H_2O/CO_2 ratio and so the SEOC can in principle be used to provide the feed required for a number of different commercially important reactions such as:

the synthesis of methanol:

$$CO + 2H_2 \rightarrow CH_3OH,$$

methanation:

$$CO + 3H_2 \rightarrow CH_4 + H_2O$$

and the *Fischer Tropsch Process*:

$$nCO + (2n + 1)H_2 \rightarrow C_nH_{2n+2} + nH_2O$$

The consequence of such advances, currently still largely in the development stage, is that it is, at least in principle, possible to produce many important products that are largely currently produced from natural gas by using instead a mixture of water and CO_2 as feedstock. Even though the current costs of such an approach are still relatively high compared with traditional methods, there has been a steady improvement in the performance of suitable SEOC cells so that, if the cost of the renewable electricity used becomes steadily lower (as is currently occurring), the production of totally green products will become commercially attractive. The economic aspects will also become more favourable relatively quickly if CO_2 emissions are taxed at higher levels than at present. Fig. 8.7 shows a schematic representation of the complete technology required, this being based on wind and solar energy, to produce chemicals and 'electrofuels'. The more conventional uses of renewable electrical power are also included for completeness.

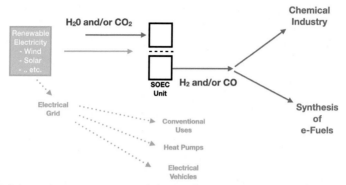

Fig. 8.7 Schematic representation of the production and uses of hydrogen and/or carbon monoxide using an SOEC system.

The 'chemical production' routes shown in Fig. 8.7 include the synthesis of ammonia, methanol and hydrocarbons. For the processes requiring the use of CO such as methanol or Fischer Tropsch synthesis, the CO_2 required as a feed for the SEOC system could be sourced from many different industrial processes if these are available at the required site. However, if the process is to be carried out at a site where such CO_2 is not conveniently available, an alternative route is to capture the required CO_2 from the atmosphere using technology such as that shown in Fig. 8.8. This process has been developed by Climeworks (www.climeworks.com) in collaboration with the company Carbfix (www.carbfix.com). (The latter company is concerned with the permanent storage of CO_2 as carbonate minerals in underground rock structures.) In the example shown, the system is used in conjunction with a geothermal power plant of the type described in Chapter 3 and this supplies the operational energy. The Climeworks process extracts the CO_2 from ambient air by passing it through filters that adsorb it selectively at a temperature around room temperature. When it becomes saturated, the collector is then closed and the temperature is raised to between 80°C and 100°C to desorb the CO_2 (with a purity of 99.7%). This is then cooled to 45°C and collected before delivery for storage or further reaction. Climeworks claims that their filters are fully recyclable and that a single unit can capture approximately 50 t of CO_2 per year directly from the atmosphere.

Use of green hydrogen in steel production

Two major industrial sources of greenhouse gases discussed in Chapter 2 that have traditionally come from the use of coal as fuel are the production of iron

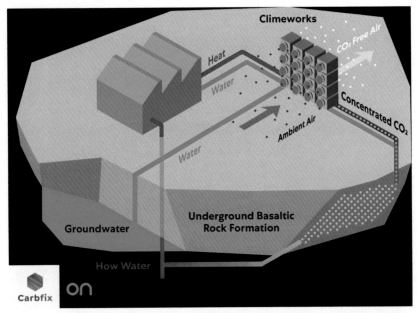

Fig. 8.8 The Carbafix process for the extraction of CO_2 from the atmosphere. *(From (N.d.-b). Climeworks. Reproduced with kind permission of Carbafix and Climeworks.)*

and steel and that of cement. Current progress towards applying green hydrogen as fuel in these processes to minimise the very large emissions of CO_2 that are involved are now discussed.

Fig. 8.9 shows the stages involved in the steel–making process. Iron ore is admitted to the blast furnace together with coke (formed in a coking oven

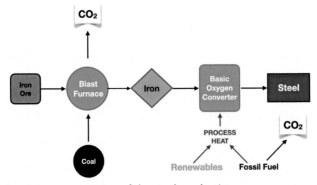

Fig. 8.9 Schematic representation of the steel production process.

prior to admission to the blast furnace) and the latter is converted to carbon monoxide by the residual traces of oxygen present in the feed:

$$2\,C + O_2 \rightarrow 2\,CO$$

The CO reduces the iron ore (assumed here to be predominantly Fe_3O_4) to produce pig iron:

$$Fe_3O_4 + 4\,CO \rightarrow 3\,Fe + 4\,CO_2$$

The pig iron is then fed, together with any other required additives (e.g. Cr_2O_3), to the basic oxygen converter where the mixture is converted to steel; the CO formed from the feedstuff is used to provide the process heat as well as a source of the carbon needed for the steel.[1] Scrap iron may also be fed in various proportions at this stage, thus reducing the need for heat energy and therefore cutting the total emissions of CO_2.[m] The emission of CO_2 from the process therefore comes from both the reduction process in the blast furnace and from the combustion of the fossil fuel needed to provide both the process heat and the additional carbon needed in the basic oxygen converter.

It is possible to operate a blast furnace by using natural gas instead of coal and this already gives some improvement in greenhouse gas emissions. Some steel works have already been converted for this purpose. A small number have also been converted to using hydrogen both as a fuel and as a reducing agent and it has been claimed that this already gives 'green steel'. However, as pointed out by the Bellona Foundation (see Box 8.4), the product is only truly green if the hydrogen is produced by renewable electricity and any claims of the production of a 'climate-neutral steel' should be treated cautiously. Most of the plants in Europe using hydrogen for reduction purposes do not appear to be using fully renewable hydrogen.

---

[1] If a solid oxide electrolysis cell (SEOC) such as that shown in Fig. 8.7 were to be used and the feed used for it was water plus CO_2, the system could become largely self-sufficient in carbon.

[m] Scrap iron can be used for the production of steel for many purposes but the product is not pure enough for specialised purposes such as automobile manufacture due to the factthat scrap iron often adds undesirable metal concentrations to the alloy produced. Some 40 years ago, there was a serious problem with the corrosion of automobile bodywork due to the use of recycled steels and this led to the introduction of galvanising by the addition of ZnO to the mix. See for example: https://www.hagerty.com/media/automotive-history/galvanization-sensation-how-automakers-fought-off-the-scourge-of-rust/.

BOX 8.4 Current European use of hydrogen in the steel industry

The Bellona Foundation (https://bellona.org/about-bellona) is a Norwegian organisation with offices in Brussels and also in Russia which aims to provide 'a solution-oriented approach to the environmental challenges'. They have recently published two reports considering the use of hydrogen in steel production in Europe.[n] The first of these considers the use of hydrogen as an auxiliary reducing agent in the 'Blast Furnace - Basic Oxygen Furnace' (BF-BOF) route and the second considers hydrogen as the sole reducing agent in the Direct Reduction of Iron (DRI) route. The first report lists ten companies in Europe who are using hydrogen reduction in BF-BOF plants and it concludes that only three of them are using hydrogen produced by electrolysis while most of the others use grey hydrogen (from natural gas). Bellona queries the claims that these companies are producing green steel, pointing out that using grid electricity in the current situation would actually increase the output of greenhouse gases by 36.7% if a plant was using German electric power. The second report is more hopeful and discusses the situation if iron ore were to be reduced to sponge iron in either a shaft furnace or a fluidised bed reactor using hydrogen produced with renewable electricity. The sponge iron would then be fed to an Electric Arc Furnace (EAF) to which carbon is also fed. Some 14 such plants are planned for Europe in the next ten years or thereabouts and a further ten are in operation. Of the latter, most are most probably currently using natural gas while only two or three are already using hydrogen from electrolysis. Even in the latter cases, it is not yet clear that renewable electricity is being used.

[n] Hydrogen in Steel Production: What is happening in Europe, Parts 1 and 2: https://bellona.org/news/climate-change/2021-03-hydrogen-in-steel-production-what-is-happening-in-europe-part-one and: https://bellona.org/news/industrial-pollution/2021-05-hydrogen-in-steel-production-what-is-happening-in-europe-part-two.

A 2020 briefing prepared for the European Parliament[o] has pointed out that replacing coal with hydrogen would at current prices drive the cost of steel up by one-third. In order to produce the hydrogen by electrolysis would require a 20% increase in the current generation of electricity and hence an even greater increase of renewable production, going beyond the replacement of current fossil–fuel–based electricity generation systems. The briefing also points out that an increased price of steel is likely to cause a shift towards the use of alternative greener materials and that the steel industry would no longer be tied to geographic regions in which coal is plentiful.

[o] https://www.europarl.europa.eu/RegData/etudes/BRIE/2020/641552/EPRS_BRI(2020)641552_EN.pdf.

The situation in Europe regarding a wish to use renewable energy in steel production (Box 8.4) is reflected in many other countries. For example, there has recently been significant interest in reports from Australia that renewable electricity (produced using solar power) will be used to produce hydrogen for steel production by electrolysis.[P] This change would enable Australia's steel production to be concentrated there rather than retaining the current practice of exporting iron ore to China for the production of steel that is then returned to Australia for its domestic use. Not only would such a change allow significant decreases in China's CO_2 emissions but there would also be substantial reductions arising from lower associated maritime emissions during transportation in both directions.

Use of green hydrogen in cement production

Chapter 2 gave a brief description of the process involved in cement production and showed that this industry is responsible for between 8% and 10% of the world's annual emissions of CO_2, a total of almost 900 kg of CO_2 being emitted for every 1000 kg of Portland Cement produced. The main producers are China, India and the US. Coal combustion is generally used to provide the energy needed to bring about the endothermic decomposition of the $CaCO_3$ feedstock, the combustion itself causing significant CO_2 emissions. However, the decomposition reaction itself:

$$CaCO_3 \rightarrow CaO + CO_2$$

is responsible for almost 70% of the CO_2 emission. Hence, even if sustainable energy is used to give green hydrogen that is then burnt to give the necessary energy for the decomposition reaction, only approximately 30% of the total CO_2 emissions will be removed.[q] The use of sustainable energy can also give significant decreases in the substantial quantities of CO_2 emitted in the transportation and handling of materials to and at the cement plant. However, there is no way of cutting down the emissions from the decomposition reaction. Hence, one partial solution is to find other materials to replace at least part of the CaO in the cement needed for construction purposes worldwide by some other constituent. There is currently significant effort aimed at finding alternative materials for that purpose.

[P] See, for example: https://www.forbes.com/sites/kensilverstein/2021/01/25/we-could-be-making-steel-from-green-hydrogen-using-less-coal/.

[q] The excess CO_2 could also be converted in a SOEC system of the type described in an earlier section by feeding it with a mixture of steam and the CO_2 emitted from the calciner. The product syngas could then be used for methanol production or Fischer Tropsch synthesis.

However, as there is always likely to be some CaO used in the product, some method of using the emitted CO_2 in a sustainable fashion must be found. Although the CO_2 formed could be trapped for storage underground, the available storage capacities would soon be exhausted if such a method was used for all global cement production. A possible use for the CO_2 would be to use it as a source of syngas by the combined electrolysis of water and CO_2 discussed above, synthesising either methanol or hydrocarbon fuel from the syngas. This could be done either at each cement plant by establishing suitable facilities there (photovoltaic or wind energy generation plus a fuel synthesis plant) or by setting up a suitable pipeline to an existing fuel-synthesis facility. The fuels resulting would not have been made using 'green materials'; however, since the CO_2 used would otherwise have been emitted during the cement preparation process, there would be an all-over reduction in CO_2 emission.

Other areas for energy savings and for the reduction of greenhouse gas emissions

There are many other industrial processes which make use of fossil fuels for the provision of energy. The vast majority of these could instead use hydrogen or a sustainable fuel such as bio-methanol as a source of energy. Although the oil-processing and the petrochemical industries both use a significant amount of self-produced hydrogen (e.g. that formed in cracking processes), improved savings could result in both industries from the use of significantly more renewable energy.

Substantial quantities of electrical energy are used in data-storage facilities and it is clear that many such users are currently installing solar or wind generation systems to provide the majority of their energy needs. Such facilities also most frequently have substantial battery storage facilities.

A great deal of energy is used for both heating and lighting processes in both commercial and domestic situations and considerable effort is being made to achieve significant improvements. Currently, the vast number of users in both of the latter categories still use fossil fuels for heating purposes, these fuels ranging from coal in some areas to kerosine and natural gas in others. All of these energy sources must be phased out and renewable electricity or heat exchange systems such as those described in Chapter 3 must be used instead. There are major moves in many countries to improve the use of energy in commercial and domestic buildings. As an example, Ireland has a very ambitious programme to improve the energy efficiencies of existing housing and has a significant grant programme administered by the Sustainable Energy Authority of Ireland (https://www.seai.ie/grants/home-energy-grants/). Scotland has recently announced an ambitious programme for the installation of heat pumps (https://www.pv-magazine.com/2021/06/ 14/scotland-announces-massive-plan-for-heat-pump-deployment/) and a similar plan has been announced for the Netherlands (https://www. pv-magazine.com/2021/04/23/massive-plan-for-hybrid-heat-pump-deployment-in-the-netherlands/). The main problem in all of such programmes is that it is difficult to convince existing homeowners that they should upgrade their energy use. However, significant advances are being made in that regard (see Box 8.5) and much more effort of this sort is planned.

Many countries have plans to promote the widespread use of electric vehicles and the use of renewable electricity. However, one of the major problems associated with such vehicles is that of recharging when on an extended journey and, despite the steadily increasing ranges possible on a single charge as discussed in Chapter 6, this is a drawback when it comes to encouraging everyone to change to such vehicles. Further, unless a substantial proportion of the electricity supplied from the grid is renewable, the use of hybrid vehicles gives better figures than those for electric vehicles in many countries. If the subsidiary recharging motors used in hybrid vehicles were powered by biomass-derived fuels, then their range is likely to continue to exceed that of fully electric vehicles and to be significantly

BOX 8.5 Irish household energy savings

Ireland has a very significant national programme aimed at encouraging the use of sustainable energy in all regions of the country, both urban and rural. Individual householders are encouraged to upgrade their home energy systems in order to decrease greenhouse gas emissions and, in parallel, to cut their energy costs. Many households have installed photovoltaic (PV) systems and heat exchange systems of the type discussed in Chapter 3 and have also improved their home insulation in order to improve the all-over efficiency of their energy systems. In the case now discussed, the homeowner has recently installed PV panels as well as a heat exchange system that is used for water heating. He also uses a battery vehicle which is now charged using electricity generated with the PV panels when this is available. (His PV system includes battery storage which allows for some load balancing.) Fig. 8.10 shows a recent record of the operation of his PV system that illustrates very clearly how such a system operates effectively even in a country at a relatively northerly latitude (ca. 53°N).

Continued

BOX 8.5 Irish household energy saving—cont'd

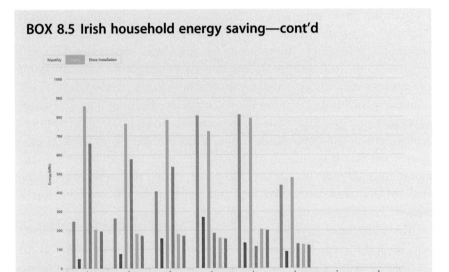

Fig. 8.10 Data for the operation of a domestic Irish photovoltaic system from January to mid-June 2021. *(Data kindly provided by F. Sheehan. Shown are the monthly figures for power generation (orange), feedback to the grid (red), household energy use (blue), draw-down from the grid (brown) and feed to (green) and from (grey) the battery storage system. The maximum household energy use in January 2021 was about 770 kWh.)*

These data, which record the operation of the solar panels on a monthly basis in the period January to mid-June 2021, show that the panels only generate a relatively small proportion of the energy needed for the household (averaging approximately 800 kWh per month) in the early months of the year but that they provide a large proportion of the household load from April onwards. (Data for the period following the commissioning of the system at the end of July 2020 that are not included in this figure show that the PV generation fell off significantly only from November onwards.) The battery storage included in the system has the advantage that when the use of grid electricity is necessary during the winter months, it can mostly be downloaded and stored during the night when the electricity provided is less expensive and also very largely renewable.

more sustainable. Biomass–derived fuels will also play an increasing part in transport, particularly in aviation.

Another area is of major concern: agricultural emissions. Ireland has a particularly significant problem in this regard as it depends very significantly on its agricultural activity, particularly in relation to cattle and dairy products. Recent reports show that Ireland has fallen significantly behind its targets in reducing the emissions from that sector, these contributing 34% of the county's total greenhouse emissions; these agricultural emissions have risen significantly over the last few years rather than falling. Only two other European countries, Hungary and Poland, have shown increases in agricultural emissions over the same period, while Greece, Croatia and Lithuania have shown decreases, these as a result of reductions in their dairy herds. Some methods of decreasing the Irish agricultural emissions are discussed in a very recent report,[r] these including improvements in the efficiency of nitrogen utilisation as well as the use of protected urea products and low emission slurry spreading. However, several other possibilities are not included in that report. For example, it would seem that increasing use should be made of anaerobic digestion methods (see Box 8.6) to reduce the quantities of agricultural residues such as slurry since

---

BOX 8.6 Methanation of biogas

The production of biogas by anaerobic digestion of biological waste was discussed in Chapter 5. The gas from a simple biogas plant contains a mixture of CO, CO_2, H_2 and CH_4 and this can be used directly for heating purposes or can be added in relatively low proportions to natural gas pipelines. However, if green hydrogen can be produced at competitive prices at each plant using the electrolysis methods as described earlier in this chapter (see, for example, Fig. 8.7), it could be used to methanate both the CO and CO_2 components of the biogas mixture to produce high purity methane that could be distributed undiluted in existing gas pipelines as a totally green product.

One of the main sources of greenhouse gas emissions is agriculture and there is therefore a great need to reduce these emissions. An acceleration of the installation of anaerobic digestion systems to provide biogas on farms should therefore be a priority for countries such as Ireland whose agricultural greenhouse gas emissions are currently very high. Ireland has at present just above 30 biogas plants and it has been estimated that it would need to have an additional 900 such plants by 2050. Unfortunately, there are signs of a reluctance to accept such plants by the inhabitants in many areas, it being argued that such plants cause local environmental and safety issues: a NIMBY approach.[s]

[s]NIMBY = 'Not in my back yard'.

---

[r] https://www.epa.ie/publications/monitoring–assessment/assessment/state-of-the-environment/irelands-environment-2020—an-assessment.php.

the spreading of this slurry for fertilisation purposes is a very major contributor to greenhouse gas emissions. As the climate in Ireland is particularly suited to the growth of grass, it would be possible for many farmers to concentrate more on the production of silage intended as a feedstock for such anaerobic digestion plants. Further, the report does not address the problem of the steady continued increase in the cattle population in Ireland that has occurred since the removal of milk quotas in 2015. An alternative method of decreasing the agricultural greenhouse emissions, particularly those of biogenic methane, would be to reduce the size of cattle herds and to encourage increased attention to the growth of suitable sustainable energy crops as an alternative to dairy and beef husbandry. As with many of the actions needed in relation to all the sectors discussed in this chapter, decisive government action is needed to attain the necessary reductions in emissions.

Conclusions

The objective of this book is an attempt to give a brief introduction to many aspects of the production and uses of energy as well as an indication of some current approaches to the improvement of energy efficiency in a number of different applications, the aim throughout being to highlight approaches to the reduction of emissions of greenhouse gases. The situation changes from day to day and new reports on governmental and industrial activities in emission reduction are appearing very regularly. What is very clear is that each government must engage fully in the process and must apply a 'carrot and stick' approach to reducing its own emission figures. It is hoped that this book will supply some of the background information needed by those involved in making governmental decisions as well as indicating to academic and industrial researchers some of the areas which require the greatest scientific and technological inputs.[t]

---

[t] It is the author's intention to add in due course short reports on new developments related to some of the topics discussed in this book on his blog at www.contemporarycatalysis.wordpress.com.

Tailpiece

The concept for this book emerged several years ago when I began recognising that various reports outlining plans for the reduction of greenhouse gas emissions failed to include significant details of the chemistry of how these reductions might be achieved. A trivial example was a statement that hydrogen would be produced by reforming natural gas; the report that I read implied that hydrogen production from methane was a straightforward and simple process involving an uncomplicated decomposition of methane. In another item, it was stated that a fleet of city buses in Ireland would be powered by hydrogen produced entirely from indigenous resources; however, no feasible explanation was given as to how that would be achieved, and Ireland does not currently have any appropriate technology. I therefore felt that it would be helpful to write a book that outlines some of the possible methods for controlling greenhouse gas emissions, with the emphasis being placed on the chemistry associated with such methods. My initial intention was that I would first consider only the chemistry behind currently operating processes for the production of energy and then go on to discuss new technologies. My plan was to concentrate on areas in which I had particular knowledge and that were closely related to my own research activities, namely processes involving heterogeneous catalysis. However, I soon realized that I also needed to cover topics about which I have had somewhat less hands-on experience. Hence, the writing process also became a learning experience. I found that I had to expand my prior knowledge of other related topics such as the construction and operation of batteries; the increasing use of photovoltaic, wind, hydroelectric and tidal energy; and even of improvements in domestic heating systems. I hope that at least some part of this rather wide range of content will prove informative to the reader. Further, I hope that specialists in any of the topics that I have covered who feel that my treatment of their subjects is relatively trivial will excuse my attempts at producing easily understandable and readable explanations. Further information on many of the topics handled is easily accessible by following the many links included.

During the final stages of preparation of this book, I exchanged some interactive emails with my friend Prof. Dr. Miguel A. Bañares of CSIC-Instituto de Catalisis, Madrid, Spain, regarding the subjects to be included in my book. During this correspondence, he introduced the analogy of

'Mount Sustainable', representing the huge barrier that must be conquered before the nations of the world can reach "Net Zero" by 2050. This analogy of climbing a mountain is very appropriate: we are not going downhill towards a target; instead we are going to have to surmount a whole series of very significant barriers and even dead ends in our efforts to achieve the summit and the final goal. There are groups of mountains in almost all the countries of the world, some relatively simple to climb, some requiring good guidebooks and some requiring experienced guides, but to climb them all requires dedication and planning. Extending the analogy to climate change, each nation of the world has a number of differing 'Mount Sustainables'. The task faced by each is to overcome a series of their own 'Mount Sustainables', each of these peaks requiring a different approach to the summit. This must be done in a collaborative manner such that each nation ends up by the year 2050 having conquered all of its summits. Teamwork is required for each ascent, especially if all are to be conquered in this relatively short period. The nations possessing less experience will require both financial and technological assistance from those fortunate enough to have much more.

A group of some of Ireland's highest mountains, the MacGillicuddy's Reeks, situated in the Killarney National Park, County Kerry.

Climate change mitigation is on everyone's minds and there are daily reports highlighting one or the other aspect of the subject. The United Nations Climate Change Conference of the Parties (COP 26), being held in Glasgow (31 October–12 November 2021), is about to start. It is to be hoped this gathering will be more than just a talking shop and that agreement may be reached there between the participating nations on the steps that now need urgently to be taken to reach the targets set out in the Paris Agreement of 2015.

Even during the months since I finished writing this book, many documents relevant to its content have appeared, many related to the COP 26 objectives. The latter reports describe in some detail how different nations and industrial groupings hope to achieve the essential reductions in greenhouse gas emissions. It is my intention to attempt to provide updates on some of the recent developments related to the content of the book on my blog at https://contemporarycatalysis.wordpress.com/. I encourage you–the reader–to register and login to this blog and even to consider contributing additional items for inclusion.

Julian R.H. Ross, October 2021

Postscript

During the final stages of publication of this book, it has proved possible for me to write a brief additional note on the outcome of the COP 26 gathering. This meeting, held in Glasgow at the beginning of November 2021, was attended by the leaders of most of the nations of the world and received much attention from the world press. Towards the end of the meeting, it appeared that full consensus had been reached regarding all the items for which a unanimous decision had been required. However, in the final moments of the closing session, three countries—Australia, China and India—withdrew their support for a decision to cease the use of coal by 2030. (Following the mountaineering analogy of the Postscript, an insurmountable ravine had been encountered just below the summit of 'Coal Mountain'.) Nevertheless, some very important decisions were reached during the meeting, not only within the conference walls but also through parallel international negotiations, regarding how the objectives of the Paris Agreement might be reached. It was agreed in Glasgow that most of the Paris objectives should be significantly accelerated between now and 2030. All the participating nations have agreed to submit updated plans as to how these new objectives will be achieved in their own countries. (A good summary of the main conclusions of COP 26 is to be found at https://ec.europa.eu/commission/presscorner/detail/en/ip_21_6021.)

It is abundantly clear that now is the time for action on all fronts and that we must all now move more firmly from words to action. It is also clear that such action will be costly to all involved. Governments must help and encourage all their citizens and corporations to work towards the reduction of greenhouse gas emissions on all fronts and must impose sanctions when necessary on those who do not contribute to the required reductions. Many of the methods of reducing greenhouse gas emissions described in this book are well advanced and are immediately applicable when appropriate, but some are still under development. Much supporting research and development in the field of energy will therefore be required over the next decade.

Julian R.H. Ross, November 2021

Index

Note: f = figure; t = table; b = box

Printed in the United States
by Baker & Taylor Publisher Services